ELEMENTARY FOOD SCIENCE

Food Science Texts Series

Series Editor
Dennis R. Heldman, University of Missouri

Ernest R. Vieira, *Elementary Food Science, Fourth Edition*

Food Science Texts Series

Cameron Hackney, Merle D. Pierson and George J. Banwart,
Basic Food Microbiology, 3rd Edition (1997)
Dennis R. Heldman and Richard W. Hartel,
Principles of Food Processing (1997)
Hildegarde Heymann and Harry T. Lawless, *Sensory Evaluation of Food* (1997)
James M. Jay, *Modern Food Microbiology, 5th Edition* (1996)
Norman G. Marriot, *Essentials of Food Sanitation* (1997)
Norman N. Potter and Joseph H. Hotchkiss, *Food Science, 5th Edition* (1995)
Romeo T. Toledo,
Fundamentals of Food Process Engineering, 3rd Edition (1997)
Vickie Vaclavik and Elizabeth W. Christian, *Essentials of Food Science* (1997)

ELEMENTARY FOOD SCIENCE

FOURTH EDITION

ERNEST R. VIEIRA
Department of Food Science, Nutrition and Culinary Arts
Essex Agricultural and Technical Institute
Hathorne, Massachusetts

CHAPMAN & HALL

I(T)P® International Thomson Publishing
New York • Albany • Bonn • Boston • Cincinnati • Detroit • London • Madrid • Melbourne
Mexico City • Pacific Grove • Paris • San Francisco • Singapore • Tokyo • Toronto • Washington

Cover Design: Andrea Meyer, emDASH inc.

Printed in the United States of America

For more information, contact:

Chapman & Hall
115 Fifth Avenue
New York, NY 10003

Chapman & Hall
2-6 Boundary Row
London SE1 8HN
England

International Thomson
Publishing-Japan
Hirakawacho-cho Kyowa
Building, 3F
1-2-1 Hirakawacho-cho
Chiyoda-ku, 102 Tokyo
Japan

Thomas Nelson Australia
102 Dodds Street
South Melbourne, 3205
Victoria, Australia

Chapman & Hall GmbH
Postfach 100 263
D-69442 Weinheim
Germany

International Thomson
Editores
Campos Eliseos 385, Piso 7
Col. Polanco
11560 Mexico D.F.

International Thomson
Publishing Asia
221 Henderson Road #05-10
Henderson Building
Singapore 0315

2 3 4 5 6 7 8 9 10 XXX 01 00 99 97

Library of Congress Cataloging-in-Publication Data

Vieira, Ernest R.
Elementary food science / Ernest R. Vieira -- 4th ed.
p. cm. -- (Food science texts series)
Third ed. by Louis J. Ronsivalli
Includes bibliographical references and index.
ISBN 0-412-07961-5 (hb : alk. paper)
1. Food handling. 2. Food industry and trade. I. Ronsivalli, Louis J. Elementary food science. II. Title. III. Series.
TX537.R66 1996
641.3--dc20 96-14761
CIP

British Library Cataloguing in Publication Data available

To order this or any other Chapman & Hall book, please contact **International Thomson Publishing, 7625 Empire Drive, Florence, KY 41042.** Phone: (606) 525-6600 or 1-800-842-3636.
Fax: (606) 525-7778, e-mail: order@chaphall.com.

For a complete listing of Chapman & Hall's titles, send your requests to
Chapman & Hall, Dept. BC, 115 Fifth Avenue, New York, NY 10003.

To Louis J. Ronsivalli

Contents

	Preface	**xvii**
	Acknowledgements	**xix**

Part I Interrelated Food Science Topics

Chapter 1	**Why Food Science?**	**3**
	The Beginning	3
	Food Science Pioneers	3
	Entering the Twentieth Century	4
	The Interdisciplinary Years	5
	Advances in Technology and Quality	6
	The 1980s to Present	6
	The Future	8
	Food—The Largest of All Industries	8
	The Impact of Food Science on Society	8
	Food Science as a Profession	9
Chapter 2	**Composition and Nutritional Value of Foods**	**11**
	Establishing a Healthy Diet	12
	Proteins	13
	Some Properties of Proteins	18
	Carbohydrates	23
	Sugars	23
	Polysaccharides	28
	Lipids	33
	Fats and Oils	33
	Fatty Acids	34
	Phospholipids, Waxes, Shingolipids, and Sterols	36
	Vitamins	37
	Fat-Soluble Vitamins	37
	Water-Soluble Vitamins	39
	Minerals	41
	Iron	41
	Calcium	42
	Other Major Minerals	42
	Trace Minerals	43
	Natural Toxicants	45

Chapter 3	**Microbial Activity**	**47**
	Characteristics of Microbes	47
	Structure and Shape of Microbes	47
	Size of Microbes	48
	Reproduction in Microbes	49
	Motility in Microbes	49
	Effect of pH on Microbial Growth	50
	Nutritional Requirements of Microbes	50
	Effect of Temperature on Microbial Growth	50
	Water Requirements of Microorganisms	51
	Oxygen Requirements of Microorganisms	51
	Effect of Oxidation–Reduction Potential on Microbes	52
	Effects of Microbes on Foods	52
	Undesirable Changes in Foods	52
	Food Infections	54
	Food Intoxications	62
	Desirable Changes in Foods	66
Chapter 4	**Food Safety and Sanitation**	**68**
	Personal Hygiene	68
	Sanitation in the Home	69
	Foodservice Sanitation	72
	Food Plant Sanitation	74
	Plant Exterior	75
	Plant Construction	76
	Equipment	76
	Personnel Facilities	78
	Storage	79
	Cleaning the Plant	79
	Water Supply	81
	Sewage Disposal	81
	Sanitation in Retail Outlets	81
	The Hazard Analysis and Critical Control Point System (HACCP)	84
	I. Assessing Hazards	85
	II. Identifying Critical Control Points (CCPs)	86
	III. Establish Requirements to Be Met at Each CCP	86
	IV. Monitoring Critical Control Points	86
	V. Taking Corrective Action	86
	VI. Setting up a Record Keeping System	87
	VII. Verify that the System is Working	88
Chapter 5	**Quality and Sensory Evaluation of Foods**	**92**
	Appearance	92
	Texture	93
	Flavor	93
	Consumer Sensory Testing	97
	New Technologies	99

Chapter 6	**Regulatory Agencies**	**100**
	The Food and Drug Administration	100
	Adulteration	101
	Misbranding	101
	The U.S. Public Health Service	102
	The Meat Inspection Bureau	103
	The Poultry Inspection Service	104
	Other Regulatory and Inspection Agencies	104
	Trade and Professional Organizations	106
	The Institute of Food Technologists (IFT)	106
	The American Dietetic Association	106
	The National Restaurant Association (NRA)	107
	The Educational Foundation of the National Restaurant Association (EF)	107
	The Center for Science in the Public Interest	107
	The American Council on Science and Health	107
	The National Health Officers Association (NEHA)	107
	The International Association of Milk, Food, and Environmental Sanitarians (IAMFES)	107
	The Frozen Food Industry Coordinating Committee	107
	The National Pest Control Association (NPCA)	108
	NSF International	108
	Underwriters Laboratories, Inc. (UL)	108
Chapter 7	**Food Labels**	**109**
	Basic Information	109
	Ingredients	110
	Grades	111
	Nutrition Labeling	112
	Serving Sizes	114
	Exemption and Exceptions	114
	Daily Values	115
	Raw Foods	117
	Health Claims	119
	Other Claims	120
	Open Dating	121
	Other Information	
Chapter 8	**Enzyme Reactions**	**123**
	The Nature of Enzymes	123
	Proteolytic Enzymes (Proteases)	125
	Oxidizing Enzymes (Oxidases)	126
	Fat-Splitting Enzymes (Lipases)	126
	Enzymes That Decompose Carbohydrates (Carbohydrases)	127
	Applications	128

Chapter 9 **Chemical Reactions** **130**
Oxidation 131
Nonenzymatic Browning 133
The Strecker Degradation 134
The Aggregation of Proteins 135

Part II Food Processing Methods

Chapter 10 **Thermal Processing** **139**
Pretreatment of Foods 140
Vacuum in Cans 141
The Need for Vacuum 141
Obtaining the Vacuum 143
Liquid in Cans 143
Filling the Cans 144
Sealing the Cans 145
The Heat Process 146
The *D*-Value 147
Estimating Processing Times 147
The z- and F_0-Values 149
The Conventional Heat Processing Chamber (Retort) 150
Cooling Heat-Processed Foods 151
Other Methods of Heat Processing 152
Continuous Agitating Retort 152
The Hydrostatic Cooker 153
The High-Temperature–Short-Time (HTST) Process 153
Aseptic Fill Method 154
Cook Under Pressure 155
The Sous-Vide Process 155
Modified-Atmosphere and Controlled Atmosphere Packaging (MAP) 155
Microwave Processing 156
Containers 156
Cans 156
Flexible Pouches 156
Glass Containers 157
Microwavable Containers 158
Plastic, Flexible Containers 158
Canning of Acid Foods 158
Warehouse Storage of Canned Foods 158
FDA Regulations and Safety 159

Chapter 11 **Drying** **160**
Pretreatment 160
Methods of Drying 160
Sun Drying 160

	Hot-Air Drying	161
	Fluidized-Bed Drying	162
	Drum Drying	163
	Spray Drying	163
	Freeze-Drying	164
	Puff Drying	166
	Microwave Drying	166
	Reconstitution of Dried Foods	166
	Packaging of Dried Foods	166
	The Effect of Drying on Microorganisms	167
	Deterioration of Dried Foods	167
	Oxidative Spoilage of Dried Foods	167
	Nonenzymatic Browning in Dried Foods	168
	Enzymatic Changes in Dried Foods	168
Chapter 12	**Low-Temperature Preservation**	**169**
	Refrigeration at Temperatures Above Freezing	169
	Mechanical Refrigeration	170
	Refrigeration Practices	171
	Other Refrigeration Methods	171
	Freezing	172
	The Preservation Effect of Freezing	173
	Freezing Methods	173
	Air-Blast Freezing	173
	Fluidized-Bed Freezing	175
	Dehydrofreezing	175
	Plate Freezing	175
	Liquid Freezers	175
	Slow Freezing	176
	General Considerations of Freezing Preservation of Foods	176
	Preparation of Foods for Freezing	177
	Packaging	178
	Problems with Freezing or Thawing of Bulk Foods	179
	Quality Changes During Frozen Storage	179
	Shelf-Life of Frozen Foods	181
	Thawing	181
Chapter 13	**Food Additives**	**183**
	Philosophy of Food Additives	185
	Antioxidants	186
	Nutrient Additives	187
	Flavorings	188
	Flavor Enhancers	188
	Acidulants	189
	Alkaline Compounds	190
	Sweeteners	190
	Fructose	190

Molasses 191
Honey 191
Maple Sugar 191
Lactose 192
Maltose 192
Xylitol 192
Sorbitol 192
Mannitol 192
Aspartame 193
Saccharin 193
Starches 193
Gums 193
Enzymes 194
Invertase 194
Pectinase 194
Cellulases 194
Proteases 195
Lipases 195
Glucose Oxidase 195
Catalase 195
Sequestrants 196
Polyhydric Alcohols 196
Surface-Active Agents 196
Leavening Agents 197
Ionizing Radiation 198
Chemical Preservatives 199
Sodium Chloride 200
Fatty Acids 201
Sulfur Dioxide 201
Sorbic Acid 202
Sodium Nitrite 202
Oxidizing Agents 203
Benzoates 203
Colorants 204

Part III Handling and Processing of Foods

Chapter 14 Meat 209
Beef/Veal 211
Pork 214
Lamb/Mutton 215
Cured Meat Products 217
Sausage Products 219

Chapter 15 Dairy Products 224
Fluid Milk 224
Microorganisms in Milk 224

	Sources of Bacteria in Milk and Methods of Limiting Bacterial Contamination	225
	Handling of Milk on Farms	226
	Transportation of Fluid Milk	226
	Processing of Fluid Milk	228
	Other Dairy Products	232
	Ice Cream	232
	Yogurt	236
	Cheese	238
	Butter	243
	Whey	243
	Buttermilk	243
	Sour Cream	243
	Dairy Product Substitutes	244
Chapter 16	**Poultry and Eggs**	**245**
	Poultry	246
	Chicken	246
	Turkey	249
	Ducks and Geese	250
	Eggs	250
	Processing of Eggs	253
	Egg Consumption	255
	Egg Substitutes	255
Chapter 17	**Fish and Shellfish**	**256**
	Important Families of Fish and Shellfish	259
	The Herring Family *(Clupeidae)*	259
	The Cod Family *(Gadidae)*	262
	The Mackerel Family *(Scrombridae)*	265
	The Salmon Family *(Salmonidae)*	266
	The Flatfish Family *(Pleuronectidae)*	267
	Other Fish	268
	Shellfish	268
	Bivalve Molluscs (Class Pelycopoda)	268
	Crustaceans (Class Decapoda)	272
	Aquaculture	277
	Surimi	281
Chapter 18	**Cereal Grains**	**283**
	Wheat	284
	Corn	287
	Oats	288
	Barley	289
	Rye	291
	Rice	291
	Other Cereal Grains	292
	Sorghum	292
	Buckwheat	292

	Cottonseeds, Soybeans, and Peanuts	293
	Millet	293
	Triticale	294
Chapter 19	**Bakery Products**	**295**
	Bread	295
	Cakes and Cookies	299
	Doughnuts	299
	Crackers	300
	Pie Crusts	300
	Low- and No-Fat Bakery Items	300
Chapter 20	**Fruits and Vegetables**	**302**
	Fruits	302
	Grapes	303
	Apples	306
	Bananas	307
	Oranges	309
	Aseptic Packaging	310
	Coffee	310
	Other Fruits	313
	Vegetables	313
	Green and Wax Beans	315
	Broccoli	316
	Cucumbers	316
	Lettuce	316
	Olives	317
	Potatoes	318
	Tomatoes	319
	Cabbage	321
	Tea	322
	Other Vegetables	323
Chapter 21	**Sugars and Starches**	**324**
	Sugar from Cane	324
	Beet Sugar	327
	Other Sucrose Sources	328
	Corn Syrup and Sugar	328
	Beverages	329
	Candy	330
	Starch	332
Chapter 22	**Fats and Oils**	**337**
	Lard	339
	Oils	340
	Hydrogenation	342
	Margarine	343
	Lipid Emulsifiers	343
	Salad Dressings	343

	Chocolate	344
	Lipid Substitutes	344

Part IV Food Science and the Culinary Arts

Chapter 23	**Equipment Used in Food Preparation**	**349**
	Cutting Equipment	349
	Heating Utensils	352
	Types of Heat and Heat Sources	354
Chapter 24	**Food Preparation—An Important Application of Basic Chemistry and Physics**	**358**
	Meat	358
	Dairy Products	360
	Poultry and Eggs	361
	Fish and Shellfish	364
	Cereal Grains	365
	Bakery Products	366
	Vegetables	369
	Fruits	371
	Sugar	372
	Fats and Oils	373

Part V Food Science Laboratory Exercises

Laboratory Safety Recommendations	**377**
Cleaning and Sanitizing Procedures	**381**
Manual Cleaning and Sanitizing Policy	**382**
Acidity in Foods	**383**
Salt (Sodium Chloride) in Foods	**385**
Microorganisms in Food	**388**
Sensory Evaluation of Food	**390**
Thermal Processing of a Low-Acid Product	**392**
Production of Wine and Beer	**395**
Production of Sauerkraut	**398**
Production of Hot Dogs	**400**
Production of Surimi	**402**
Production of Yogurt	**404**

Appendix	407
Suggested Readings	409
Index	413

Preface

The scientific study of food is one of human society's most important endeavors. Food is necessary for survival, growth, physical ability, and good health. Food processing, handling, preparing, and serving comprise the world's largest industry. Many factors require that those scientists who choose to study foods be prepared in many of the physical and life sciences. Among these are the chemistry of food, its vulnerability to spoilage, its role as a disease vector, its quality attributes, its nutritional content, and its varied sources.

Prior to the establishment of university curricula in food science, the first food science practitioners were drawn from such fields as chemistry, chemical engineering, bacteriology, agriculture, and biology. Many of the earlier activities of food scientists emphasized the application of the latest technological developments to all aspects of food production, the activity was largely described as food technology, and many of the newly formed departments of universities that offered courses in the study of foods were called food technology departments. But much basic work was required to understand the characteristics of foods and the factors that affected these characteristics. Subsequently, there developed a distinction between food science (which involved basic studies) and food technology (which involved the application of technological developments to all aspects of food handling). The term food technology is used less frequently now and many of the colleges and universities have renamed their departments food science departments.

This text in elementary food science is written for a wide variety of students studying food science and related concentrations. It should be used in one- and two-year certificate and associate degree programs in food science. It is also an excellent choice for those Baccalaureate candidates who are studying food science for the first time.

Many related areas of study such as nutrition, dietetics, culinary arts, foodservice, and hotel and restaurant management at present include or are planning to add a required food science course. This text is especially useful for these courses, as it was written for students who do not yet have an extensive science background.

The text has many practical applications and can serve as a reference book for people in any part of the industry that is involved with food processing, labeling, testing, food quality, and safety in general.

Parts I and II have been designed to reach a broad audience, including food handlers who may not have had the necessary educational preparation to enable them to handle foods in a safe manner. The author believes that the improper handling of foods is the primary cause for the tens of thousands of food poisonings that occur annually. Yet there is no reason for even one case to occur. In fact, it is reasonable to propose that the safety of foods and the factors that affect it should be required instruction at the high school level, because everyone will, sooner or later, either prepare foods for others or be confronted with a situation in which the safety of a food is in question.

The commercial preparation of food, whether in a large food processing plant or a small restaurant, should not be done by anyone other than qualified personnel. Qualified is defined as those who have met two criteria. The first is the assimilation of the required knowledge regarding the safe handling of foods and the second is certification by an appropriate organization recognized by regulatory authorities that the person has demonstrated the ability and the attitude to handle foods properly. A major aim of the book is to help its readers to gain the necessary knowledge to handle foods safely. The expanded section on the Hazard Analysis Critical Control Point (HACCP) method of food handling stresses this.

As with any introductory text, it is not possible to cover all of the elements of a subject as broad as food science—from the many scientific disciplines that one must absorb in order to be an effective food scientist, to the many applications of food science in government, university, and industry activities. Therefore, it can be seen that whereas the text covers many major topics in food science, it does not cover every aspect of food science. The Suggested Readings at the end of the volume should help fill the need for additional information on those subjects that were not covered completely or were omitted.

Part IV of the Fourth Edition is based on the growing awareness and importance of the principles of food science and technology in today's food management and preparation methods. An understanding of these principles is invaluable to any cook, chef, baker, or foodservice manager and this knowledge should be passed on to the students of this very large and essential segment of the food industry, to prepare them properly for what lies ahead. The information here will also be useful to Food Scientists and their students because it will give them an insight into the needs of this large segment of the industry, without whose existence food science would have little purpose. This section of the book will also show the practicality of the scientific principles as they are put to use in today's society.

To allow hands-on experience for the students of elementary food science, Part V has been added. This section contains laboratory exercises that demonstrate some of the theories and methods described in other sections. Students will be able to take their knowledge of the sciences into the laboratory or pilot plant and produce and test products that are familiar but possibly not thought of as "science experiments." The application of science to foods can be a fun, enlightening, and very practical experience. The inclusion of this section was supported by members of the industry as it gives students practical processing and testing experience.

Acknowledgments

The author would like to thank the following people for their technical and psychological support in the production of this text:

Peggy Vieira, who read, and changed, the book many times;

Doreen Iovanna, my former student and very competent transcriber;

Ted Novakowski, Technology Manager at the Essex Agricultural and Technical Institute and *Dave Luca,* art director at Georgetown (MA) High School for their assistance and creativity in developing new tables and figures;

Hannah Luca for her assistance, patience, and cooperation

ELEMENTARY FOOD SCIENCE

Part 1

Interrelated Food Science Topics

1

Why Food Science?

The scientific study of food is one of human society's most important endeavors. Food is necessary for survival, growth, physical ability, and good health. Food processing, handling, preparing, and serving comprise the world's largest industry. Many factors require that those scientists who choose to study foods be prepared in many of the physical and life sciences. Among these are the chemistry of food, its vulnerability to spoilage, its role as a disease vector, its quality attributes, its nutritional content, and its varied sources.

THE BEGINNING

The origin of food science is unclear but history reports that the Egyptians, Greeks, and Romans were able to preserve a variety of foods in vinegar, brine, honey, or pitch. Some foods were dried in the sun or with fire, and cheeses and wines were also produced. Drying, freezing, smoking, fermenting, and heat processing (cooking and baking) have been practiced for centuries. It is believed that, until the latter part of the eighteenth century, food preservation evolved as an art handed down from generation to generation and its development was slow, depending on accidental discoveries. The first food technologists were actually discoverers, inventors, and improvers of the preservation methods mentioned previously. Later, science improved upon discovery techniques and invention, resulting in great gains to human health with provision of a safe, nutritious, varied, and economical food supply.

FOOD SCIENCE PIONEERS

Food Science is an interdisciplinary field that evolved first from chemistry, then microbiology and medicine. Since then biochemistry, nutrition, toxicology, mathematics, physics, engineering, psychology, genetics, biotechnology, and law have become integral parts of this fascinating profession.

It is generally agreed that food science actually advanced from accidental discovery and trial and error invention to the "dawn-of-science" era in 1795, when Nicholas Appert responded to the need of Napoleon's army to be less dependent on local foods. Through a heating process, Appert developed the first canning methods and is credited with the discovery of the first thermal processing techniques for food preservation. A few more people warrant mention in connection with these early days of food science. Shortly after Appert's work, the primitive analytical chemistry done by Frederick

Accum targeted the detection of adulterants in food and the first efforts to ensure a safe and honest food supply were initiated. In 1820, he published "Treatise on the Adulteration of Food, Culinary Poisons, etc."

Another chemist, Louis Pasteur, was a pioneer in food science and his interest in fermentation led to research on the production of vinegar, wines, beers, and alcohol. He also heated milk and milk products to destroy disease-causing and spoilage organisms. The process, known as pasteurization, is used today on juice and other beverages as well as dairy products. Pasteur, the chemist, became the father of bacteriology and of much modern preventive medicine. The work of Pasteur and others led to the recognition of the existence of enzymes, at that time meaning "an unorganized ferment." Scientists were beginning to realize that when people became sick after the consumption of food, it was often due to the presence of microorganisms, most of which had not yet been identified.

ENTERING THE TWENTIETH CENTURY

During the years from 1895 to 1938 many more significant advances were made in food science and nutrition. Essential amino acids and their role were explained, most vitamins were discovered and named, and the means of preventing rickets, pellegra, goiter, and beri-beri were found. Canning was now a technology under scientific control, and the use of refrigeration in the storage and transport of perishable food was becoming common. Frozen foods had reached the retail level and home refrigerators were replacing iceboxes.

Some of the people responsible for these advances in food science include Samuel Prescott, William Underwood, Edwin Hart, Clarence Birdseye, and Carl Fellers.

Samuel Prescott integrated biology and chemistry and from this, the growth of biochemistry and nutrition was reinforced. In 1895, he started work with William Underwood at the Massachusetts Institute of Technology (MIT) on canning research, which transformed that industry into a scientifically controlled technology. The growing knowledge of bacteriology was soon applied to the food, beverage, and dairy industries. Prescott's interest in sanitation led to his crusade in pasteurization and production of a sanitary milk supply. His interests widened and he became involved in food dehydration, refrigeration, food additives, and the chemical composition of foods. Prescott's work certainly broke down many barriers between the basic sciences, enabling the development of interdisciplinary sciences such as food science.

Edwin Hart was a pioneer nutritionist who did much work on animal feeding studies. His classic "single grain" experiment, using different grain sources with equal calories, protein, carbohydrate, fat, and ash content showed that gross chemical analysis alone was not enough to adequately define a diet. This experiment, started in 1906, led to much productive work on vitamins and minerals and improved nutrition through food science.

Clarence Birdseye is known for his contributions to the development of frozen foods. He was quite disappointed with the poor and unsanitary way in which fish were handled. These observations led to the development of several methods including quick freezing to capture quality that would be lost if fish were handled improperly and frozen by slow methods. Birdseye was patient and worked many years in the 1920s and 30s to get frozen foods accepted at the retail level. His discoveries, combined with the advancements in freezing technologies, allowed substantial penetration into this market by 1939.

Carl Fellers had an early interest in the nutritional quality of canned and other processed foods. He served with the National Canners Association from 1921 to 1924 and his work led to the present high level of safety in canned foods. In 1926 he moved to the Massachusetts Agriculture College (now the University of Massachusetts), where he remained. Among his accomplishments are the invention of methods for pasteurizing dried fruits, canning Atlantic crab, the use of ascorbic acid as an antioxidant, the fortification of apple juice, and the use of cranberry culls to develop cranberry juice and other cranberry products that are so popular in today's market. Dr. Fellers was a founder of IFT and served as president in 1949–1950.

THE INTERDISCIPLINARY YEARS

Up until about 1939, most of the early food scientists mentioned were chemists. The next four scientists of major influence in food science, Bernard Proctor, Bernard Oser, Loren Sjöström, and Amihud Kramer, came from varied and interdisciplinary backgrounds.

Bernard Proctor conducted early research in food microbiology and was one of the earliest to study the use of radiation for the preservation of foods. During World War II, the needs of the times led him to do research in the compression of dehydrated foods and the scientific investigation of effective food packaging. Many of the foods present-day soldiers carry on missions are of this dehydrated variety, packaged in flexible laminated containers. This technology has also been implemented in foods used by hikers and campers.

Bernard Oser was a biochemist who brought to food science the needed knowledge and insights from two related areas, nutrition and toxicology. His work, involving bioassays, using test animals, often gave different results from in vitro "chemical" tests. The recognition of these differences led to the concepts of bioavailability, ingredient interactions, absorption, and excretion rates. Oser was a pioneer who also was one of the first to press for the need to examine the effects of exogenous substances on reproduction. Oser was quite active in the 1950s as concern over food additives grew. He took an active role in educating the food industry to meet new requirements for toxicological studies and safety evaluation.

Loren Sjöström, while working with the consulting firm of Arthur D. Little (ADL), established a flavor laboratory and was one of the developers of sensory analysis methods that are used internationally today. This organized, scientific approach to "the art of taste-testing" became the flavor profile method for descriptive testing of food, beverages, and similar products. The first paper on this method was presented at the IFT annual meeting in 1949. ADL trains people in this method of analysis and it is used on hundreds of foods in hundreds of locations.

Amihud Kramer can be credited with bringing statistics to food science. This application to quality control helped industry meet increasing quality and cost requirements. His book, *Quality Control for the Food Industry* (Kramer and Twigg, 1970; 1973), played a significant role in this influence. He was active in devising procedures for production of desirable nutritional and sensory quality at a reasonable cost.

ADVANCES IN TECHNOLOGY AND QUALITY

Through the mid 1950s many other pioneers joined those already mentioned to develop food science as we know it today. Since then, many more significant discoveries, applications, and methodologies have developed. During the 1960s advances in technology allowed for improvement in quality. Freeze drying and products such as freeze-dried coffee hit the market, computer control in the processing plant was introduced, and product quality and efficiency of production resulted. Rigid and flexible plastic containers became common in the marketplace and many new ingredients such as oil blends and flavorings were developed. Proteins from products such as fish, whey, and soy were used to engineer and formulate new foods. Plant sanitation was improved by clean-in-place methods and foods with longer shelf-lives resulted. Dried milk flavor was improved by advances in dehydration technology, and the use of enzymes produced some unique products such as meat tenderizers. Automation developed for single-product plants such as beer, milk, and baked products, resulting in substantially increased production rates.

The 1970s brought a need for energy saving processes, and consumption of energy dropped by 25% in most plants. The era of health foods and organic foods was born during this decade. This gave the opportunity for many small-scale processors to become competitive with the bigger companies. By the end of the 1970s, smaller, quicker, user-friendly computers were introduced into food plants, contributing to more energy-efficient plants producing higher quality products at a faster rate.

THE 1980S TO THE PRESENT

With the 1980s came great optimism on the part of the food processor and increased importance of food safety and quality demanded by the consumer. The "new generation" gourmet type products were in and the advanced technologies in freezing systems, frozen distribution, and retail and home freezers led to the development of new, high-quality, frozen entrees. Aseptic packaging also was introduced during this time and products such as fruit juices soon were on the shelves in their now familiar box containers. The use of microwave ovens in almost every home put a great demand on the food industry. The consumer demanded high-quality, attractive, nutritious, safe food that could be made with minimal preparation. The high-speed life style in much of America had developed a demand for these products and as we entered the 1990s, the food industry was presented with this challenge.

Nutrition, health, convenience, and food safety are high-priority concerns in the 1990s and will undoubtedly carry an elevated level of importance into the twenty-first century. We have seen the development and widespread use of the food pyramid in an effort to make it easier for the consumer to understand the importance of proper nutrition and a balanced diet. The Nutrition Labeling and Education Act (NLEA) of 1990 required the Department of Health and Human Services to propose new nutritional labeling regulations, and by 1994 these were in effect and consumers now had information to help them achieve their nutrition goals and better utilize the food guide pyramid.

Food Science is also responding to present day and expected future health and safety trends. The use of sciences such biotechnology, genetic engineering, computer technology, microbiology, and chemistry will allow the food scientist to help bring to the marketplace new foods that will meet the needs and desires of the consumer. Examples of other challenges that will be faced are convenience foods, low-calorie and low-fat foods, quick methodologies such as gene probes or biosensors for detection of harmful microbes, inventory computerization to control and maintain quality, and computer "shopping" through the Internet lines. All these factors will have an effect on how foods are processed, packaged, labeled, distributed, and consumed.

In order for consumers to gain the benefits of the expanding technologies, they must understand the importance of science in the larger picture. Too often the media sensationalizes certain published works and the consumer may get information based on opinion rather than scientific facts and evidence. In the future, it must be stressed that scientific conclusions can be drawn only from data, not from consensus. Scientific evidence cannot be drawn from a "popular vote." Educators must play a significant role in getting the message to all consumers, starting with preschool and continuing through the ever-growing elderly population. Education should show the clear advantages that technology can provide but also expose the risks involved. This will help develop a consumer able to make informed decisions. Education in the "Science of Food" as opposed to simply cooking must be stressed in elementary, middle, and high schools. Teachers should be encouraged to take courses in the areas of nutrition, food science, and food safety. Many colleges offer such programs through night schools, and college faculty and students of food science and nutrition often visit schools to give demonstrations and "mini-laboratories" on the subjects. Some colleges offer courses to qualified high school students, enabling them to earn college credits while still in high school. Also, with the increased availability of satellite-transmitted information, discussions among groups in different parts of the world are possible. The educator in food science will play a significant role in the future of the world's food supply and the health and quality of life of its human inhabitants.

As a result of some of the catastrophes that occurred in the 1990s such as the death of young children who ate undercooked hamburger purchased at a major fast-food restaurant, food safety has become a major concern of the consumer. Again, the media has led much of the public to concentrate on some of the areas that scientific data show to be low-risk categories such as intentional additives (colors, flavors, preservatives, and functional aids), and unintentional additives (pesticide residues). These are certainly areas that deserve attention and regulation but food poisoning from microbial hazards pose the higher risk. Proper sanitation and food handling at all levels, including the home, would significantly lower the numbers of food poisoning outbreaks.

Advances in molecular biology are providing the food scientist with a vast array of new tools such as gene probes to detect bacteria, viruses, and toxins. Methods are also available to improve and develop foods that are more stable and can have "built in" nutrition and health benefits. These methods will soon be routine and the result will be a safer food supply. Biotechnology and the use of biosensors to detect drugs, pesticides, and indicators of freshness and quality are being used for quick detection of problems in food. Consequently, fewer foods that may pose a safety problem will ever get to wholesale or retail distribution.

THE FUTURE

Future food scientists will have many problems to solve. Some will be new, such as the development and use of genetically engineered food sources. Others, such as the prevention of food spoilage and poisoning, have always been with us and probably always will. In the future, we may have foods in packages with digital readouts of degrees of freshness, risks due to presence of pesticides or other compounds, and nutritional information that can be quickly analyzed by home computers to incorporate the food into a specially designed diet. There will be products from plants and animals at present considered waste but containing components that will be isolated, extracted, and purified to become food additives or medical supplements that will contribute to our health and longevity. Not too long ago, some of these concepts were pure "science fiction" but, with our technological advances, many are already being used and soon others will be developed. With all this futuristic advancement, it must be stressed that none of it would be possible without the basic knowledge of the sciences. We still must depend on the scientific method which involves understanding the nature of the problem, the nature of the materials, and the physical and chemical laws controlling the process. As was mentioned earlier, food science is an applied science and utilizes many others. Food science is certainly one of most exciting and potentially rewarding careers of the future.

FOOD—THE LARGEST OF ALL INDUSTRIES

Food is not only our most important need, but the food industry is the largest of all industries, employing tens of millions of people in all areas such as growing, fishing, processing, transportation, storage, preparation, inspection, distribution, sales, and marketing. Many more people are required to service the needs of the industry, as well as the people who operate it by providing required facilities, housing, equipment, transportation, financing, utilities, and communications. The retail value of the U.S. food industry is the largest of all and sales are in the hundreds of billions of dollars.

THE IMPACT OF FOOD SCIENCE ON SOCIETY

Perhaps the greatest impact of food science on society has come in the area of agriculture. The earliest developments changed nomadic societies to settled ones. Once the idea of farming the same piece of land was conceived, permanent shelters became more practical and the forerunners of modern-day houses were built. The protection of farms was important enough to cause farmers to group together and to establish boundaries and tiny hamlets, the forerunners of modern-day towns and cities. As agriculture improved, farmers became more efficient and fewer and fewer farmers were able to supply food for more and more people. This freed a large portion of the population to engage in other efforts such as manufacturing, construction, transportation, education, research, and medical services.

This transformation of efforts allowed our country to grow from a fledgling nation to the largest food-producing, industrial, political, and military power in the world.

We are now in an era where food and what is done with it are often the topics of many newspaper and other publication media articles. How food is grown, harvested, transported, processed, packaged, stored, prepared, and served is frequently the subject of media scrutiny and often the resulting information is authoritative and authentic. Other times, however, it seems that the public is not given all of the facts, only those that make sensational headlines and "sell newspapers." As mentioned previously, good scientific judgment cannot be made without reviewing all the data. The consumer should be made aware of all of the available facts and allowed to make an informed judgment, not directed to make judgments based on opinions of the authors. Many publications are available giving information on food and how it should be handled, prepared, served, and stored. Local boards of health can be contacted with questions on food safety and the United States Department of Agriculture (USDA), Food and Drug Administration (FDA), State Departments of Public Health, State Cooperative Extension Services, and the American Dietetic Association have available information on food handling and safety. Many colleges and universities have public information programs and offer continuing education for the general public, providing information about nutrition, sanitation, and other food science related areas such as food processing, microbiology, and food safety. The professional organization for food technology also has accurate information through the science communications program regarding food questions. Their address is:

The Institute of Food Technologists
221 North LaSalle St.
Chicago, IL 60601
Phone: 312-782-8424 Fax: 312-742-8348

Always check the sources of information in the printed media. Look for credentials of the authors and references as to where the information was obtained. If you have questions on the credibility of any information, contact one of the above mentioned organizations.

FOOD SCIENCE AS A PROFESSION

Food science may well be among the most important, the most timely, and the most relevant professions of our time. It is one of the most important because it is the device by which we can control the availability, the nutrition, and the wholesomeness of food—the one commodity that is critical to human survival. It is timely because we are in an era of significant technological advances and with these advances come a need for an understanding of the way our food is grown, harvested, transported, stored, processed, packaged, inspected, distributed, advertised, marketed, sold, displayed, prepared, and served. Food safety is the top priority of all portions of the food distribution chain and food science and its understanding is becoming increasingly necessary in all steps of food processing and distribution. Its relevance to other disciplines is apparent and colleges and universities offer special practical courses in food science as science electives to nonscience majors. These types of courses should be included in all elementary, middle, and high school programs as well, teaching food safety and nutrition at various levels to the different age groups. This will lead to a more informed, healthier society that will have a positive impact on the economy in general.

To prepare for a career in food science, it is a good idea to take science courses at the high school level and apply to one of the colleges or universities that offer curricula in food science or food technology. If you haven't done this, however, keep in mind that nonscience majors have transferred to food science and have done well. There are also two-year technical colleges that offer Associate Degree programs, with graduates of these programs choosing to either obtain employment or to transfer their credits to universities or colleges to earn a Bachelor or more advanced degree. There are a number of universities that offer degrees in food science through the Ph.D. level.

Food Science is not a discipline such as chemistry or mathematics but is rather a combination of disciplines, with emphasis placed on food-related aspects of each discipline. Thus, the food science student concentrates on subjects such as the microbiology of foods, the biochemistry of foods, the rheology of foods, the engineering principles of food processing, and food preservation. The student of food science also studies some disciplines and takes some courses without emphasis on food-related aspects. The latter include mathematics, inorganic chemistry, and basic physics. Food science students will take laboratory courses that will expose them to much equipment that will heat process, freeze, dry, homogenize, package, and seal foods. Engineering ability is quite useful for aspiring food scientists. Some schools have started programs incorporating the business aspect of the food industry into the curriculum. This blending of business acumen and scientific training is indeed an interesting combination. The reasons for the broad academic preparation required by food scientists are as follows:

- Foods, originating from animals and plants, are complex biochemical systems that continue to undergo change (mainly deteriorative) at rates that depend on such environmental conditions as temperature, humidity, and presence of oxygen.
- Foods are generally contaminated by a variety of microorganisms that subsist on the food components, creating changes in proteins, fats, and carbohydrates resulting in the formation of offensive, toxic, or desirable byproducts.
- The food scientist may want to employ certain additives to solve problems such as rectifying nutritional deficiencies, preventing the development of certain microbial toxins (e.g., botulism toxin), preventing or slowing spoilage, and improving texture, flavor, or appearance.
- The food scientist needs to be proficient in industrial plant operations and have a knowledge of process equipment and processes.
- The food scientist must be concerned with pesticide and fertilizer residuals, as well as compounds of mercury and other elements in levels as low as parts per million or less.
- The food scientist needs to be familiar with parasites, insects, and other foreign materials.

When research is done in chemistry or biology, a search of the published abstracts in either discipline is usually sufficient. The food scientist, however, may find relevant information in both of these areas or in others. The food scientist must be a well-rounded applied scientist, able to work within many areas to solve the problems that develop in this ever-changing, exciting field.

2

Composition and Nutritional Value of Foods

Food is a complex mixture of chemicals that an organism takes in and assimilates to promote growth, to expend energy, to replace worn or injured tissue, and to prevent some diseases. Nutrition, however, encompasses many processes, and thus it may be given many definitions. Gregor Mendel, the Austrian botanist and founder of the science of genetics, defined it as "The Chemistry of Life." This definition may be the most appropriate for the food scientist because the process by which food components are assimilated, converted, and utilized are understood and properly managed only when the chemistry is understood. Most foods are extremely complex mixtures of thousands of chemicals. Proteins, carbohydrates, and lipids are organic substances that, with water, usually make up more than 97% of a food's mass. The remainder consists of thousands more compounds, which exist in small amounts measuring in concentrations of parts per million or less. These compounds are often important in the taste, odor, and color of the food. Vitamins and minerals also exist in minute amounts and are extremely important in many of the body's functions. Most food constituents contain chemically active groups that can enter into complicated series of reactions with each other or with surrounding materials such as air, water, packaging, and equipment. Heat, moisture, and concentration changes in food processing together with biological catalysts called enzymes also may induce reactions. The resulting changes in the food may be desirable or undesirable and the food scientist must be able to predict and control the changes to gain the desired effects.

The physical structure of the foods also plays a very important role in the food's acceptability. Changes in textural characteristics may not directly affect the nutritional constituents in a food but if the consumer does not enjoy eating it, the valuable nutrients may never be ingested. With very few exceptions foods are not well-defined physical systems such as crystalline solids (salt, sugar), but more often occur in solutions, mixtures, or colloidal suspensions.

Nutrition has become a discipline in itself and many textbooks are entirely dedicated to the subject. In this chapter, the importance of nutrition is covered briefly and the chemistry of the major nutrients and their reactions in certain food processing systems are discussed.

Biochemistry is defined as the branch of chemistry relating to vital processes, their mode of action, and their products. The composition of foods and the reactions that lead to changes in their constitution and characteristics certainly fit the definition of biochemistry. The biochemical changes that occur in foods may directly or indirectly affect the nutritional quality of foods. The chemistry or biochemistry of life is an area with which all food scientists must be familiar.

ESTABLISHING A HEALTHY DIET

The high rates of obesity and complications resulting from consumption of too many calories or too many calories from saturated fats have led to the concept of "Western Disease." The United States has set national health goals and has implemented programs to improve health. Among these are guidelines developed by the United States Department of Agriculture (USDA) such as the "Dietary Guidelines for Americans" which were designed to help in planning and evaluation of dietary intake.

Another guide released by the USDA in April of 1992 was the "Food Guide Pyramid" (see Fig. 2.1), which replaced the basic four food groups as the recommended guide for designing healthy diets. The main objective of both guides is to lower the intake of fats to the level of 30% of total caloric intake and to keep the level of saturated fats to less than 10%. The nutritional labels (see Chapter 6) help consumers to calculate these percentages as they give the total calories, calories from fat, and grams of fat and saturated fat per serving. With this information, it requires simple arithmetic to determine if guidelines are being met. An example is a food that gives 200 calories per serving and contains 5 grams of fat and 2.5 grams of saturated fat. The calculation is shown below:

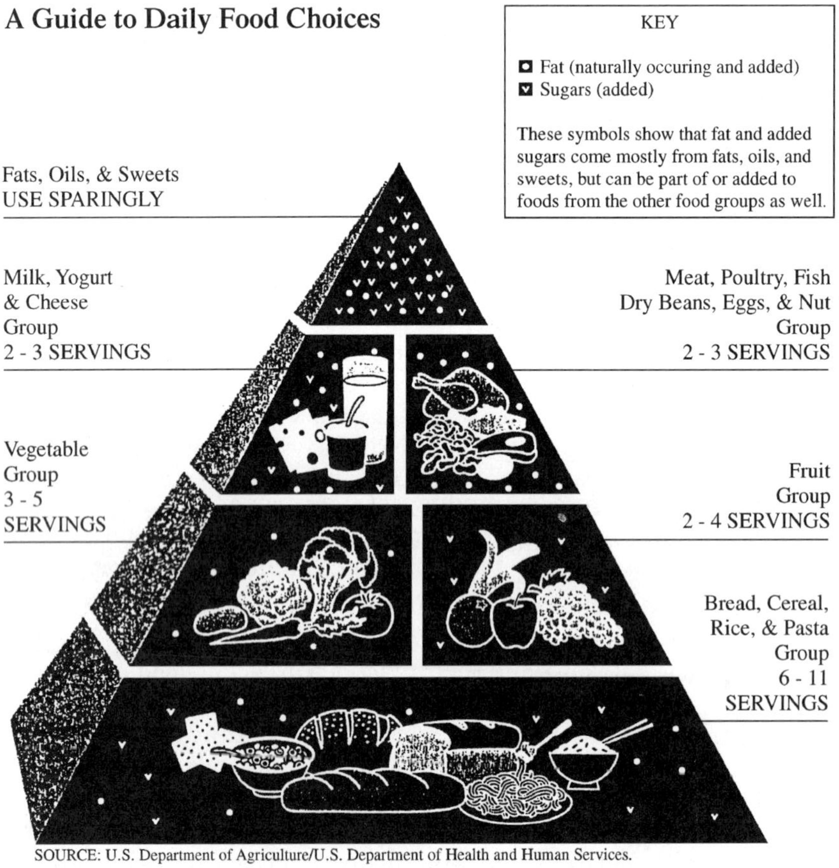

Figure 2.1. The food guide pyramid.

1. Multiply grams of fat times 9 calories per gram to get calories from fat:

$$5 \times 9 = 45$$

2. Divide calories from fat by total calories and multiply by 100 to get percent calories from fat:

$$(45/200) \times 100 = 22.5\%$$

3. Multiply grams of saturated fat times 9 calories per gram to get calories from saturated fat:

$$2.5 \times 9 = 22.5$$

4. Divide calories from saturated fat by total calories and multiply by 100 to get percent calories from saturated fat:

$$(22.5/200) \times 100 = 11.25\%$$

This food meets the guidelines for total fat calories but is slightly high in saturated fat. To use the food in a balanced diet, it must be supplemented with other foods that are low in saturated fats to reduce the total dietary intake to the desired total of less than 10%.

When developing a diet, dietitians use the food pyramid and the nutritional labeling requirements as aids to ensure the recommended dietary allowances (RDAs) (Table 2.1) are met.

PROTEINS

Proteins are the chief organic constituents of muscle and other tissues. Proteins are major components of the enzymes that regulate and carry out the general metabolism and functional processes of living organisms. Proteins are part of the intracellular and extracellular structures of animals; they make up the structure of many hormones and antibodies (disease-resisting components), and are concerned with many other factors involved with body functions. Proteins contain nitrogen, carbon, hydrogen, oxygen, and sometimes sulfur and phosphorus. Most proteins contain nitrogen at a level near 16%. The procedure for finding total protein in a food is determined by analyzing for protein nitrogen, and then multiplying the results by 6.25 to determine the actual amount of protein in the food analyzed. All proteins are composed of amino acids having the general formula:

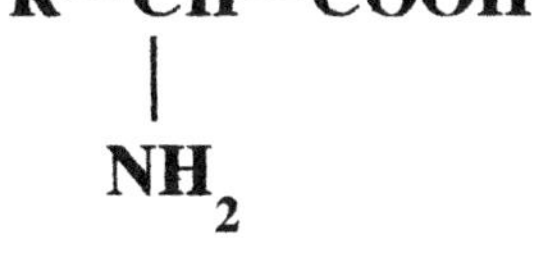

R could represent any one of a variety of chemical structures, the simplest being a hydrogen atom in glycine, whereas other amino acids have much more complex structures (Table 2.2) The R-groups of the amino acids in this table are classified as either

Table 2.1. Recommended Dietary Allowances, Revised 1989[a]
(Designed for the maintenance of good nutrition of practically all healthy people in the United States)

							Fat-Soluble Vitamins			
		Weight[b]		Height[b]						
Category	Age (years) or Condition	(kg)	(lb)	(cm)	(in)	Protein (g)	Vitamin A (μg RE)[c]	Vitamin D (μg)[d]	Vitamin E (mg α-TE)[e]	Vitamin K (μg)
Infants	0.0–0.5	6	13	60	24	13	375	7.5	3	5
	0.5–1.0	9	20	71	28	14	375	10	4	10
Children	1–3	13	29	90	35	16	400	10	6	15
	4–6	20	44	112	44	24	500	10	7	20
	7–10	28	62	132	52	28	700	10	7	30
Males	11–14	45	99	157	62	45	1,000	10	10	45
	15–18	66	145	176	69	59	1,000	10	10	65
	19–24	72	160	177	70	58	1,000	10	10	70
	25–50	79	174	176	70	63	1,000	5	10	80
	51+	77	170	173	68	63	1,000	5	10	80
Females	11–14	46	101	157	62	46	800	10	8	45
	15–18	55	120	163	64	44	800	10	8	55
	19–24	58	128	164	65	46	800	10	8	60
	25–50	63	138	163	64	50	800	5	8	65
	51+	65	143	160	63	50	800	5	8	65
Pregnant						60	800	10	10	65
Lactating	1st 6 months					65	1,300	10	12	65
	2nd 6 months					62	1,200	10	11	65

Source: From Recommended Dietary Allowances, Revised 1989. Food and Nutrition Board, National Academy of Sciences–National Research Council, Washington, D.C.

[a]The allowances, expressed as average daily intakes over time, are intended to provide for individual variations among most normal persons as they live in the United States under usual environmental stresses. Diets should be based on a variety of common foods in order to provide other nutrients for which human requirements have been less well defined.

[b]Weights and heights of Reference Adults are actual medians for the US population of the designated age, as reported by NHANES II. The use of these figures does not imply that the height-to-weight ratios are ideal.

Summary Table: Estimated Safe and Adequate Daily Dietary Intakes of Selected Vitamins and Minerals[a]

		Vitamins	
Category	Age (years)	Biotin (μg)	Pantothenic Acid (mg)
Infants	0–0.5	10	2
	0.5–1	15	3
Children and adolescents	1–3	20	3
	4–6	25	3–4
	7–10	30	4–5
	11+	30–100	4–7
Adults		30–100	4–7

Source: From Recommended Dietary Allowances, Revised 1989. Food and Nutrition Board, National Academy of Sciences–National Research Council, Washington, D.C.

[a]Because there is less information on which to base allowances, these figures are not given in the main table of RDA and are provided here in the form of ranges of recommended intakes.

Water-Soluble Vitamins							Minerals						
Vitamin C (mg)	Thiamine (mg)	Riboflavin (mg)	Niacin (mg NE)[f]	Vitamin B_6 (mg)	Folate (μg)	Vitamin B_{12} (μg)	Calcium (mg)	Phosphorus (mg)	Magnesium (mg)	Iron (mg)	Zinc (mg)	Iodine (μg)	Selenium (μg)
30	0.3	0.4	5	0.3	25	0.3	400	300	40	6	5	40	10
35	0.4	0.5	6	0.6	35	0.5	600	500	60	10	5	50	15
40	0.7	0.8	9	1.0	50	0.7	800	800	80	10	10	70	20
45	0.9	1.1	12	1.1	75	1.0	800	800	120	10	10	90	20
45	1.0	1.2	13	1.4	100	1.4	800	800	170	10	10	120	30
50	1.3	1.5	17	1.7	150	2.0	1,200	1,200	270	12	15	150	40
60	1.5	1.8	20	2.0	200	2.0	1,200	1,200	400	12	15	150	50
60	1.5	1.7	19	2.0	200	2.0	1,200	1,200	350	10	15	150	70
60	1.5	1.7	19	2.0	200	2.0	800	800	350	10	15	150	70
60	1.2	1.4	15	2.0	200	2.0	800	800	350	10	15	150	70
50	1.1	1.3	15	1.4	150	2.0	1,200	1,200	280	15	12	150	45
60	1.1	1.3	15	1.5	180	2.0	1,200	1,200	300	15	12	150	50
60	1.1	1.3	15	1.6	180	2.0	1,200	1,200	280	15	12	150	55
60	1.1	1.3	15	1.6	180	2.0	800	800	280	15	12	150	55
60	1.0	1.2	13	1.6	180	2.0	800	800	280	10	12	150	55
70	1.5	1.6	17	2.2	400	2.2	1,200	1,200	320	30	15	175	65
95	1.6	1.8	20	2.1	280	2.6	1,200	1,200	355	15	19	200	75
90	1.6	1.7	20	2.1	260	2.6	1,200	1,200	340	15	16	200	75

[c]Retinol equivalents. 1 retinol equivalent = 1 μg retinol or 6 μg β-carotene. See text for calculation of vitamin A activity of diets as retinol equivalents.

[d]As cholecalciferol. 10 μg cholecalciferol = 400 IU of vitamin D.

[e]α-Tocopherol equivalents. 1 mg D-α tocopherol = 1α-TE. See text for variation in allowances and calculation of vitamin E activity of the diet as α-tocopherol equivalents.

[f]1 NE (biacin equivalent) is equal to 1 mg of niacin or 60 mg of dietary tryptophan.

Category	Age (years)	Trace Elements[b] Copper (mg)	Manganese (mg)	Fluoride (mg)	Chromium (μg)	Molybdenum (μg)
Infants	0–0.5	0.4–0.6	0.3–0.6	0.1–0.5	10–40	15–30
	0.5–1	0.6–0.7	0.6–1.0	0.2–1.0	20–60	20–40
Children and adolescents	1–3	0.7–1.0	1.0–1.5	0.5–1.5	20–80	25–50
	4–6	1.0–1.5	1.5–2.0	1.0–2.5	30–120	30–75
	7–10	1.0–2.0	2.0–3.0	1.5–2.5	50–200	50–150
	11+	1.5–2.5	2.0–5.0	1.5–2.5	50–200	75–250
Adults		1.5–3.0	2.0–5.0	1.5–4.0	50–200	75–250

[b]Since the toxic levels for many trace elements may be only several times usual intakes, the upper levels for the trace elements given in this table should not be habitually exceeded.

Table 2.2. Amino Acid Structures

Polar uncharged R-groups	Charged R-groups:

Polar uncharged R-groups:

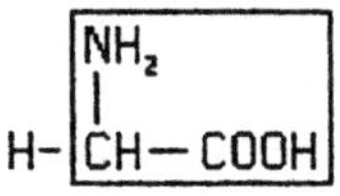

Glycine

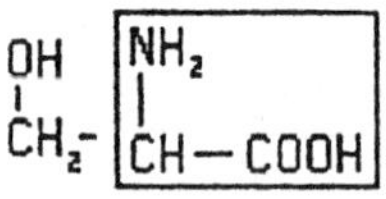

Serine

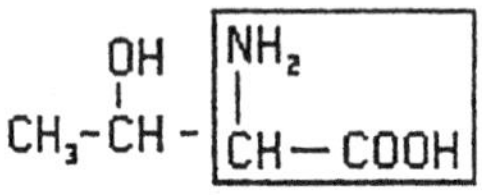

Threonine

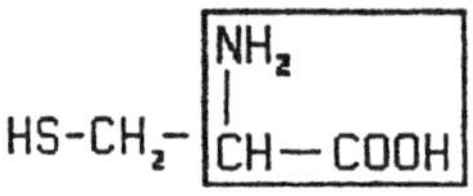

Cysteine

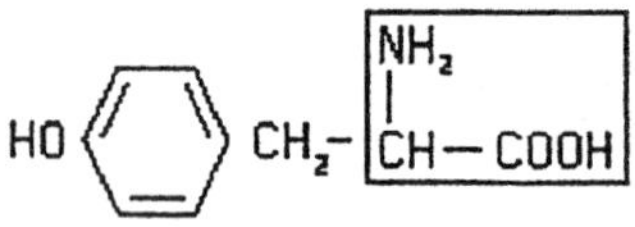

Tyrosine

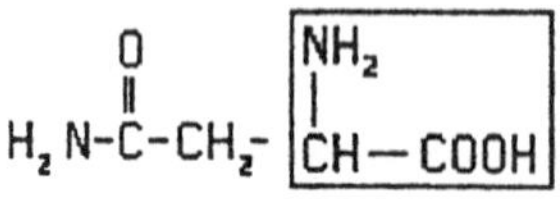

Asparagine

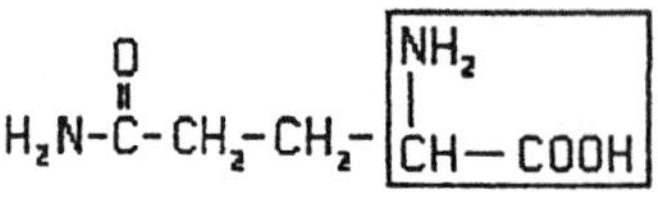

Glutamine

Charged R-groups:

Negatively charged

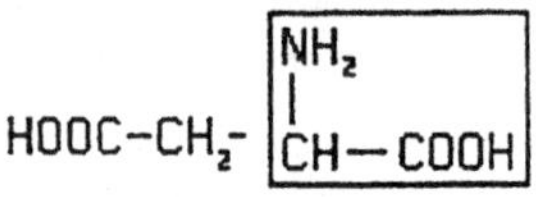

Aspartic acid

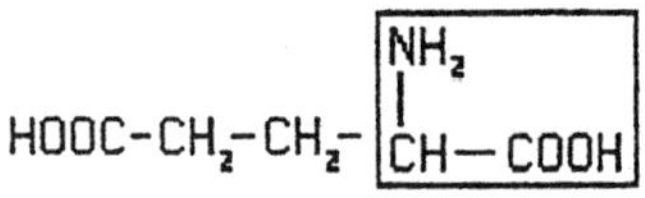

Glutamic acid

Positively charged:

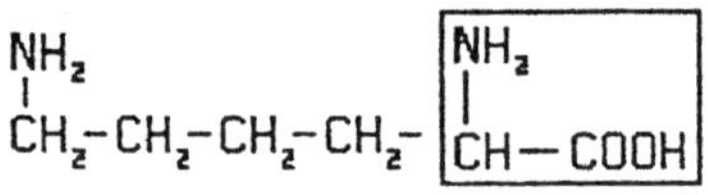

Lysine

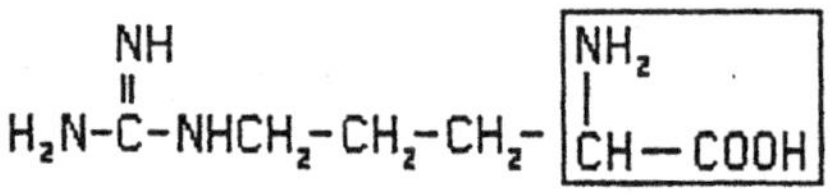

Arginine

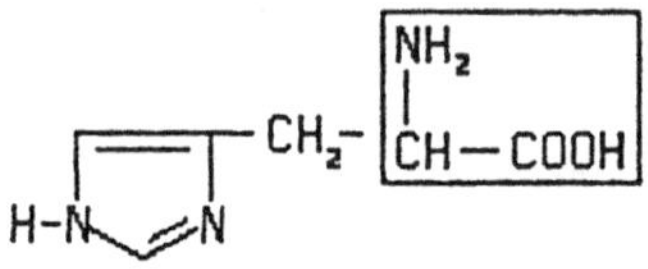

Histidine

Table 2.2. Amino Acid Structures *(continued)*

Essential Nonpolar R-groups

Essential Nonpolar R-groups:

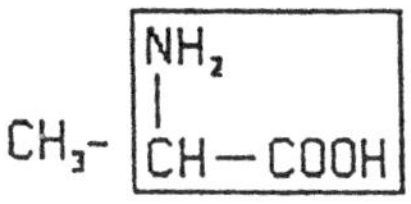

Alanine

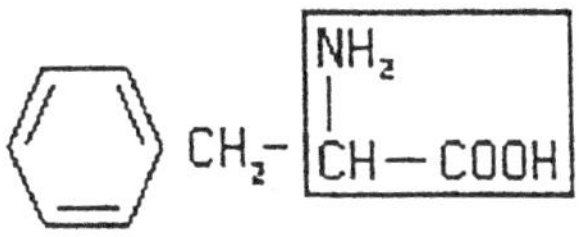

Phenylalanine

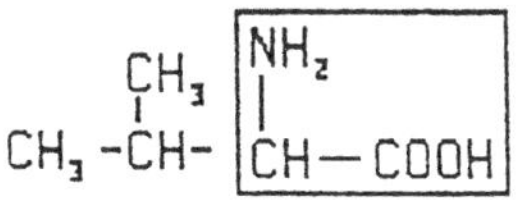

Valine

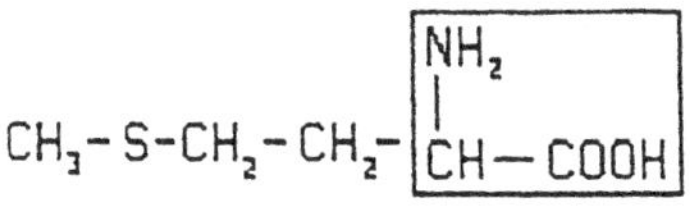

Methionine

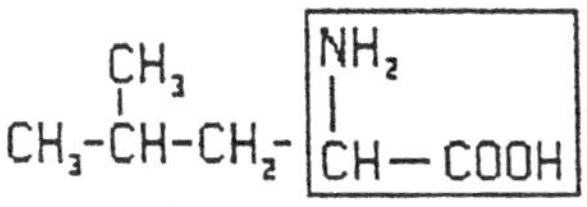

Leucine

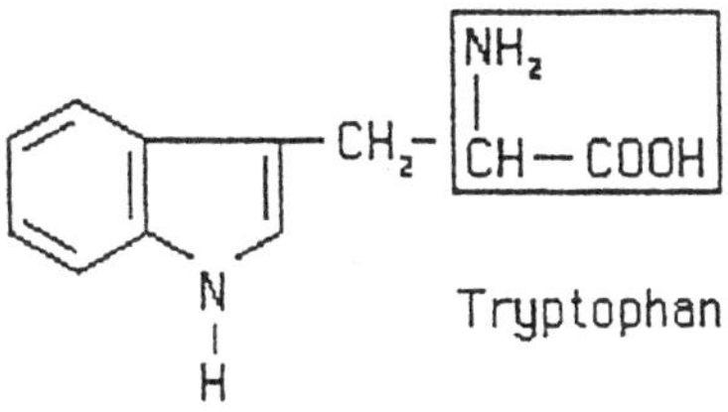

Tryptophan

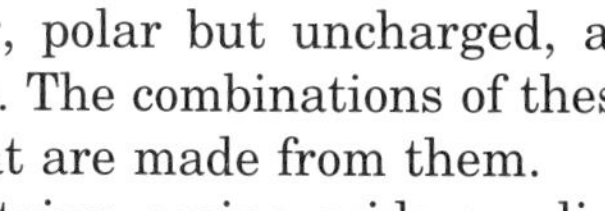

Isoleucine

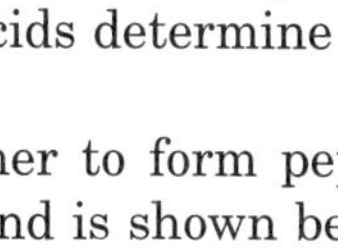

Proline

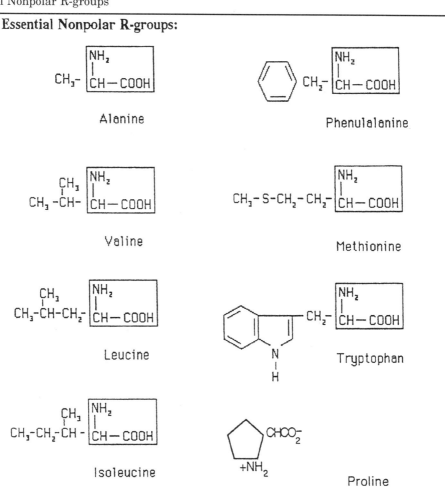

nonpolar, polar but uncharged, acidic (negatively charged), or alkaline (positively charged). The combinations of these amino acids determine the properties of the proteins that are made from them.

In proteins, amino acids are linked together to form peptides by a peptide bond (–CO–NH–). The formation of the peptide bond is shown below:

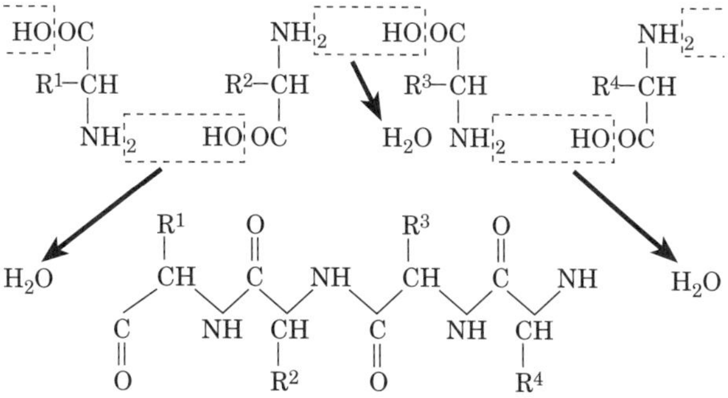

The bond links the carboxyl group (COOH) of one amino acid with the amino group (NH_2) of another amino acid, releasing one water molecule.

Because proteins contain carbon, they can be used as fuel because part of the molecule can be oxidized, sometimes involving deamination, to supply energy.

Proteins are required by humans for growth (protein synthesis) and for repair and maintenance of cells. Because mature adults have in essence ceased to grow, their protein requirement is less, per unit weight, than that of those who are still growing.

Although humans require proteins, not all proteins are of a suitable composition to supply the needs of the body, especially that of the growing child. Some proteins, such as those in animal products such as meat, milk, and eggs, are considered "high quality." Proteins of high quality contain all the essential amino acids in the amounts needed to support protein tissue formation by the body. Most plant proteins are not considered complete although soybeans are very nearly so and are actually sometimes believed to be complete. Essential amino acids are those that cannot be synthesized by the body in adequate amounts and, therefore, must be supplied by the diet. They are sometimes referred as "indispensable amino acids." The other amino acids are referred to as nonessential or "dispensable" and can be made by the body if the diet has supplied sufficient amounts of the essential amino acids. If any of the essential amino acids are missing from the diet, proteins are not formed, even those that could be produced from available amino acids. This may seem inefficient but if the body did not cease all protein formation, cells would end up with an imbalance of proteins which would seriously affect cell functions. In all, there are 20 amino acids, nine essential and eleven nonessential. These are listed below:

Essential	Nonessential
Histidine	Alanine
Isoleucine	Arginine
Leucine	Asparagine
Lysine	Aspartic acid
Methionine	Cysteine
Phenylalanine	Glutamic acid
Threonine	Glutamine
Tryptophan	Glycine
Valine	Proline
	Serine
	Tyrosine

Amino acids cannot be stored in the body for very long so a fresh supply of essential amino acids is needed daily.

Some Properties of Proteins

Solubility

There are great differences in solubility among proteins. These differences are determined by the amino acid content and sequence. Some proteins, such as histones, which are involved in protein replication, and albumins such as ovalbumin in egg white, lactalbumin in milk, and albumins of blood are soluble in pure water. Some proteins are insoluble in water, for example, keratin in hair and fingernails and collagen in bones, cartilage, connective tissue, and epidermis. Others are generally insoluble in water but are soluble in weak salt solutions. These include myosin in muscle tissue and lactoglobulins in milk. Others (wheat glutenin, oryzenin in rice) are not soluble in water but are soluble in acids or alkali. Still others such as wheat gliadin and corn zein are soluble in 50% to 80% ethanol.

Water Binding

Proteins such as those in the muscle tissue of meats are able to bind water molecules through forces such as hydrogen bonding. This is a great advantage when making ground, comminuted products such as hot dogs and bologna. An emulsion is made with the fat, protein, and water and when the product is cooked (smoked), it will retain much of the moisture and a palatable, attractive product is obtainable.

Structure

Protein structures are quite different than those of other food constituents. Molecular weight, amino acid composition, and sequence along the polypeptide chain constitute the "primary structure" of the molecule. The spatial arrangement of the various atoms and groups in a molecule is responsible for a number of other structures which make up the "conformation" of the molecule including secondary, tertiary, and quaternary structures.

The secondary structure is in the shape of a coiled helix. This is believed to be formed as the result of each C=0 group forming a hydrogen bond with each NH group in the chain (Fig. 2.2).

Tertiary structures are a result of the folding of a chain over itself. The result is a three-dimensional state that is very important to the level of protein activity. Amino acid residues that are far away from each other in the sequence come closer together because of the folding of the structure. In a water medium, the polar side chains of the structure tend to be pushed toward the surface, while the hydrophobic chains are pushed toward the middle of the protein molecule. The result is a structure that may be either fibrous or roughly spherical, whichever gives it better stability in water.

A quaternary structure is also possible if two or more of these polypeptide chains join together. This final molecule is a huge tangled, complicated chain of amino acids, bristling on the surface with positive and negative charges. These large protein molecules are quite fragile and the food scientist must be aware of reactions due to exposure to heat, acids, salt, or other conditions that may disturb their stability.

There are more stable, less fragile structures in proteins also. The pleated sheet structure is a result of hydrogen bonds between each C=0 and NH of different chains running side by side (Fig. 2.3). This is a secondary structure and is common in the fibrous proteins such as keratin and collagen.

Denaturation

As mentioned earlier, the conformation or the secondary, tertiary, and quaternary structures can be quite fragile and may be altered by a number of factors that may be utilized in food processing. This change in molecular structure without breaking covalent bonds or altering amino acid sequence is referred to as denaturation. With denaturation usually comes the loss of biological activity and significant changes in

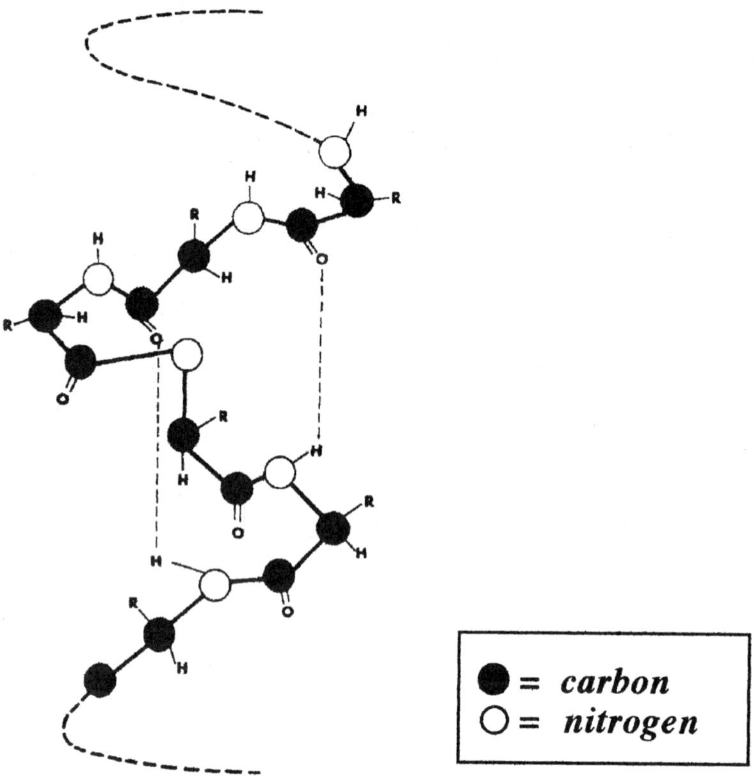

Figure 2.2. Helical structure of a protein.

some physical or functional properties such as solubility. Denaturation occurs only in proteins and proteinlike substances and has different levels of effect on individual proteins. Enzymes, which are primarily protein substances, can be denatured, causing the biochemical functions that they catalyze to cease.

Denaturation may be caused by a number of factors such as heat, acids, solvents such as ethyl alcohol, concentrated solutions of some salts, and by surface forces. Besides the loss of biological activity and solubility mentioned earlier, irreversible gels may be formed, their optical rotation (how polarized light rotates when sent through a solution containing the protein) is altered, they may become more susceptible to enzymatic hydrolysis (making them more digestible), and the sulfhydryl groups of some amino acids are more exposed. Very often, the result of denaturing of a globular protein is the unfolding of the polypeptide chain to a randomly coiled polymer (Fig. 2.4). Denaturation, in most cases, is said to be "practically irreversible," which means that the renaturation process, if any, will be much slower than the denaturation process.

Denaturation and its control is essential to the food technologist. Some results of denaturation include the inactivation of enzymes, which may cause problems in storage. The "blanching" step in freezing and canning of many fruits and vegetables is done to denature enzymes. Denaturation contributes to the flavor and texture of foods. Cheeses and yogurt are direct results of denaturation of milk proteins. The flavor of cooked protein-rich foods such as eggs and milk largely results from the exposure of some of the –SH (sulfhydryl) groups in the conformation of the molecule when it unfolds upon being cooked. The hardening of an egg white in a hot frying pan is an example of heat

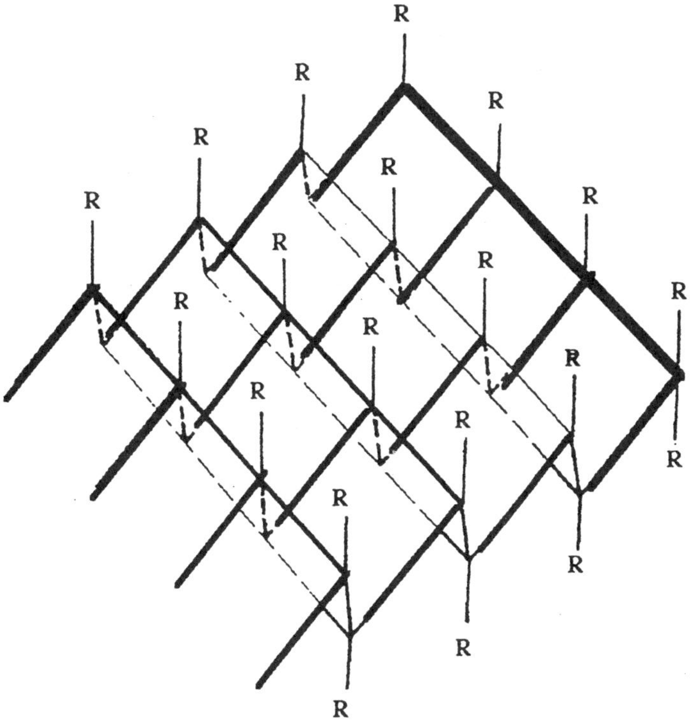

Figure 2.3. Pleated-sheet structure of a protein.

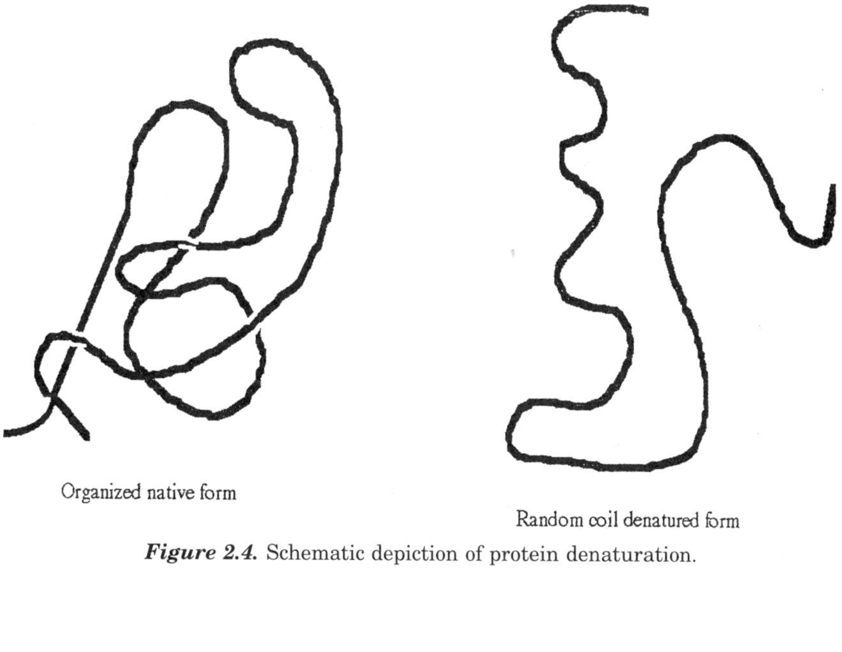

Figure 2.4. Schematic depiction of protein denaturation.

denaturation and the whipping of egg whites to a foam results from exposure to surface forces. Many meat proteins that are cooked to temperatures of about 134 to 167°F (57 to 75°C) will have profound changes in texture, water holding capacities, and shrinkage.

Reduction

As shown in Table 2.2 the amino acid cysteine contains –SH groups and in a protein molecule these are the most susceptible to oxidation (loss of electrons). Even in the presence of mild oxidizing agents the following reaction takes place:

$$\mathrm{R{-}SH + HS{-}R} \underset{(2H)}{\overset{(O)}{\rightleftharpoons}} \mathrm{R{-}S{-}S{-}R}$$

The formation of disulfide bonds from two sulfhydryl groups of the same molecule or of different molecules results in the increase of rigidity in the protein texture. This is accomplished in bread flour when bromates or other oxidizing agents are added and the physical strength of the wheat protein gluten is increased. Cysteine itself acts as a reducing agent, causing the gain of electrons, and it is added to foods when oxidation is to be reversed or prevented.

Hydrolysis

Hydrolysis in proteins is the cleaving of the peptide bonds by the addition of a water molecule:

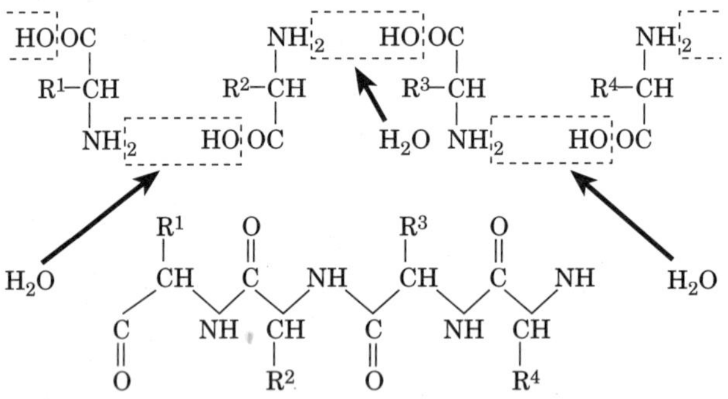

This can be caused by strong acids, alkalis, and certain enzymes.

Hydrolysis may be complete or partial. When complete hydrolysis is accomplished, the protein is broken down to its amino acid components. In partial hydrolysis, smaller peptide chains containing few amino residues are formed. These peptides are used in the food industry as protein hydrolysates. In a method to make the protein hydrolysates, plant proteins are hydrolyzed by strong hydrochloric acid and then neutralized with alkali. The result is a mixture of peptides and amino acids that is concentrated or dried. The protein hydrolysates in this form can be used as flavor components in soup, sauce, gravy and similar mixtures.

CARBOHYDRATES

To carry out its day-to-day physiological functions and maintain a constant body temperature (invariably in an environment of changing temperatures, usually less than that of body temperature), the body requires a constant source of energy. Beyond its continuing maintenance need for energy, the body periodically needs relatively larger amounts of energy to do work or to engage in other vigorous physical activities. Humans derive their energy mainly from carbohydrates (55% to 65%), although they can also utilize fats and proteins for this purpose.

The carbohydrates are a class of compounds that consist of carbon, oxygen, and hydrogen. The carbohydrates that are important in nutrition include sugars, starches, dextrins, and glycogen. Others that are not digestible (the fibers) do not supply calories but are very important in the overall health and well being of the human body.

Sugars

The sugars are the simplest of the carbohydrates and the simplest of the sugars are those that cannot be broken down any further by hydrolysis. These are the monosaccharides and the most common ones in foods are the hexoses or six-carbon sugars. Among the many theoretically possible monosaccharides, five that occur free in nature are glucose, mannose, galactose, fructose, and sorbose. The open chain formulas for these are shown below:

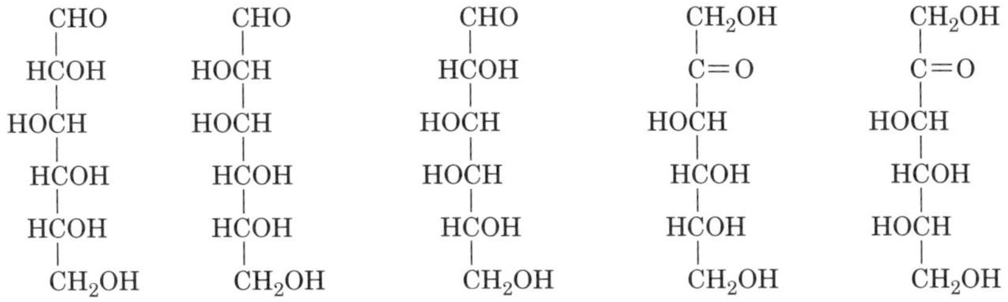

D–glucose D–mannose D–galactose D–fructose L–sorbose

You can see that all but one (L-sorbose) has a "D" prefix. This refers to the configuration of the hydroxyl group on the last carbon before the end or the fifth carbon as is the case in hexoses. If the sugar has the D configuration, the OH appears on the right (dextro). The L configuration is the opposite, with the OH on the left (levo). Sorbose is the only one shown that has this configuration. Each monosaccharide can exist in either the D or L configuration, and these are called enatiomorphs or mirror images.

As you can also see in the structures, two of the sugars have a different location of the carbonyl group (C=0) in the molecule. The first three, having the carbonyl group at carbon-1, are called polyhydroxyaldehydes or aldoses and the other two, with the carbonyl carbon at carbon-2, are called polyhydroxyketones or ketoses. They all have the same molecular weight and general formula of $C_6H_{12}O_6$. Sugars, like proteins, are optically active and, in solution, will rotate polarized light to either the left or right. This is determined only by testing with a laboratory apparatus that measures optical rotation (a polarimeter). The direction of the optical rotation cannot be predicted by the structures alone and, in some cases, for example, D-fructose, polarized light is described as D-(−) fructose with the D denoting the configuration of the OH group at carbon-5 and the (−) telling that polarized light is rotated to the left when a solution of the sugar is tested with the polarimeter.

The open chain structure, designating sugars as true aldehydes or ketones, is inadequate in fully describing monosaccharides. Aldehydes, for example, tend to condense with alcohols to form hemiacetals:

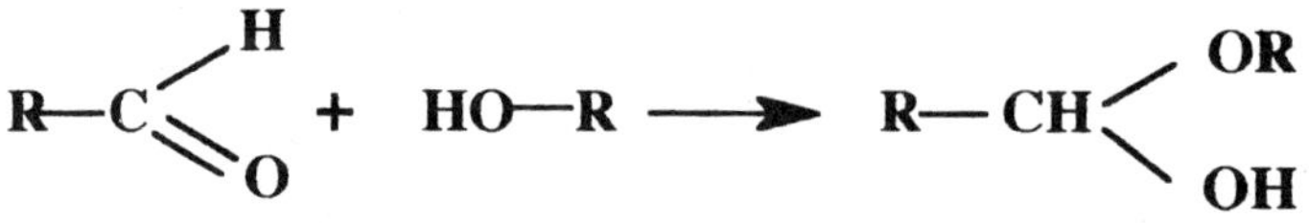

In the open chain structures, it can be seen that the carbonyl and hydroxyl groups needed for this reaction are in the same molecule. An "inner hemiacetyl bridge" is formed and the closed six-membered ring structure results:

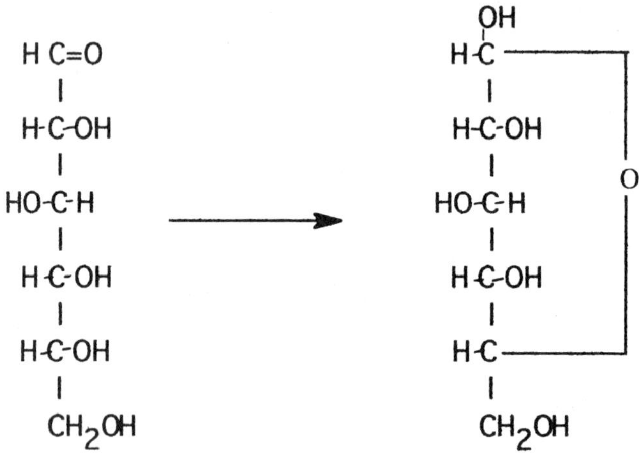

This can also happen with the ketoses as a five-membered ring is formed:

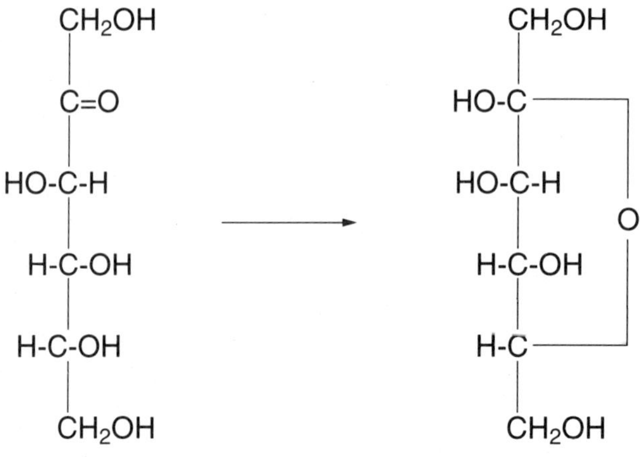

With the formation of the hemiacetyl ring, carbon-1 becomes asymmetric and two stereoisomers are possible:

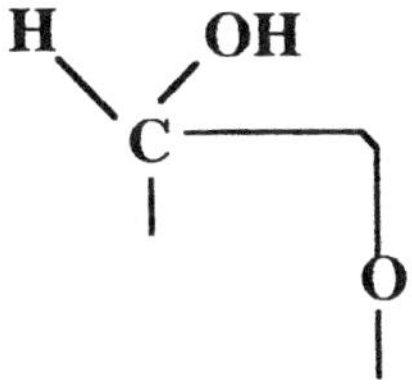

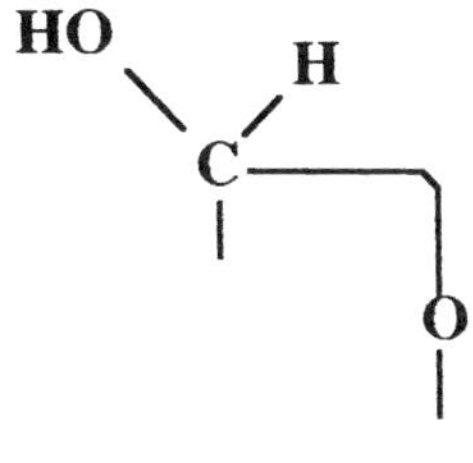

Most monosaccharides can exist in either manner and the configuration with the OH on the right is referred to as the alpha configuration and when the OH occurs on the left, it is referred to as the beta configuration. Glucose in solution exists as a mixture of the alpha and beta configurations.

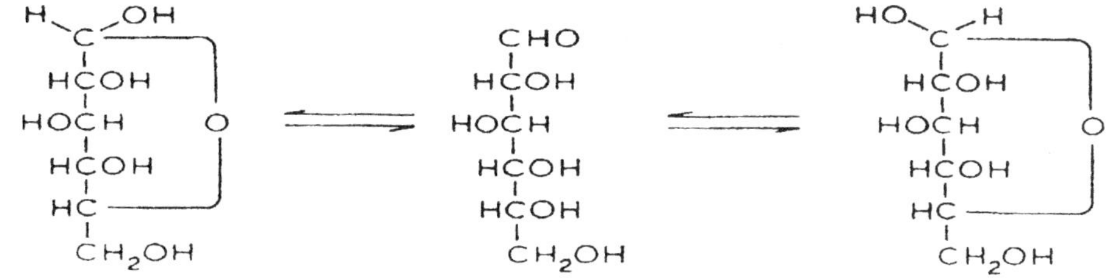

α-D(+) glucose open-chain glucose β-D(+) glucose

In solution, the sugars may convert from the alpha to beta configurations until equilibrium is reached (the ratio of alpha to beta does not have to be equal for this to occur). This conversion is referred to as mutarotation.

Monosaccharides that have six-membered rings are referred to as pyranoses and those with five-membered rings are furanoses. These configurations for two sugars in their alpha and beta forms are shown below:

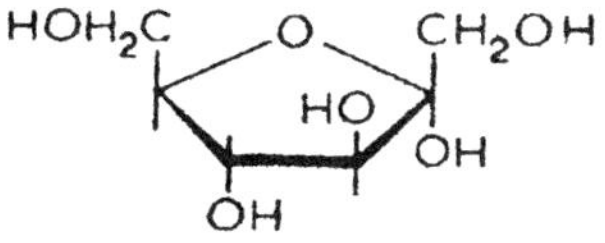

α-D(-) **fructose**

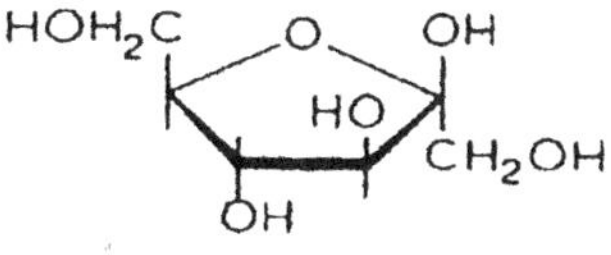

β-D(-) **fructose**

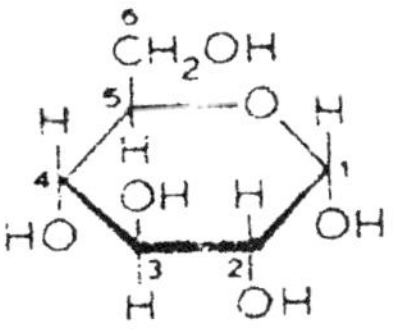

α-D(+) **glucose**

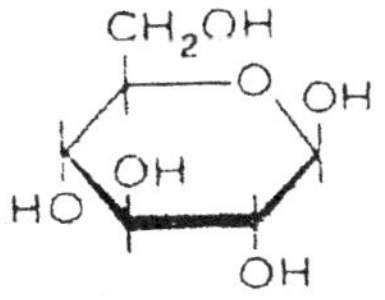

β-D(+) **glucose**

These configurations show the OH groups either in the "up" or "down" configuration with "down" being the alpha configuration and "up" denoting the beta. These correspond to "left" and "right" in the open chain presentation. The locations of these hydroxyl groups at carbon-1 in the pyranoses and carbon-2 in the furanoses will be very important in the digestibility of polysaccharides that are formed from monosaccharides.

More complex sugars may be formed as two or three molecules of monosaccharides join together. When this happens, water molecules are formed:

glucose + glucose → maltose + water

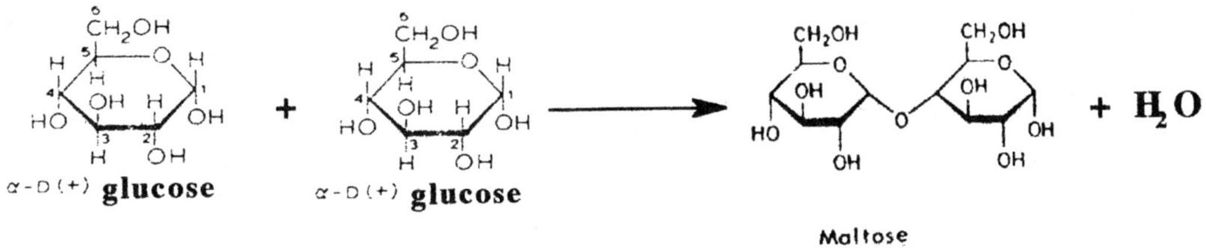

This explains why monosaccharides are shown by the formula $C_6H_{12}O_6$ and disaccharides (made from two monosaccharide units) are shown as $C_{12}H_{22}O_{11}$.

Disaccharides can be composed of two units of the same monosaccharide or of two different monosaccharides. Some common disaccharides are shown below:

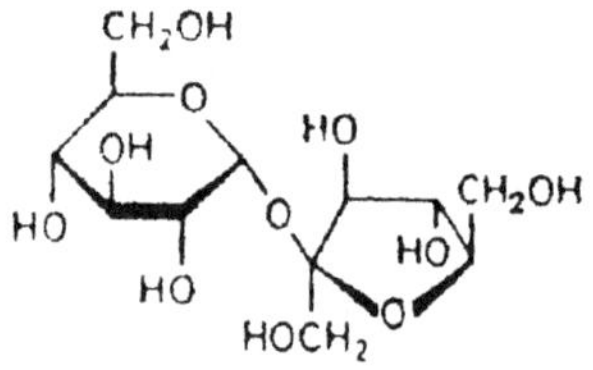
sucrose

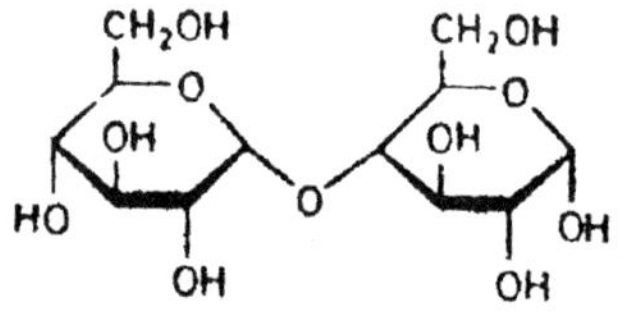
maltose

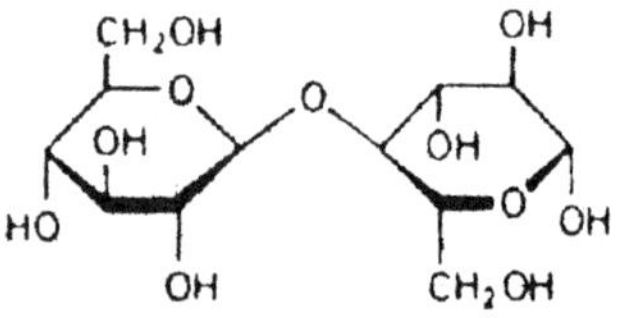
cellobiose

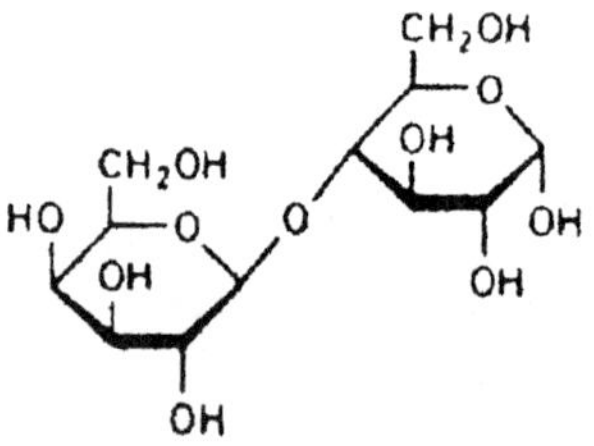
lactose

These show two disaccharides composed of two molecules of the same monosaccharide and two disaccharides composed of two different monosaccharides. Sucrose, made from glucose and fructose, is found in nature and is quite common. Cane sugar is mainly sucrose and is refined and sold as table sugar. Whenever you see "sugar" on the ingredient label of a food, sucrose is the chemical that is present. Lactose, the other sugar made from two different monosaccharides (glucose and galactose), is a major sugar in milk. The other two sugars, maltose and cellobiose, are each composed of two molecules of glucose. Neither of these exist free in nature but are byproducts of either metabolism or synthesis. One might ask why these disaccharides are different if they both have the same molecular weight and both are made from two glucose molecules.

Closer examination of the configurations will show that the linkages between the two glucose molecules are different. Both link between carbon-1 of the first molecule and carbon-4 of the second molecule but the OH's of the number 1-carbons in the maltose molecule are in the alpha configuration and the OH's on the number 1-carbons of the glucose molecules in the cellobiose molecule are in the beta configuration. Thus, it is said that maltose is made from two glocuse molecules joined with an alpha-1,4 linkage and cellobiose is made from two glucose molecules joined with a beta-1,4 linkage. Both of these molecules are used as building blocks for the polysaccharides starch and cellulose, respectively. This one difference (linkage) yields one product that has 4 calories per gram (starch) and one that is not digestible (cellulose).

Oligosaccharides (*oligo* = few) are described as those molecules that have between 2 and 10 monosaccharide units. The disaccharides are obviously included in this group by definition but were given separate attention because of their importance in food processing and nutrition. There are several trisaccharides that are found in nature and found in trace amounts in crystallized commercial beet sugar. It is made up of one molecule each of D-galactose, D-glucose, and D-fructose, linked with an alpha-1,6 and alpha-1,2 linkage.

A tetrasaccharide called stachyose made from two D-galactose units, one D-glucose, and one D-fructose unit is linked by two alpha-1,6 and one alpha-1,2 linkage. Stachyose is one of the main carbohydrates in soybeans, and because of its poor digestibility is thought to be a cause of flatulence and other gastrointestinal disturbances experienced after eating soy and other beans.

Sugars are used extensively in food processing for a number of reasons. The most obvious is use of sugar as a sweetener but, as shown in Table 2.3, not all sugars are very sweet. Some (lactose) are only 16% as sweet as sucrose.

Sugars may also be added for texture and appearance qualities. Sugar's contribution to viscosity is important in the consistency, body, and mouth feel of some foods. Water in food may be available for microbiological use or it may be chemically bound and add a moist texture to food but not offer a source of nutrients for microorganisms. The availability of water to microorganisms is measured as water activity and sugars, with their affinity for water, depress water activity in some foods such as jams and jellies, giving a preservative effect. (See Chapters 2 and 13 for more on water activity.) Sugar also has a high refractive index and is responsible for the shiny appearance in high-sugar products such as syrups, jellies, and dried fruits.

Table 2.3. Degree of Sweetness of Various Sugars and Other Sweeteners

Sweetener	Degree of Sweetness
Sucrose	100
Fructose	173.3
Glucose	74.3
Lactose	16
Maltose	32
Galactose	32
Saccharin	30,000–50,000
Sodium cyclamate	10,000
Neohesperidin dihydrochalcone	1,000,000

Polysaccharides

Polysaccharides are polymers (compounds of many smaller molecules) of the simple monosaccharides joined together by glycosidic bonds. They may contain repeated structures of the same monosaccharide (homopolysaccharides) or may contain two or more different monosaccharide units in their long chains (heteropolysaccharides). When the chains total ten or more units, the molecule is defined as a polysaccharide but most polysaccharides in nature consist of hundreds to several thousand of monosaccharide units. Polysaccharides, unlike many sugars, are not sweet, but are responsible for another sensory characteristic important in foods, texture. Viscosity, mouth feel, consistency, gelatin, smoothness, and toughness are some characteristics for which starches are responsible. Nutritionally, starch is the most important polysaccharide, providing the main source of calories in the human diet. Other polysaccharides that are mentioned in this section are glycogcn, polydextrose, and fibers.

Starch

Starch is a polymer of alpha-D-glucose and is found as a storage carbohydrate in plants. It is composed of two different polymers: amylose, which is a linear compound, and amylopectin, a branched compound. The glucose molecules in amylose are joined together with the alpha-1,4 linkage and whenever the branches occur to form an amylopectin molecule, the alpha-1,6 linkage occurs (Fig. 2.5). Upon hydrolysis (Fig. 2.6), shorter chained polysaccharides, oligosaccharides, disaccharides, and finally monosaccharides are made by introducing water at the position of the glycosidic bonds. This reaction can be catalyzed by the presence of acid, alkali, or various enzymes. Five of the major enzymes that can catalyze starch hydrolysis are alpha-amylase, beta-amylase, pullulanase, amyloglucosidase, and maltase. Alpha-amylase is widely distrib-

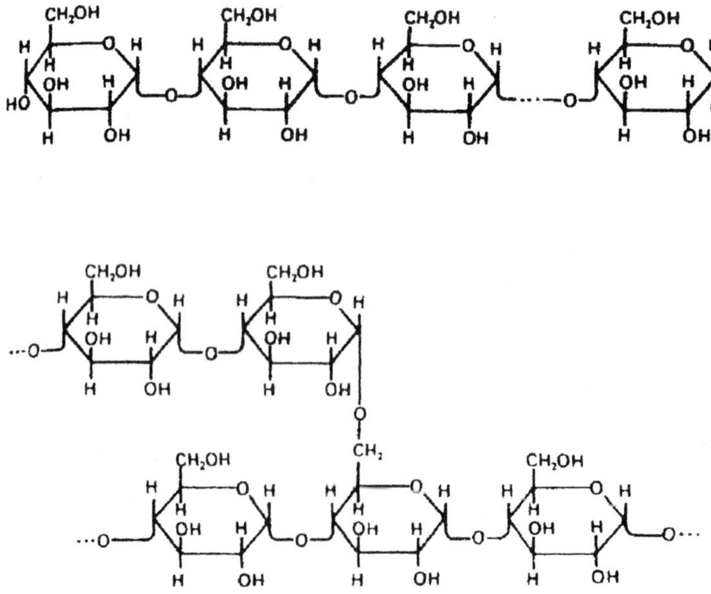

Figure 2.5. Starches.

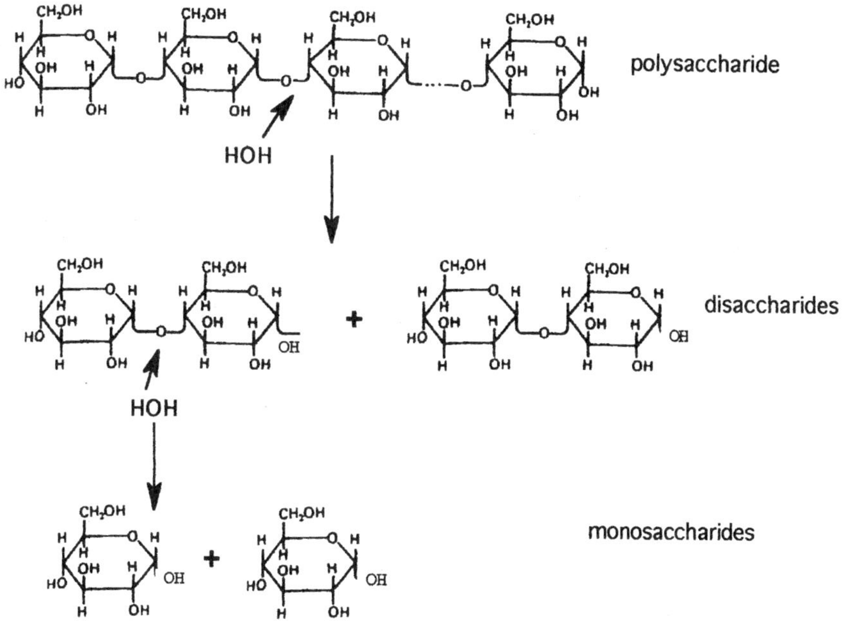

Figure 2.6. Hydrolysis of polysaccharides.

uted in nature, occurring in saliva and pancreatic excretions in mammals as well as in plants and microorganisms. Alpha-1,4 linkages are hydrolyzed in the presence of this enzyme, resulting in a mixture of glucose, maltose, and some short polysaccharides which, because of the presence of branches, are not hydrolyzed further. Beta-amylase is found almost exclusively in higher plants. The products of beta-amylase hydrolysis are similar to those of alpha-amylase except mostly maltose molecules and the branched polysaccharides remain. Pullulanase is specific to the alpha-1,6 linkage and is needed to hydrolyze a branched molecule to allow for the eventual release of the monosaccharide units. Amyloglucosidase is found mainly in molds and can hydrolyze both alpha-1,4 and alpha 1,6-linkages. Maltase is capable of hydrolyzing any remaining maltose molecules into two glucose molecules.

Starches generally are not readily soluble in cold water but, when heated, as a result of water uptake, the starch molecules swell and gelatinize. The viscosity increases, forming a paste, and upon cooling, a gel is formed. Starch pastes are often used to thicken foods and gels can be modified when combined with sugar or acid and used in puddings.

Gel strength in starches is interesting and very important. Starches used in products such as canned soups or stews, where very high temperatures for long periods of time are needed in the canning process, must be able to withstand the abuse and continue to hold water and maintain a smooth, but thick texture. Generally speaking, those starches with more amylose in their structure will form a stronger gel. The analogy of a cut tree is used. A stack of logs represents amylose, and a pile of branches represents amylopectin. The pile of branches is larger but is much less stable than the stack of logs.

Food technologists have been able to modify starches, changing their properties. These modifications make it possible to achieve different textural characteristics in a variety of products that undergo various processing procedures. One way to reduce the viscosity of starches is to treat them with acid or enzymes, resulting in hydrolysis of some of the bonds. This low viscosity is desired in some sauces, toppings, and gravies. Starches can also be treated with oxidizing agents such as sodium hypochlorite to reduce viscosity and achieve paste clarity. These starches can be used as emulsion stabilizers and thickeners in dressings and spreads. Starches may also be cross-bonded by reacting starch suspensions with products such as sodium trimetaphosphate in the presence of an alkaline catalyst. This process inhibits disintegration of the swollen cooked granules which helps keep high viscosity in the presence of acid or high shear. Freeze–thaw stability is needed in some products such as frozen meat pies and some frozen desserts. To achieve this, starch derivatives such as starch acetate, starch phosphate, and hydroxypropyl starch are used. Starch granules may also be precooked to produce a starch that will swell in cold water. This type of starch is necessary in products such as instant pudding.

Retrogration is a characteristic of starches that involves primarily the association or attraction of amylose molecules. Results of this can be varied depending on the processing procedures the food product is exposed to. One example occurs in frozen products containing gravies thickened with high-amylose-containing starches. After freezing, the amylose molecules undergo retrogration rather rapidly, resulting in exuding of water from the starch gel. Upon thawing, the gravy loses its viscosity and seems completely liquid. To avoid this, modified starches or those high in amylopectin such as "waxy" maize are used. Another result of retrogration is the staling of bread and other baked goods such as low-fat cakes. Again, the unbranched amylose molecules are the first to get involved and when they associate, a hardening of the crumb (staling) results. Amylopectin retrogrades very slowly, and in its presence, the retrogration of amylose is inhibited. Stale bread can be "softened" and retrogration temporarily reversed. This can be accomplished by slightly moistening and heating for a few minutes in a hot oven or by wrapping in a moist paper towel and microwaving for a short time.

Glycogen

Glycogen is an animal starch that is a highly branched system of glucose units resembling amylopectin. Glycogen is produced in the liver from glucose and is stored in the liver, as well as in the muscles, where it is available for immediate use as energy. Both the liver and the muscles can store only a limited amount of glycogen; therefore, when an excess of carbohydrates is ingested, there will be a tendency to develop an excess of glycogen. The excess carbohydrates will be converted to fat and stored in the body as such. The body maintains an equilibrium between glucose, the energy-producing sugar, and glycogen, which can be converted to glucose as the glucose in the blood is used up to produce energy. The nutritional value of glycogen in foods is minimal and animal foods are not considered a good source of carbohydrates.

Energy is produced from glucose, whether from glycogen or from plant sources, by oxidation of the sugar with the release of water and carbon dioxide, which are easily removed from the body.

$$C_6H_{12}O_6 + O_2 \rightarrow E + 6\ CO_2 + 6\ H_2O$$

glucose + oxygen yields energy + carbon dioxide + water

Polydextrose

Polydextrose is used as a food additive for texture. It is a randomly bonded condensation polymer of glucose synthesized in the presence of minor amounts of sorbitol and

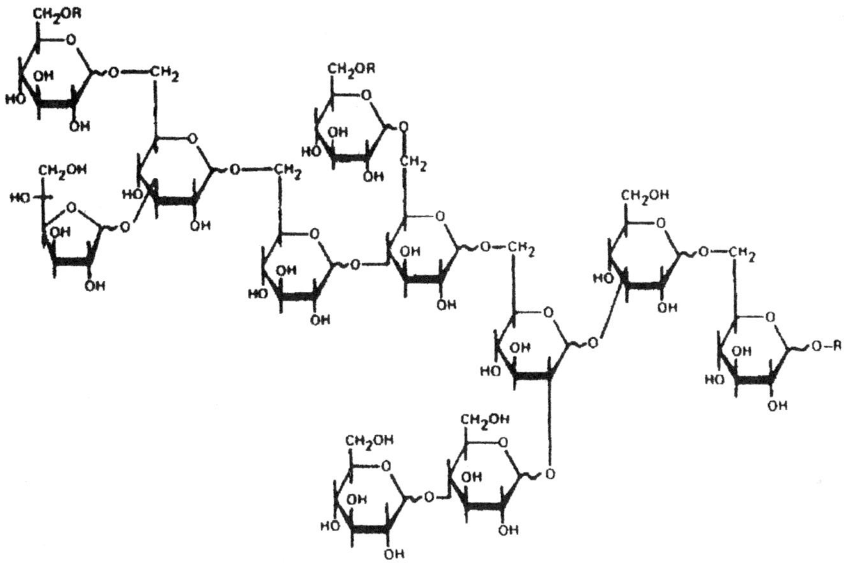

Figure 2.7. Polydextrose chemical bonding.

citric acid (Fig. 2.7). It is highly branched and, because of its unusual structure, is not readily digested and contains only one calorie per gram. It can be used as a "bulking agent" in low-calorie products to replace sugar, bind water, and add to textural attributes without greatly increasing calories.

Fibers

Dietary fiber includes the nondigestible carbohydrates. These may be either water-soluble or water-insoluble. Both have nutritional significance. The water-insoluble group, which includes wheat products and wheat bran, is believed to reduce chances of colon cancer by increasing bulk and diluting the effect of secondary bile acids. The water-soluble fibers, such as some cereal brans and pectin, are believed to lower serum cholesterol levels by binding with bile acids and causing removal of cholesterol in the feces. Short discussions of cellulose, pectin, and gums are given here.

Cellulose comprises most of the structural material in plants and is the main component of many industrially important substances including wood, paper, and fibers such as cotton. Chemically, cellulose is a linear polymer of D-glucose molecules with beta-1,4 bonds:

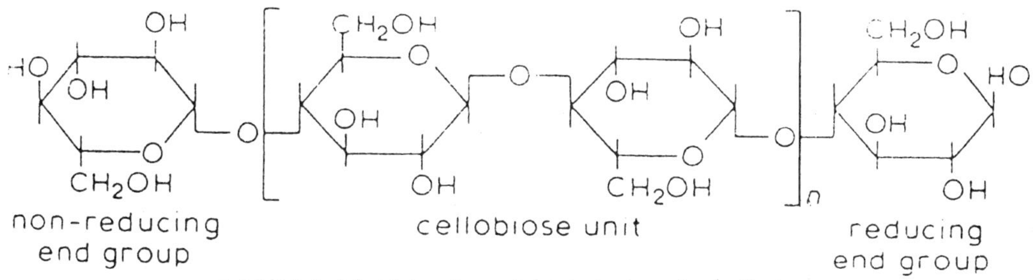

section of structural formula of cellulose

In nature, cellulose occurs in fibers that contain very high mechanical strength and are insoluble in water. The cellulose molecule is elongated and rigid. The hydroxyl group protruding from the chain may readily form hydrogen bonds, giving cellulose this tough texture. Upon hydrolysis, cellobiose and finally glucose are produced. Beta-glucosidase is an effective cellulase enzyme and is able to catalyze cleavage of the beta-1,4 linkages. These enzymes occur in fungi and in some bacteria. Man and most animals are unable to digest cellulose because they lack cellulases in their digestive tracts. Ruminants such as cows and sheep are able to utilize cellulose as energy sources because of the presence of cellulase-excreting microorganisms in the rumen. Similarly, termites utilize microorganisms living in their digestive tracts also to hydrolyze wood cellulose.

Treating cellulose with dilute acid yields "microcrystalline cellulose" which consists of very fine particles that can be used as an indigestible noncaloric food additive. Many low-calorie muffins and other baked products use this type of ingredient to increase volume and mass without adding calories.

Pectin is a water-soluble fiber that exists in the intercellular spaces, the middle lamella, of plant tissues. Pectins are heteropolysaccharides, as their structure includes more than one type of molecule. The unbranched chains consist of polygalacturonic acid molecules joined with a 1,4 linkage. The galacturonic acid units may be esterified (an acid is combined with an alcohol) with methyl alcohol:

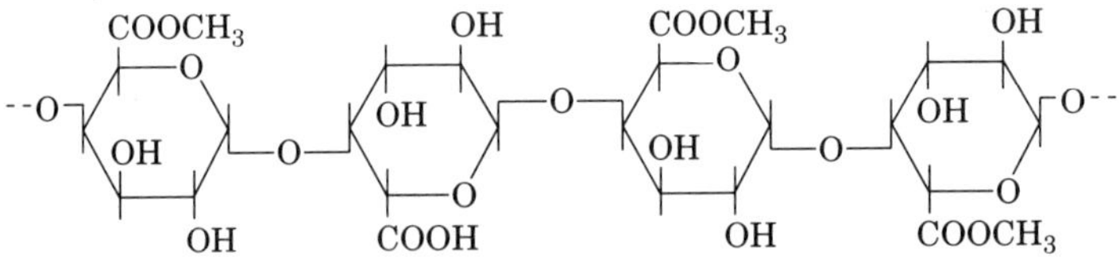

This structure shows 75% esterification of methylation.

Upon acid, alkali, or enzyme hydrolysis, the methoxy groups are first removed, leaving an unesterified chain of galacturonic acid called pectic acid. Further hydrolysis results in individual galacturonic acid units.

Depending on the degree of esterification and the chain length, pectins vary in their jelly-forming properties. With normal degrees of esterification of about 50% to 70%, pectins will form gels with addition of acid and sugar and will do so rather slowly as the jelly cools. Increasing the esterification to above 70% will cause the gel to form quicker and this "rapid set" pectin is chosen for products such as jams and preserves, where a gel is needed before the particles of fruit sink. Another variation is the low-methoxy pectin, which has less methoxylation (less than 50%) and will form a gel with less sugar and the addition of a divalent ion such as calcium (Ca^{a+}). This pectin is used in the low-calorie fruit spreads.

Gums, because of their ability to give highly viscous solutions at relatively low concentrations, are chosen as gelling, stabilizing, and suspending agents. Gums all have hydrophilic properties and are therefore included in the water-soluble fibers. They can be obtained from several sources including exudate gums (gum arabic), bean gums (locust bean gum), seed gums (guar gum), and seaweed gums (agar, algin, and carrageenan). They are used in many foods such as candies and other confections, fruit sauces, syrups, toppings, spreads, baked goods, salad dressings, and beverages.

All are heteropolysaccharides and their sources and composition are outlined in Table 2.4.

Table 2.4. Some Common Gums

Gum	Source	Components
Gum arabic	Acadia trees	L-Arabinose, L-rhamnose, D-galactose, D-glucuronic acid
Locust bean gum	Carob bean	D-galacto-D-mannoglycan, pentoglycan, protein, cellulose, ash
Guar gum	Guar plant seed	D-galacto-D-mannoglycan chain with single galactose branches
Agar	Algae (*Rhodophyceae*)	Agarose, agaropectin
Algin	Giant kelp (*Macrocystis pyrifera*)	Anhydro-1, 4-beta-D-Mannuronic acid, L-glucuronic acid
Carrageenan	Irish moss (*Chondrus crispus*)	—[a]

[a]Three fractions have been isolated and all consist of salts or sulfate esters with a ratio of sulfate to hexose units of close to unity.

LIPIDS

The term lipid is generally defined as the heterogeneous group of substances, associated with living systems, which have the common property of insolubility in water but solubility in nonpolar solvents such as hydrocarbons or alcohols. The major members of this group are the fats and oils, which are given the most emphasis in this section. Other members of the lipid group include waxes, phospholipids, sphingolipids, and sterols which all have important roles in food biochemistry and nutrition and will be mentioned briefly. Nutritionally, the fats and oils contain 9 calories per gram while proteins and carbohydrates contain only 4 calories per gram.

Fats and Oils

When an alcohol and an acid are joined producing a water molecule, the result is an ester.

$$\text{R–OH} + \text{R}_1\text{–COOH} \rightarrow \text{R–00CR}_1$$
$$\text{alcohol} + \text{acid} \rightarrow \text{ester}$$

Glycerol (glycerine) is a polyhydric alcohol that is capable of forming three esters with different acid components.

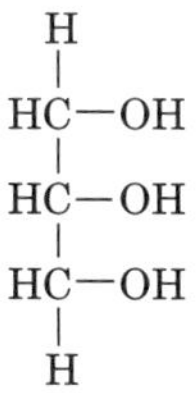

glycerol

Fats and oils are glyceryl esters of fatty acids. The fatty acids esterify with the glycerol molecule to form triglycerols. There are a number of fatty acids in nature and some of them are discussed in this section.

Fats and oils, as do carbohydrates, contain the elements carbon, hydrogen, and oxygen, but the proportion of oxygen in fats is far less. It can be said that fats and oils are fuel foods of a more concentrated type than carbohydrates. It can be shown, by calorimetry, that fats and oils produce more than twice the heat energy produced by carbohydrates. As mentioned previously, 1 gram of fat yields 9 calories, whereas 1 gram of carbohydrate yields 4 calories. An additional advantage of fat from the aspect of energy availability is that it stores well in large amounts in adipose tissues. Thus fat, considered to be a reserve form of fuel for the body, is an important source of calories. Paradoxically, this is not advantageous in affluent societies where the problem is not the availability of food for energy, but rather the health hazard of obesity.

Fats may occur in foods as lipid materials that are solid at room temperature or as oils that are liquid at room temperature. The reason for the variation in the melting points of these is the fatty acid components in the molecules.

Fatty Acids

Fatty acids are aliphatic (open chain) carboxylic acids. Natural fatty acids most commonly found in fats and oils almost always contain an even number of carbon atoms ranging from 4 to 28. The fatty acids may be saturated or unsaturated. Those whose structures have the maximum number of hydrogen atoms are called saturated and those which are lacking hydrogen atoms at certain points in the structure are classified as unsaturated.

$$H-\underset{H}{\overset{H}{C}}-\underset{H}{\overset{H}{C}}-\underset{H}{\overset{H}{C}}-\underset{H}{\overset{H}{C}}-\underset{H}{\overset{H}{C}}-\underset{H}{\overset{H}{C}}-\underset{H}{\overset{H}{C}}-\underset{H}{\overset{H}{C}}-\underset{H}{\overset{H}{C}}-\underset{H}{\overset{H}{C}}-\underset{H}{\overset{H}{C}}-\underset{H}{\overset{H}{C}}-\underset{H}{\overset{H}{C}}-\underset{H}{\overset{H}{C}}-\underset{H}{\overset{H}{C}}-\underset{H}{\overset{H}{C}}-\underset{H}{\overset{H}{C}}-COOH$$

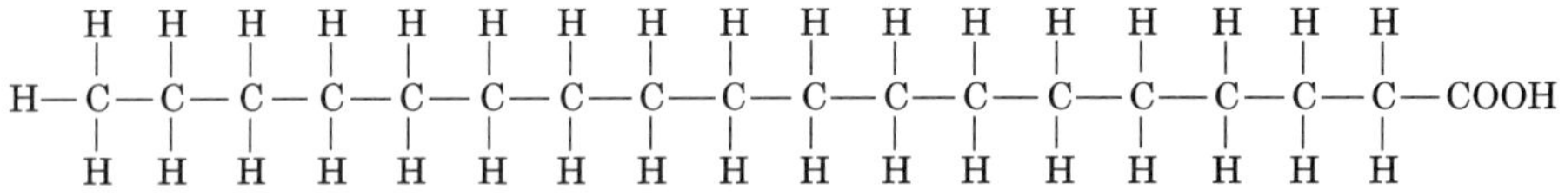

a saturated fatty acid

$$H-\underset{H}{\overset{H}{C}}-\underset{H}{\overset{H}{C}}-\underset{H}{\overset{H}{C}}-\underset{H}{\overset{H}{C}}-\underset{H}{\overset{H}{C}}-\underset{H}{\overset{H}{C}}-\underset{H}{\overset{H}{C}}-\underset{H}{\overset{H}{C}}-\overset{H}{C}=\overset{H}{C}-\underset{H}{\overset{H}{C}}-\underset{H}{\overset{H}{C}}-\underset{H}{\overset{H}{C}}-\underset{H}{\overset{H}{C}}-\underset{H}{\overset{H}{C}}-\underset{H}{\overset{H}{C}}-\underset{H}{\overset{H}{C}}-COOH$$

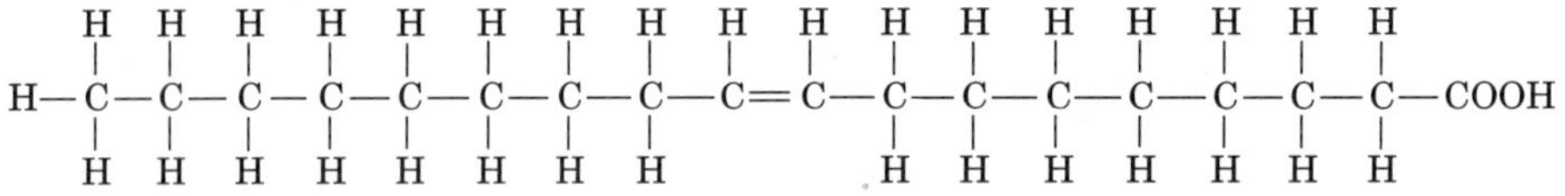

an unsaturated fatty acid

The points of unsaturation (double bonds) show the carbon atoms here are sharing more electrons (2) with the adjacent carbon in the chain than the carbons in the saturated fatty acid (1). The unsaturated fatty acid also has a slightly smaller molecular weight (2 less for each point of unsaturation).

Some common saturated fatty acids are listed in Table 2.5.

Unsaturated fatty acids occur in both plant and animal sources but are more prevalent in most oils from plant sources such as corn oil, safflower oil, and canola oil. Some of the most important unsaturated fatty acids are shown in Table 2.6.

Fatty acids with one double bond are monounsaturated and those with two or more double bonds are known as polyunsaturated fatty acids. As with amino acids, those fatty acids that must be supplied by the diet are designated as essential. Which fatty

Table 2.5. Some Common Saturated Fatty Acids

No. of Carbons	Common Name	M.P. (°C/°F)	Formula
4	Butyric acid	–7.9/17.8	$CH_3(CH_2)_2COOH$
6	Caproic acid	–3.4/25.9	$CH_3(CH_2)_4COOH$
8	Caprylic acid	16.7/62.1	$CH_3(CH_2)_6COOH$
10	Capric acid	31.6/88.9	$CH_3(CH_2)_8COOH$
12	Lauric acid	44.2/111.6	$CH_3(CH_2)_{10}COOH$
14	Myristic acid	53.9/129	$CH_3(CH_2)_{12}COOH$
16	Palmitic acid	63.1/145.6	$CH_3(CH_2)_{14}COOH$
18	Stearic acid	66.9/157.3	$CH_3(CH_2)_{16}COOH$

acids are essential is controversial but the most recent edition of the RDAs defines linoleic and alpha-linolenic as essential.

In fats and oils, all three hydroxyls of the glycerol molecule are involved in ester linkages, hence the chemical name triglyceride.

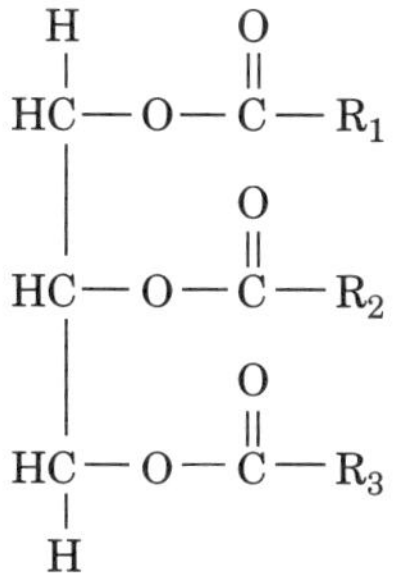

The fatty acid components of fats and oils not only affect the melting point but also have some nutritional implications. It has been shown that saturated fats in the diet can lead to increased serum cholesterol levels. It has been estimated that at least 85% of the deaths in this country are, in some way, diet related. Leading causes are heart attack, stroke, and other coronary problems. Obesity and serum cholesterol levels are contributory factors in these diseases. With these facts in mind, it is very important to be aware of the amounts and types of fat that are ingested daily. It is recommended by both the American Heart Association (AHA) and the American Dietetic Association (ADA) to keep the percentage of calories from fat in our diet to less than 30% with less than 10% from saturated sources.

Table 2.6. Some Common Unsaturated Fatty Acids

No. of C	Common Name	Melting Point (°C/°F)	Formula
18	Oleic acid	10.5/50.9	$CH_3(CH_2)_7CH{=}CH(CH_2)_7COOH$
18	Linoleic	–5/23	$CH_3(CH_2)_4CH{=}CHCH_2CH{=}CH(CH_2)_7COOH$
18	Linolenic	–11/12.2	$CH_3CH_2CH{=}CH\text{-}CH_2\text{-}CH{=}CH\text{-}CH_2\text{-}CH{=}CH(CH_2)_7COOH$
20	Arachidonic	–49.5/–57.1	$CH_3(CH_2)_4CH{=}CH\text{-}CH_2\text{-}CH{=}CH\text{-}CH_2\text{-}CH{=}CH\text{-}CH_2\text{-}CH{=}CH(CH_2)_3COOH$

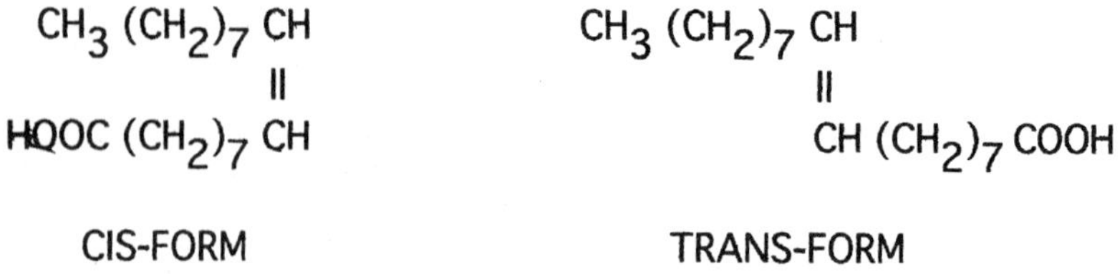

Figure 2.8. Oleic acid in the *cis* and *trans* configurations.

When oils are hardened, as in the production of hard cooking oils and margarines, the melting point is obviously raised, making it easier to transport the oils or to spread them on foods. The hardening process is described in Chapter 22. Unsaturated fatty acids, because of their points of unsaturation, can exist in different configurations or forms. In nature, most occur in the *cis*-form (Fig. 2.8) but when the oils are hydrogenated or hardened, as in some margarines, a change occurs in the configurations of some of the unsaturated fatty acids that remain. These changed fatty acids now take on a new configuration, the *trans*-form (Fig. 2.8). This change was once thought to be "the best of both worlds" because not only did the fatty acids that had been saturated have an increased melting point, but any of the remaining unsaturated fatty acids that converted to the *trans*-form also showed an increase in melting point.

Studies have shown a link in the ingestion of *trans*-fatty acids and the increase of serum cholesterol. The fact that the fatty acids "straighten" converting from the *cis* to *trans*-form is one explanation for this but more studies are being done. Some consumer advocate groups and newspapers have published much information on the controversial subject of whether consumption of margarine containing *trans*-fatty acids is worse than consuming butter. The recommendation from the AHA and ADA are agreed with by most health professionals and if it is followed, the ingestion of *trans*-fatty acids will be lowered also.

When fats are ingested, they can take one of two general paths. In one, they are hydrolyzed to glycerine and fatty acids by lipase enzymes in the small intestine and then will eventually be oxidized at the cellular level for energy, carbon dioxide, and water. In the other, they are emulsified and carried to adipose cells where they will be stored.

Phospholipids, Waxes, Sphingolipids, and Sterols

Other lipid components include phospholipids, sterols, waxes, and sphingolipids. Phospholipids are made from glycerol, fatty acids, phosphoric acid, and often an amino alcohol. The structure of a common phospholipid, lecithin (phosphatidyl-choline) is shown. Lecithin from various sources contains different saturated or unsaturated fatty acid groups.

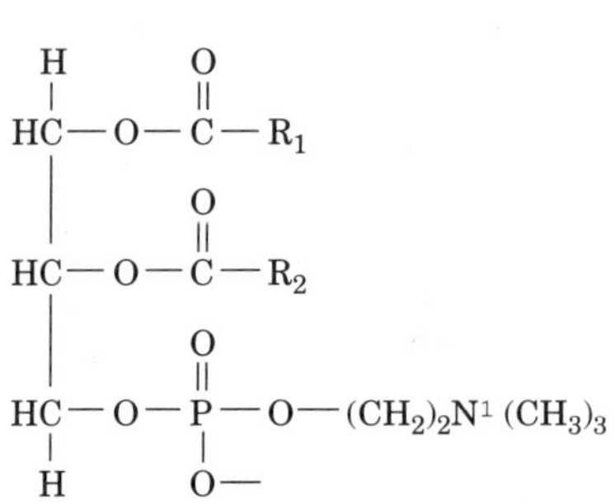

Phospholipids, because of their polarity (+ and – charges) at one end of the molecule and nonpolarity at the other, are very good emulsifiers and are used in the food industry in such products as chocolate, salad dressings, and mayonnaise (as a component of egg yolks) to help hold polar and nonpolar components together. In the body, they are an important constituent of cell membranes and are involved in the building of such.

Waxes are not made from glycerol, but from fatty acids and monohydric alcohols of 24 to 36 carbons in length. An ester linkage is needed to combine the two molecules. Waxes serve in the body as protective, water repellent coatings on tissue surfaces. Their function is to prevent overevaporation of moisture or invasion of the tissue with water from the environment. In the food industry, waxes are used in some packaging and as ingredients in some confections and candies for texture or appearance. They are not digestible.

Sphingomyelin is a sphingolipid and is an important constituent of nerves and brain tissue. In this lipid, glycerol is replaced by a long-chained nitrogen-containing alcohol.

There are a number of sterols that have important functions in the body. These are very complex chemical compounds containing an alcohol group to which fatty acids can be esterified. The sterol cholesterol is involved in the composition of bile salts, which play a role in the emulsification of fats in the intestine, hence, in the digestion of fats. Ergosterol, another sterol, may be converted to vitamin D in the body under the influence of sunlight or ultraviolet light.

More information on fats and oils, how they are used in industry, and their oxidation is found in Chapters 8, 9, and 22.

VITAMINS

There are a number of vitamins required in small amounts by the human body for sustaining life and good health. Some are fat-soluble; others are water-soluble. The recommended dietary allowances (RDA) for the vitamins and minerals described here are shown in Table 2.1.

Fat-Soluble Vitamins

Vitamin A

Vitamin A is found only in animals, although a number of plants contain carotene, from which vitamin A can be produced in the body once the plants containing carotene are eaten. Vitamin A may be formed in the body from the yellow pigments (containing carotene) of many fruits and vegetables, especially carrots. Vitamin A, which is required for vision and resistance to infection, is also found in fats and oils, especially in the liver oils of many saltwater fish. Epithelial cells (those cells present in the lining of body cavities and in the skin and glands) also require vitamin A. Deficiency in vitamin A may cause impairment in bone formation, impairment of night vision, malfunction of epithelial tissues, and defects in the enamel of the teeth.

Vitamin A previously was expressed in international units (IU). Because of the different biological activities of vitamin A sources such as beta-carotene, other carotenes, and preformed vitamin A (retinol), vitamin A is now expressed in terms of retinol equivalents (RE). To convert, 1 IU = 0.3 RE. The definition of RE is given by the Food and Nutrition Board as:

1 RE = microgram all-*trans*retinol = 6 micrograms of
beta-carotene = 12 micrograms of other provitamin A carotenoids.

The Reference Daily Intakes (RDIs), on which the % daily values on the nutritional labels are based, for vitamin A are 5000 IU based on the 1968 RDAs. This corresponds to 1500 REs. Vitamin A is routinely added to milk. The 1989 revised RDA for vitamin A is 1000 REs for adult males and 800 for females (see Table 2.1). Nutritional labels are discussed further in Chapter 7.

Vitamin D

Vitamin D (calciferol or activated ergosterol) is necessary for normal tooth and bone formation. Deficiencies in vitamin D result in rickets (deformities of bone, such as bow-legs and curvature of the spine) and tooth defects. Fish oils, and especially fish liver oils, are excellent sources of vitamin D. The human body is also able to make vitamin D by converting sterols such as cholesterol with ultraviolet light from the sun or an artificial source. The RDA for vitamin D was in the past expressed in IUs and 400 was the RDS. It is now expressed in micrograms (μg) as cholecalciferol and 10 μg = 400 IU. Vitamin D, as vitamin A is also routinely added to milk.

Vitamin E

Vitamin E has four different forms (the tocopherols) that have the same name except with the prefixes alpha-, beta-, gamma-, and delta-. The four compounds are closely related, with slight differences in structure, but alpha-tocopherol is the most common and serves as an antioxidant that serves to prevent the oxidation of body components, such as unsaturated fatty acids. Vitamin E has also been shown to enhance the absorption of iron. Diets excessive in polyunsaturated fats can lead to the formation of peroxidized fatty acids that could reach dangerous levels. Vitamin E may prevent this. Good sources of Vitamin E are corn oil, cottonseed oil, and peanut oil.

While the symptoms of vitamin E deficiency in humans are not clearly established, experiments with various animals have shown that vitamin E deficiency has an adverse effect on reproduction, with apparent irreversible injury to the germinal epithelium. Other symptoms noted in animal studies include injury to the central nervous system, growth retardation, muscular dystrophy, and interference with normal heart action.

The RDA for vitamin E is now given in milligrams of alpha-tocopherol equivalents (α-TE). The old measurement was in IUs and 30 was the RDA (10 α-TE = 30 IU).

Vitamin K

Vitamin K is essential for the synthesis of prothrombin, a compound involved in the clotting of blood. Cabbage, spinach, cauliflower, and liver are especially good sources and it is also synthesized by bacteria in the human intestine. Antibiotic therapy that destroys intestinal organisms can produce deficiencies of vitamin K.

The significant symptom of vitamin K deficiency in humans and in animals is the loss of the ability of the blood to clot which is, of course, a dangerous condition that can result in death whenever bleeding occurs. It is believed that humans ordinarily receive adequate amounts of vitamin K in the diet. The RDA ranges from 5 μg in infants to 80 μg in males over 25 years of age.

Water-Soluble Vitamins

Vitamin B_1 (Thiamin)

Thiamin is involved in all bodily oxidations that lead to the formation of carbon dioxide. It is also necessary for nerve function, appetite, and normal digestion as well as growth, fertility, and lactation. The symptoms of thiamin deficiency are retardation of growth, palpitation and enlargement of the heart, hypertension, and beri-beri. The various effects of a disturbance of the nerve centers such as forgetfulness or difficulty in thinking are other manifestations of vitamin B_1 deficiency.

Important to the food scientist is the sensitivity of vitamin B_1 to sulfur dioxide (SO_2) and to sulfite salts. SO_2 destroys vitamin B_1 and should not be used as a preservative in foods that are a major source of the vitamin. FDA and meat inspection laws prohibit such use. Thiamin is stable to heat in acid foods but is less so in neutral and alkaline foods. This should be taken into account when processing foods.

Good sources of the vitamin are fresh pork, wheat germ, and cereals containing bran and some fair sources are beef and lamb.

Vitamin B_2 (Riboflavin)

Vitamin B_2 makes up a part of enzyme systems involved in the oxidation and reduction of different materials in the body. Deficiency of riboflavin generally results in growth retardation and may result in vision impairment, scaling of the skin, and lesions on mucous tissues. Neuritis is another effect of deficiency. Milk, liver, and eggs are good sources while meats and leafy green vegetables are moderate sources. Riboflavin is sensitive to light, so packaging such as cardboard and other light-resistant containers for milk are used.

Niacin (Nicotinic Acid)

Niacin is part of enzyme systems regulating reduction reactions in the body. It is also a compound that dilates blood vessels. It is part of the coenzyme nicotinamide adenine dinucleotide (NAD) which is involved in glucose breakdown. Deficiency of niacin causes pellagra (a disease that causes diarrhea, dermatitis, nervous disorders, and sometimes death). Good sources are yeast, meat, fish, poultry, peanuts, legumes, and whole grain.

Vitamin B_6 (Pyridoxine)

Other substances closely related to this vitamin are pyridoxal and pyridoxamine. Vitamin B_6 is part of the enzyme system that removes CO_2 from the acid group (COOH) of certain amino acids and transfers amine groups (NH_2) from one compound to another in the body. Although the vitamin is needed for processes such as those mentioned, it seems a deficiency does not cause a well-recognized disease. Bananas, barley, grain cereals with bran, muscle meat, liver, and green vegetables are all among sources of this vitamin. Vitamin B_6 has been used by women taking steroid contraceptive pills, and it has been used in treatment of such ailments as premenstrual syndrome (PMS) but this is not recommended without valid blood or urine tests to show low B_6 levels.

Biotin

Biotin is reported to be a coenzyme in the synthesis of aspartic acid, which plays a part in a deaminase system and in other processes involving the fixation of carbon dioxide. Deficiency of this compound is unusual, although it can be tied up by a substance in raw egg whites called aviden and the deficiency has been demonstrated in feeding studies with mice. Because of the production of biotin by microbial flora of the intestines, the dietary requirement for this compound is unknown. Liver is an excellent source of biotin and peanuts, peas, beans, and whole cooked eggs are good sources.

Pantothenic Acid

Pantothenic acid, a vitamin required for normal growth, nerve development, and normal skin, is a component of coenzyme A and others involved in metabolism. Pantothenic acid is widespread in foods, so obvious symptoms of a deficiency are rare in humans. Organs of animals (liver, heart, kidney), eggs, whole wheat products, and peanuts are excellent sources.

Choline

Choline is generally listed with the B vitamins and is consumed in adequate amounts in a normal diet as well as being produced by intestinal microbial flora (other growth factors such as biotin, para-aminobenzoic acid, and inositol are also produced in this manner). It is a component of cell membranes and brain tissue and it functions as part of the substance acetylcholine, which is one of the brain's principal neurotransmitters.

Vitamin B_{12} (Cyanocobalamin)

This vitamin is a very complex chemical compound and is the largest vitamin molecule. It is required for the normal development of red blood cells, and a deficiency in vitamin B_{12} causes acute pernicious anemia and a variety of other disorders. Cobalt is part of Vitamin B_{12}'s structure, thus giving rise to the requirement for this mineral in nutrition. Some vitamin B_{12} is synthesized by bacteria in the intestine but intake of 2 μg per day for adults is recommended.

Folacin

Folacin is required for the formation of blood cells by the bone marrow and is involved in the formation of the blood pigment hemoglobin. It is also required for the synthesis of some amino acids. Deficiency results in some types of anemia including pernicious anemia. The RDA is increased from 180 μg to 400 μg for pregnant women, as it is believed the vitamin may act in prevention of some birth defects. It is part of the coenzyme system tetrahydrofolate (THF), which is required for the synthesis of new genetic material and therefore new cells. Liver, leafy vegetables, legumes, cereal grains, and nuts are all sources of the vitamin.

Vitamin C (Ascorbic Acid)

Ascorbic acid is required for the formation of intercellular substances in the body including dentine, cartilage, and the protein network of bone. Hence, it is important in tooth formation, the healing of broken bones, and the healing of wounds. It may be important to oxidation–reduction reactions in the body and to the production of certain hormones. Vitamin C, like vitamin E, enhances the absorption of iron.

There are several claims that have been attributed to vitamin C such as prevention of colds and removal of cholesterol from the blood as demonstrated in a study conducted on rats. The significance of the rat study for humans is yet to be determined and the cold prevention claim is not supported by the medical profession or the FDA. It has been shown that exceedingly high doses of the vitamin taken for extended periods of time can cause some problems. Infants born to mothers who have been on high doses of the vitamin for a long time have been born with a dependence on the higher doses. A deficiency of the vitamin causes scurvy (spongy, bleeding gums, loss of teeth, swollen joints), fragile capillary walls, and impaired healing of wounds. Excellent sources of the vitamin are orange juice, tomato juice, green peppers, broccoli, cabbage, and brussels sprouts. Potatoes are a fair source because we eat large amounts. Many fruits contain fair amounts of vitamin C.

The food technologist must be aware that vitamin C is easily destroyed by oxidation and heat. It also can be lost in cooking water during processing. Steps must be taken to minimize the losses, and fortification may be necessary before or after processing to ensure sufficient nutrient amounts.

MINERALS

When testing for minerals in the food laboratory, the first step is to remove all water and then all organic matter from the product being tested. After these steps, most of the food (often more than 97%) is gone. What remains is the inorganic material or ash. The human body consists of about 3% minerals, most of which is in the skeletal system. Although the minerals exist is minute amounts, their functions are necessary for normal growth and reproduction. Minerals are generally categorized as "major" and "trace" based on the amounts in the body. Here, a very brief overview of the importance of each mineral is discussed. Further reading on the subject is recommended for students of nutrition. All minerals are important but only two are included on the nutritional labels; thus these are mentioned first.

Iron

Iron is considered a trace mineral because it is needed in such small amounts. Of all the required nutrients, iron may be consumed in insufficient amounts in the industrialized world more commonly than any other. One reason may be the poor absorption of iron sources such as the iron phosphates and iron phytates found in plants. Animal sources and those from soluble salts used in food fortification are generally absorbed more efficiently. Iron is an essential part of both the blood pigment hemoglobin and the muscle pigment myoglobin. It is included in some enzymes. Deficiencies of iron cause anemia. The amount of iron needed is related to growth rate and blood loss. Women who are of menstruating age should take special caution to ensure sufficient intake. Iron toxicities are rare but care should be taken when storing iron supplement tablets. As few as 6 to 12 tablets could prove fatal if taken by a small child. Good sources of iron are liver, meats, eggs, oatmeal, and wheat flour. As mentioned previously, both vitamin E and vitamin C have been shown to aid in iron absorption.

Calcium

Calcium is one of the minerals that humans require in the greatest amounts (phosphorus is the other). It is required for bone and tooth structure and is necessary for the function of nerves and muscles. Calcium is also needed in the blood clotting mechanism. Deficiencies can lead to osteoporosis, especially in older women. It is suggested that increased calcium intake, especially when young, may help prevent this in later years. Vitamin D is essential for calcium absorption and lactose has been shown to aid in this also. This makes vitamin D fortified milk an excellent, if not the best source of calcium. Calcium deficiencies may be widespread in our society, but unlike most other deficiencies, symptoms are not apparent until later in life. Care should be taken to ensure sufficient calcium intake.

Other Major Minerals

Sodium

In humans, sodium is required in the extracellular fluids to maintain osmotic equilibrium and body-fluid volume. As salt is a major food ingredient there is little evidence of deficiencies except in diseases involving prolonged vomiting or diarrhea. The consumption of salt and other sodium sources should be constrained. Sodium has been shown to aggravate hypertension (high blood pressure) and it is recommended that daily consumption levels be kept between 1100 and 3300 mg. The average in the typical American diet is closer to 6000 mg.

Chlorine

Chlorine is also involved in extracellular fluids as the major negatively charged ion. It has a role in controlling blood pH and is necessary in the production of hydrochloric acid of gastric juice. It is a component of table salt and is never really lacking in the diet. If great losses of body fluids occur, as in vigorous exercise, the amount of chlorine may have to be replenished.

Potassium

Potassium is present in the body cells as the chief intracellular cation and is associated with the function of muscles and nerves and with the metabolism of carbohydrates. It is important in maintaining the fluid volume inside cells, and the acid–base balance. Good sources of potassium are meats, eggs, oranges, bananas, and fresh milk. Cell membranes are quite permeable to potassium, but as it leaks out, a highly active membrane pump returns it to the cell in exchange for sodium. This is critical because if as little as 6% of the potassium contained in the cells were to escape into the blood, the heart would stop.

Phosphorus

About 85% of the phosphorus in the body is found combined with calcium. It is also part of the body's major buffers (phosphoric acid and its salts) and it is part of both DNA and RNA, the genetic code material present in every cell. Some lipids contain phosphorus in their structure and these phospholipids help transport other lipids in the blood. Phosphorus also plays a key role in energy transfer as it is part of the energy carrier of the cells, ATP. Some sources of phosphorus are meats, fish, eggs, and nuts.

Magnesium

Magnesium is a minor component of bones and is present in soft tissue cells, where it is involved in protein synthesis. Deficiencies of magnesium are unusual and good food sources are most vegetables, cereals and cereal flours, beans, and nuts.

Sulfur

Sulfur is present in virtually all proteins and plays a most important role in forming the cystein bridges that are essential for protein conformation. Deficiencies of sulfur are associated with protein deficiency and if foods containing sufficient protein are eaten then sulfur amounts will be adequate.

Trace Minerals

Iodine

Iodine is part of the hormone thyroxine which regulates metabolic levels. Deficiency of iodine leads to low-level metabolism, lethargy, and goiter, which results in an enlarged thyroid gland. There is rarely a deficiency of iodine when saltwater fish are available and eaten. Today, iodized salt prevents the deficiency but there is some concern about overconsumption of the mineral. Fast food operations use iodized salt liberally, and iodates are used by some bakeries as dough conditioners. Symptoms of toxic levels are similar to those resulting from a deficiency, that of an enlarged thyroid gland. With the identification of the problem, food industries are reducing use of iodine-containing products but care must still be taken.

Flourine

The flourine ion is present in body tissue in trace amounts and helps to prevent tooth decay. Drinking water is the chief source of flourine and fish is also a good source. If flourine is taken in excessively high doses for long periods of time, it is toxic. Too much flourine in the form of supplements can cause flourosis, a mottling of the tooth enamel.

Copper

Copper aids in the utilization of iron in hemoglobin synthesis and is required by some body enzyme systems. It can be toxic in high concentrations and these can be reached if copper utensils are used for storage or distribution of acid foods (copper tubing in machines that dispense lemonade or other acid beverages). Deficiencies are virtually unknown and some food sources are fruits, beans, peas, eggs, liver, fish, and oysters.

Cobalt

Cobalt is a component of vitamin B_{12}, the only component present in the body known to contain this element. Sufficient amounts of cobalt are present in most foods and some may be absorbed into food from cooking utensils. Even though it is part of vitamin B_{12}, it does not replace the need for the vitamin.

Zinc

Zinc is a cofactor in more than 70 enzymes that perform specific tasks in the eyes, liver, kidneys, muscles, skin, bones, and male reproductive organs. These include carbohydrate and protein metabolism and nucleic acid synthesis. Deficiencies are rare but dwarfism, gonadal atrophy, and possible damage to the immune system have been attributed to a deficiency of the mineral. Good sources are protein-rich foods such as shellfish, meat, and liver.

Manganese

Manganese is needed for normal bone structure, fat production, reproduction, and functioning of the central nervous system. Deficiencies result in bone disorders, sexual sterility, and abnormal lipid metabolism. Meats are a source of the mineral but adequate supplies are found in most human diets.

Selenium

Selenium acts as an antioxidant in conjunction with vitamin E. Its major food sources are meat, seafood, and grains. A deficiency results in anemia, muscle pain, and sometimes heart failure.

Vanadium

Vanadium deficiencies in humans are not known but animal and bird studies have shown growth retardation, deficient lipid metabolism, impairment of reproductive function, and bone growth retardation.

Silicon

Silicon is found in unpolished rice and grains. Certain diseases involving connective tissue are believed to result when it is not present in adequate amounts.

Tin

Tin, occurring naturally in many tissues, is necessary for growth in rats. It is believed essential to the structure of proteins and possibly other biological components. As it is present in many foods, a deficiency has not been noted.

Chromium

Chromium plays a physiological role thought to be related to glucose metabolism, perhaps by enhancing the effectiveness of insulin. Although it is a normal body component, its content decreases with age. Depleted tissue concentrations in humans have been linked to adult-onset diabetes. It has been shown to remedy impaired carbohydrate metabolism in several groups of older people in the United States. Whole, unprocessed foods are the best sources whereas refined foods have less of the mineral.

Aluminum, Boron, and Cadmium

Aluminum, boron, and cadmium are also found in trace amounts in the human body, but neither their roles nor the effects of deficient or excessive amounts are known. Although the affected areas of the brain of those afflicted with Alzheimer's disease have been found to contain excessive amounts of aluminum, the effect of this abnormality on the disease or vice versa is not yet clear. It is recommended, however, to avoid long-term storage of acid foods in aluminum pans.

NATURAL TOXICANTS

Some plants are able to produce compounds that serve as protectants or help ensure reproduction. These compounds may attract pollinating insects or repel animals or insects that may eat the plant. Some of these metabolites are quite toxic to man. Some mushrooms, for example, produce specific nitrogen-containing bases or alkaloids that cause severe physiological effects. Other plants, such as potatoes, may produce alkaloid

Table 2.7. Some Natural Toxicants and Their Food Sources

Toxin	Food Source
Cyanide-generating compounds	Lima beans
Safrole	Spices
Prussic acid	Almonds
Oxalic acid	Spinach, rhubarb
Enzyme inhibitors and hemagglutinins	Soybeans
Gossypol	Cottonseed oil
Goitrogens (interfere with iodine binding by the thyroid gland)	Cabbage
Tyramine	Cheese
Avidin (antagonistic to the growth factor biotin)	Egg white
Thiaminase (destroys vitamin B_1)	Fish and shellfish
Vitamins A and D and methionine (exhibit toxic effects in excessive concentrations)	many foods

solanine. Consuming large amounts of potato skins which contain these alkaloids may have toxic effects to humans.

Heavy metals such as lead, mercury, and arsenic are often found in soil and water and these can make their way into foods. Usually, they occur in such small amounts that they are considered harmless and some, such as zinc and selenium, are essential nutrients.

Some other natural toxicants and their food sources are shown in Table 2.7. Many of these are largely removed or inactivated when foods are processed; for example, enzyme inhibitors, hemagglutinins, avidin, and thiaminase are destroyed by the heat of cooking; water soaking and fermentation removes many cyanogenic compounds; and removal of gonads, skin, and parts of certain fish eliminates toxins present in these tissues. Over the years, man has also developed the ability to detoxify low levels of many potentially dangerous chemicals.

3

Microbial Activity

Foods are normally contaminated with microbes (or microorganisms) that are so small that a microscope is required to see them. Microbes include bacteria, yeasts, molds, virus, algae, and protozoans. However, the organisms that normally contaminate and spoil food are the bacteria, with yeasts and molds of secondary importance.

Under normal conditions, microbes feed on the food in which they live and reproduce. During their life cycles, they cause a variety of changes in food, most of which result in a loss of the food's quality. In some cases, the controlled growth of specific microbes can produce desirable changes, such as the change that results in the formation of sauerkraut from cabbage and wine from grapes. Microorganisms function through a wide variety of enzymes that they produce, and the changes in food attributed to microorganisms are actually brought about by the chemical action of these enzymes.

CHARACTERISTICS OF MICROBES

The physical and biological abilities, requirements, and tolerances of microbes are factors that determine the effects of microbes on food.

Structure and Shape of Microbes

Bacteria, molds, and yeasts have rigid cell walls enclosing the cell materials and the cytoplasm, but they differ greatly in their other properties. The microbes most important in the spoilage and controlled changes in food have various shapes (see Fig. 3.1).

Many are rod-shaped and occur either as single cells, two adjoining cells, or short chains of cells. Other bacteria are spherical in shape (cocci). Some cocci exist mainly as a grapelike cluster of cells, the staphylococci, others as a cubelike cluster of spherical cells, the sarcina. Some cocci occur in groups of two, the diplococci, or in chains, the streptococci. Some disease-causing bacteria are included in the cocci group.

Some bacteria are curved or comma-shaped rods, the vibrios, and others are long, slender, corkscrew-shaped cells, the spirochaetes. Other bacterial groups either are not important to changes in food or are disease-causing types.

Molds are multicellular (bacteria and yeasts are single cells) and are made up of branched threads (hyphae), consisting of chains of cylindrical cells united end to end. Some of the hyphae serve to secure nutrients from the material in which they are growing while others produce the spores that provide for reproduction, or for new mold growth. Some molds produce a mycelium (mass of hyphae) that has cross walls (septa) while others do not have cross walls in the mycelium.

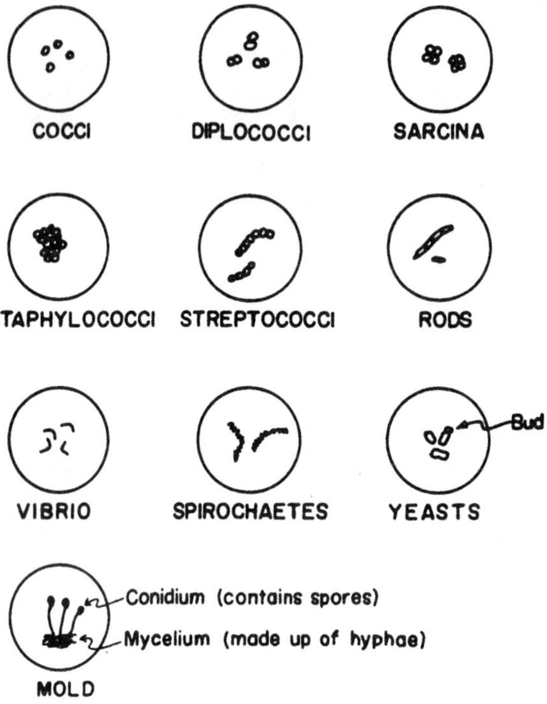

Figure 3.1. Shapes of various microorganisms.

Yeasts form single cells or chains of cells that may be spherical or of various shapes between the spherical and the cylindrical.

One common characteristic of bacteria, yeasts, and molds is that they can exist either as active, vegetative cells or as spores. As vegetative cells, they metabolize, reproduce, cause food spoilage, and sometimes cause disease. They generally have a considerable effect on the environment. Their activity, however, is dependent on several environmental conditions, which are discussed later on. When conditions are unfavorable (such as high temperatures), the vegetative cells begin to die before they can reproduce, and soon all the vegetative cells die. Many microorganisms, however, also exist in the spore state, sometimes because it is a necessary step in reproduction, and sometimes because it is necessary for the microbe to survive unfavorable conditions. In other cases, certain microorganisms, such as some bacteria, exist in both the vegetative and the spore states and when conditions become unfavorable, the spores survive.

In order for spores to survive the conditions that are destructive to their vegetative counterparts, they must have special properties. In general, spores are concentrated forms of their vegetative counterparts; thus, spores contain less water, and are more dense and smaller than the vegetative cells. The spore wall is thicker and harder than the cell wall.

Size of Microbes

Bacteria are comparatively small. The single cell of many bacteria is about 1/40 millionth of an inch (1 micron) in diameter. There are, however, some types of bacteria that may be 50 or more times larger than this. Because of their small size, bacteria cannot be seen with the naked eye, and when evidence of their growth can be seen, such as when slime forms on meat, the organisms will have multiplied to very large numbers, billions of cells to each square inch (6.5 cm^2) of the meat surface.

The single cell of the mold, although not visible without magnification, is much larger than the single bacterial cell, and any significant growth of molds on foods can be seen. The visible mold comprises mycelia with or without spore heads. The spore head carries the spore that will give rise to more growth.

Yeasts vary in size from that of the more common spherical bacteria to a form several times larger. Like bacteria, when grown in solution, the individual cells cannot be seen, but eventually, cell numbers will accumulate to the point where the solution becomes cloudy. There are, however, some instances in which groups of cells will form visible clumps (colonies) on foods or on the surfaces of solutions.

Reproduction in Microbes

Bacteria usually reproduce by fission, a transverse division across the cell to form two new cells. Under favorable conditions, this form of reproduction continues until there are billions of cells per ounce (29.6 ml) of solution or per square inch (6.5 cm^2) of surface of food material. (It should be noted that microorganisms must live in liquid material, and when they grow on food surfaces, they are actually living in the liquid available on or in the food.)

There are some instances in which sexual reproduction occurs in bacteria, one cell uniting with another before division occurs, but this is not the usual circumstance and need not be discussed further here.

Under favorable environmental conditions, bacterial multiplication for some species occurs at an exceedingly rapid rate, a doubling of the number requiring about 20 min. From this standpoint, the number of cells with which a food is contaminated is very important, for if the number is high, only a few hours may be required for bacterial multiplication to reach a level that will cause food spoilage.

Molds usually reproduce by means of spores, each organism producing spores in great numbers. Mold spores are of two types: sexual spores produced by the fusion of two sex cells, or asexual spores that arise from the fertile hyphae. Most molds produce asexual spores. Asexual spores are formed on the sides or ends of hyphae threads (a conidium) or produced in a special spore case called a sporangium. Conidia contain many spores, and when a spore comes in contact with a suitable growth medium under favorable conditions, it grows to become the adult mold.

Yeasts reproduce by budding, by fission, or by spore formation. In budding, a projection is formed on the original cell that eventually breaks off to form a new cell. As is the case with bacteria some yeast may undergo fission, the process in which one cell divides into two cells. Sometimes, yeasts reproduce by spore formations that may be sexual or asexual.

Motility in Microbes

Many types of bacteria are motile, that is, they are able to move about in the solutions in which they live by means of the movement of flagella, which are thin, protoplasmic, whiplike projections of the cell. Some bacteria, such as the cocci, have no flagella and are not motile.

Molds are not motile, but they spread in vinelike fashion by sending hyphae outward.

Yeasts are not motile; consequently, they exist in relatively compact clusters, unless distributed by agitation or by some other external dispersing force.

Effect of pH on Microbial Growth

Both the growth and the rate of growth of microbes are greatly affected by pH. Thus, microorganisms have an optimum pH at which they grow most rapidly and a pH range above or below which they will not grow at all. Generally, molds and yeasts grow best at pHs on the acid side of neutrality, as do some bacteria. Many species of bacteria grow best at neutral or slightly alkaline pH. Some bacteria will grow a pH as low as 4, whereas others can grow at a pH as high as 11. At least part of the reason why fruits are usually spoiled by molds or yeasts and flesh-type foods (meats, fish, poultry, and eggs) are usually spoiled by bacterial growth is that fruits have a low (acidic) pH and flesh-type foods have a near-neutral pH.

Nutritional Requirements of Microbes

Microorganisms, especially bacteria, vary greatly in nutritional requirements from species to species. In the presence of particular inorganic salts, some bacteria can utilize the nitrogen in air to form proteins and utilize the carbon dioxide in air to obtain energy or to form compounds from which they can then obtain energy. Others can utilize simple inorganic salts, such as nitrates, as a source of nitrogen and relatively simple organic compounds, such as lactates, as a source of energy. Nearly all yeasts can derive all their nitrogen from lysine, an amino acid. Some bacteria require complex organic compounds for growth, including amino acids (primary units of proteins) and vitamins—especially those belonging to the B group—and traces of certain minerals.

It has been shown that, in some cases, not only are trace minerals necessary, but their careful control is needed to sustain an optimum growth rate of microorganisms. There is some evidence that demonstrates the ability of at least some microbes to utilize substitute elements for required ones. Sometimes, one trace element may protect microbes from the toxic effects of the presence of other elements; for example, the presence of zinc has been reported to protect yeasts against cadmium poisoning.

Molds and yeasts, like bacteria, may require basic elements (carbon, hydrogen, nitrogen, phosphorus, potassium, sulfur, etc.) as well as vitamins and other organic compounds.

Although sugar is a nutrient important to microbes, some molds and yeasts can grow well in high concentrations that inhibit bacterial growth. In fact, yeasts grow extremely well in the presence of sugar.

Effect of Temperature on Microbial Growth

The optimum growth rate of microorganisms depends on temperature, as they do not grow above or below a specific range of temperatures. Again, growth temperatures vary with species, and bacteria are arbitrarily classified according to the temperatures at which they grow. Bacteria classified as psychrophiles grow fastest at about 68 to 77°F (20 to 25°C), but some can grow, although more slowly, at temperatures as low as 45°F (7.2°C), while others grow at temperatures as high as 86°F (30°C). Some will grow at temperatures as low as 19°F (–7.2°C), as long as the nutrient solution for growth is not frozen. Bacteria classified as mesophiles grow best at temperatures around 98°F (36.7°C), but some will grow at temperatures lower than 68°F (20°C) and others at temperatures as high as 110°F (43.3°C). Bacteria causing diseases of animals are mesophiles. Bacteria classified as thermophiles grow best at temperatures as low as 113°F (45°C) and others at temperatures as high as 160°F (71.1°C) or slightly higher.

It should be pointed out that although microorganisms grow within a given range of temperatures, their growth rate decreases greatly at the low or high temperatures within that range.

When microbes are subjected to temperatures higher than those of their growth range, they are destroyed at rates that depend on the degree to which the temperature is raised above the growth range. Some bacteria form heat-resistant spores whose degree of heat resistance varies with the species. Some types may be destroyed by raising the temperature (in the presence of moisture) to 212°F (100°C) over a period of several minutes while others can survive boiling many hours. Some may even survive temperatures as high as 250°F (121.1°C) for many minutes. Bacterial spores are the most heat-resistant living organisms known to humans.

In general, there are bacteria that can grow and can survive under more extreme conditions than those tolerated by any of the molds or yeasts. Molds, as a class, can grow and survive under more extreme conditions than can yeasts. On the other hand, although the growth and survival of the many species of bacteria cover a broad range of conditions, each species is highly selective, and the range of conditions under which it can metabolize is generally narrower than that of molds and yeasts.

Water Requirements of Microorganisms

Microorganisms grow only in aqueous solutions. A term, water activity (a_w), has been coined to express the degree of availability of water in foods. This term is applied to all food, with the ordinary fresh-type food having an a_w of about 0.99 to 0.96 at ambient temperatures. Low water activities, which limit the growth of microorganisms in foods, may be brought about by the addition of salt or sugar, as well as by the removal of water by drying. Under such conditions, the remaining water has been tied up by chemical compounds added to or concentrated in the food or bound to some food component, such as protein.

$$a_w = \frac{\text{equilibrium relative humidity}}{100}$$

Equilibrium relative humidity is reached in a food when the rate at which it loses water to its environment is equal to the rate at which it absorbs water from the environment.

Oxygen Requirements of Microorganisms

Some bacteria are aerobic, that is, they require oxygen (in air) for growth. Others grow best when the oxygen concentration is low (microaerophiles). Still others can grow either in the presence or absence of oxygen (facultative aerobes or facultative anaerobes). A number of bacterial species will not grow in the presence of oxygen because it is toxic to them, and these are called anaerobes.

It should be pointed out that although oxygen is toxic to anaerobic bacteria, they sometimes can grow under conditions that appear to include the presence of oxygen. The explanation for this is that organic materials (animal and vegetable matter) contain compounds that themselves tie up the available oxygen. Also, in many instances, aerobic bacteria first grow and consume the oxygen and produce reducing compounds that also combine with available oxygen, thus producing conditions suitable for the growth of anaerobic bacteria. In contrast to this, certain aerobic bacteria can utilize the oxygen in certain oxygen-bearing compounds, such as nitrates. In general, molds require oxygen for growth. Yeasts grow best in aerobic conditions, but some grow anaerobically (e.g., fermentative types) although more slowly.

Effect of Oxidation–Reduction Potential on Microbes

The oxidation–reduction potential, expressed as E_h and measured in millivolt units, is defined as the tendency to yield electrons (become oxidized) or to capture electrons (become reduced). The greater the degree to which a substance is oxidized, the more highly positive will be its E_h value, and the greater the degree to which a substance is reduced, the more highly negative will be its E_h value. In general, aerobes require a substrate having a positive E_h, and anaerobes require a substrate having a negative E_h. The optimum E_h values for the growth of microbes vary with species.

EFFECTS OF MICROBES ON FOODS

As stated earlier, bacteria, molds, and yeasts are the main causes of the spoilage of unpreserved foods, with bacteria playing the major role. The changes in food caused by microbial action can be classified into two types: undesirable changes and desirable changes.

Undesirable Changes in Foods

Undesirable changes can be further subdivided into (1) those that cause food spoilage, not usually associated with human disease, and (2) those that cause food poisoning whether or not the food undergoes observable changes.

Food Spoilage

Food spoilage can be detected organoleptically. That is, we can either *see* the spoilage, *smell* the spoilage, *taste* the spoilage, *feel* the spoilage, or experience combinations of the four sensations. Quite often, the evidence of microbial growth is easily visible as in the case of slime formation, a cottonlike network of mold growth, irridescence and greening in cold cuts and in cooked sausage, and even obvious large colonies of bacteria. In liquids, such as juices, microbial spoilage is often manifested by the development of a cloudy appearance or curd formation. The odors of spoiled protein foods are very objectionable and, in some cases, when they are intense enough, even toxic. Some of these obnoxious odors are common enough, and the compounds responsible for them are ammonia, various sulfides (as in the smell of clam flats), and hydrogen sulfide (the typical smell of rotten eggs). The changes in taste properties of spoiled foods range from loss of good characteristic taste to the development of objectionable tastes. Thus, when a pear or an orange spoils, the sweet characteristic taste of either is lost, and when milk spoils, it develops an acidic, sometimes also bitter, taste. The feel of spoiled foods reflects the spoilage in different ways, depending on the type of food and the microbe involved in the spoilage, so some spoiled foods feel slimy, while others may feel mushy.

Food Poisoning

Food poisoning may result from a variety of hazards including those from chemical, physical, or biological sources.

The chemical hazard is a danger posed by chemical substances contaminating food all along the food chain. The public controversies concerning pesticides in foods and water contamination caused by chemical spills have been well publicized and have initiated regulatory action. Chemicals must be handled very carefully by the food processor, and training in their proper use and storage must be stressed. In food processing and foodservice, there are a number of concerns regarding chemicals that must be controlled by foodservice managers and food processors. These include: (1) contamination with chemicals such as detergents and sanitizers; (2) excessive use of food additives, preservatives, and spices; (3) chemical action with containers such as acids with metal-lined containers; and (4) toxic metals such as lead (not to be used in food production areas), copper from water lines exposed to carbonated or other acidic beverages, and cadmium from uncovered meats that come in extended contact with refrigerator shelves which contain the metal.

The physical hazard is the danger from particles or items that are not intended to be part of the food product. Chips of broken glass, metal fragments such as staples from packaging or metal curls from can openers, or even cherry pits in a "depitted" canned cherry are included in this group. Strict quality control and examination of raw and finished products are essential to control this hazard.

The biological hazard includes the food poisoning resulting from the activity of microbes on the food. In many instances, grain used for animal feeds becomes contaminated with mold growth during storage, the molds producing toxic materials that cause diseases in the animals to which the grain is fed.

Bacteria, and especially molds, cause many diseases of vegetables and cereals as they are grown, causing economic losses, but this subject is beyond the scope of this discussion.

Almost all the factors that contribute to foodborne diseases result from ignorance of proper food-handling procedures or the unwillingness of some individuals in food industries (including food processors) to comply with the basic guidelines for proper food handling. Thus, foodborne diseases will continue to occur at unnecessarily high rates as long as (1) food handlers do not employ strict sanitation in both their personal habits and in the maintenance of their work area and equipment, (2) foods are not properly refrigerated, (3) foods are not adequately processed, (4) cross-contamination situations are not avoided, and (5) management does not realize the importance of incorporating practices that prevent foodborne diseases. Proper food-handling procedures include rather simple techniques, such as holding at specified temperatures, but they also include complex procedures, such as calculating processing times and predicting certain biochemical reactions that might result from processing modifications. Ordinarily, therefore at some level in the food-processing stage, the services of a professional food technologist are required. Once a food has been properly processed, the remaining handling it undergoes does not require a professional food technologist, but it does require periodic quality control checks by personnel qualified in microbiology and sanitation.

Foodborne diseases are mostly caused by several species of bacteria, although viruses, parasites, amoebas, and other biological, as well as chemical, agents may be responsible. The disease-causing bacteria in foods, or their end products that cause disease, are transmitted by consumption of the food. Foodborne diseases are classified as food infections and food intoxications, although the distinction between the two categories is not clear in some cases (e.g., Bacillus cereus and Clostridium perfringens).

Food Infections

Food infections are those in which the disease organism is carried through foods to the host (human or animal), where it actually invades the tissues and grows to numbers large enough to cause disease.

Salmonellosis

Salmonellosis is caused when foods contaminated with *Salmonella* bacteria are eaten. At the present time, approximately 42,000 cases of salmonellosis are reported yearly. About 150 deaths annually are due to this disease. However, it is considered that only about 1% of acute digestive illnesses are reported in this country, so that actually there may be many more cases of this disease.

Typhoid fever, of which there are fewer cases than of salmonellosis, and hence fewer deaths, is caused by an organism belonging to the *Salmonella* species, but this disease is usually not considered to be the ordinary salmonellosis for three reasons: (1) The ordinary *Salmonella* organisms will infect other animals as well as humans, whereas the typhoid germ is known to infect only humans; (2) typhoid fever is usually more severe than the ordinary salmonellosis; and (3) in healthy adults, several hundred thousand to several million ordinary *Salmonella* bacteria (cells) must be ingested (eaten) to cause salmonellosis, while the ingestion of even one typhoid cell may cause typhoid fever.

The ordinary symptoms of salmonellosis are abdominal pain, diarrhea, chills, frequent vomiting, and prostration. However, there are instances in which much more severe symptoms may be encountered. The incubation period (time after ingesting the organisms until symptoms are evident) is 7 to 72 hours. In typhoid fever, the incubation period is 7 to 14 days.

Persons with salmonellosis often become carriers of the organism for a period of time after they have recovered from the disease. That is, they continue to discharge the organisms in their feces. Because of this, carriers often contaminate their hands with these organisms, which may not be removed completely after thorough washing. Hence, if carriers handle foods that are to be eaten by others, they may contaminate them with these bacteria, and in this manner, transmit the disease to others. In most salmonellosis cases, the carrier stage does not persist longer than 12 weeks after symptoms, and with typhoid fever, for shorter periods. However, there are isolated cases in which the carrier stage lasts much longer than 12 weeks, and 2% to 5% of those ill with typhoid fever may become permanent carriers.

The *Salmonella* bacteria are rod-shaped; they do not form spores, and thus they are not especially heat-resistant. They are motile (can move about in the water, in foods, or other materials in which they are found) and will grow either with or without air (oxygen). At the present time, more than 2000 types of *Salmonella* bacteria are known, all of which are considered to be infective to man. Obviously, these organisms are very widespread.

Whereas it is considered that many *Salmonella* bacteria must be consumed to cause the disease in a normal adult, it is known that the very old and especially the very young may contract the disease after ingesting only a few of these organisms. Therefore, any food, especially a food that can be eaten without cooking, should be kept essentially free of these bacteria.

The *Salmonella* bacteria grow at temperatures near 95°F (35.6°C), but they will also grow more slowly at both higher and lower temperatures than this. It has been found that many foods are suitable for the growth of these organisms and that in some foods they will grow slowly at temperatures as low as 44°F (6.7°C) or as high as 114°F (45.5°C). Moreover, since the destruction of bacteria by heat is a matter that involves both time and temperature as well as the degree to which a food protects bacteria; temperatures as high as 140°F (60°C) may be required to bring about marked decreases in levels of these bacteria during the cooking of some foods.

Regarding the destruction of bacteria by heat, it should be explained that as they are heated and a temperature is reached at which they are destroyed, they are not all destroyed at once. For instance, if at about 140°F (48.9°C), 90% of the organisms would be destroyed in a period of 5 minutes, it would take 10 minutes to kill 99% of the organisms, 15 minutes to kill 99.9% of the organisms, and so on. It should be noted that if some of these *Salmonella* organisms in foods survive whatever heating they receive during cooking, and the food is thereafter held at temperatures at which they will grow [44 to 110°F (6.7 to 43.3°C], especially at room temperatures, the organisms may grow again to large numbers. The following is the USDA, FSIS regulation for the production of cooked beef, roast beef, and cooked corned beef. These times and temperatures will ensure destruction of *Salmonella* organisms:

Degrees Farenheit	Degrees Centigrade	Minimum processing time in minutes after temperature is reached
130	54.4	121
131	55.0	97
132	55.6	77
133	56.1	62
134	56.7	47
135	57.2	37
136	57.8	32
137	58.4	24
138	58.9	19
139	59.5	15
140	60.0	12
141	60.6	10
142	61.1	8
143	61.7	6
144	62.2	5
145	62.8	instantly

Some types of cooking are not sufficient to destroy all *Salmonella* bacteria that may be present in foods. Examples of cooked foods in which these organisms may survive are scrambled, boiled, or fried eggs; meringue; turkey stuffing; oysters in oyster stew; steamed clams; and some meat dishes. Foods eaten raw or without further cooking, such as clams, oysters, milk powder, cooked crabmeat, and smoked fish, may be infective should they be contaminated with *Salmonella* bacteria, especially because they may be eaten by young people who are quite susceptible to infections of this kind.

Shellfish, egg products, prepared salads, and to some extent, raw and cooked meats have often been associated with the transmission of salmonellosis. Raw shellfish may be taken from waters contaminated with *Salmonella* bacteria and cooked shellfish meats may be contaminated by humans because they are usually removed from their shell by hand. Poultry of all types, beef cattle, and hogs may have salmonellosis or may be carriers of the organism causing the disease; hence under conditions of cooking in which these organisms are not destroyed, they may be transmitted to humans.

It has been reported that 2.6% to 7% of all fresh eggs are contaminated with *Salmonella*. The USDA recommends cooking all eggs until the yolks are solid to ensure destruction of all *Salmonellae*.

Pets, such as cats and dogs, can have salmonellosis and be carriers of the causative organism. As this is the case, children, especially the very young, who handle materials contaminated by pets and who are very susceptible to contracting the disease, may contract salmonellosis.

It may seem curious that animals can have salmonellosis. The reason appears to be that their feeds, especially fish meal, meat meal, and bone meal, fed to them as supplements, often contain *Salmonella* bacteria. This is also true of some of the dried types of food used for feeding pets.

Methods or procedures that would help to eliminate salmonellosis or greatly decrease the number of cases are:

1. Good sanitation methods and procedures in food manufacturing plants, in restaurants and institutions serving food, and in the home are essential. This includes not only cleaning and sanitizing of equipment and utensils; elimination of insects and rodents; and cleaning and sanitizing of floors, walls, and so on, but also the personal cleanliness of workers who prepare or serve food. All personnel must wash and sanitize their hands prior to handling foods once having left their work stations for any reasons.
2. All foods should be held at temperatures of 40°F (4.4°C) or below when not being cooked, prepared for cooking, or being served. This would not eliminate *Salmonella* organisms from foods, but it would prevent salmonellosis in healthy adults.
3. Foods that can be eaten without further cooking should be produced under the best conditions of sanitation, and some foods of this type, such as milk powder, should be subjected to frequent bacteriological examination to determine that they are essentially free of *Salmonella* bacteria.
4. Where possible, foods (poultry stuffings, etc.) should be cooked to temperatures of at least 150 F° (65.6°C), at which it can be assumed that all *Salmonella* bacteria have been destroyed.
5. Egg products (dried or frozen) should be pasteurized (heated to 140°F [60°C] for 3 to 4 minutes), then cooled prior to drying or freezing.
6. Flocks of poultry known to have salmonellosis (this can be determined by testing) should be eliminated as egg producers. This can be done without economic loss because such poultry may be used as food.

7. The food, especially protein supplements, given to pets and other animals, should be treated to eliminate the *Salmonella* bacteria that infect these animals. This can be done, for instance, by pelletizing the food or supplements, a treatment that raises the temperatures sufficiently to destroy many of the bacteria that might be present.
8. Animals that have died from disease should not be eaten. (This rule is sometimes not observed on farms.)

Shigellosis

Shigellosis, sometimes called bacillary dysentery, is a food infection caused by bacteria of the genus (group) known as *Shigella*. Each year, there are about 17,000 cases of shigellosis reported. There is a higher mortality rate with shigellosis than with salmonellosis.

The ordinary symptoms of shigellosis are diarrhea with bloody stools (feces), abdominal cramps, and some fever. In severe cases, the symptoms are much more complex. There are fewer known species or types of *Shigella* bacteria as compared with more than 2000 for *Salmonella*. However, one of these organisms, *Shigella dysenteriae*, usually causes a much more severe disease than either *Salmonella* or other *Shigella* bacteria. The incubation period (time after ingesting the bacteria before symptoms are evident) is said to be as long as 7 days with an average of 4 days.

The *Shigella* bacteria are nonmotile, rod-shaped cells that will grow in both the presence and absence of oxygen. They do not form spores.

As in the case of salmonellosis, humans may become carriers of the *Shigella* organism after they have had the disease, and hence may become a source of contamination or infection to others who eat food that they have handled. However, the carrier stage with shigellosis is shorter than with salmonellosis.

Shigellosis is transmitted chiefly through water or milk, but the disease has also been transmitted through the eating of soft, moist foods, such as potato salad. At the present time, it is believed that when foods become the source of the organism in *Shigella* infections they have been contaminated, directly or indirectly, with small amounts of human fecal discharges.

It has been reported that the optimum growth temperature for *Shigella* bacteria is about 98.6°F (37°C) and that the range in which growth can occur is 50 to 104°F (10 to 40°C). They are relatively sensitive to heat and can tolerate up to 6% salt.

Certain control methods are noted based on the knowledge of how food and drink should be handled.

1. Since water is a known source of the organism causing shigellosis, all water used for drinking, for adding to foods, or for the cleaning and sanitizing of equipment and eating utensils should be potable (drinkable). The potability of water is determined by sanitary surveys and by bacterial tests made on samples of the water. Any food-manufacturing or food-serving operation should use only water known to be potable. Municipal water supplies should be checked periodically to determine that they are not polluted. In some cases, water from deep wells from sources not connected with the municipal water supply is used for foods and cleaning. Also, sea water is sometimes used for cleanup in food-serving or food-manufacturing operations. In such instances, management should arrange for bacterial tests to be made on such water supplies at frequent intervals to determine that they are not polluted.

2. Because it is good practice in food handling to refrigerate foods not in use, another method of controlling shigellosis is to hold foods at 40°F (4.4°C) or below at all times when they are not being served or prepared for serving.
3. It is known that humans may become *Shigella* carriers; therefore, in food-handling operations such as food manufacturing, the preparing of food, or the serving of food, personnel known to have had an intestinal disease should be excluded from any tasks that would bring them into direct contact with any food.

Vibriosis

Vibriosis, a disease caused by the bacterium *Vibrio parahaemolyticus*, was first identified in Japan where there were outbreaks involving the infection of many people. A few cases have been reported in the United States. This disease, unlike others that may be transferred from a variety of foods, is contracted almost solely from seafood. With the increased consumption of seafood in the United States, this organism may gain importance in its implications in foodborne disease.

The symptoms of vivbriosis are abdominal pain, nausea, vomiting with diarrhea, and occasional blood and mucus in the stools (feces). A fever involving a 1 to 2°F (0.55 to 1.1°C) rise in temperature is experienced in 60% to 70% of the cases. The period of incubation after ingestion of the organism is 15 to 17 hours and the symptoms last for 1 to 2 days. It is questionable whether or not people who have had this disease become carriers.

The organism causing vibriosis is a short, curved, rod-shaped, motile cell. It grows with or without oxygen and is believed to require 2% to 4% sodium chloride for growth. The organism is found naturally in the ocean, and, because it grows fastest at 86 to 104°F (30 to 40°C), is found in highest concentrations near the shoreline during the summer months.

Raw fish and molluscs (squid, octopus) have been the foods most often involved in the transmission of vibriosis, but it is known to be present sometimes in shellfish, such as clams and oysters, and has also been found in cooked crab meat. The extent to which the latter foods may have transmitted the organism and caused the disease is not known.

The control of vibriosis involves the following precautions:

1. Because the *Vibrio* is found in high concentrations in seawater only during the period in which coastal waters are quite warm, it is preferable not to eat raw molluscs (squid, octopus, clams, and oysters) during the months of July, August, and September in temperate climates or in whatever months the coastal waters of a particular region are warmest. Some believe it is too risky to eat raw molluscs at any time!
2. Because *Vibrio parahaemolyticus* is quite heat-sensitive and would be destroyed by whatever cooking is necessary to remove crab meat from the shell, the presence of the organism in cooked crab meat must have been due to contamination after cooking. This indicates that good food plant sanitation is a method of controlling vibriosis. The use of potable fresh water that has been adequately chlorinated for the cleaning and sanitizing of food plants and food-plant equipment is a method of controlling vibriosis.

Cholera

Cholera is rarely encountered in the United States; however, sporadic cases occur during summer months along coastal areas. The outbreaks generally result from consuming inadequately cooked crabs or oysters. The risk of contracting cholera is markedly decreased by thoroughly cooking shellfish and preventing their recontamination from surfaces or containers holding raw shellfish.

In the Near and Far East, cholera causes, from time to time, much sickness and death. The symptoms of cholera include diarrhea with an abundance of watery stools, vomiting, and prostration. Eventually, because the patient is unable to retain water taken by mouth, dehydration becomes a weakening factor in the disease. As a result of the rundown condition of the patient, secondary infection may set in.

The organism causing cholera is *Vibrio cholerae.* It is a short curved rod that is motile. This bacterium is aerobic (requires oxygen to grow).

Cholera is ordinarily transmitted through drinking water, but it can be transmitted through contaminated foods that have been washed in polluted water or handled by persons having the disease.

The control of cholera appears to require mainly pure water for drinking, adding to foods, and cleaning and sanitizing utensils and equipment used for the manufacture, preparation, and serving of foods.

Trichinosis

Trichinosis is not a bacterial disease. It is caused by the microscopic roundworm, *Trichinella spiralis.* Approximately 100 cases of the disease are reported in this country each year, with an indeterminate number of unreported cases. There are also a few deaths attributed to this disease each year. Trichinosis is transmitted through the eating of pork, although some wild game, such as bear, has also been known to cause trichinosis. Many of the reported cases resulted from eating pork processed by farmers or local butchers. However, some of the products of larger packers have also transmitted this disease.

The symptoms of trichinosis vary with the number of organisms ingested. If large numbers of larvae (stage of the life cycle of the worm) are eaten, nausea, vomiting, and diarrhea develop 1 to 4 days after the food is eaten. If only a few organisms are eaten, initial symptoms may be absent. On the seventh day after they are taken in, the larvae (produced by the adult worms) migrate from the intestines to the muscles, causing an intermittent fever as high as 104°F (40°C) that can last for a few weeks. The upper eyelids may swell owing to the accumulation of fluids. Once the larvae have become embedded in areas between the muscle fibers, a cyst is formed that becomes calcified. In this state, the larvae remain dormant in the body over a period of years.

The *Trichina* organisms in meat are present as the larval stage of a roundworm, and in the human intestine, develop into adult roundworms. The adult worms unite sexually and the females produce larvae that migrate from the intestine to the muscle tissues.

The following paragraphs describe a number of ways in which trichinosis can be controlled or prevented:

1. Since the *Trichina* larvae are destroyed by a temperature of 137°F (58.3°C), fresh pork should not be eaten unless all parts have been heated to this temperature or

higher. It is a good general rule never to eat fresh pork that shows any pink or red color or that has not been heated to the point where no indication of uncooked meat fluids can be seen.

2. The USDA has several rules governing the processing of pork sold as cured products. (a) The fresh pork shall have been frozen and held at 5 to –20°F (–15 to –28.9°C) for a period of 6 to 30 days, the time of holding depending on the temperature at which it is held and the size of the portion of pork; (b) all parts of the cured product shall have been heated to at least 137°F (58.3°C) in ready-to-eat products; (c) for dried, summer-type sausage (Italian salami, cervelat, etc.), the product shall have curing compounds added, after which it must be held for at least 40 days at temperatures not lower than 45°F (7.2°C).
3. The cooking of garbage to be fed to hogs is considered to be a method of controlling trichinosis because uncooked pork may be present therein. Such pork might contaminate the hog eating it, thus causing it to be a source of infection to humans.

Amebiasis

Amebiasis, like trichinosis, is not a bacterial disease. A single-celled animal, an amoeba, causes amebic dysentery in humans. The organism is called *Endamoeba histolytica*.

The symptoms of amebiasis vary greatly from patient to patient, and periodically the severity can vary in the same patient. Diarrhea is a common symptom of disease, and it may be persistent and severe, or mild and occasional. Abdominal pain, fatigue, and fever are sometimes encountered. Incubation lasts from 2 days to several months, but is usually 3 to 4 weeks.

The control of amebiasis is essentially a matter of good sanitary procedures:

1. Drinking water, water added to foods, and water used for washing and sanitizing of equipment and utensils and for irrigation of crops should be potable.
2. Water to be used for foods or for equipment or utensils that contact foods, taken from deep wells, lakes, or ponds, should be tested periodically for bacteriological safety. Although this is not a bacterial disease, it appears to occur when bacteria indicative of pollution with human discharges are present.
3. Persons known to have had amebic dysentery should be rigidly excluded from food handling of any kind.

"Red Tide" Shellfish Poisoning

Paralytic shellfish poisoning is caused by consuming shellfish that have fed on the algae *Gonyaulax catenella* or *G. tamarensis*. These toxic dinoflagellates produce a toxin (saxitoxin) that may cause death in humans within an hour or two after consumption of contaminated shellfish. The amount of poison required to cause illness varies considerably. The toxicity is measured in mouse units (MU). One MU is the minimal amount of poison needed to kill a 20-gram white mouse 15 minutes after intraperitoneal injection of an acidic extract of mussel (or other shellfish) tissue. Deaths in humans have resulted from a range of 3000 to > 20,000 MU of the toxin.

When environmental conditions are right, concentrations greater than 20,000 organisms per milliliter can occur and the water becomes brownish red, hence the term *red tide*. Proper surveillance and control by governmental agencies is needed to ensure contaminated shellfish will not be harvested. One to three weeks after the "red tide" bloom has disappeared, mussels and clams gradually destroy or excrete the poison.

Protozoa

A waterborne disease, prevalent where surface waters serve as drinking water sources, has reached a high level of incidence in the United States. *Giardia lamblia,* a flagellate one-celled animal (protozoa), is the causative agent. Symptoms including abdominal cramps, diarrhea, fatigue, weight loss, and nausea may persist for 2 to 3 months if untreated. The organism may be spread through human feces, and animals such as beavers have been found to harbor the organism and it can be spread through their feces also. Methods to prevent the transmission of the disease include appropriate disposal of fecal matter, wastewater treatment, and inclusion of a filtration step before chlorination in the preparation of drinking water from surface sources. Boiling the water will also kill the organism.

Other Food Infections

Tuberculosis (caused by *Corynebacterium tuberculosis*) and brucellosis (caused by *Brucella melitensis* and also known as undulant fever) are diseases that in the past were transmitted through milk but have been controlled in recent years by heat pasteurization of milk and the testing of dairy herds, which eliminated infected animals.

The ingestion of foods may be involved in infectious hepatitis when infected food handlers who are careless in their personal habits are involved in preparing or serving food to others, or when contaminated shellfish are eaten raw or without adequate cooking.

Streptococcal infection is quite rare and can be prevented by pasteurization.

Taenia solium (a tapeworm that may infest pork), *Taenia saginatta* (a tapeworm sometimes found in beef) and *Diphyllobothrium latum* (a tapeworm sometimes found in fish) all cause illness in humans, but none of these pose any hazard when foods are thoroughly cooked. Small worms of the genus *Anisakis* may also infect fish and cause illness in humans when the infected fish is not properly cooked.

Listeria monocytogenes has attracted attention because of two cheese-related outbreaks that resulted in cases of meningitis, miscarriage, and perinatal septicima. *L. monocytogenes* has been implicated in well over 300 reported cases of listeriosis that resulted in approximately 100 deaths. These outbreaks also had a significant economical impact on the dairy industry, which suffered more than $66 million in losses due to product recalls. The ability of this organism to grow at refrigeration temperatures in combination with its appearance in raw and processed meats, poultry, vegetables, and seafood make it a serious threat to susceptible consumers and to the entire food industry. It has also survived the minimum high-temperature–short-time treatment for pasteurized milk (161°F [71.7°C] for 15 seconds) and the standard treatment for cottage cheese manufacturing (135°F [57.2°C] for 30 minutes).

Yersinia enterocolitica is another psychrotropic organism that has been a causative foodborne pathogen in raw and pasteurized milk, and contaminated water. Unlike *L. monocytogenes,* it is destroyed by normal pasteurization. Yersiniosis usually results in enterocolitis, characterized by diarrhea, fever, and severe abdominal pain in the lower right quadrant. This illness sometimes has been incorrectly diagnosed as appendicitis and has resulted in unnecessary appendectomies. The organism is widely distributed and outbreaks have been traced to a wide variety of foods such as milk that had contaminated chocolate syrup added to it after pasteurization and tofu that was packed in contaminated spring water.

Campylobacter jejuni outbreaks have been linked to consumption of raw milk, cake icing, eggs, poultry, and beef. These underscore the need for thorough cooking and proper handling of raw products. *Campylobacter* is emerging as a leading cause of gastroenteritis in humans. Reduced oxygen levels are required for growth (microaerophilic) and the bacteria are destroyed easily by heat. *C. jejuni* is part of the intestinal microflora of mammals and birds and has been isolated from feces of healthy swine, cattle, dogs, cats, rabbits, rodents, chickens, turkeys, and wild birds. Fecal matter from healthy mammals and birds is probably the major source of this organism in our foods.

A strain of *Escherichia coli* (serotype 0157:H7) has become the causative agent in hemorrhagic colitis, resulting in severe abdominal cramps and watery, grossly bloody diarrhea and hemolytic uremeic syndrome, which is the leading cause of acute renal failure in children. Vehicles of infection for this organism have been raw or undercooked ground beef, undercooked ground beef sandwiches, raw milk, pork, lamb, and chicken. In 1993, an outbreak of *E. coli* food poisoning resulted in 475 cases, including deaths of three children.

As a direct result of the outbreak, the USDA's FSIS began a sampling program testing raw ground beef for *E. coli* 0157:H7. It was also proposed that meat and poultry plants be required to adopt HACCP programs to identify and control potential food safety hazards. It is recommended to cook all ground beef to the well-done stage, which can be reached when the meat is held at 155°F (68.3°C) for at least 15 seconds. Prevention of this disease can be accomplished by proper cooking of ground beef and prevention of cross-contamination from raw foods and food handlers.

Other food infections have implicated *Bacillus cereus*, *Aeromonas hydrophilia*, *Arizona hinshawaii*, and *Plesimonas shigelloides*.

Viruses are obligate parasites and do not grow on culture media as do molds and bacteria. Viruses also do not replicate in foods and therefore numbers will not increase during storage, preparation, and transportation. Sanitary methods must be imposed to prevent contamination of foods with intestinal viruses. Foods can serve as a vehicle of infection for viruses even though the organisms do not proliferate in the food. Viral hepatitis of food origin is a good example of what can happen if proper handling methods are not carefully practiced. If contaminated shellfish are eaten raw or without adequate cooking, they could also be a vehicle of transportation of the virus. Another virus, the Norwalk virus, was implicated in a food poisoning outbreak that affected over 200 students in 1994 who had all eaten at the same college cafeteria. A foodservice worker was found to have the virus and foods such as cold salads were suspected as the vehicles of infection. According to the American Red Cross, "You Won't get AIDS in a Restaurant." There has never been a scientifically documented case of acquired immune deficiency syndrome (AIDS) transmitted through food, so therefore it is believed that there is no threat from AIDS as a foodborne disease.

Food Intoxications

Food intoxications are those diseases in which the causative organism grows in the food and produces a chemical substance in the food that is toxic to humans and other animals.

Staphylococcal Poisoning

The symptoms of staphylococcal poisoning are nausea, vomiting, abdominal cramps, diarrhea, and prostration. While the symptoms last, suffering may be acute, but this is usually for a period of only a few hours. Generally, the patient recovers without complications. The incubation period after ingestion of food containing the toxin is from 1 to 7 hours, usually 3 to 6 hours. Staphylococcal poisoning was once often wrongly called ptomaine poisoning. Ptomaines are produced by bacteria in some foods when extreme decomposition occurs. Most ptomaines are not poisonous, and it is unlikely that many people would eat foods decomposed to this extent. It is probable, therefore, that ptomaine poisoning rarely occurs.

The bacterium that causes staphylococcus poisoning is *Staphylococcus aureus*, the same bacterium that causes white-head pimples, infections, boils, and carbuncles. These cells are spherical or ovoid in shape, nonmotile, and in liquid cultures arrange themselves in grapelike clusters, in small groups, in pairs, or in short chains. They grow best in the presence of air (oxygen), but they will also grow in its absence. They will grow in media or in foods that contain as much as 10% salt (sodium chloride). When these organisms grow in foods they produce a toxin that can be filtered away from the food and from the bacterial cells. The toxin is not destroyed by normal cooking.

Almost any food except acid products is suitable for the growth of *Staphylococcus aureus*, but certain foods have been most often the cause of staphylococcal poisoning. The foods most frequently involved in this disease are ham and ham products, bakery goods with egg, custard fillings, chicken products (especially chicken salad), potato salad, and cheddar cheese. The reason why ham products are frequently involved is that in their preparation they may become contaminated with *Staphylococcus aureus,* and as this product contains 2% to 3% salt, other bacteria that might grow and inhibit the growth of staphylococci are themselves inhibited by the salt. Also, people who handle ham and its products are apt to believe that such foods are not perishable. However, ham and ham products are perishable and should always be held at 40°F (4.4°C) or below.

The toxin produced by *Staphylococcus aureus* is not readily destroyed by heat. With most cooking methods, the organism itself is destroyed, but if the organism has grown and produced toxin prior to cooking, the toxin is still present. Therefore, such foods as milk powder have caused staphylococcal poisoning, yet no living staphylococci could be isolated from them. In the case of milk powder, the bacteria were destroyed during the heating required for drying the milk.

Humans are the main source of the organism causing staphylococcal poisoning. It has been found that approximately 40% of normal human adults carry *Staphylococcus aureus* in their noses and throats. Thus, the fingertips and hands often become contaminated with the organism. Also, any person having infected (containing pus) cuts or abrasions is a definite source of the organism. Cows may also be a source of *Staphylococcus aureus*, particularly if the animals have mastitis (an infection of the udder).

In foods, *Staphylococcus aureus* will grow at temperatures as low as 44°F (6.7°C) and as high as 112°F (44.4°C). In some foods (turkey stuffing), temperatures as high as 130°F (48.9°C) have sometimes been necessary to destroy the organism.

Staphylococcal poisoning usually occurs after a food has been held at temperatures that allow the organism to grow at relatively fast rates.

Methods of controlling staphylococcal poisoning are:

1. Hold all foods, when not being eaten or prepared for eating, at temperatures of 40°F (4.4°C) or below.

2. Prohibit all persons having boils, carbuncles, or pus abrasions or cuts on their hands from handling foods.
3. Require all personnel handling food in food manufacturing or food serving establishments to wash and sanitize their hands with solutions such as chlorine or one of the iodophors prior to performing their particular tasks.
4. Eliminate use of milk from cows with mastitis.

Botulism

Botulism is an unusual disease occurring only rarely (about 15 cases per year in the United States) but with a high mortality rate (in the past, over 50%, but more recently, about 30%). The symptoms of botulism are vomiting, constipation, difficulty of eye movement, double vision, difficulty in speaking, abdominal distension, and a raw sore throat. In severe cases, and this disease is usually severe, the breathing mechanism and eventually heart action are affected, often resulting in death.

There are several types of the bacterium *Clostridium botulinum* that may cause botulism in humans or other animals. The types that affect humans most often are A, B, and E. Humans are most susceptible to the toxin of types A and E. The botulinum organisms are spore formers that have some heat resistance, and they grow only in the absence of oxygen or under conditions in which oxygen is quickly taken up by chemical substances (reducing compounds) present in some foods. The toxin is produced in the food by the bacteria before the food is eaten. Whereas the various types (A through F) of *Clostridium botulinum* are classified mainly on the basis of the different antitoxins required to neutralize each of the different toxins, there are some differences in the growth (cultural) characteristics of the cells of the different types. The types of *Clostridium botulinum* have different minimum temperatures for growth. Types E, F, and some B will grow at temperatures as low as 38°F (3.3°C), although growth will be comparatively slow at this temperature, while types A and B will not grow at temperatures below 50°F (10°C). The maximum growth rate for all types is at 86 to 95°F (30 to 35°C).

Unlike the toxin produced by *Staphylococcus aureus*, that produced by *Clostridium botulinum* is readily destroyed by heat. In foods, all botulinum toxins present would be destroyed by bringing the temperature to that of the boiling point of water (212°F [100°C]). Destruction of the toxin starts at temperatures well below 212°F (100°C). This is a fortunate circumstance, as most cases of botulism are caused by home-canned foods, many of which are heated prior to serving. This heating has no doubt saved many lives, because although the organism itself may not be destroyed by such heating, the toxin is usually destroyed, and it is the toxin, not the organism, that causes the disease.

Clostridium botulinum will not grow in acid foods (pH 4.5 or below), yet there have been some acid foods (pears, apricots, tomatoes) that have been involved in causing the disease. In such cases, it is believed that some other organism (mold, yeast, or bacterium) has first grown and raised the pH of the food to the point where *Clostridium botulinum* would grow.

Botulinum toxin affects the nerves associated with the autonomic functions of the body (contraction and dilation of blood vessels, breathing, heart action, etc.) At least in the case of type E poisoning, even when symptoms are recognized, the patient may be treated with antitoxin and eventually recover. With type A botulism, the situation is not so certain, and it is considered by some that once symptoms have occurred, the toxin is fixed, and antitoxin treatment does little good. Because the toxin blocks the function of the nerve synapses, the voluntary nervous system also becomes affected.

Because most cases of botulism are caused by home-canned foods, one method of control would be to make sure that all vulnerable home-canned foods are pressure cooked at times and temperatures suitable to destroy all spores of *Clostridium botulinum* that may be present. Pamphlets are available from the USDA that specify times and temperatures for sterilizing different products in different size containers.

Although extremely rare, there have been some commercially canned products that have caused botulism. All manufacturers of such products should have technical advisors who can determine that the heat processes given their products are sufficient to destroy all spores of *Clostridium botulinum* that may be present.

Since types E and F *Clostridium botulinum* will grow at temperatures as low as 38°F (3.3°C), all flesh-type foods, especially fish (since type E is often present), should be stored at temperatures below 38°F (3.3°C).

A good control precaution is to heat all canned vegetables, especially home-canned, to the boiling point and to hold them for several minutes prior to serving.

Infant botulism, first recognized in the 1970s, is not a foodborne disease. However, it is caused by *Clostridium botulinum* that grows in the intestine, possibly due to the lack of the development of normal intestinal flora, and it produces toxin there.

Perfringens Poisoning

Perfringens poisoning has sometimes been classified as a food infection and at other times as a food intoxication. Research indicates that when this organism grows in foods, it produces a substance (an enzyme or other compound) that causes an intestinal disturbance in humans.

Perfringens poisoning is caused by *Clostridium perfringens*, which, like *Clostridium botulinum*, grows only in the absence of oxygen. Also, as in the case of botulism organism, perfringens is a spore-forming bacterium, although the spores of this organism are not as heat-resistant as those of some types of the botulinum spores. The symptoms of perfringens poisoning are diarrhea and abdominal pain or colic. The illness occurs 8 to 22 hours after the food has been eaten, and the symptoms are of short duration (1 day or less). The number of people involved in an outbreak is quite often comparatively large. In perfringens poisoning, what ordinarily happens is that meat or poultry is cooked, then held at comparatively high temperatures, then served. The spores survive the cooking because they have some heat resistance; then when the meat or gravy is held at room temperature or on a steam table where the temperature is below 140°F (60°C), the spores grow and reproduce to large numbers, causing illness.

To control perfringens poisoning, meats and gravies should be refrigerated at temperatures of 40°F (4.4°C) or below shortly after cooking if not immediately eaten, or should be held at temperatures not lower than 140°F (60°C) in preparation for serving.

Other Food Intoxications

Other diseases may occur from the accidental accumulation of toxins in foods exposed to unusual environmental concentrations of chemical or biological toxins from polluted areas. In such cases, the toxins are not easily removed or destroyed; therefore, we must rely on our regulatory and public health agencies to make periodic inspections of foods that are suspect, as well as the area from which they are produced.

Bacillus cereus gastroenteritis is listed by some as a food intoxication because it is believed that toxin is released in the food as a result of cell autolysis (enzymatic destruction of cells after death). Surveys of foods and ingredients have shown high percentages containing *B. cereus.* High numbers of the viable cells are required for symptoms to develop. These include abdominal cramps, watery diarrhea, and some vomiting. *B. cereus* is a spore former and can survive some cooking temperatures. Quick cooling of foods, proper holding temperatures for hot foods (140°F [60°C] or higher), and proper reheating of leftovers to at least 165°F (73.9°C) are necessary for prevention.

Mycotoxins are produced by molds. Some are highly toxic to many animals and potentially to humans. Aflatoxins are the most widely studied of the mycotoxins. As of 1980, 14 mycotoxins were recognized as carcinogens. The FDA has recognized the potential danger of mycotoxins and has established allowable action levels.

Desirable Changes in Foods

During the development of civilization, humans have learned how to utilize some of the products produced by the growth of certain microorganisms. The bacterial products utilized by humans are mostly produced by the bacteria that form lactic acid from sugars, although other types may form useful materials. Cultured milks (soured cream, yogurt, buttermilk, etc.) are produced through the growth of the lactic acid bacteria.

To obtain the curd in the manufacture of cheese, milk is cultured with bacteria that produce lactic acid, which in turn precipitates the casein, although this can be accomplished in a different way. The particular flavors and textures of many cheeses are attributed to bacterial growth during or after curd formation. Flavor and texture are influenced especially during aging or during the period in which the cheese is held in storage at a particular temperature for purposes of maturing.

The particular flavor of butter is the result of the formation of small amounts of a chemical compound from sugars or citrates by the growth of lactic acid bacteria in the cream prior to churning. Pickles and olives are at least partially preserved by acid formed by bacteria when the raw materials are allowed to undergo a natural fermentation. The particular flavor and texture of sauerkraut are due to acid and other products produced by the growth of lactic acid bacteria in shredded cabbage to which some salt has been added. The typical flavor of dried-cured sausage (Italian salami, cervelat, etc.) is produced by the growth of lactic acid bacteria in the ground meat within the casing during the period in which the sausages are stored on racks and allowed to dry.

Acetic acid (vinegar) is produced from ethyl alcohol by the growth of *Acetobacter* that oxidizes the alcohol to acetic acid.

$$\underset{\text{ethyl alcohol}}{2(CH_3CH_2OH)} \xrightarrow{O_2} \underset{\text{water acetaldehyde}}{2HOH + 2(CH_3CHO)} \xrightarrow{O_2} \underset{\text{acetic acid}}{2(CH_3COOH)}$$

Various types of bread and certain other bakery products are leavened (raised) by yeasts. The yeasts, in this case, not only produce carbon dioxide, a gas that causes the loaf to rise, but also produce materials that affect the gluten (protein) of flour, causing it to take on a form that is elastic. The elasticity of the gluten is necessary to retain the gas and to support the structure of the loaf.

All alcoholic drinks are produced through the growth of different species of yeasts. These include beer, ale, wines, and whiskeys. Whiskeys, brandies, gins, and so on are produced by distilling off the alcohol from materials that have been fermented by yeasts.

Molds are sometimes allowed to grow on sides or quarters of beef for purposes of tenderizing the meat and possibly to attain particular flavors.

Some cheeses, such as Roquefort, Gorgonzola, blue, and Camembert, owe their particular flavor and texture to growth of molds. Under present-day manufacturing procedures, these cheeses are inoculated with molds of particular species, and incubated to allow the molds to grow and produce the desired textures and flavors.

The utilization of bacteria, molds, and yeasts to produce particular foods or beverages enjoyed by humans in many cases results from accidental discoveries in which foods or food materials have undergone natural fermentations or changes due to the growth of microorganisms. In many cases, these natural fermentations were first used as a method of food preservation, artificial refrigeration being unavailable at the time. Today, these foods have become special products valued for their own particular taste and texture, even though food-handling and food-processing methods have done away with the need for preservation by natural fermentation, except in very primitive areas.

In general, it can be said that microorganisms, including bacteria, molds, and yeasts, are essential to the existence of humans and other animals on earth. Microorganisms can cause economic losses in that they cause diseases and the decomposition of animal foods. However, such decomposition is a part of the cycle that returns elements, especially nitrogen, carbon, hydrogen, and oxygen, to a form that can be utilized by living things to form new organic materials.

Antibiotics, a group of compounds successful in combating microbial infections in humans, are produced by other microbes.

4

Food Safety and Sanitation

In the preface, it was mentioned that food safety must be the premier concern at all levels of food processing, preparation, and service. Every step in the process including harvesting, storage, and transportation of raw materials; processing; packaging; storage and transportation of the processed product; receiving of the processed product at the wholesale and retail level; storage of this product; use of it in preparation of another product; storage and display of the newly prepared product; service of the new product; preparation and storage of leftovers; reservice of the leftovers; and disposal of unused product must be carefully executed and monitored for safety and quality. This chapter covers some of the hazards involved in food processing and gives methods used to ensure safety and quality. Many colleges offer certification courses in food safety and sanitation and it is highly recommended that anyone aspiring to a career in the food industry take such a course.

Food must be handled in a sanitary manner to prevent the growth of microorganisms already present and to prevent further contamination if we are to stop the estimated millions of cases of food poisoning that occur each year in the United States. Sanitary handling involves personal hygiene as well as sanitary procedures.

PERSONAL HYGIENE

Rules of personal hygiene that must be strictly observed by food handlers are reasonable and require little more than common sense and awareness.

1. Persons with communicable diseases, including skin infections, should never be allowed to handle foods to be consumed by others. Obviously, this means that all food handlers must undergo periodic physical examinations to ensure this criterion.
2. Food handlers should observe physical cleanliness and should wear clean (preferably white) uniforms and no dangling jewelry.
3. Food handlers should keep the head covered, and fingernails short and clean (no nail polish).
4. When possible, gloves should be worn, but whether or not gloves are worn, the hands should be washed and dipped in a disinfectant prior to handling food. Whenever the hands are used for other activities, they should be rewashed any time prior to handling food again.
5. When handling foods, the hands should not touch the mouth, nose, or other part of the body, especially body openings, since these are possible sources of pathogens.

Remember that the hands are involved in most instances of personal contamination of foods.
6. During work, food handlers should neither smoke, drink, nor eat in the work area.
7. Pets and other animals do not belong in the food-processing area.
8. Sneezing and coughing should be confined to a handkerchief, and in fact, the individual should leave the work area when either is imminent. After using a handkerchief, the hands are to be rewashed.
9. Cloths should not be used for cleaning.
10. Foods that appear to be unwholesome, or that may contain unacceptable contaminants, should not be handled.

SANITATION IN THE HOME

The home should be kept clean by periodic cleaning and by setting rules of conduct for members of the household such as discouraging litter and encouraging use of ash trays (if discouraging smoking is unsuccessful). The surfaces should be kept dust-free, preferably by vacuum cleaning, and surfaces that contact foods should be constructed of easily cleanable material such as plastics and stainless steel. Tableware, such as knives, and pots and pans should have handles of metal or plastic rather than wood. Tableware and utensils should be cleaned as soon after use as possible and should not be left uncleaned for long periods either in the sink or on the counter, as bacteria will grow and may become airborne to contaminate other food. Tableware and utensils should be washed in hot water and detergent and rinsed, then immersed in hot water (that has been heated to a minimum of 170°F [76.7°C]) for at least 30 seconds.

In foodservice establishments, it is a violation of federal law to use a product as a chemical sanitizer unless it is Environmental Protection Agency (EPA) registered. Some household bleaches have EPA registration and the accuracy of all claims on the label has been proven. Follow the directions for sanitizing carefully. To sanitize dishes and other food utensils, the Food and Drug Administration (FDA) recommends a solution of 50 ppm available chlorine as a hypochlorite. After washing and rinsing, the utensils must be submerged in the warm sanitizing solution (75°F [24°C] or higher) for 10 seconds. If other approved sanitizers such as iodophores or quarternary ammonium are used, 30 seconds of submersion is required. For counter tops, cutting boards, or equipment that cannot be submerged, spray or wipe with the proper strength solution and allow to air-dry.

Automatic dishwashers may be used with confidence, because they clean effectively and the temperature of the water can be raised to a higher level than can be tolerated in hand washing. Keep the refrigerator and freezer clean and free from odors. Keep a clean house and enforce the habits that keep it clean, prompting members of the household to help. The home should be kept free from rodents and other pests, such as roaches and flies, by a strict preventive program and by a continuing check for indications of the presence of these undesirable elements. When they are present, concerted efforts to eliminate them should not be spared.

Care of food in the home involves precautions in a number of areas. Approximately 88% of all cases of food poisoning can be traced to handling of food at improper temperatures. Food must be kept out of the temperature "danger zone" (Fig. 4.1) in all phases of preparation (purchase, transport, preparation, service, and storage).

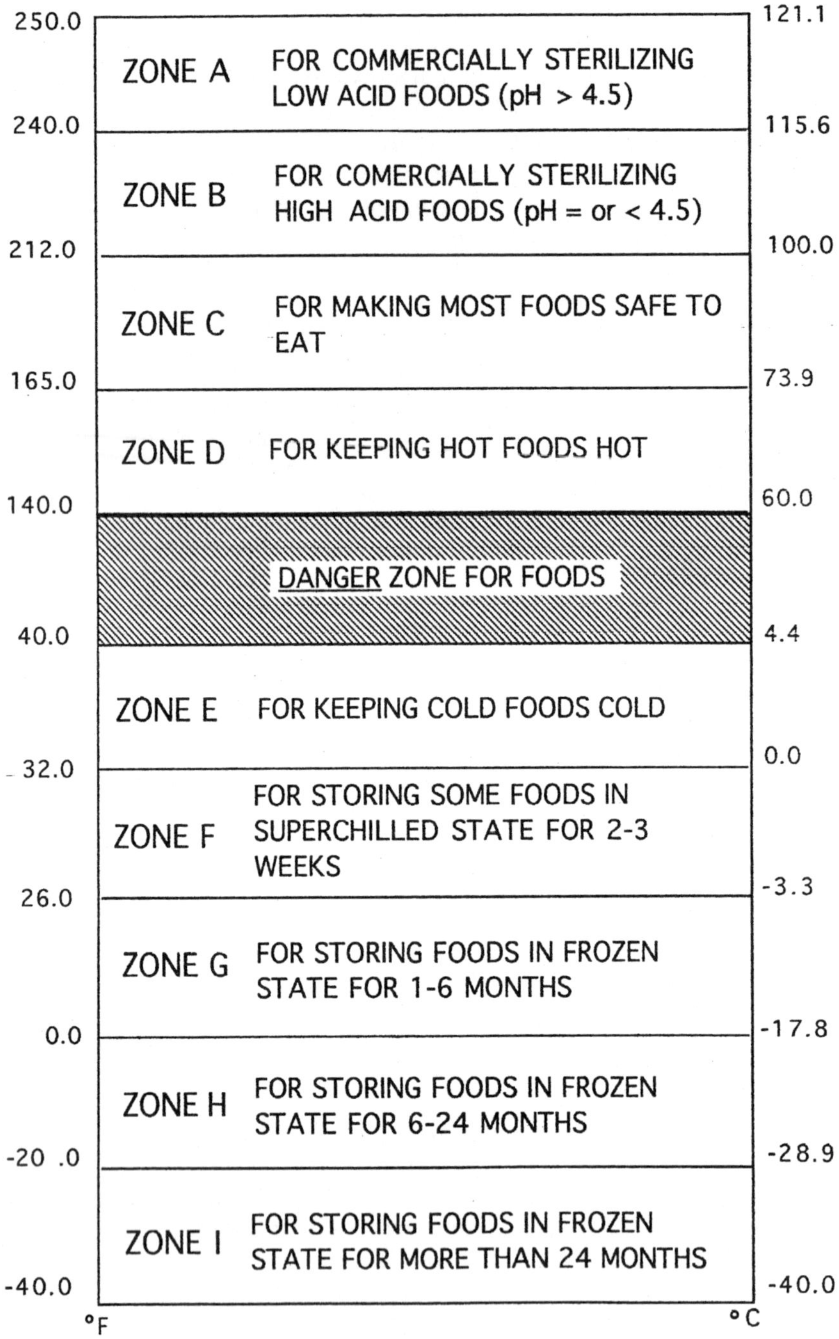

Figure 4.1. Temperature zones used for processing, holding, and preserving foods.

When shopping for foods, shop at clean stores where employees observe the rules given earlier in this chapter. Shop for the perishable foods (e.g., milk, meat, fish) last, and do not delay returning home once the shopping has been completed. Remember that bacteria (which spoil foods) grow to very large numbers in just hours while in a warm car. Once home, the perishable foods should be unpacked immediately and transferred quickly to the refrigerator or freezer.

Perishables should be kept in a refrigerator held at as low a temperature (above freezing) as possible. Remember that the most important deterrent to food spoilage is low temperature. The refrigerator should be kept at temperatures in the range 32 to 38°F (0 to 3.3°C), preferably at the lower side of the range. Frozen foods should be held at 0°F (–17.8°C) or below. Neither the refrigerator nor the freezer can lower the temperature of bulk foods quickly enough, and spoilage occurs during the period of cooling; therefore, when the bulk of foods can be reduced to smaller sizes it should be done. Large pots of stew, bulk hamburger, and large cuts of meat could all be divided and put in small containers. Cooling food quickly helps to avoid spoilage (see Fig. 4.2).

Packaging of the foods, especially in impervious films, is important to minimize freezer burn, oxidative deterioration, and dehydration. Refrigerated foods should be covered, except for ripe fruits and vegetables. All cooked foods, meats, fish, cold cuts, bacon, and frankfurters should be packaged before refrigeration. Nuts should be packaged and refrigerated to slow down the oxidation of their fats, which leads to rancidity. Greens (e.g., spinach) should be refrigerated unwashed because if water is retained they may spoil more quickly. They should, however, be washed before using. It should be remembered that fresh foods should be used as soon as possible. Fresh foods, with some exceptions, deteriorate rapidly. Some foods are best held at room temperature and these include baked goods (to hold for more than 2 days, they should be frozen), unripe fruit, and bananas. When meats, eggs, poultry, fish, and other perishable foods are to be held, it should be remembered that the safe holding temperatures are those below 40°F (4.4°C) and those above 140°F (60°C).

Meals that are consumed following their preparation are not likely to be responsible for food poisoning. On the other hand, meals that are consumed hours after preparation, such as is customary at picnics and outings, should be supervised by knowledgeable people. This is especially true of salads containing eggs, chicken, or turkey, or other products made from them. Prepared foods that are not to be used for long periods should be refrigerated immediately and later reheated, if necessary, just prior to use. Some products such as puddings, custards, and eclairs should be held under refrigeration at all times. Home freezers are not of sufficient capacity to flash-freeze stuffed poultry, chicken, and especially turkey, and therefore poultry should not be stuffed prior to freezing. Instead it should be stuffed prior to cooking. Leftovers should be used as soon as possible; this is especially true of salads, chicken, and other perishables. If, because of odor or other indicator, there is any doubt about the safety of a food, it is best to discard it unless advice can be obtained from a food scientist.

Extreme care must be taken with foods to be taken on a picnic or outing.

1. Precool all perishable foods, including ingredients for salads, preferably in a refrigerator.
2. Heat all foods, to be served hot, to at least 165°F (73.9°C).
3. Keep cold foods cold (40°F [4.4°C] or below) and hot foods hot (140°F [60°C] or above) until served.

It is best to refrain from handling pets while preparing foods.

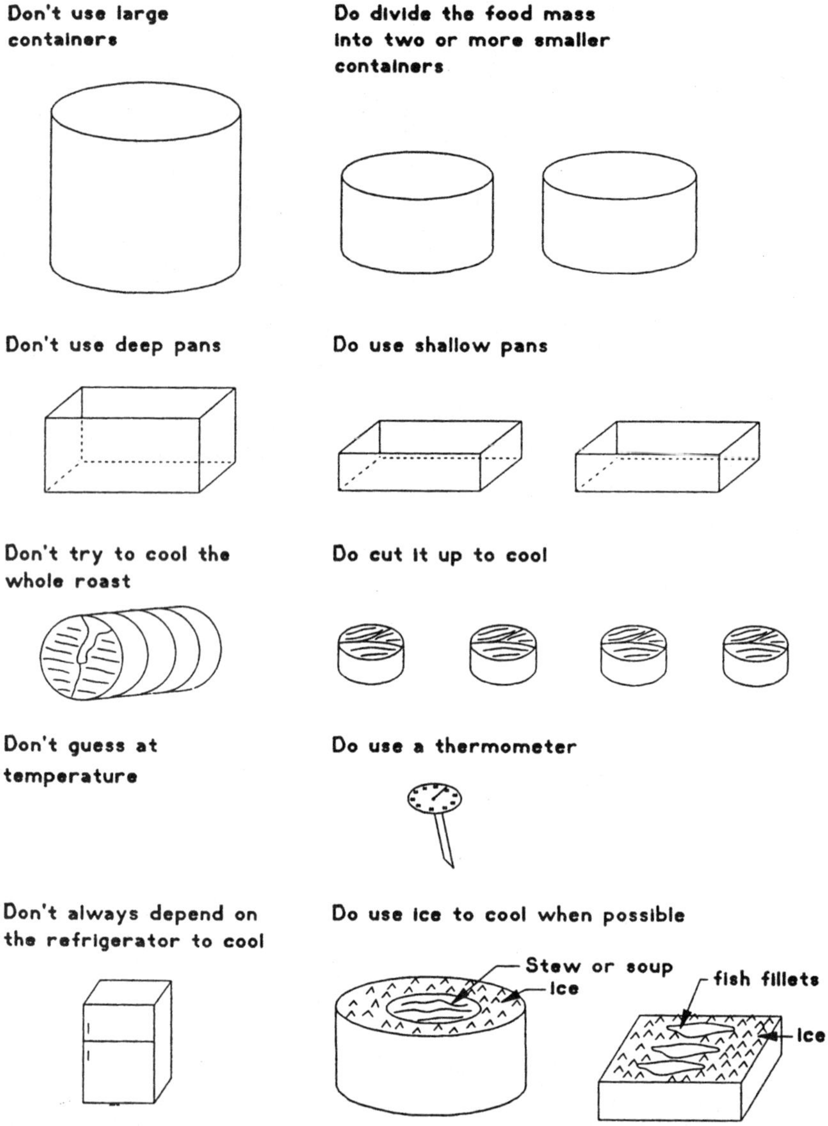

Figure 4.2. Do's and don'ts for the safe cooling of foods.

FOODSERVICE SANITATION

It has been estimated that over 100 million meals per day are served in approximately over half a million restaurants in the United States. The U.S. Public Health Service has reported that about two-thirds of all reported food poisonings result from meals served in restaurants. In terms of numbers, no one can be sure, as only a small percentage of all food poisonings is ever reported. One thing is known, and that is the number is very high, with some estimates indicating that it is about several million per year. Thus, sanitation in restaurants requires considerable improvement, as it is known that nearly all food poisonings are avoidable. The main reason for food poisoning is the poor attitude of some workers and the ignorance of others. Therefore, it must be the responsibility of the managers and owners of restaurants to employ only those with healthy attitudes and to make sure that they have been properly educated. Attitude has to be the most important criterion for employment, as the education part is simply a matter of a little time and effort. On the other hand, no amount of education in sanitation can improve an unhealthy attitude. People with unhealthy attitudes must be removed from food handling responsibilities. The problem is aggravated by the fact that restaurants are visited by large numbers of people over short periods (taxing the cleaning and food handling efforts of the facilities) and by the fact that some of the customers are bound to have communicable diseases.

Whereas restaurants outnumber other feeding establishments (e.g., hospitals and school and industrial cafeterias), in the latter, over a million meals are served daily.

The number of people involved in serving meals in restaurants and institutions is estimated to be over 8.0 million. Whether food is to be prepared and served in restaurants or in institutional cafeterias, safety precautions and sanitary procedures are very similar.

In general, the area where people are fed need not be special except that it can be easily cleaned and kept clean. The area should be pleasant; the floor may be carpeted, but in any case, it must be easy to clean thoroughly.

It should be mandatory to use only potable (drinkable) water in foodservice establishments. If nonpotable water is used for cooling refrigeration units in walk-in refrigerators, for example, this water should not be connected with the potable water system, and the pipes should be painted with an identifying color.

In the food preparation and utensil-cleaning areas, the floors should be constructed of acid-resistant unglazed tile or of epoxy or polyester resin on a suitable base material. Also, the floor should be sloped to drains to facilitate cleaning and prevent the accumulation of water. Drains should empty into lines separated from toilet sewer lines to a point outside of the building and should be so constructed as to prevent the possibility of backup into the building. The walls of food preparation and utensil cleaning areas should be constructed of smooth glazed tile to distances equivalent to splash height, since this type of construction allows for easier and more effective cleaning. The junction of the walls and floor should be coved or curved, which also facilitates cleaning, there being no angled corners where food materials can lodge.

The surfaces of benches and tables in food preparation areas should be of stainless steel or plastic, because such materials are impervious, noncorrosive, and easy to clean. For the same reason, cooking utensils, including steam kettles, should be constructed of stainless steel.

Areas in which steam cooking, steam cleaning, or deep-fat frying is done should be provided with hoods and exhaust fans to the outside. Hoods should be constructed of stainless steel for ease in cleaning and should be equipped with traps to prevent condensed moisture from running back into foods being prepared.

Cutting boards should be constructed of plastic, preferably Teflon, as such material is easy to clean, does not absorb water, and does not foster bacterial growth.

Walk-in refrigerators should have floors constructed of unglazed tile that are sloped to drains to facilitate cleaning. The drains should not empty directly into the sewerage system. Instead, they should empty into a sink or other container that empties in turn into a drainage system, so there is no possibility of backup into the refrigerator. The walls and ceilings of walk-in refrigerators should be constructed of glazed tile to facilitate cleaning.

Wash basins, soap, hot and cold water, and a container of disinfectant (preferably a solution of one of the iodophors) together with paper towels should be present in the food preparation, utensil washing, and food exit areas to make certain that anyone having left his or her particular operation will wash, disinfect, and dry his or her hands prior to resuming work.

The management of foodservice establishments should obtain raw food materials only from reliable sources. It should also determine that the water to be used for drinking, cooking, and cleaning is potable.

The precautions to be used in cooking foods are no different from those that have been given for household food safety, but if foods, including gravies, are to be held on steam tables, the temperature of all parts of the food should never fall below 140°F (60°C), and preferably not below 150°F (65.6°C). Also, any food container held on steam tables should be emptied, removed, and replaced with a new container of the particular food instead of being partially emptied and refilled with more food.

If leftover cooked foods are to be refrigerated, they should be placed in covered impervious containers (plastic or metal) and labeled. A card catalog of leftover foods should be maintained so that they may be thrown away if held for periods of more than 4 days at temperatures of 38°F (3.3°C) or above.

If cream-filled pastries, such as eclairs or pies, and salads, such as potato, tuna fish, crab meat, or chicken, are to be held in the refrigerator or in display cases, the temperature of such storage areas should be 38°F (3.3°C) or below.

Personnel with boils or pus-producing infections on their hands should not be allowed to handle foods or to clean utensils or equipment. Any personnel known to have had a recent intestinal ailment should be excused from work until such a period as it can be determined that he or she is not infective.

All garbage and waste materials should be held in leakproof metal or plastic containers with tight-fitting covers when held on the premises of the foodservice establishment. Rubbish should also be held in this manner. After emptying, each container should be cleaned inside and outside, and the water used for such cleaning disposed of in the sewerage system. Such containers should be washed in an area well separated from that used for washing utensils; containers of garbage and trash should be stored in verminproof rooms, well separated from the food preparation and serving areas. The floors of such storage rooms, and the walls up to distances of splash heights, should have an impervious easy-to-clean surface. Storage areas of this type should be cleaned periodically. Care should be exercised to see that rodents and insects do not become established in waste storage areas.

An effective method of sanitizing glasses, chinaware, and utensils is to wash in water at least at 120°F (48.9°C) with detergent followed by rinsing, then immersing in clean water at 170°F (76.7°C) for 30 seconds. Immersion for 10 seconds in a solution containing 50 ppm available chlorine or 30 seconds in a solution containing 12.5 ppm of an iodophor, in each case at a temperature not below 75°F (23.9°C), may be substituted for the hot-water dip for purposes of sanitizing utensils and tableware.

For the floor plan of a typical foodservice facility, see Figure 4.3.

FOOD PLANT SANITATION

Food plant sanitation is necessary, first because it is a law (see the Food, Drug and Cosmetic Act), and second because it is ethical, economical, and expected.

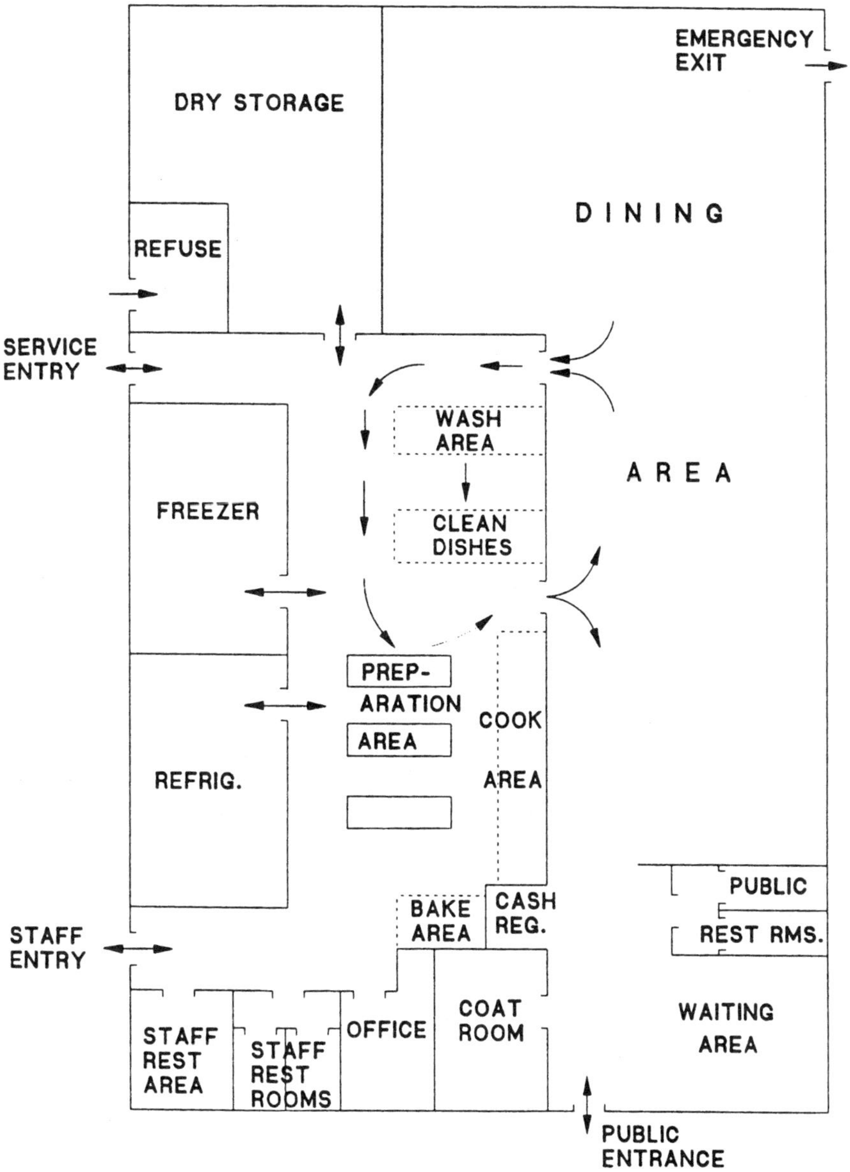

Figure 4.3. Simplified floor plan for a well-organized foodservice operation.

An important part of plant sanitation is the establishment of a strict quality control over the incoming raw materials. No amount of plant sanitation can remain effective if incoming materials are allowed to bring in pests or contaminants.

Many of the factors concerned with food safety in plants manufacturing food products are the same as those indicated for food preparation and serving establishments. There are some additional considerations, however.

Plant Exterior

The surroundings for food plants should be neat, trim, and well landscaped. There are several reasons for this. Nice surroundings have a good psychological effect on those who work within. If the environs are well kept, the personnel working there are much more apt to try to keep things neat and clean on the inside. If the surroundings are dirty or cluttered, those working in the plant are apt to become careless in matters concerned with general sanitation. All parking spaces, roadways, and walks should be paved so that dust contamination of the air will be minimized, and contamination, such as animal droppings, will be washed away with each rain rather than be soaked into the ground to be airborne during dry spells. The area surrounding a food plant, including platforms, should not be used for storing crates, boxes, or machinery, as these materials may harbor rodents that could eventually find their way into the plant. There should be no area around the plant where the landscaping allows potholes or depressions of any kind in which water may accumulate and become a breeding place for insects which then could become established within the plant. Food materials, ensilage piles, or other organic wastes should not be present in any exposed area near the plant, as they attract and become breeding places for insects, especially flies, which are difficult to control in food plants, even under the best conditions. There should be no neighboring plants, such as chemical, sewage, poultry, or tanneries, that may transfer bacteria or chemicals to the food plant.

Plant Construction

Food manufacturing plants are best constructed of brick or concrete, as wood is difficult to maintain in a clean and sanitizable condition and is more vulnerable to invasion by rodents, birds, and other pests. If the plant is constructed of wood, the foundation should be "rat-stopped" (constructed of cement to a distance of several feet below and above the ground). The walls and roof junction of the food plant should be weatherproof and impenetrable by insects. In food-processing or utensil-washing areas, the junction of the wall and floor should be curved and have no angled corners, to facilitate cleaning. Window ledges should be slanted to prevent their use by personnel for storage of materials. The floor should be made of acid-resistant unglazed tile or of epoxy resin material, an epoxy tile grout laid on cement, for instance. The floor should be sloped to drains so that water does not accumulate. A cement surface is undesirable, as it tends to become pitted, leaving areas where water and food scraps accumulate and where bacteria may grow to large numbers, thus becoming sources of contamination and putrid odors. The walls should be covered with glazed tile at least up to distances of splash height, to facilitate cleaning.

Raw materials should be separated from areas producing the finished product by a solid nonleaking wall (no openings, no doors) or by using separate buildings. The boiler room must also be separate and closed off. Each entryway to the area producing the finished product should be equipped with a shallow pan containing a disinfectant, so that those who enter the area must step into the pan, thus disinfecting their shoes or boots.

Equipment

Food-processing equipment should be so designed that all surfaces contacting foods are smooth, relatively inert, nonabsorbent, easily reached for cleaning, and of materials that are easily cleaned and sanitized. Moreover, it is desirable that as much of the equipment as possible can be cleaned without disassembling, but, where required, disassembly is possible and easily done. These specifications are desirable because they ensure that sanitation of equipment is possible and that sanitation can be accomplished quickly, effectively, and inexpensively. They are also desirable because they do not foster delaying tactics on the part of employees. The NSF International (formerly the National Sanitation Foundation) has developed widely accepted standards for food processing and foodservice equipment. These grade equipment design, construction, and installation.

The equipment, benches, and machinery used in food-processing and utensil-cleaning areas should be of such materials and design as to make cleaning as easy as possible. Surfaces should be smooth (about 150 grit). Cutting boards, knife handles, and shovel handles should be of hard plastic or other materials impervious to water. Black and cast irons and ordinary steel may be used for equipment that has no contact with foods, such as retorts and can sealers. However, because these materials have rough surfaces and are subject to corrosion and are difficult to clean, their surfaces should not come in contact with food materials. While the surface of new galvanized iron is corrosion-resistant, the zinc covering soon wears off, exposing the iron, which can corrode. Also, zinc may cause a discoloration of certain foods. For these reasons, galvanized iron is not acceptable as a surface for contact with foods. Copper is used in steam kettles for the manufacture of certain foods such as jams and jellies, because it is a good conductor of heat. When used in processing other foods, it must be kept scrupulously clean, however; otherwise, oxides accumulate, accelerating the destruction of vitamin C and the oxidation of fats.

Alkaline materials used on copper equipment may cause a discoloration of foods. It is generally undesirable to use cooper in food-processing equipment, even though it is among the best conductors of heat available. It is especially important to avoid use of copper with acid foods as the copper is absorbed in acid foods to levels that can be toxic. Monel metal, an alloy consisting mainly of nickel and copper, is suitable for food-processing equipment but is expensive. Aluminum conducts heat well but is subject to corrosion when contacting alkaline materials or fruit acids. Glass may be used for piping in the transportation of liquids, such as milk, and such piping can be cleaned in place without dismantling. It is not suitable for lining metal equipment, nor is enamel, being subject to chipping, thus exposing the metal, which may then corrode. Glass or enamel chips may also become incorporated into the food. Rubber is used for conveyor belts that transport materials, but rubber is not easily cleaned and belts lined with Teflon or a suitable metal are preferable to those lined with rubber. Stainless steel is doubtless the most suitable material for the construction of equipment that contacts food. It is noncorrosive and easily cleaned and sanitized. A special stainless steel may be required for areas contacted by chlorides, such as those occurring in seawater or brine.

Piping and pumps should have no threaded joints where food can accumulate. Sanitary design for such equipment calls for flush joints held together by clamps, allowing thorough cleaning and sanitizing. Pipes that carry food materials should have no dead ends that cannot be cleaned and where food materials can accumulate and decompose. With such construction, surges on the line cause decomposed material to enter the mainstream of the food material passing through the pipes. Pipes should not be joined to tanks or hoppers, so that the pipe end extends into the tank itself. In such cases, when the liquid falls below the level of the pipe, some food remains in the pipe end, where it may decompose and eventually contaminate new material entering the tank. Tanks, flumes, thermometer wells, pots, and pans should have only curved corners and junctions of sides and bottom to facilitate cleaning.

Personnel Facilities

Locker rooms should be provided, separate for male and female personnel, and a sufficient number of lockers should be available to provide one for each worker so that outside clothing may be stored. These locker rooms should be kept clean and tidy.

Separate toilet rooms with self-closing doors (that do not open directly into processing area) should be provided for men and women workers. Toilet rooms should have wash bowls with hot and cold running water, soap dispensers, paper towels, and containers for refuse. The suggested minimum number of toilets and wash basins is given in Table 4.1. Urinals may be substituted for toilets in the men's rooms on a one-to-one basis, but the number of toilets must never be fewer than two-thirds of the number given in Table 4.1. The hand-washing units should be of the foot-activated type.

If personnel are to eat lunches at the plant, a room separate from other rooms, including locker rooms, should be provided for this purpose. This room should be kept clean and sanitary. Drinking fountains should never be located in toilet rooms.

All personnel working in food-processing or utensil-cleaning areas should be provided with clean outer uniforms daily.

In processing or utensil-cleaning areas, there should be wash basins, a container of disinfectant (preferably a weak solution of an iodophor), and paper towels so that those in charge can make sure that workers wash and disinfect their hands before returning to work once having left their particular area of operation. This is of great importance to good food plant sanitation.

Food processing plants must have adequate light to ensure that employees can perform their duties effectively. Table 4.2 cites the minimum lighting requirements to carry out specific operations in the plant.

Smoking in food processing areas is prohibited and many plants now are "smoke free" environments. If a foodservice or food processing worker does smoke in designated areas, he or she must wash his or her hands thoroughly as one would after using the toilet. Signs should be conspicuously placed in the area to remind the workers to wash their hands.

Food processing plants should have good ventilation with filtered air, to reduce the atmospheric humidity, and hence the amount of condensation, which if allowed to occur, promotes growth of bacteria and molds on many surfaces including those of walls, ceilings, floors, equipment, utensils, and foods.

Table 4.1. Suggested Laboratory Facilities

Number of Employees	Minimum Number of Toilets	Minimum Number of Wash Basins
1–9	1	1
10–24	2	1
25–49	3	2
50–74	4	3
75–100	5	4
> 100	5 + 1/30 additional employees	4 + 1/50 additional employees

Table 4.2. Suggested Minimum Light Requirements for Food Processing Plants

Operation	Minimum Light (Ft-Candles)
Sorting, grading, inspection[a]	50
Processing, active storage	20
Instrument panels, switchboards	10
Toilet rooms, locker rooms, etc.	10
Dead storage	5

[a]Local lighting for inspection may have to be as high as 100–150 ft-candles, depending on the type of inspection performed.

Storage

Dry materials such as breadings and flour, to be stored, should be held in a room constructed of materials, such as brick or cement, that do not allow the entrance of insects and rodents. Such rooms should be refrigerated to about 50°F (10°C) to prevent the hatching of eggs and the development into adult insects, should viable insect eggs be present.

Cleaning the Plant

All floors, walls, benches and tables, conveyors, hoppers, fillers, kettles, and utensils used for processing foods should be thoroughly cleaned and sanitized at least once, and preferably twice, per 8-hour working shift. Large plants should have cleaning crews with a foreman. A list of approved cleaning compounds was published by the U.S. Department of Agriculture in 1987, and it is available from that agency. The water used for cleaning should be potable and have a temperature of about 130 to 140°F (54.4 to 60.0°C). The type of detergent used in cleaning should be suitable to remove the type of soil that will be encountered. Depending on the application, detergents should have certain properties. In general, detergents should not be corrosive. When used in hard water, they should not form precipitates. They should have good wetting, and in many cases emulsifying, properties. All detergents contain surfactants (surface-active agents) that reduce tension where the detergent meets the soil surface, allowing the detergent to penetrate in order to loosen and disperse the soil. Most detergents used in the food industry are alkaline because most food soils are acidic in nature and will disperse more easily in the alkaline solutions. Detergents should also be good solvents for both organic and inorganic soils. They should saponify fats and have good dispersal and deflocculating properties. They should not form residual films on surfaces. High-pressure water and high-pressure steam can be used to flush hard-to-reach places with detergent.

To emulsify effectively, detergents act much as soaps, having an ionic and a nonionic end. The ionic or hydrophilic (water loving) "head" attracts water and the hydrophobic

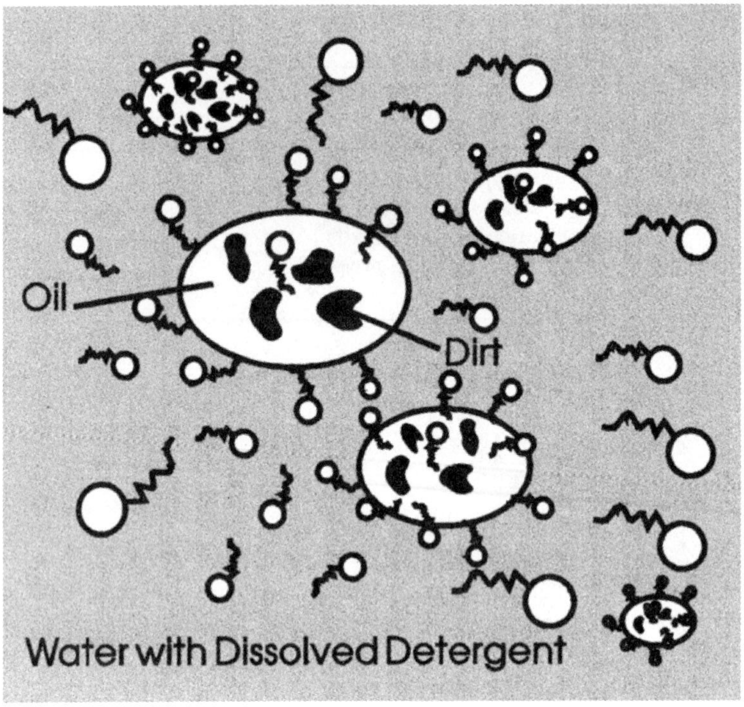

Figure 4.4. Action of detergents in removing soil.

"tail" will "dissolve" in grease and oil and the soil can be removed in the water medium (see Fig. 4.4).

Chemical agents used for controlling microbes include bacteriostats, which prevent the growth and spread of bacteria, and bactericides, which not only stop bacterial growth, but also destroy the bacteria. Some agents are bacteriostatic in small amounts and bactericidal in large amounts. Effective sanitation programs should be using mainly, if not completely, bactericides that include compounds in the following chemical classes: halogens, phenolics, quaternary ammonium compounds, alcohols, carbonyls, and miscellaneous others. The halogens chlorine and iodine are considered to be the most important sanitizing agents known. Such common uses such as chlorination of drinking water and iodine treatment for cuts make these compounds familiar. Phenolics, such as cresol, are considered to be also very good sanitizers; however, they have some disadvantages, among which are high irritation characteristics and relatively high cost. Quaternary ammonium compounds, although effective in the control of algae and some bacteria, are relatively ineffective against a variety of microbes that are not resistant to the halogens and phenolics. Alcohols are not as effective as generally believed, and while some carbonyls, such as formaldehyde, are effective, they are also hazardous to use. There are other miscellaneous sanitizers, but they are generally low in effectiveness and high in cost. The three most commonly used sanitizers for food contact surfaces accepted by the FDA are chlorine, iodine, and quarternary ammonia.

After cleaning and rinsing with hot water, equipment should be immersed for at least 30 seconds or rinsed with a solution of iodophor containing 12.5 ppm of available iodine. Iodophors, or "tamed iodines," are combinations of iodine and surface-active agents that have the sanitizing advantages of iodine with minimized disadvantages.

Iodine by itself is not very soluble in water, has a high vapor pressure, is corrosive, and leaves a stain. In combination with surfactants, these undesirable properties are minimized. A solution containing 50 ppm of available chlorine at a temperature not below 75°F (23.9°C) may be substituted. All equipment so constructed as to hold liquid should be thoroughly drained after cleaning and sanitizing, and containers, such as pans, should not be nested after cleaning and sanitizing, as this prevents drainage and evaporation of moisture and thus provides moisture in which bacteria may grow. For a summary of the properties of sanitizing agents see Table 4.3.

Water Supply

The water supply should be adequate for filling the plant's needs. All water that may contact either foods or surfaces that may be contacted by foods must be potable quality (safe for drinking). The water used for cleaning should be of adequate temperature and pressure and supplied via a plumbing system of adequate capacity and in conformance to building codes.

Sewage Disposal

Sewage disposal must be through a public sewerage system or through a system of equal effectiveness in carrying liquid disposable waste from the plant. The sewerage system must conform to building codes and in no way be a source of contamination to the products, personnel, equipment, or plant. Drains must be sufficient to ensure the rapid and complete transfer of all wash water, spilled liquids, and so on, to the sewage system.

SANITATION IN RETAIL OUTLETS

Most regulations applicable to foods in retail stores are the same as those that are applied in food manufacturing plants, but there are some precautions especially applicable to retail outlets. Fresh uncut meats should be stored in a walk-in refrigerator with walls and ceilings of glazed tile to facilitate washing and cleaning. The floor should be constructed of unglazed tile and sloped to drains. The temperature of the meat storage room should be held at 32 to 37°F (0 to 2.8°C). Sawdust should not be used on the floor because this creates dust, which is, to some extent, a source of contamination.

The room where meat is cut into retail portions should have the same wall, ceiling, and floor construction as does the meat storage area. Benches used for cutting meat should have surfaces of stainless steel, and cutting boards should be made of a plastic material, preferably Teflon, which is impervious and easily cleaned and sanitized. The temperature of this room should be about 50°F (10°C), as personnel find it difficult to work at lower temperatures. However, a low relative humidity must be maintained in the cutting area; otherwise the meat will condense moisture (sweat), which will facilitate the growth of spoilage bacteria. Also, neither cut nor uncut meat should be allowed to accumulate in this area but should be moved back into the storage area or into display cases as soon as possible. The meat-cutting area should be cleaned and

Table 4.3. Three of the Most Commonly Used FDA Approved Chemical Sanitizers

Minimum Concentration	Hypochlorite (Chlorine) (50 ppm)	Iodophor (Iodine) (12.5 ppm)	Quarternary Ammonium (200 ppm)
Temperature of solution	75°F/23.9°C+	75–120°F/23.9–48.9°C Note: iodine leaves solution at 120°F/48.9°C and is no longer effective as a sanitizer.	75°F/23.9°C+
Time needed for sanitizing	10 seconds for immersion. Allow to air-dry when spraying a surface or piece of equipment.	30 seconds for immersion. Allow to air-dry when spraying a surface or piece of equipment.	30 seconds for immersion, but some products require more time. Read instructions on label.
Effective pH range note: detergent residues will raise pH	Below 10	Below 5.5	7 but varies with compounds—read label
Corrosiveness	Corrosive to some substances such as aluminum, silver plate utensils, and stainless steel	Noncorrosive	Noncorrosive
Reaction to organic substances in water	Quickly inactivated	Less effective	Minimally affected
Reaction to hard water	Not affected	Not affected	Read label—hardness of 500 ppm undesirable for some quats
How to recognize or measure strength	Test kit or strip	Test kit or strip note: loses amber color as weakens	Test kit or strip

sanitized at least once per 8-hour working period and should follow essentially the same methods as indicated earlier for food manufacturing plants.

In the grinding of meat, such as hamburger, separate grinders (or grinder heads) should be used for beef and pork. The reason for this is that pork may contain an infective roundworm, which causes the disease known as trichinosis in humans. In pork, this is usually taken care of by cooking to a temperature that destroys the worm cysts. Hamburger, however, is sometimes eaten undercooked or only lightly heated, in which case the cysts would not be destroyed. If, therefore, the same grinder is used for beef and pork, small pieces of pork containing cysts may contaminate the ground beef. It seems reasonable that a grinder head be used for pork, then washed and used for beef without any hazard. However, there is always the chance that the grinder head will not be thoroughly washed.

Retail display cases for cut meats, chicken, cooked and fresh sausage products, bacon, and cold cuts should be held at temperatures of 32 to 38°F (0 to 3.3°C) and should be of the closable variety. Also, personnel in charge of display cases should make certain that products move more or less in the order of "first in, first out" and that no item remains in the display case for long periods of time.

Large or small canned hams of the type requiring refrigeration should be held, at all times, at temperatures of 38°F (3.3°C) or below. They should never be displayed in general store areas, aisles, or windows where there is no refrigeration.

Fresh produce is oftentimes poorly handled in retail outlets. The cells of such foods continue to respire after they are harvested, and the higher the temperature at which they are held the faster the rate of respiration. Respiration brings about chemical changes in fresh produce that cause a deterioration of quality. Loss of sweetness, loss of succulence, toughening, and the development of off-flavors are some of the changes that may take place in fresh produce because of respiration. For instance, sweet corn on the cob, freshly picked and then held at 35°F (1.7°C), is perfectly good in taste and texture after 15 days of storage. At high temperatures, its quality will be lost in a few hours. Storage areas for fresh produce should be clean and held at 32 to 37°F (0 to 2.8°C). Potatoes, turnips, and cabbages should be held at about 50°F (10°C), as cabbage and turnips stand up well at this temperature, and potatoes convert starch to sugar at 40°F (4.4°C) or below—becoming sweet. Most fresh produce should be displayed in an area or cabinet held at 32 to 37°F (2.8°C) or partially surrounded with ice, since such temperatures maintain quality. A possible exception is lettuce which, if held in ice, may freeze, causing the leaves to wilt.

Fish and shellfish, such as shucked clams, oyster, scallops, and shrimp, should be held in a display case surrounded by ice. This provides a temperature of about 33°F (0.6°C) and is the best way to maintain this low temperature, without freezing, for products that are extremely perishable.

Milk, cream, sour cream, cheeses, and butter should be held in an open-top display case, the temperature of which is held at 32 to 37°F (0 to 2.8°C).

Frozen foods, which are not indefinitely stable, are rarely handled under satisfactory conditions in retail stores. At –30°F (34.4°C), frozen foods deteriorate at an extremely slow rate. At 0°F (–17.8°C), many foods will have a storage life (no noticeable loss of quality) of at least 6 months, and some foods have a storage life of at least 1 year at this temperature. As the storage temperature is raised above 0°F (–17.8°C), for each 5 degrees F (2.8 degrees C), the rate of deterioration is approximately doubled. Thus, a product that has a storage life of 6 months at 0°F (17.8°C) will have a storage life of only 3 months at 5°F (–15°C). When frozen foods are delivered to the retail store, they should not be allowed to stand on platforms or in a room at high temperature but should be immediately placed in the frozen storage room. The frozen storage room should always be held at a temperature of 0°F (–17.8°C) or below, preferably at –20°F (–28.9°C). Display cases used for holding frozen foods in the retail area should be of the open-top type, or of the enclosed shelf type. Shelf-type display cases for frozen foods are not suitable unless they are enclosed by doors. The reason for this is that cold air is heavier than warm air; hence, in the open-top case the cold air tends to remain in the area where the foods are held. Shelf-type frozen food cases must be refrigerated by blowing cold air out and over the product. Because this air is heavier than warm air, it tends to flow outward and downward into the room. In display cases of this kind it is therefore difficult to maintain temperatures of 0°F (–17.8°C) or below around the entire product. Those items in front where warm air has access are usually surrounded by air temperatures much higher than 0°F (–17.8°C), and hence are subjected to an accelerated rate of deterioration.

Canned foods require some consideration in retail handling. To begin with, the buyer of canned foods for a retail outlet should have the knowledge, or employ personnel with the knowledge, to determine whether or not these products have been sufficiently heat processed for safe consumption. That means that the foods must be heated to the point where all spores of the bacterium known as *Clostridium botulinum* have been destroyed. Actually, to prevent spoilage (not disease) by other bacteria, canned foods should be heated beyond the point at which all disease-causing bacteria will have been destroyed.

An adequate backlog of canned foods must be available to the retail store. This means that there must be a warehouse where canned foods are stored. Such warehouses should be held at temperatures not above 75°F (23.9°C). For this reason, the temperature of the warehouse where canned foods are stored should be regulated as previously stated. Nor should the storage warehouse for canned foods be held at temperatures below 50°F (10°C), for if this is done, when higher temperatures are reached in this area or when the cans are placed in the retail outlet at higher temperatures, moisture may condense on the surface (the cans may sweat) and cause rusting of the outside surface, which discolors the label and is otherwise unsightly, and it may eventually weaken the can to permit microbial invasion of the contents.

Canned foods, both those stored in the warehouse and those held in retail stores, should be handled on a first-in, first-out basis. The reason for this is that although most canned foods have a relatively long storage life, they are not indefinitely stable. The usual type of deterioration after long periods of storage is due to internal corrosion of the container, which results in a swelled can or in leakage. Deterioration of this type is most often encountered in acid foods, such as tomato products, in which case internal corrosion produces hydrogen gas, causing the can to swell. The food in this case may be perfectly edible, but consumers would be running a risk to eat the food from a swelled can because they cannot be sure that the cause of the swelling was not due to gas produced by some disease-causing bacterium such as *Clostridium botulinum*. Swollen cans should always be returned to the supplier to be discarded properly.

THE HAZARD ANALYSIS AND CRITICAL CONTROL POINT SYSTEM (HACCP)

A foodborne disease is one that is carried or transmitted to humans by food. In Chapter 3, a number of foodborne diseases from microbiological sources are discussed. If an outbreak of food poisoning occurs, there is much suffering, especially obvious in the person or people who are actually stricken with food poisoning and experience physical and emotional pain as well as possible loss of income as a result of doctor's bills and absence from work. The restaurant, other foodservice facility, or food processing company that is responsible for the outbreak will also suffer, with the consequences of legal fees, a scarred reputation, and often forced closure and possible bankruptcy. An outbreak is defined by the International Association of Milk, Food and Environmental

Sanitarians (IAMFES) and described as an incident where two or more people have the same disease, similar symptoms, or excrete the same poisonous microorganisms (pathogens). A time, place, or person association between these people must be established. Other outbreaks include a single incident of botulism or an illness caused by a chemically contaminated food.

In the processing of a food product, one must consider a number of potential hazards that could result in contamination. These can be generally classified as biological, chemical, and physical. Biological hazards pose the biggest threat to food safety. They include pathogenic microorganisms such as bacteria, viruses, parasites, fungi (molds, yeast, and some mushrooms), and certain toxin-carrying fish. Chemical hazards can also be a problem. Examples include pesticides, chemicals in cleaning supplies, and toxic metals from equipment that get into foods accidentally. Food additives and preservatives are purposely added to the food, but some people may have allergic reactions to them, making them chemical hazards in certain situations. The third major hazard, the physical hazard, includes many materials that just do not belong in the food but, through negligence or chance, get in. Examples of these are broken glass or crockery, metal fragments, bits of packaging including nails and staples, dirt, or hair.

Processing and preparing can be a risky business and precautions must be implemented to prevent problems and to correct them if they do exist.

HACCP, a system for ensuring the safety of foods, was developed in 1971 in a cooperative effort by the U.S. Army Natick Laboratories, the National Aeronautics and Space Administration, and the Pillsbury Co. The system employs seven elements:

1. Assess potential hazards in all stages of food production, from growing to the finished products.
2. Determine critical control points (CCP) where controls are necessary to eliminate or reduce hazards.
3. Establish requirements to be met at each CCP.
4. Establish procedures to monitor each CCP.
5. Establish corrective actions when monitoring uncovers deviations from plan.
6. Establish record-keeping procedures.
7. Establish procedures to monitor effectiveness of HACCP.

When organizing and setting up a HACCP program, each step is important and necessary for the assurance of a safe, high-quality finished product.

I. Assessing Hazards

1. Potentially Hazardous Foods

Review the recipe or ingredient list for potentially hazardous foods (PHF's) which may be either separate items or ingredients. (see Table 4.4)

2. Flow Chart

Set up a flow chart showing the path the food travels in your procedure (see Fig. 4.5).

3. Estimate Risks.

A number of risks can develop based on the number of PHFs you have in your product. The following factors should be considered in this risk assessment:

Table 4.4. Potentially Hazardous Foods (PHFs)

Any food that contains in whole or in part of:

Milk or milk products	Edible crustaces (e.g., shrimp, lobster, crab)	Raw seed sprouts
Shell eggs	Baked or boiled potatoes	Sliced melons
Meats	Tofu or other soy protein products	Synthetic ingredients (e.g., textured soy proteins)
Poultry	Plant foods that have been heat treated (e.g., beans)	
Fish		
Shellfish		

NOTE: the term PHF does not include foods with a pH of 4.6 or below, or a water activity (a_w) of 0.85 or less.

a. Customers with special needs—will they be old (nursing home), hospitalized and possibly immunodeficient, or have some other specific condition?
b. Suppliers—make sure all suppliers are reputable, listed with the FDA or an individual state as an approved shipper, as in the case of shellfish suppliers.
c. Size of operation—be sure your facility can handle the project.
d. Employees—will your employees need further training, equipment, or materials for the project?

II. Identifying Critical Control Points (CCPs)

1. Adapt Flow Charts.

Critical control points can now be added to the flow charts (see Fig. 4.6).

III. Establish Requirements to Be Met at Each CCP.

1. Establish the criteria or standards that must be met at each CCP. These will be based on proven methods; research data; or federal, state, or local food regulations.
2. Be specific and use exact criteria such as temperatures, times, and recognized procedures whenever possible.

IV. Monitoring Critical Control Points

1. Determine whether criteria and standards are being met at all CCPs and throughout the process.
2. Involve employees responsible for the procedures in monitoring.
3. Remember, having CCPs without monitoring is a waste of time.

V. Taking Corrective Actions

1. *Immediate* corrective action must be incorporated if the standards at any CCP are not being met.

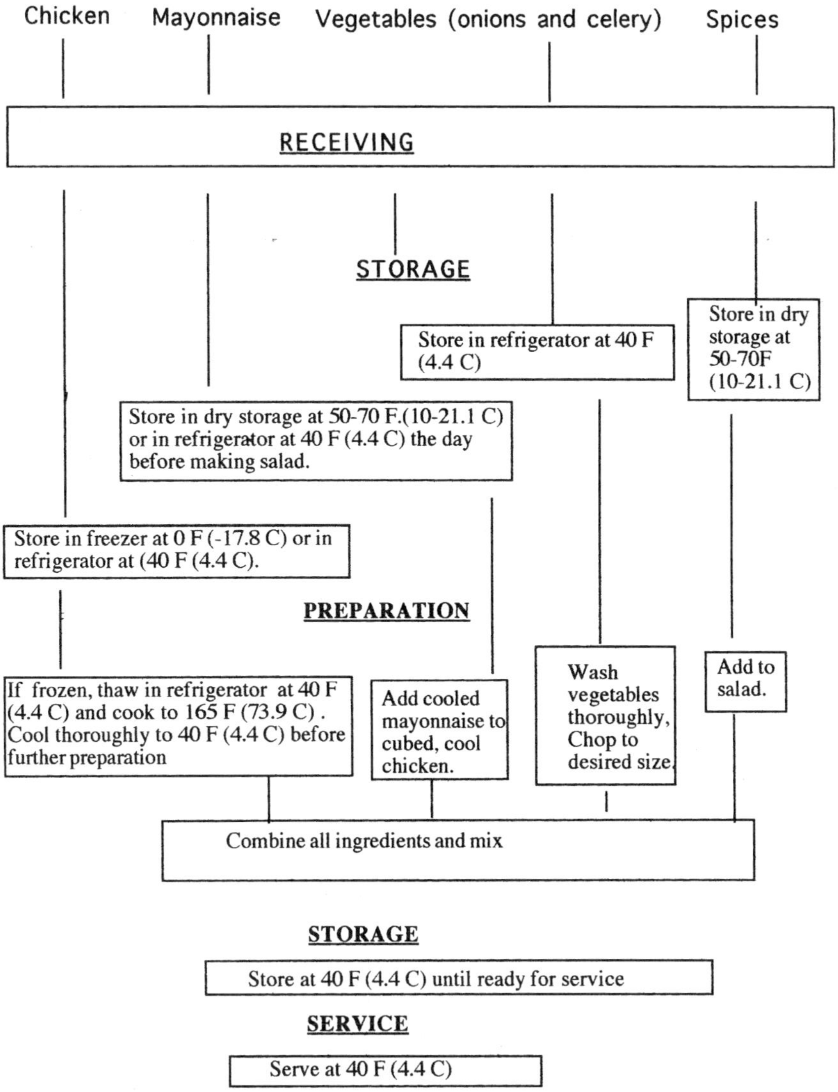

Figure 4.5. Flow chart for chicken salad.

2. Corrective actions must be clear and based on the individual situation. For example, a corrective action for a food not being held at a proper warming temperature of 140°F (60°C) may be different depending on how long the food has been at the improper temperature. If it has been held improperly (less than 140°F) for less than 2 hours, reheat to 165°F (73.9°C) before serving. If it has been held at below 140°F (60°C) for more than 2 hours, discard.

VI. Setting up a Record Keeping System

1. These should be simple, easy to do quickly, and should contain information needed to ensure safety.

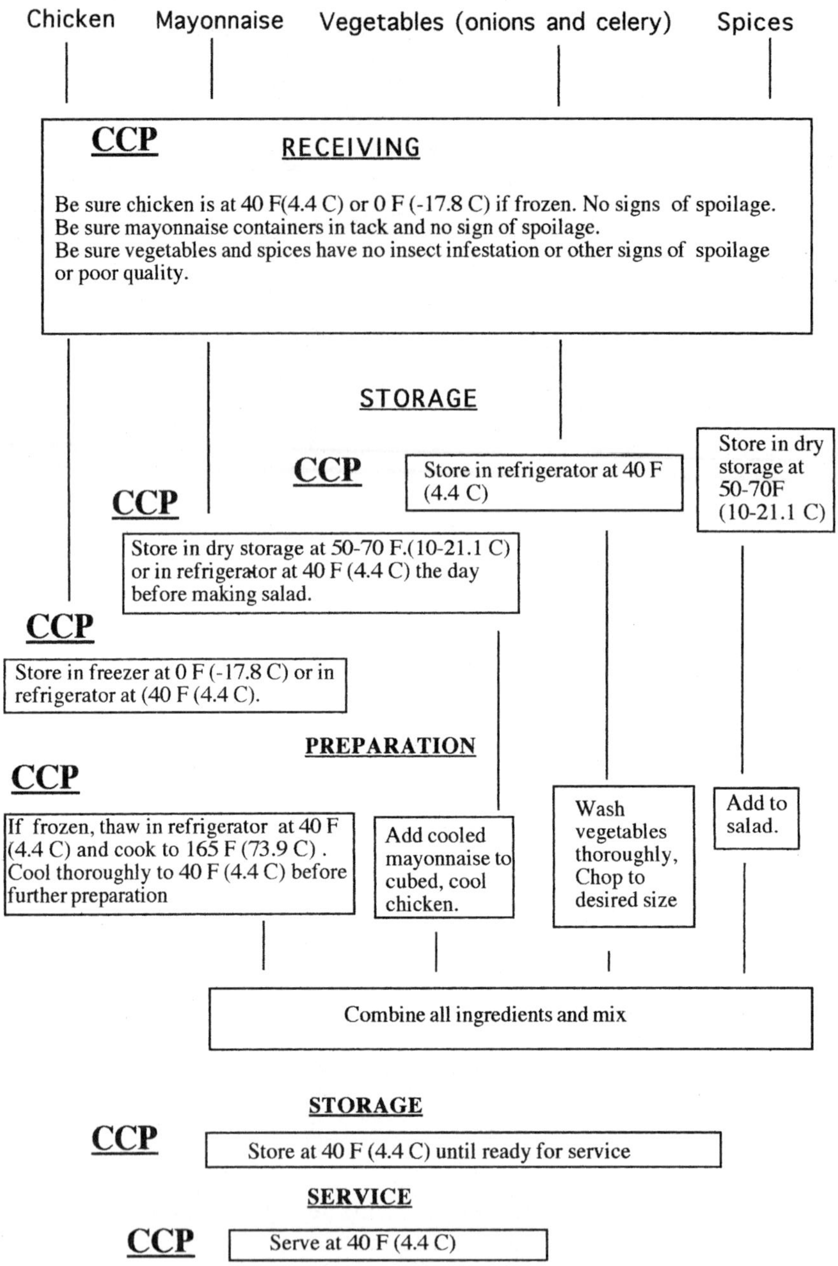

Figure 4.6. Flow chart for chicken salad with CCPs.

2. Check lists to examine all CCPs are useful here. Employees doing the job may be very helpful in setting up the form to be used.

VII. Verify that the System Is Working.

1. This may be done by laboratory analysis (for microbiological or chemical hazards, depending upon the product).

Critical Control	Hazard	Standards	Corrective Action if Below Standard
		RECEIVING	
Receiving raw chicken	Contamination and spoilage	Accept at 40°F (4.4°C)—check with thermometer	Reject delivery
		Package intact	Reject delivery
		No off color, stickiness, etc.	Reject delivery
Receiving vegetables	Contamination and spoilage	Package intact	Reject delivery
		No signs of insects or rodents	Reject delivery
		No cross-contamination from other foods on delivery vehicle	Reject delivery
Receiving spices	Contamination and spoilage	Package intact	Reject delivery
		No signs of insects or rodents	Reject delivery
		No cross-contamination from other foods on delivery vehicle	Reject delivery
		No moisture inside package package	Reject delivery
Receiving mayonnaise	Contamination and spoilage	Package intact	Reject delivery
		No separation of product	Reject delivery
		STORAGE	
Storing raw chicken	Cross contamination of other foods	Store on lower shelf	Move to lower shelf away from other foods
	Bacterial growth and spoilage	Chicken temperatures must stay below 40°F (4.4°C)	Discard if temperatures have been too high for too long (judgments must be made here by people responsible for quality. If any signs of poor quality exist, e.g., off odors or stickiness, *discard*.) If quality meets standards, use as soon as possible.
		Label and date package. Use first in, first out (FIFO) method of rotation	Discard product past rotation date
Storing vegetables	Cross-contamination from PHFs	Store above raw PHFs	Move to upper shelf if not damaged. If damaged or spoiled, discard.
		Label and date package. Use first in, first out (FIFO) method of rotation	Discard product past rotation date
Storing spices	Contamination and spoilage	No signs of insect or rodent infestation	Discard
		No moisture spoilage	Discard
			Keep temperature between 50 and 70°F (10–15.6°C) and relative humidity at 50–60%
Storing mayonnaise	Contamination and spoilage	No separation	Return to supplier
		Package intact	Discard

Figure 4.7. A HACCP flow chart for chicken salad.

Critical Control	Hazard	Standards	Corrective Action if Below Standard
		SERVICE	
Serving salad	Contamination and spoilage	Servers must have clean hands and preferably gloves	Clean hands and wear gloves
		Temperature of salad must remain at 40°F (4.4°C) or below.	Lower temperature
		All plates, bowls, and other utensils and silverware must be clean and sanitary	Clean and sanitize properly. Discard any food contacting unsanitary surfaces
		STORAGE	
Storing prepared salad	Contamination and spoilage	Container must be covered properly and stored on upper shelves away from raw foods.	Cover and move to proper shelves
		Temperature must remain at 40°F (4.4°C) or lower	Cool rapidly to proper temperature if food has not been in danger zone for too long (more than 2 hours)—good judgment needed here.
IF RESERVICE OF LEFTOVERS IS DONE, ALL STANDARDS DESCRIBED IN THE SERVICE SECTION MUST BE OBSERVED.			
		PREPARATION	
Cooking chicken	Bacterial survival	Cook to 165 °F (73.9°C) check with thermometer	Continue cooking until proper temperatures met
Cooling chicken	Spoilage	Cool to 40°F (4.4°C) with 2–4 hours	Cut chicken to smaller pieces to quicken cooling
Cubing chicken	Contamination	Use clean, sanitary knives, cutting board, and containers. Avaoid metal containers other than stainless steel for preparation and storage	Clean and sanitize all food contact surfaces
Cubing vegetables	Contamination	Use clean, sanitary knives, cutting board, and containers. Avoid metal containers other than stainless steel for preparation and storage	Clean and sanitize all food contact surfaces
Combining ingredients (chicken, vegetables, spices, and mayonnaise)	Contamination	Use clean, sanitary utensils, avoid handling food, wear approved plastic gloves. Avoid metal containers other than stainless steel for preparation and storage	Clean and sanitize all food contact surfaces, wash hands
	Reaching spoilage temperatures in danger zone	Minimize time in danger zone.	Make smaller batches if necessary. Use cool mayonnaise [40°F (4.4°C)] Use cool stainless steel bowls and utensils.

Figure 4.7. *Continued*

2. Internal quality control to check temperatures, texture, color, taste, absence of defects, or other quality and safety aspects.

Figure 4.7 has a flow chart with CCPs, hazards, standards to be met, and corrective actions to be taken. This type of chart is very useful for both monitoring and verifying that the system is working.

HACCP is widely accepted and recommended for all food production, handling, and distribution.

5

Quality and Sensory Evaluation of Food

When evaluating the quality of foods, many attributes must be considered. Wholesomeness of the food should be included as a quality factor and is most important to food safety (see Chapter 4). The nutritional value (see Chapter 2) is considered by some to be a quality factor in many foods, especially those purchased as low-calorie or low-fat. Quality has been defined as "degree of excellence" with appearance, texture, and flavor becoming major factors. If a food doesn't look, feel, and taste good, most consumers in countries where food is abundant will not purchase it a second time, even if the price is right, and the food is wholesome and nutritious. Testing these sensory attributes in food is quite subjective but methods have been developed to standardize this measurement.

APPEARANCE

Color, and the shade or intensity of such, the size, and the shape of a portion of food all contribute to its visual perception. Many different foods have different characteristics that could add to or subtract from its quality as perceived by the consumer. Examples include the browness of a pie crust, the shape of a pickle, the wholeness of a product such as canned pineapple, for which the price decreases as the size changes from rings to chunks—same product, different shape.

Color is perhaps the most recognizable of a food's visual attributes. It is the first level of quality control performed by the consumer. Color indicates freshness or spoilage of products such as bread, fruits, and vegetables. It may also indicate the strength of a product as in the case of coffee and tea or completeness of cooking, as in the case of french fries or meats.

To the trained food technologist, color can indicate many quality factors. The color of chocolate can give clues as to how it has been stored, and the color of a batter changes with density and can indicate an error in mixing. If too much oxygen is present in a package of dried foods such as tomatoes, a bleached color results.

There are many laboratory apparatuses such as spectrophotometers and colorimeters that can measure color and its gloss, intensity, and opaqueness. These are valuable and necessary tools for ensuring consistent quality but foods still must pass the human visual perception test to be accepted by the consumer.

TEXTURE

Textural factors are those qualitites felt by the tongue, palate, teeth, and even the fingers. Texture can be an indicator of freshness in many foods. With aging, many foods undergo water changes. In fruits and vegetables, for example, cell walls break down with aging and water is released. The result is a softer, soggy product that indicates low quality and spoilage. Other products, such as crackers and pretzels, must be protected from water gain rather than loss. If not packaged properly, they will absorb water and become soft.

In food processing, the food scientist adds ingredients specifically for textural attributes. The baker adds lipids to tenderize cakes. Starches, pectins, and gums are added to thicken, gel, and increase viscosity. Sugar is added for texture as it can add body to beverages, improving mouth feel or, in higher concentrations, can add chewiness or even brittleness if concentrations are high enough for crystallization to occur.

Viscosity and consistency are interesting characteristics that can have both visual and textural quality attributes. Syrups are visually more appealing if thicker, yet this increased viscosity also adds appeal to the feel of the syrup on the food. In some foods, such as ketchup, thickness is a major advertising and marketing tool.

In the food laboratory, texture can be measured in a number of ways. Generally speaking, however, texture is the measure of resistance to force. Examples of the different forces are numerous. "Compression" is tested on foods that, when squeezed, remain in one piece, for example, bread. "Cutting" is a force that goes through the food, thus dividing it, for example, an apple. A force applied away from a material resulting in tearing or pulling apart is termed "tensil strength," for example, donuts or breads. A force where one part of the food passes another is "shearing," as in the chewing of some candies and gum. When grading some foods, such as beef steak, all of these factors can come into play.

The food technologist has many tools to measure texture and control quality during processing. These include the "consistometer" for viscosity, the "squeeze-tester" for softness of bread, the "penetrometer" or "tenderometer" for tenderness in raw beef or baked goods or spreadability in margarine, and the "succulometer" which measures succulence by squeezing juice out of food.

FLAVOR

Flavor is a combination of both taste and smell but is also influenced by a textural composite of sensations referred to as "mouth feel." Some foods, such as sauces or gravy, that are thicker will often be judged to have the fuller or richer flavor. This perception may be either purely psychological or real. The change in texture may actually affect the volatility or solubility of a flavor compound, increasing the flavor intensity.

Color can also influence the perceived flavor in foods. Taste tests have been done in which the colors of fruit sherbets had been altered but the flavors were not changed. Only 43% of the flavors in the altered samples were identified correctly whereas just under 70% of the unaltered samples received correct responses. One tester even commented on how easy one sample was to identify because there were pieces of cherry in it. He was actually tasting a sample of pineapple sherbet (with pieces of pineapple) that had been colored red!

Although color and texture have an effect, flavor is mainly controlled by the mixture of tastes (salt, sour, bitter, and sweet) and the numerous combinations of compounds that are responsible for aroma. The flavor of food is very complex and has not been completely described for many foods. Flavor measurment is largely subjective and thereby very difficult to measure. Differences of opinion on flavors and their contribution to quality result because of this. People have wide ranges of sensitivity to detect different tastes and odors and even when these are detected, individuals differ in their preference. The cultural and biological differences in people make flavor measurement quite difficult.

FORMAT FOR A FLAVOR PROFILE

AROMA

Amplitude Rating

Character Note (in order of appearance)	Intensity
Character Note (in order of appearance)	Intensity
Character Note (in order of appearance)	Intensity
Character Note (in order of appearance)	Intensity
Etc.	

Others:

FLAVOR

Amplitude Rating

Character Note (in order of appearance)	Intensity
Character Note (in order of appearance)	Intensity
Character Note (in order of appearance)	Intensity
Character Note (in order of appearance)	Intensity
Etc.	

Others:

AFTERTASTE (TIME OF MEASUREMENT IF NOT ONE MINUTE)

Character notes (intensities optional)

FOOTNOTES

Color, Appearance, Texture (optional)

____
Signature of Panel Leader

____
Date

Figure 5.1. Example of a flavor profile score sheet.

Some food companies utilize a method of flavor analysis developed in 1948 by Messrs. Sjöström and Cairncross of Arthur D. Little, Inc. in Cambridge, Massachusetts. "The Flavor Profile" method is based on the concept that flavor consists of identifiable taste, odor, and chemical feeling factors (sensations in the mouth and throat, such as pepper bite) plus an an underlying complex of sensory impressions not separately identifiable.

Trained panelists are needed for this method and they are selected on the basis of their abilities to discriminate odor and flavor differences and communicate their perceptions. The panelists must take a series of tests to determine their abilities to notice flavor and odor differences and to identify basic tastes, rank, intensities, and identify common odorants.

After the panelists are selected, they must go through a training period of approximately 60 hours of training and 100 hours of practice sessions. The time of training varies depending on whether the panel will be testing any food or beverage or testing a specific food. Training for a single type of product requires less time. The training generally covers the nature of taste and smell, basic requirements for panel work, and techniques and procedures for reproducible odor and taste work. The development of sensory terminology through exercises and reference standards is also stressed. As the panelists advance, more difficult products are used and flavor situations of a more complex nature are covered. Interpretation and utilization of panel data are covered also.

When the panel is in operation, a quiet, well-lighted, odor-free environment removed from external distractions must be provided. All samples must be presented to the panel under identical conditions and should be analyzed and scored in an identical manner at the same time. A product's aroma is analyzed first, followed by the flavor, and finally the aftertaste.

Each panel has a leader and four or more members. The leader conducts a product orientation before the panel does the testing. In the orientation, the objectives of the product are outlined and samples to be tested are introduced with similar products to establish a framework for comparison. At this time the panel will establish a list of character notes for the sample such as vanillin, peppermint, chalky, or buttery. Reference materials will also be chosen (pure compounds that display particular flavor and aroma notes). The panel then decides the best way to present and examine the samples.

During the test, each panel member evaluates the sample independently using standardized Flavor Profile techniques and records the findings on a blank sheet (Fig. 5.1). The panel leader, who is also a member of the panel, now has each panelist present his or her findings, records them, and leads a panel discussion to reach a consensus on each component of the description. The final profile may take three to five sessions and after this, the panel leader interprets and reports the results.

The components of the Flavor Profile method are:

1. Overall impression of aroma and flavor, called amplitude.
   Amplitude measures balance (blend) and body (fullness) of flavor and aroma in the product. It is scored on a point system from 0 to 3 (in 1/2-point increments) with 0 representing no blend or fullness and 3 representing a high degree of blending and fullness.
2. Identification of perceptible aroma and flavor notes.
   These are called character notes and are objective rather than subjective (i.e., a note may be described as "vanilla" but not "good"). These include aromatics, basic tastes, and chemical sensations or feeling factors (cool, burn).

3. Intensity of each character note.
   The degree to which each character note is perceived is "intensity" and the scale is constant over all product categories. The point system such as the one used with amplitude is used, with 0 representing "not present" and 3 being "strong."
4. Order in which these character notes are perceived.
   The "order of appearance" is determined through standardized techniques for tasting and smelling because order of appearance will be influenced by the location of tastebuds on the tongue, volatility of the components, and texture of the product.
5. Aftertaste
   Aftertaste can include tastes, aromatics, and/or sensations. It is an important part of the flavor of a product. The sensations are noted at a specific time after tasting has been completed, usually 1 minute. If aftertaste is a major factor, intensity ratings may be given.

When color or texture is important to the product's description, it is also noted in the panel session.

When all data are collected and analyzed, a final test report, including complete identification of the sample(s) and the objectives and duration of the study is given. The methods used to prepare the product, the reference standards used, a summary of the amplitude ratings, and major character notes with their intensities are also given. The presence of any off-notes should be mentioned and the order of appearance and aftertaste noted. Any observations of visual and textural qualities are important and should also be described.

Another method is The Profile Attribute Analysis (PAA), which is an objective method of sensory analysis that uses an extensively trained panel to numerically describe the complete sensory experience through profile attributes. These are a limited set of

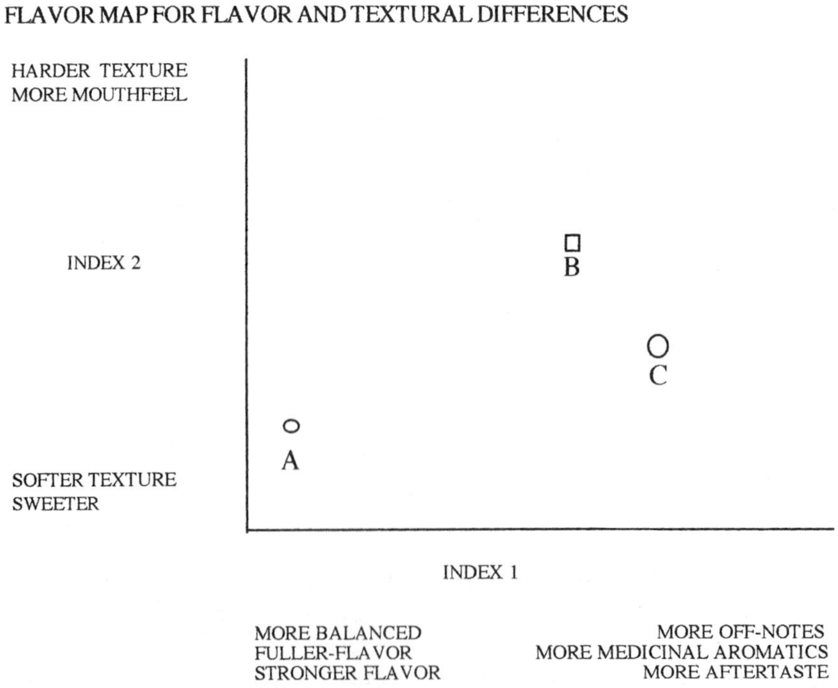

Figure 5.2. Example of a flavor map.

characteristics that, when properly selected and defined, provide a complete description of the sensory characteristics of a sample with little descriptive information lost. PAA is based on the Flavor Profile Method in concept and implementation but by limiting the number of profile attributes the panelist measures, it is possible to evaluate four or more samples per session as compared with one per Flavor Profile session.

PAA relies on numerical measures to describe differences among samples, and "Flavor Maps" can be used to display summary statistics (Fig. 5.2). In the sample flavor map, it can be seen that product A has a more balanced, fuller-flavored, stronger flavor and has less off-notes and aftertaste.

CONSUMER SENSORY TESTING

Because of the differences in people's sensitivity, perceptions, and preferences, it is virtually impossible to create a food that everyone would choose as his or her favorite. The food scientist does have some methods of testing foods to get a consensus of what

Two of the samples are identical and the other different. Please check the duplicate samples and score your preference for either the identical or odd samples.

SCORING : Designate whether you prefer the pair or the odd sample by (P).

Sample #	Identical Samples (indicate by check mark)	Score
________	______________________________	_____
________	______________________________	_____
________	______________________________	_____

Plate # _______ Date: __________

Note: This test sheet was used in a triangle test done on two fish species prepared with identical recipes. To avoid problems with tendencies of testers to choose based on order of testing, samples were given in different sequences. This is shown below:

Plate #	Recipe #I	Arrangement of Samples (left to right on plate)
1		A A B
2		A B A
3		A B B
4		B A B
5		B B A
6		B A A

Figure 5.3. Triangle test score sheet.

a majority of the prospective consumers will accept and therefore desire to purchase. Examples of sensory testing procedures are "difference" or "discrimination" testing and "acceptance" or "consumer" testing.

The "difference" test is used for various reasons. One of the most common types of this test is the "triangle" test (Fig. 5.3). Testers are given three samples, two of which are identical, and are told to identify the odd sample. If the consumer cannot detect the difference, it should indicate that there is no detectable difference in the flavor of the samples. This test may be used to convince consumers that a particular product tastes as good as a similar brand-name product that may be more expensive. It has also been used to try to convince people to purchase lesser utilized fish species rather than those whose numbers are being depleted as a result of over-fishing. If consumers are convinced that there is no difference in flavor and taste, they may demand the lesser utilized species, the fishing boats will now seek out this species because there is a market for it, and the numbers of the over-fished species will be replenished. All of this, of course, is utopian and seems easy. In reality, it is very difficult to convince consumers and even more difficult to get them to bring about a change in established eating habits even when the test proves a difference in the samples could not be detected.

Another use for the "difference" test is a situation in which a change in formulation is desired to reduce production costs. The food scientist can use the "triangle" test to determine if the new formulation results in a detectable change in flavor and quality. Small test panels trained in sensory evaluation are usually used in this instance.

The "acceptance" test usually requires a large number of individuals who represent the target group of potential consumers. For example, if you are testing a flavored cereal advertised by a cartoon character, you would use a panel of children. A gourmet frozen entree would require a more mature panel.

The panel here requires little training and is usually given a score sheet (Fig. 5.4) which rates different sensory attributes from "excellent" to "inedible." This method, the "hedonic rating," uses a numerical scale that gives a score of 9 for "excellent" and

	Appearance	Odor	Texture	Flavor
Excellent				
Very Good				
Good				
Fair				
Borderline				
Slightly Poor				
Poor				
Very Poor				
Inedible				

COMMENTS:

Figure 5.4. Hedonic rating of food score sheet.

a score of 1 for "inedible." A mathematical average can now be determined for the food's acceptability.

NEW TECHNOLOGIES

In the mid-1980s researchers at Britain's Hartwick University used work done by the U.S. military in developing polymers that could conduct electricity to help planes evade enemy radar, and applied it to the development of the first "electric nose." In the nose, the polymers act as spongy sensors that absorb sense vapors and match them with computer models. These are useful in matching particular odors or fragrances.

In medicine, for example, ailments such as diabetes and some liver and kidney diseases cause the production of certain odors that appear in the breath. Because the machines are more precise than the human nose in finding the specific odor components, scientists see its use as a diagnostic tool increasing in the future.

In the food industry, these noses could be used to detect specific aromas indicating spoilage, off quality, or use of a particular additive.

This "technology of the future" is being used now in a number of applications and wherever the same component is being tested for in many samples, it is very useful. The use of this technology along with flavor panels gives the food scientist an additional tool to ensure a consistent, safe, high-quality food supply.

6

Regulatory Agencies

A major evolution in modern societies is the widespread use of food produced and often preserved in areas remote from the consumer. Because consumers do not know how the food was handled, they do not know if it is safe to eat. (In ancient times, slaves and animals were sometimes compelled by their masters to eat food of questioned safety before it was eaten by their masters.) Thus, as the entire population presently requires protection, it is a proper government function to determine the wholesomeness and purity of foods and to protect the consumer against economic fraud as well as health hazard. Yet government seldom assumes this responsibility spontaneously, and protective regulatory legislation is passed and enforced only after consumer-oriented persons or groups stimulate broad public support for government action.

In the United States, the basic regulation covering the safety and legitimacy of commercial foods was not enacted until several states had already passed laws to protect consumers against adulterated and misbranded food and drugs within their own borders. It was not until 1906 that the Federal Food and Drug Act was passed, owing to the efforts of dedicated people, such as Dr. Harvey W. Wiley. Dr. Wiley is given credit for the enactment of the first federal regulation covering safe and pure foods and drugs. Subsequent additions to federal regulations have followed.

There are several federal agencies for regulating foods and food products sold in the United States, but only three have enforcement authority. These are the Food and Drug Aministration (Department of Health and Human Services) and the Meat Inspection Division and Poultry Inspection Service, both of which come under Food Safety and Inspection Service (FSIS) of the U.S. Department of Agriculture.

These agencies use *standards, regulations, ordinances*, and *codes* to conduct their enforcement.

A *standard* is a measure used to compare quality
A *regulation* is a governmental control with the force of law.
An *ordinance, law*, or *statute* gives the legal authority for governmental agencies to publicize and put regulations into effect.
A *code* often equals a regulation or statue and is a collection of regulations, statutes, and procedures.

The role of regulation is to protect the public from foodborne illness. Food processors, retailers, wholesalers, and foodservice establishments are required to meet standards for critical areas of their operations that require effective control measures.

THE FOOD AND DRUG ADMINISTRATION

The Food and Drug Administration (FDA) is probably the most important enforcement agency because it regulates all our foods except meat and poultry and, in some instances, can even regulate these products. There are two general regulatory categories: adulteration and misbranding. A food is adulterated if (1) it is filthy, putrid, or decomposed; (2) it is produced in unsanitary conditions; or (3) it contains any substance deleterious to health. A food is misbranded if (1) it is a food for which standards of identity have been written and it fails to comply with these standards, (2) it is wrongly labeled, or (3) it fails to meet the regulations for fill of container.

Adulteration

Adulteration is not difficult to determine because there are tests that can be made to detect sources of contamination such as that from rodents (hairs, pellets, or urine), insects, dirt, and other detritus. Also, if a food is putrid, this can be detected by the ordinary human senses, a fact that is well known and accepted. However, the detection of decomposition is not easy and often scientists do not agree on what constitutes decomposition of a particular food. Citations based on the decomposition of a food, therefore, frequently have to be settled in court.

Regarding additives that may be present in foods, the FDA and the industry know that certain chemicals are toxic and cannot be added to foods at all. The FDA has a GRAS (generally recognized as safe) list that specifies which chemicals may be added to foods and, in many instances, how much may be added to a particular food. Many compounds on this list come under what is called the Grandfather Clause, these chemicals having been used in foods for years with no apparent ill effect. For some chemicals that can be added to foods and for any new chemical that will be added, tests have been or will be made by feeding several kinds of animals (e.g., rats, dogs, mice, guinea pigs) a diet containing the chemical over a period of several generations. The results of such tests are determined by observations on the weight and general health of animals, as well as their ability to breed, and on autopsies and chemical tests for specific enzyme activities, and so on. Time, trained personnel, and special facilities are required for testing a new food additive. This is a very expensive process, requiring, as a rule, the outlay of several hundred thousand dollars, and no producers of such a new compound are apt to initiate such testing, which must satisfy the FDA, unless they are certain that the new additive will provide specific advantages and have great utility.

It is difficult to enforce the section of the law prohibiting substances that may be deleterious to health of the consumer. For example, pathogenic bacteria, such as the *Salmonella* organisms often present in food, can cause disease and even death. The FDA and food scientists know that poultry and other foods generally contain these organisms but are unable to control the situation. If enforcement were attempted, for instance, in the case of poultry, the whole industry would be shut down, as approximately 25% of the product contains salmonellae. The testing for this group of organisms requires at least 2 days to obtain results, another factor that serves to impede enforcement. This regulation, therefore, should probably be changed or reworded, because laws that cannot be enforced are useless and lead to unsatisfactory practices. Foods found to be adulterated are seized by the FDA and destroyed.

Misbranding

If a standard of identity has been set up for particular food, the food can contain only the ingredients specified in the standard. If the food is found to contain other ingredients or additives, it will be seized and destroyed. As an example, sulfur dioxide is allowed in some food and might be used to provide good color in ketchup. However, there is a standard of identity for ketchup in which sulfur dioxide is not included. Therefore, if ketchup were found to contain sulfur dioxide, it would be seized and destroyed. Food for which a standard of identity has been established and that does not conform to this standard is destroyed on the premise that it has no identity. Standards of identity have been set up for some bakery products, cacao products, cereal flours and related products, alimentary pastes, milk and cream, cheeses, processed cheese, some cheese foods and spreads, some canned fruits, fruit preserves and jellies, some canned shellfish, eggs and egg products, oleomargarine, some canned vegetables, canned tomatoes, tomato products, and other foods.

A food that is misbranded may be wrongly labeled as to weight, portions, or ingredients if no standard of identity has been established for it. If wholesome, wrongly labeled products need not be destroyed. Instead, they can be relabeled to comply with ingredients, weight, and so on, and sold. If the product fails to meet the "fill of container" requirement, it may be relabeled to specify this fact and sold under the new label.

It should be noted that, theoretically, the FDA can regulate only those foods that are shipped interstate (from one state to another), but there are ways to get around this. For instance, if a company ships one food interstate but not another, both foods may come under the jurisdiction of the FDA.

In response to need the FDA developed a new "Model Food Code" released in 1993 that assists health departments in developing regulations for foodservice inspection programs. Another updated edition, called simply the "Food Code," was released in 1995 to provide additional clarification and to further facilitate the processes for improving food laws. Some state and local agencies develop their own codes but they all must be at least as strict as the FDA code. State and local laws can be tighter, but not looser than the FDA code.

THE U.S. PUBLIC HEALTH SERVICE

The U.S. Public Health Service has regulatory authority over the sanitary quality of drinking water and foods served on interstate and international carriers (airlines, trains, etc). This agency carries out research and surveillance on foodborne diseases (infections and intoxications) and sanitary processing and shipping of foods. It is presently part of the FDA. In cooperation with state regulatory agencies, it sets up standards for coastal waters from which bivalve shellfish, such as oysters and clams, may be harvested for consumption. It also sets up specifications for waters from which bivalves may be harvested and then subjected to purification procedures or relocated in approved waters. It also helps set up standards for the bacteriological quality of the bivalve shellfish and for waters in which shellfish are grown. Shellfish dealers are registered and must keep records indicating the area from which bivalves were harvested, from whom they were purchased, and to whom they were sold. The state and the Public Health authorities keep a list of approved shellfish dealers. It is up to state authorities to make sanitary surveys of bivalve growing areas and bacteriological tests of the waters of shellfish-growing areas and of the shellfish. If a particular dealer does not comply with regulations, he is taken off the approved list and any product that he ships interstate will be seized by the FDA. It is also probable that state authorities would seize his product shipped intrastate. Also, if the state program for shellfish sanitation does not meet the requirements of the Public Health Service, all producers within the state are taken off the approved list and no bivalve product can be shipped interstate.

The Public Health Service, together with state authorities, also sets up standards for milk and cream. This includes the control of disease in dairy herds; the management and milking of herds; the bacteriological quality of raw milk, raw cream, and certified milk; pasteurization procedures including time and temperatures; and the bacteriological quality of pasteurized milk and cream.

The FDA has the power to inspect any food processing or food handling plant and to close any plant it considers to be unsanitary or to be adulterating foods in any way. FDA inspectors inspect some food plants but do not have the personnel to inspect them all even annually. Most enforcement results from chemical or bacteriological analyses of some product. If according to their analysis a product is found to be adulterated, it will be seized and destroyed, either with or without court hearings. In a case where the FDA learned that a certain boat, without freezer capacity, had been fishing for an extraordinarily long time prior to landing its catch, the FDA inspectors waited until the product was filleted, packaged under a particular label, and frozen, then seized the product. This resulted in a court case in which the FDA convinced the judge that the product was decomposed. The product was destroyed and the producers fined.

The U.S. Centers for Disease Control (CDC), located in Atlanta, GA, are field agencies of the U.S. Public Health Service. Their function is to investigate outbreaks of foodborne illnesses, study the cause and control of the disease, and publish statistical data and case studies in the *Morbidity and Mortality Weekly Report* (MMWR). CDC also provides educational services in sanitation.

THE MEAT INSPECTION BUREAU

The Federal Meat Inspection Act, made law in 1906, is administered by the United States Department of Agriculture (USDA) through its Meat Inspection Bureau, a branch of the Agriculture Research Service. This law regulates the safety of meats (beef, pork, lamb) entering into interstate commerce. The Meat Inspection Bureau is also an enforcement agency. It differs from the FDA in that most of the regulation is carried out by inspectors stationed at the plant processing the food. They deal with any food that contains more than a small percentage of meat. When cattle, hogs, or sheep are slaughtered, one or more inspectors (who are veterinarians) must be stationed at the slaughtering plant if any parts of such products are to be shipped interstate. The animals may be inspected prior to slaughter, and if diseased, are destroyed. However, the main inspection comes after slaughter. As the animals are slaughtered, the carcass, entrails, and organs are tagged and identified with a particular carcass. The veterinarian inspectors test the viscera to determine whether or not the animals were diseased. Diseased animal carcasses including organs and entrails are covered with a dye and must be disposed of as tankage and not used for human consumption. (Tankage is slaughterhouse waste that is heat processed and dried and used as fertilizer.)

The Meat Inspection Bureau also has "lay" inspectors. These inspectors have training but are not veterinarians. Lay inspectors deal primarily with those plants or areas of plants where meat is cut into portions for further processing or for shipment as fresh cuts; where sausages (fresh sausage, frankfurters, bologna and other cooked sausage, dried sausage, etc.) are produced; and where hams, shoulders, and bacon are cured, smoked, and so on. Lay inspectors determine that processing rooms and equipment are clean and sanitary before work is started, and if not satisfactory, the room is tagged and cannot be used for processing until cleaned and sanitized to the satisfaction of the inspector.

The Meat Inspection Bureau maintains lists of approved ingredients for cured and processed products, and it is part of the inspectors' job to determine that nothing is added to processed products that is not on the approved list. Inspectors also check the amount of some approved materials added to processed products. The Meat Inspection Bureau has laboratories that do some analyses. For instance, a certain amount of water can be added to some cooked sausages or to cured hams. The amount of water allowed is specified (e.g., 10% of the finished product in frankfurters). In frankfurters, the added water gets into the product as ice during the cutting of the meat, one of the reasons for adding water in this case being to keep the ingredients (the meat emulsion) cool during cutting, which improves the quality of the finished product. The inspector cannot tell exactly how much water is being added during the cutting, but if he or she becomes suspicious he or she will take samples of the finished product, ship them to a bureau laboratory, and have them analyzed for added water. In any case, samples for analysis may be taken periodically from processing plants. The same kind of inspection and analysis may be used for other processed products, including hams.

States also have inspectors who regulate the slaughter and processing of meats that are used intrastate but not shipped interstate. However, such inspection in the past has been far from adequate. In recent years federal authorities decided improvement in the inspection and regulation of local slaughtering and meat-processing plants was essential, and this improvement is presently being implemented.

The USDA periodically inspects (at least twice a year) "custom" slaughterhouses (those that slaughter but do not sell the meat). If the plants start selling meat wholesale or retail, the USDA will inspect more frequently. The inspections will be done for sanitation, adherence to USDA standards, and a USDA veterinarian will inspect the carcasses for disease.

The USDA also has a Total Quality Control (TQC) program in which a plant sets up its own regulations that are at least as stringent as those set by the USDA. The plant personnel conduct inspections and keep the records. USDA inspectors will inspect the plants regularly and will check all records closely.

THE POULTRY INSPECTION SERVICE

The Poultry Inspection Service, similar in scope and operation to the Meat Inspection Bureau, is responsible for the inspection of poultry products. An agency within the USDA, it ensures that all poultry sold in interstate commerce is processed in U.S. government-inspected plants and is wholesome. Although poultry inspection is mandatory, the grading of poultry products is voluntary on the part of the processor. Poultry is inspected prior to dressing, during evisceration, during packing, and after packing.

OTHER REGULATORY AND INSPECTION AGENCIES

There are some inspection agencies that perform activities of a regulatory nature but these are not necessarily enforcement agencies. One such is the Voluntary Inspection

Service of the U.S. Department of Commerce. It is administered by the Department's National Marine Fisheries Service, National Oceanic and Atmospheric Administration. If a fish processor so desires, he may have an inspector stationed in his plant on a permanent basis to assess the premises for sanitary conditions, determine that wholesome ingredients are used, and grade the product. As previously stated, this is an entirely voluntary service, and the processor must pay whatever costs are involved in such an inspection. With this type of inspection, the product is evaluated and graded as A, B, or C in quality, the last grade being the lowest acceptable quality. The packer can label the product accordingly.

Continuous inspection is available for certain canned and frozen fishery products. While the inspectors do not inspect every fish, they do inspect a sufficient number so that all lots of fish are sampled, and because each lot is relatively uniform in quality, unacceptable fish are bound to be identified. This is also true for canned fish. The inspection of frozen fish products does not permit the same degree of confidence, because frozen products are manufactured from blocks of frozen fish or shellfish, usually prepared and frozen outside the United States. It is true that the inspection service can and does sample the finished product to determine quality, but this can only be a spot check that is not adequate to evaluate the quality of the whole pack. There is another reason why grading canned fish is more effective than grading frozen fish. Once the canned product has been processed, the quality does not change to any extent. This is not the case with frozen foods. Frozen fishery products are not indefinitely stable even at 0°F (−17.8°C). At higher temperatures they deteriorate at a much faster rate. It is unfortunate that while manufacturers tend to freeze and store fishery products to a temperature of 0°F (−17.8°C) or below, those who transport such products and especially retailers, during the handling and display of fishery product, do not. Thus, a product labeled grade A may actually be grade C or substandard by the time that it reaches the consumer.

The USDA has a grading service for fruits, fruit juices, and vegetables, both canned and frozen. Again, for canned fruits and vegetables, the grading is quite effective, but for frozen products the same problems exist as do for frozen fishery products. To make the grading of frozen foods effective, the temperatures at which these products are shipped, the temperatures at which they are handled and stored at the retail level, and especially the temperatures at which frozen foods are held during display at the retail level must be regulated.

The Environmental Protection Agency (EPA) authorizes and regulates the use of pesticides and other environmental contaminants, and it monitors compliance and provides technical assistance to the states. It sets standards for air and water quality and regulates the use of sanitizers and the handling of wastes.

The Internal Revenue Service (IRS) enforces, with FDA collaboration, the Federal Alcohol Administration Act and other pertinent regulations that control the commerce of alcoholic beverages (e.g., whiskey, wine, beer, brandy).

The National Bureau of Standards (NBS) is responsible for setting the official standards for units of weights and measures for all commercial products, including food.

The Office of Technical Services (OTS) issues voluntary "Simplified Practices Recommendations" to limit types and sizes of packages, bags, and jars used as food containers.

The Federal Trade Commission (FTC) enforces the provisions of the FTC Act, which prevents unfair and deceptive trade practices.

State and local authorities such as Boards of Health regulate inspections in establishments such as restaurants, cafeterias, and other foodservice establishments that conduct local sales. Many state and local ordinances are patterned after the "Model Food Code" but some are different and many boards of health issue *guidelines* that explain the laws, making them easier to understand.

The regulation and standardization of food in international trade represent a prodigious and nearly impossible task. Yet the benefits to be derived are of such magnitude that they merit the necessary efforts to achieve them. The Codex Alimentarius Commission, an international organization, was formed by over 90 countries to establish food standards. Its importance is recognized when we are told that the U.S. FDA detains about 40% of the imported foods that it checks. What of the food shipments that escape inspection? What proportion of the imported foods is mislabeled? What proportion contains harmful substances or is contaminated or adulterated? It is because of these considerations that the FDA detains so much of the imported foods that it checks. The member countries of the Codex Alimentarius send their experts to the international meetings to help formulate the quality standards, which are more strict in some aspects than those of many individual countries. For example, the international standards require the listing of all ingredients in food formulations. The work done by this international body, slow though it is, should facilitate trade among the member countries, and the risks of foodborne illness and deceptive practices should be substantially reduced.

TRADE AND PROFESSIONAL ORGANIZATIONS

The Institute of Food Technologists (IFT)

The professional organization for food technologists and other food professionals, the IFT is a nonprofit scientific society founded in 1939. Since 1947, the IFT has printed a monthly publication (*Food Technology*) that presents information regarding the development of new and improved food sources, products, and processes, their proper utilization by industry and the consumer, and their effective regulation by government agencies. The IFT also has a Communications Division that comments on and provides information on proposed regulations.

The IFT has a number of sections that have monthly meetings and the national IFT also has an annual meeting that is attended by about 20,000 people. The IFT also has a student organization and encourages the teaching of food science at various levels from elementary school through college.

The American Dietetic Association (ADA)

The ADA is the professional organization for dietitians and dietary technicians. The ADA has a registration program for qualified dietitians and dietary technicians who obtain education from an approved college or university program and successfully complete a certification examination. The ADA has over 60,000 members and has local sections and does much work with educational institutes to educate the public in proper nutrition and healthy life styles.

The National Restaurant Association (NRA)

The NRA is the national trade association for retail and institutional foodservice establishments. It promotes sanitation with governmental, scientific, commercial, and educational institutions.

The Educational Foundation of the National Restaurant Association (EF)

The EF provides educational services to the food industry which include certifications in areas such as foodservice sanitation and personnel administration. EF also offers a comprehensive risk management series called SERVSAFE that includes the sanitation certification.

The Center for Science in the Public Interest (CSPI)

The CSPI is a nonprofit public interest membership organization that advocates improved health and nutritional policies. CSPI publishes a *Nutrition Action Newsletter* 10 times a year.

The American Council on Science and Health

The Council is a reputable consumer group that distributes reliable health and nutrition information. The ACSH is located in New York, New York and produces peer-reviewed reports on important health topics and publishes *Priorities*, a quarterly magazine.

The National Environmental Health Association (NEHA)

NEHA has a registered sanitarian program that has established a national standard based on education, experience, and testing for sanitarian qualifications. Some of its members are responsible for inspection services and environmental health programs.

The International Association of Milk, Food and Environmental Sanitarians (IAMFES)

IAMFES was a pioneer in the United States milk sanitation program, and also publishes information on milk and food safety and generally accepted methods on how to investigate a foodborne illness.

The Frozen Food Industry Coordinating Committee

This organization has developed a code for *Recommended Practices for the Handling of Frozen Food* that describes handling of food from processing to retail food service.

The National Pest Control Association (NPCA)

The NPCA consists of licensed and certified pest control operators (PCOs) and provides guidelines and training materials for integrated pest management (IPM) and other topics related to safety and sanitation.

NSF International

NSF International, formerly the National Sanitation Foundation, develops standards for equipment design, construction, and installation. Any manufacturer who wishes to have NSF International evaluate their equipment can do so and, if standards are met, the equipment will be included on a list and carry the NSF International seal.

Underwriters Laboratories, Inc. (UL)

UL lists all equipment that meets NSF International standards and also lists electrical equipment that passes U.L. safety requirements.

7

Food Labels

All labels associated with foods should be read at the time of procuring any food, because the contents of labels consist of important criteria for the selection of foods. These include basic information; a list of ingredients; in some cases, quality grades; nutritional information; shelf life data; and other important information. Authority for the regulation of food labels lies with the Food and Drug Administration (FDA). See Chapter 6 for more on the FDA.

In 1989, the FDA published an advanced notice of proposed rule-making and, with the Food Safety and Inspection Service (FSIS), held nationwide hearings to find out what consumers, food manufacturers, and health professionals wanted on the label. In 1990, the Nutrition Labeling and Education Act (NLEA) became law, mandating numerous changes in food labeling. Besides mandating nutritional labels on most foods and authorizing appropriate health claims on labels, it also calls for activities to educate consumers on nutritional information on the label and how to use it to maintain healthy dietary practices. The final regulations were published in the Jan. 6, 1993 *Federal Register*. Food companies were given until July, 1994 to implement the regulations.

BASIC INFORMATION

The amount and type of information on food labels differ among various food products, but all labels must bear the name of the product, the net contents (net weight or liquid measure in both common household and metric measures), and the name and address of the manufacturer, packer, or distributor. The net weight includes the liquid in which the food is packed. Canned vegetables are usually packed in a brine and canned fruits are usually packed in a syrup (see Chapter 20). The inclusion of net contents in a label is important to consumers, so that they can ensure that they buy enough to satisfy their needs and to make purchasing judgments related to unit costs (e.g., buying a large package may yield significant savings in unit cost). They can also use unit cost to select the most economical product from among different brands of similar quality. In the past, the packaging and labels of some foods gave a false impression of what they really were. For example, a product may have been packed and labeled similarly to ice cream, but it did not contain the minimum proportion of dairy fat (butter fat), making it ineligible and illegal under the standards set by the FDA to be labeled as ice cream. Now, products must be labeled clearly, avoiding any confusion. Many foods are covered by standards of identity, another authority of the FDA. Thus, there is a standard for ice cream, one for fish sticks, one for mayonnaise, one for catsup, etc. To prevent deception of consumers, the FDA has ruled that such foods be labeled under another common name that will provide consumers with an accurate knowledge of the product. Such a common name could be "frozen dessert" if the product does not meet the standard requirements for ice cream.

Another labeling requirement covers foods in which components of a recipe are missing, such as may occur in the case of some prepared entrees. Furthermore, the labels of such products must identify the entree to be prepared, and the ingredients that must be added to complete the recipe. An example of this would be a package of vegetables to be used in a "chicken stir-fry" recipe. It must be shown clearly on the label that chicken must be added.

The names of food must include the name(s) of the important ingredient(s) in descending order by weight. Thus, in a chicken entree, chicken meat must be the major ingredient, and in fish sticks, fish, not the breading, is the major ingredient. In frankfurts and beans, frankfurts comprise the major ingredient and beans must be the second major ingredient in descending order by weight. Otherwise, the product must be labeled "beans and frankfurts."

Another label requirement is the inclusion of the word "imitation" when a product resembles a standardized product but does not have an equal nutritional value. Thus a product manufactured as a no-cholesterol substitute for eggs must be labeled as "imitation eggs" if its nutritional value is not equivalent to that of eggs. If the product is formulated to have an equal nutritional value, but is missing a major ingredient (in this case, it would be egg yolk) then the product need not be labeled "imitation," but it must be given a new name such as "no-cholesterol eggs." "Eggbeaters®," the name of one such product, is an example of the latter case.

INGREDIENTS

A list of ingredients must be included on the label of many foods. In the past, foods that are covered by standards of identity (e.g., mayonnaise, ketchup, jams, and jellies) did not have to list those ingredients that were mandatory in accordance with the standard by which a particular food was covered. FDA now requires that all ingredients be listed, even for standardized foods. The order in which ingredients are listed must start with the major ingredient (by weight) and continue with the remaining ingredients in descending order. Any additive used in a food product must be listed. For more on additives, see Chapter 13. In the past, most colors did not have to be listed by name, but had to be listed within the term "colorings." With the implementation of NLEA, this has changed and they must be listed individually. Yellow Nos. 5 and 6, Red Nos. 40 and 3, Blue Nos. 1 and 2, and Green No. 3 are all examples.

Protein hydrolysates are added to food for various reasons such as thickeners, flavorings, flavor enhancers, or nutrients. Because the law doesn't require all flavors to be identified by their common or usual names, some industries have included hydrolysates in the "flavorings" or "natural flavors" declaration. This has happened when the hydrolysates have been added as flavor enhancers, a use not exempt from common or usual name declaration. The FDA concluded that when hydrolysates are added as flavors, they also function as flavor enhancers so must be declared on the label by their common or usual name. In the past, terms such as "hydrolyzed vegetable protein," "hydrolyzed animal protein," or "hydrolyzed protein" were allowed. Now, the source of the protein, such as corn, soy, or casein, must be identified as in the declarations "hydrolyzed soy protein" or "hydrolyzed casein." In the case of casein (or caseinate), it must be identified as a milk derivative. This will help casein-sensitive people or people who, for religious reasons, avoid milk products, to eliminate products containing casein from their diet. Some "nondairy" creamers do contain caseinate, added to whiten effectively.

The NLEA also has provisions concerning sweeteners, percentage ingredient information, sulfites, and requirements for juices.

Voluntary inclusion of the food source in the name of sweeteners is now allowed. For example, "corn sugar monohydrate" may be used in addition to "dextrose" or "dextrose monohydrate," which are also permitted.

Manufacturers who decide to declare ingredients by percentage of formulation must do so by weight. They may use percentage declaration in as many or as few ingredients as they choose as long as the information is not misleading. They still must declare all ingredients in descending order by weight.

Since 1986, sulfiting agents had to be listed in nonstandardized foods. Now, standardized food labels must list them as well.

Since 1990, "a food that purports to be a beverage containing juice must declare the per cent of total juice on the information panel." This alone does not eliminate confusion, as some juice manufacturers use bland juices such as apple or white grape and prominently name the juice with some of the lesser used juice. An example would be a "strawberry or raspberry juice containing 100% juice" would lead a consumer to believe that the product contains only strawberry or raspberry juice when it actually may contain those juices in very small amounts. To correct this misconception, FDA has required manufacturers to either:

- state the beverage is flavored by the named juice such as "strawberry flavored juice drink" or
- declare the amount of the named juice in a 5% range as "juice blend, 2 to 7% raspberry juice."

GRADES

Many foods carry a grade label that certifies their level of quality. All foods, except seafoods, are graded by the United States Department of Agriculture (USDA) and by health agencies of the individual states which are authorized by FDA to do so. Seafoods are presently graded by the United States Department of Commerce (USDC). See Chapter 6 for more on the USDA and USDC. The grades, in all cases, are based on organoleptic (appearance, odor, taste, and texture) criteria as well as on physical and production characteristics. More importantly, for seafoods, the grades and inspection sticker certify that the food is safe and wholesome and produced under the sanitary standards as recommended by the FDA. The grades assigned to different foods may vary. "Prime" and "Choice" may be used for meat while seafoods are graded "A," "B," or "C" in descending order of quality.

Grading is not mandatory because all graded products are already inspected for safety. Grading is usually requested by the producer, distributor, etc. One class of perishable foods presently not under mandatory inspection is that which includes finfishes. Consideration has been given, however, to make the inspection of finfishes mandatory, and it should not be unexpected that this will eventually happen. For the present, certification of the wholesomeness of finfish is ensured only when the product carries the USDC inspection sticker, which is normally accompanied by the USDC grade sticker.

While grades are not dependent on the nutritional quality of foods, the FDA has established standards that require certain levels of vitamins A and D when these vitamins are added to milk.

NUTRITION LABELING

With the growing tendency of consumers to control the consumption of various dietary components, there is a growing proclivity to read food labels mainly for the nutritional data. This consumer demand was one of the driving forces that encouraged the FDA to change its labeling regulations and, as of 1994, nutritional labeling has been required on the labels of most foods. The following is a list of mandatory (italic) and voluntary dietary components in the order in which they must appear:

- *total calories*
- *calories from fat*
- calories from saturated fat
- *total fat*
- *saturated fat*
- stearic acid (for meat and poultry products only)
- polyunsaturated fat
- monounsaturated fat
- *cholesterol*
- *sodium*
- potassium
- *total carbohydrate*
- *dietary fiber*
- soluble fiber
- insoluble fiber
- *sugars*
- sugar alcohol (e.g., xylitol, mannitol, sorbitol)
- *protein*
- *vitamin A*
- percent of vitamin A present as beta-carotene
- *vitamin C*
- *calcium*
- *iron*
- other essential vitamins and minerals

If a food is enriched or fortified with any of the optional components or a claim is made about any of them, pertinent information becomes mandatory.

The key aspects of the new nutrition label are shown in Figure 7.1.

There are rules on how the information must be presented. "Nutrition Facts" must be set in the largest type on the nutrition panel and be highlighted in some manner such as boldface to distinguish it from the other information. Highlighting is also required for headings such as "Amout per Serving" and "% Daily Value," and for the names of dietary components that are not subcomponents (calories, total fat, cholesterol, sodium, total carbohydrate, and protein).

Nutrition Facts

Serving Size 1 cup (228g)
Servings Per Container 2

Amount Per Serving

Calories 260 Calories from Fat 120

	% Daily Value*
Total Fat 13g	**20%**
Saturated Fat 5g	**25%**
Cholesterol 30mg	**10%**
Sodium 660mg	**28%**
Total Carbohydrate 31g	**10%**
Dietary Fiber 0g	**0%**
Sugars 5g	
Protein 5g	

Vitamin A 4% • Vitamin C 2%
Calcium 15% • Iron 4%

* Percent Daily Values are based on a 2,000 calorie diet. Your daily values may be higher or lower depending on your calorie needs:

	Calories:	2,000	2,500
Total Fat	Less than	65g	80g
Sat Fat	Less than	20g	25g
Cholesterol	Less than	300mg	300mg
Sodium	Less than	2,400mg	2,400mg
Total Carbohydrate		300g	375g
Dietary Fiber		25g	30g

Calories per gram:
Fat 9 • Carbohydrate 4 • Protein 4

Figure 7.1. Nutrition facts.
(Source: Focus on Food Labeling, FDA Consumer, 1993)

Serving Sizes

The declaration of "serving sizes" was addressed in the NLEA. Some uniformity of serving sizes and sizes closer to what people actually eat were the major goals of the changes here. There are now 139 FDA-regulated food product categories that list amounts of food normally consumed per eating occasion and the suggested label statement for serving size declaration. With some products, such as cookies, a reference amount is given (30 g) and number of pieces (cookies) that comes closest to that amount may be given as a serving size. For example, if the average weight of a cookie is 13 grams, a serving size of "two cookies (26 g)" may be declared. For food products packaged and sold individually, if the individual package is less than 200% of the applicable reference amount, the item will be one serving. For example, a 12 fluid oz. (360 ml) can of cola would be one serving because the reference amount for a single serving of cola is 8 fluid oz. (240 ml).

Exemptions and Exceptions

Certain situations require special stipulations for the provision of nutritional information:

- FDA-regulated food packages with less than 12 square inches of available space for nutritional labeling do not have to provide it. They must provide an address or telephone number for consumers to acquire the information.
- Packages with less than 40 square inches for nutritional labeling may present nutritional labeling in a tabular format (see Fig. 7.2), omitting footnotes with the list of daily values and caloric conversion information and abbreviating the names of dietary components.

Some foods are exempt from nutritional labeling:

- food produced by small businesses (a business with food sales of less than $50,000 a year or total sales of less than $500,000 as defined by the FDA and a business with fewer than 500 employees and producing no more than a certain amount of product per year as defined by FSIS)

Nutrition Facts
Serv. Size 1/3 cup (56g)
Servings about 3
Calories 80
Fat Cal. 10
*Percent Daily Values (DV) are based on a 2,000 calorie diet.

Amount/serving	%DV*	Amount/serving	%DV*
Total Fat 1g	2%	**Total Carb.** 0g	0%
Sat.Fat 0g	0%	Fiber 0g	0%
Cholest. 10mg	3%	Sugars 0g	
Sodium 200mg	8%	**Protein** 17g	

Vitamin A 0% • Vitamin C 0% • Calcium 0% • Iron 6%

Figure 7.2. Tabular format label.
(Source: Focus on Food Labeling, FDA Consumer, 1993)

- food served for immediate consumption such as in restaurants, cafeteria, airplanes, and vending machines
- ready-to-eat foods that are not for immediate consumption, as long as the food is prepared on site, for example many bakery, deli, and candy store items
- food shipped in bulk as long as it is not for sale in that form for consumers
- donated foods
- plain coffee, tea, flavor extracts, food colors, some spices, and other foods that contain no significant amounts of any nutrients
- products intended for export
- if a food contains no significant amount of some nutrients, an abbreviated label may be used (see Fig. 7.3).

Daily Values

Appearing on the nutritional labels is a new set of dietary references called "Daily Values" or DVs. These are made from two sets of references, the DRVs (daily reference values) and the RDIs (recommended daily intakes). The DRVs are for nutrients for which no set of standards previously existed. The RDIs replace the term "U.S. RDAs" (Recommended Daily Allowances). Despite the name change, except for protein, the values will stay the same. U.S. RDAs should not be confused with RDAs (Recommended Dietary Allowances; see Table 2.1). The FDA uses the RDAs to set up the U.S. RDAs (now the RDIs). Tables 7.1 and 7.2 show the DRVs and the RDIs. Whatever the calorie level, DRVs for energy-producing nutrients are always produced as follows:

Nutrition Facts

Serving Size 1 can (360 mL)

Amount Per Serving	
Calories 140	
	% Daily Value*
Total Fat 0g	**0%**
Sodium 20mg	**1%**
Total Carbohydrate 36g	**12%**
Sugars 36 g	
Protein 0g	**0%**

* Percent Daily Values are based on a 2,000 calorie diet.

Figure 7.3. Simplified nutrition label.
(Source: Focus on Food Labeling, FDA Consumer, 1993)

Table 7.1. Daily Reference Values (DRVs)[a]

Food Component	DRV
Fat	65 grams (g)
Saturated fatty acids	20 g
Cholesterol	300 milligrams (mg)
Total carbohydrate	300 g
Fiber	25 g
Sodium	2400 mg
Potassium	3500 mg
Protein[b]	50 g

[a]Based on 2000 calories for adults and children over 4 years of age.
[b]DRV for protein does not apply to certain populations; RDIs for protein has been established for these groups:

Children 1–4 years	16 g
Infants under 1 year	14 g
Pregnant women	60 g
Nursing mothers	65 g

Table 7.2. Reference Daily Intakes (RDIs)

Nutrient	Amount
Vitamin A	5000 International Units (IU)
Vitamin C	60 milligrams (mg)
Thiamin	1.5 mg
Riboflavin	1.7 mg
Niacin	20 mg
Calcium	1 gram (g)
Iron	18 mg
Vitamin D	400 IU
Vitamin E	30 IU
Vitamin B_6	2 mg
Folic acid	0.4 mg
Vitamin B_{12}	6 micrograms (mcg or μg)
Phosphorus	1 g
Iodine	150 mcg or μg
Magnesium	400 mg
Zinc	15 mg
Copper	2 mg
Biotin	0.3 mg
Pantothenic acid	10 mg

Table based on 1968 RDAs.

- fat based on 30% of calories
- saturated fat based on 10% of calories
- carbohydrate based on 60% of calories
- protein based on 10% of calories
- fiber based on 11.5 g per 1000 calories

Because some nutrients are linked to certain diseases, DRVs for these nutrients represent the uppermost levels that are considered desirable. Therefore, the label will show DVs for fat and sodium as follows:

- total fat: less than 65 g
- saturated fat: less than 20 g
- cholesterol: less than 300 mg
- sodium: less than 2400 mg

Raw Foods

For consumers who want nutritional information on raw foods, the FDA established guidelines for voluntary nutrition information in November of 1991. These guidelines are in accordance with the NLEA of 1990. The NLEA mandates nutritional labeling on most processed foods but allows voluntary point-of-purchase nutrition information for raw fish, fruits, and vegetables (the FSIS and USDA have similar programs for raw meat and poultry). This voluntary status will remain as long as 60% of the stores surveyed in national surveys comply. If there is not at least 60% participation, the law will become mandatory.

The FDA has set standards on what information must be provided at the point-of-purchase. For raw fish, fruits, and vegetables, the following are included:

- name of fruit, vegetable, or fish that has been identified as the 20 most commonly eaten in the United States (see Table 7.3).
- serving size
- calories per serving
- amount of protein, total carbohydrates, total fat, and sodium per serving
- percent of U.S RDAs for iron, calcium, and vitamins A and C per serving

The information may be displayed on large placecards or listed in consumer pamphlets or brochures.

For meat and poultry, the information will be on major cuts of the product (see Table 7.4) and be displayed at point-of-purchase on posters, in brochures, or on the individual package. The information will include:

- name of meat or poultry cut
- serving size based on raw or cooked weight
- calories per serving
- calories from total fat per serving
- amount per serving—by weight and by percentage of Daily Value—for total fat, saturated fat, cholesterol, sodium, total carbohydrates, and dietary fiber
- amount by weight of sugars and protein
- percent of DVs for iron, calcium, and vitamins A and C per serving

There are also a number of optional values that may be included.

Table 7.3. Most Common Fruits, Vegetables, and Fish Sold Raw

Top 20 Fruits	Top 20 Vegetables	Top 20 Fish
Banana	Potato	Shrimp
Apple	Iceberg lettuce	Cod
Watermelon	Tomato	Pollock
Orange	Onion	Catfish
Cantalope	Carrot	Scallop
Grape	Celery	Atlantic/coho salmon
Grapefruit	Sweet corn	Flounder
Strawberry	Broccoli	Sole
Peach	Green cabbage	Oyster
Pear	Cucumber	Orange roughy
Nectarine	Bell pepper	Atlantic/Pacific and jack mackerel
Honeydew melon	Cauliflower	Ocean perch
Plum	Leaf lettuce	Rockfish
Avocado	Sweet potato	Whiting
Lemon	Mushroom	Clam
Pineapple	Green onion	Haddock
Tangerine	Green (snap) bean	Blue crab
Sweet cherry	Radish	Rainbow trout
Kiwi fruit	Summer squash	Halibut
Lime	Asparagus	Lobster

Table 7.4. Major Meat and Poultry Cuts

Beef	Pork	Lamb	Veal	Chicken and Turkey
Chuck blade roast	Loin chop	Shank	Shoulder arm steak	Whole[a]
Loin top loin steak	Loin country style ribs	Shoulder arm chop	Shoulder blade steak	Breast
Rib roast large end	Loin top loin chop boneless	Shoulder blade chop	Rib roast	Wing
Round eye round steak	Loin rib chop	Rib roast	Loin chop	Drumstick
Round top round steak	Spareribs	Loin chop	Cutlets	Thigh
Round tip roast	Loin tenderloin	Leg		
Chuck arm pot roast	Loin sirloin roast			
Loin sirloin steak	Shoulder blade steak			
Round bottom round steak	Loin top roast boneless			
Brisket	Ground			
Rib steak small end				
Loin tenderloin steak				
Regular round				
Extra lean round				

[a]Without neck and giblets. Separate nutrient panels for while and dark turkey are optional.

Health Claims

Another problem the NLEA is helping to solve is the control of any false or misleading health claims on food labels. Much research was done and FDA published in the Jan. 6, 1993 *Federal Register* the claims about relationships between food and health that are allowed:

- calcium and a reduced risk of osteoporosis
- sodium and increased risk of hypertension
- dietary saturated fat and cholesterol and an increased risk of coronary heart disease
- dietary fat and an increased risk of cancer
- fiber-containing grain products, fruits, and vegetables and a reduced risk of cancer
- fruits, vegetables, and grain products that contain fiber, particularly soluble fiber, and a reduced risk of coronary heart disease
- fruits and vegetables and a reduced risk of cancer

To qualify for labeling with a health claim, foods must contain:

- a nutrient (such as calcium) whose composition at a specified level as part of an appropriate diet will have a positive effect on the risk of disease or
- a nutrient of concern (such as fat) below a specified level

The foods must contribute nutrition to the diet by naturally containing at least 10% of the DV of one or more of the nutrients vitamin A, vitamin C, iron, calcium, protein, and fiber. The food must not contain any nutrient or food substance in an amount that increases the risk of a disease or health condition. Foods bearing health claims must contain 20% or less of the DV of: fat (13 g), saturated fat (4 g), cholesterol (60 mg), and sodium (480 mg).

Other Claims

There are also a number of claims that can be made about a food and FDA and the NLEA has set standards that must be followed. These include the standard terms concerning sugar, calories, fat, cholesterol, sodium, and fiber.

Sugar

sugar free: less than 0.5 g per serving
no added sugar, without added sugar, no sugar added:

- No sugar is added during processing or packaging, including ingredients such as fruit juice or applesauce that contain sugar.
- processing does not increase the sugar content above the amount naturally present in the ingredients.
- The food that it resembles and for which it substitutes for normally contains added sugar.
- If the food does not meet guidelines for a low- or reduced-calorie food, a statement must appear stating this and direct consumers' attention to the nutrition panel for further information on sugars and calorie content.

reduced sugar: at least 25% less sugar per serving than reference food

Calories

calorie-free: fewer than 5 calories per serving
low-calorie: 40 calories or less per serving and if the serving is 30 g or less or 2 tablespoons or less, per 50 g of the food
reduced or fewer calories: at least 25% fewer calories per serving than reference food

Fat

fat-free: less than 0.5 g fat per serving
saturated fat free: less than 0.5 g per serving and the level of *trans* fatty acids does not exceed 1% of total fat
low-fat: 3 g or less per serving, and if the serving is 30 g or less or 2 tablespoons or less, per 50 g of the food
low saturated fat: 1 g or less per serving and not more than 15% of calories from saturated fat
reduced or less fat: at least 25% less per serving than reference food
reduced or less saturated fat: at least 25% less per serving than reference food

Cholesterol

cholesterol-free: less than 2 mg of cholesterol and 2 g or less of saturated fat per serving
low-cholesterol: 20 mg or less and 2 g or less of saturated fat per serving and, if the serving is 30 g or less or 2 tablespoons or less, per 50 g of the food
reduced or less cholesterol: at least 25% less and 2 g or less of saturated fat per serving than reference food

Sodium

sodium-free: less than 5 mg per serving
low-sodium: 140 mg or less per serving and, if the serving is 30 g or less or 2 tablespoons or less, per 50 g of the food
very low sodium: 35 mg or less per serving and, if the serving is 30 g or less or 2 tablespoons or less, per 50 g of the food
reduced or less sodium: at least 25% less per serving than reference food

Fiber

high-fiber: 5 g or more per serving. (Foods making high-fiber claims must meet the definition for low-fat, or the level of total fat must appear next to the high-fiber claim.)
good source of fiber: 2.5 to 4.9 g per serving
more or added fiber: at least 2.5 g more per serving than reference food

OPEN DATING

Open dating is the practice of showing a date on packaged foods that carry information that permits the consumer to ascertain the date beyond which the quality of the product may be expected to fall below normal levels. This information may appear on the label, but in many cases it is simply stamped on the label or elsewhere on the package. One example of the open dating is the "sell by" date. This defines the last day that the product should be sold or used by the retailer. It is the type used for perishable products such as dairy and meat products. Another example of open dating is the "expiration date" or "use by" date. This defines the last day that the product should be used or eaten. It is the type used for dried yeast and baby foods. Another type of open dating is the "freshness date." This defines the last day that the product will retain its fresh quality. This type is used largely for bakery products. From the consumer's point of view, the least important type of open dating is the "pack date." This defines the date when the food was manufactured, processed, or packaged. This type of dating is used for nonperishable foods such as canned foods and some packaged foods. However, this date has little meaning to the consumer unless he or she has information on the expected shelf-life of the product. It does have a practical value similar to "code dating" (see next section).

OTHER INFORMATION

Other information that may be found on the package of foods includes code dating, the universal product code, legal symbols, and religious symbols.

Code Dating

Code dating is usually used for products having long shelf lives such as canned foods, but it may be used on the other packaged products. This type of dating, as the name implies, is in code. The code supplies information regarding the date when the product was packaged as well as the place where it was packaged. This information serves a number of purposes, but a major use is to facilitate a recall of the product should circumstances make such an action necessary.

Universal Product Code

Most food labels carry the universal product code (UPC). The value of the UPC is associated with the use of computers. The UPC is readily recognized by a line of numbers at the bottom of the block. The UPC is specific for each food product, and is used to read the product and price when computerized check-out equipment is used. The computer can be programmed to record the rate of sales, thus maintaining a status of the inventory, which in turn reveals when and how much product must be procured to maintain a desired inventory level. The ordering can be done automatically if the computer program provides for this service. Furthermore, one can determine the contribution of the product to net profit, its profit per unit space, etc. As can be seen, the UPC has a significant industrial value, with the potential to be even more valuable.

Legal Symbols

The legal symbols that may appear on labels are ® and ©. The symbol ® means that the trademark used on the label is registered with the U.S. Patent Office. The ©

means that the literary and artistic content of the label is protected under the copyright law of the United States and that copies of such labels have been filed with the Copyright Office of the Library of Congress.

Religious Symbols

Two religious symbols that may appear on food labels are relevant to people of the Jewish faith. When the letter "K" appears inside the letter "O," it means that the food is "Kosher," signifying that it complies with Jewish dietary laws and that it was processed under the supervision of a rabbi. When the letter "U" appears inside the letter "O," it means that the product complies with Jewish dietary laws and is authorized by the Union of Orthodox Jewish Congregations of America, more familiarly known as the Orthodox Union.

8

Enzyme Reactions

THE NATURE OF ENZYMES

Enzymes, produced by living things, are compounds that catalyze chemical reactions. Reactions involving enzymes may be said to proceed in two steps. In step 1, $E + S \leftrightarrow ES$ (where E = enzyme, S = substrate, and ES = unstable intermediate complex that temporarily involves the enzyme). In step 2, $ES + R \leftrightarrow P + E$ (where R = a substance in the substrate that reacts with the complex, P = the final product of the reaction, and E = enzyme liberated from the complex). Enzymes are critical to life because they have the ability to catalyze the chemical reactions that are important to life. Chemical reactions take place when the necessary reactants are present, but usually an energy input (activation energy) is required to start a particular reaction. The analogy usually given to illustrate this concept is that of a boulder located at the top of a hill. The boulder has the potential energy for rolling down the hill, but must first be pushed over the edge (see Fig. 8.1). The potential energy of the boulder could be great, depending on its mass and altitude. It could start a rock slide or landslide involving a great amount of energy. Although the energy required to push the boulder over the edge is insignificant compared to the total energy involved in the rolling of the boulder down the hill, that initial energy (called the activation energy) is nevertheless important, for without it there would be no landslide.

It is well known that the rate at which reactions take place is dependent on temperature. Because reactions can proceed more rapidly at higher temperatures but quite often have to accelerate within a system of constant temperature, such as within our bodies, only by the action of enzymes can they occur. Enzymes are produced by living organisms from the lowest single-celled members to the highest, most complex members of the plant or animal kingdoms, including humans. All life depends on enzymes to convert foods or nutrients to a form in which they can be utilized, and to carry out cellular functions.

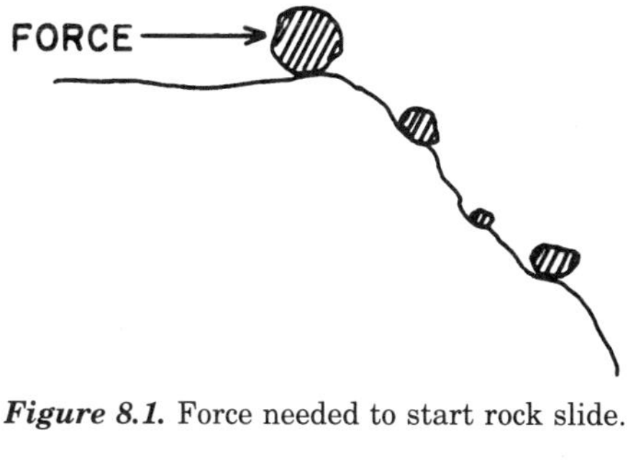

Figure 8.1. Force needed to start rock slide.

In composition, enzymes always contain a protein. They may also contain or require complex chemical compounds in order to become functional. These compounds are known as prosthetic groups (also called coenzymes) and are usually vitamins, especially those belonging to the B group. Some enzymes also require trace amounts of a metal, such as copper, to function. Enzymes therefore consist of either pure proteins, proteins with a prosthetic group, or proteins with a prosthetic group plus a metal cation.

Through their action, enzymes convert foods into less complex chemical substances that can be utilized for energy and for the building of cellular protoplasm. Proteins, fats, and carbohydrates are thus broken down to less complex compounds by enzymes in order that they may be utilized. More complex carbohydrates are broken down to glucose, a source of ready energy that can be absorbed and eventually converted to carbon dioxide and water or built up into fats that can be stored as a source of reserve energy. When reserve energy is needed, fats are first hydrolyzed (water is inserted to split the molecule), breaking down into glycerine and fatty acids. The fatty acids are then converted to acetates that can be utilized for energy. When the acetates are not completely used, those remaining may be reformed into fats and deposited as an energy reserve.

Proteins are broken down to their primary units (amino acids), in which form they may be incorporated into cellular protoplasm as proteins used for cell repair or growth. In some instances, the nitrogen portion of the amino acid may be removed and the remaining compound oxidized to provide energy.

As previously stated, enzymes always contain a protein (chain of amino acids) in which the amino acids are combined in a particular sequence, the protein itself having a particular shape or configuration.

There are approximately 20 to 22 known amino acids. Proteins consist of a large number of amino acids combined in a particular sequence. Also, chains of amino acids are crosslinked one to another and the different proteins form special configurations and shapes. Proteins are also sometimes combined with carbohydrates, lipids, or phospholipids (fatlike compounds containing phosphoric acid as part of the molecule). These are called conjugated proteins. The number and sequence of amino acids in the chain, the shape or relationship of one protein chain with another, and conjugation with carbohydrate or lipid all affect the functional properties and determine the manner win which a protein will react to physical and chemical energy.

Enzymes cause chemical reactions to occur at their fastest rates when the temperature is at an optimum level. For most enzymes, this is in the range of 60 to 150°F (15.6 to 65.5°C), but some action may occur at temperatures above or below the optimum range. Thus, some enzymes are able to react slowly at temperatures well below that of the freezing point of water and others at temperatures above 160°F (71.1°C).

Because proteins are changed chemically and physically or are coagulated by high temperatures, especially when moisture is present, enzymes are usually inactivated at temperatures between 160 and 200°F (71.1 and 93.3°C). There are some exceptions to this, however, and at least one enzyme, which splits off fatty acids from fish phospholipids, is known to remain active even after steaming at 212°F (100°C) for 20 min.

Enzymes have an optimum pH at which they cause reactions to occur at the fastest rate. Water solutions having a pH value less than 7 are said to be acidic; those having a pH value greater than 7 are said to be alkaline; and those having a pH value of exactly 7 are said to be neutral. As in the case of temperatures, some action will occur at pHs above or below the optimum, although there are low and high limits beyond which a particular enzyme action cannot take place.

PROTEOLYTIC ENZYMES (PROTEASES)

Enzymes involved in the breakdown or splitting of proteins are called proteolytic enzymes or, more simply, proteases. Proteases comprise two general classes: proteinases and peptidases. Proteinases split the protein molecules into smaller fragments called proteases and peptones, then into polypeptides and peptides. Peptidases split polypeptides and peptides into amino acids. Because most amino acids in foods are water-soluble, food proteins are essentially liquefied by proteinases and peptidases.

In meat, such as beef, pork, or poultry, held in the eviscerated state (intestines and organs removed), the proteases present in the tissue are called cathepsins. The temperatures and times under which these products are held prior to utilization are not such that extensive proteolysis can occur. Therefore, while there may be some tenderization of the tissues during holding, which may, in fact, be due to proteolysis, there is not extensive breakdown of the tissues.

In fish, proteolytic enzymes are much more active than in meats. Even when fish is held in the eviscerated state in ice or under refrigeration, there may be sufficient proteolysis to cause softening of the tissues over a period of days. In fish held in the round (uneviscerated), proteolysis is accelerated owing to a concentrated source of enzymes present in blind tubules (the pyloric ceca) attached to the intestines. Thus, even though fish in the round are refrigerated, within a few days, sufficient proteolysis may occur to dissolve the tissues of the abdominal wall, exposing the entrails. Members of the herring and mackerel families handled in the uneviscerated state are quite subject to this type of enzyme deterioration, especially if they have been feeding when caught. Other fish such as flounders and ocean perch, handled in the uneviscerated state, seem not to be especially subject to this type of deterioration.

Lobsters provide an especially good example of a deteriorative change that may take place through the action of proteolytic enzymes. As long as the lobster is alive, autolytic proteolysis does not occur. However, if the lobster dies and then is held for some hours, even under refrigeration, but especially at high temperatures, proteolysis takes place to such an extent that the lower abdominal portion will be partially liquefied. When such a lobster is cooked, the flesh will be soft and crumbly (short-meated) and part of the tail portion will have dissolved, leaving only a part of this section intact. For this reason, lobsters should never be held long after death prior to cooking. The present accepted practice is to cook lobsters from the live state. Other crustaceans (shrimp and crabs) are also subject to enzyme proteolysis, although with shrimp, this usually is not extensive, especially if the head portion (cephalothorax) is removed shortly after the shrimp are caught. The relatively high activity of enzymes in marine species is attributed to the low-temperature conditions in the marine environment. That is, in order to make reactions proceed at low temperatures, the activation energy system must be more efficient.

Plants also contain proteolytic enzymes, but these enzymes usually contribute little to deterioration, especially as long as the tissues of fruits and vegetables are not cut or damaged. Some plants provide an excellent source of proteolytic enzymes. Bromelin is found in unpasteurized pineapple juice and is so active that people who handle cut pineapple that has not been heated must wear rubber gloves; otherwise the skin of the fingers will become eroded, and could result in the exudation of some blood. Papain is a proteolytic enzyme obtained from the latex (milky liquid) of the green papaya fruit. Ficin is a proteolytic enzyme obtained from the latex of certain fig trees. Proteolytic enzymes from plants may be extracted and purified and these enzymes may be employed, for instance, to tenderize meats.

OXIDIZING ENZYMES (OXIDASES)

Oxidation has been defined as a loss of electrons in an atom or simply a chemical combination with oxygen. An example of complete oxidation is a wick of a candle burning, where the cellulose wick is converted to carbon dioxide and water. There are a number of oxidizing enzymes that bring about changes in foods that result in deterioration due to oxidation. In plants, peroxidases, ascorbic acid oxidase, tyrosinase, and polyphenolases may cause undesirable chemical reactions to occur. Peroxidases may oxidize certain phenollike compounds in root vegetables, such as horseradish, causing the prepared product to become darker in color. This does not happen while the tissues are intact but only when the vegetable has been cut up or comminuted. Ascorbic acid oxidase, present in certain vegetables, oxidizes ascorbic acid (vitamin C) to a form that is readily further oxidized by atmospheric oxygen. The resulting oxidation product is not utilized by humans as a vitamin. Therefore, the action of this enzyme may cause loss of the vitamin C content of foods. Peroxidases may also, indirectly, cause a loss of vitamin C in vegetables. In this case, the compounds formed by the action of peroxidase react with vitamin C. Phenolases are present in some fruits and vegetables. These enzymes oxidize some phenollike compounds, also present in plant product, causing brown or dark-colored compounds to be formed when the tissues are cut.

Tyrosinase oxidizes the amino acid tyrosine to form dark-colored compounds. The molecule rearranges and further oxidizes to form a red compound. Polymerization (combination of these compounds) results in the formation of dark-colored melanin compounds.

The enzyme tyrosinase, which is present in many fruits and vegetables, may cause discoloration of the cut tissue and will also oxidize compounds related to tyrosine. This enzyme is also present in shrimp and some spiny lobsters and may cause a discoloration called black spot. In shrimp, this often occurs as a black stripe on the flesh along the edges of the segments of the tail or as a pronounced band where the shell segments overlap. It is not generally recognized, but tyrosinases are also present in clams. Hence, shucked (deshelled) clams will darken at the surface if oxygen is present and if the enzymes have not been inactivated by heat.

These reactions occur only after the shrimp or clams die. In general, oxidizing enzymes do not cause deteriorative changes in tissues that are intact. In fruits and vegetables, the tissues must be cut or bruised or there must be a breakdown of cells by their enzymes before the action of oxidizing enzymes results in discoloration.

FAT-SPLITTING ENZYMES (LIPASES)

Fats are composed of glycerine (glycerol) and fatty acids. Glycerine is a polyhydric alcohol (three alcohol groups), and fatty acids are short or long chains of carbon atoms to which hydrogen is attached, either to the fullest possible extent (saturated) or to a lesser extent (unsaturated), the latter resulting in reactive groups in the chain. At one end of the fatty acid chain there is an acid group.

In the formation of fats, each one of three fatty acids combines with one of the three alcohol groups of glycerine, splitting off water in each case. On the other hand, in the breakdown of fats, water and the enzyme lipase are present, and fats are split into their original component parts, glycerine and fatty acids.

The fatty acids in most fats that are found in nature consist of a chain of more than 10 carbon atoms, and these fatty acids have no particular flavor or odor. Hence, when lipase acts on most natural fats, no bad odors are generated. However, if fats or oils high in free fatty acid content (indicating deterioration) are used for deep-fat frying, the oil may smoke during heating, which is undesirable.

$$
\begin{array}{lclcl}
CH_2OOC(CH_2)_2CH_3 & & CH_2OH & + & CH_2(CH_2)_2COOH \\
 & & | & & \text{butryic acid (odorous)} \\
CH_2OOC(CH_2)_8CH_3 & \xrightarrow[\text{lipase}]{\text{water}} & CHOH & + & CH_2(CH_2)_8COOH \\
 & & | & & \text{capric acid (some odor)} \\
CH_2OOC(CH_2)_{14}CH_3 & & CH_2OH & + & CH_2(CH_2)_{14}COOH \\
 & & & & \text{palmitic acid (no odor)} \\
\text{fat} & \longrightarrow & \text{glycerine} & + & \text{fatty acids}
\end{array}
$$

There are some fats that contain short-chain fatty acids, especially those fats present in the milk of cows or goats. These fats contain butyric (four carbons), caproic (six carbons), caprylic (eight carbons), and capric (ten carbons) acids. All these fatty acids have an odor and flavor. Butyric acid, especially, is pungent and considered to be distasteful. When lipase acts on butter, therefore, it splits off butyric acid, which gives the butter a strong, undesirable rancid taste. Actually, butter is an emulsion of water in oil and contains about 16% water, the water being present as fine droplets. Butter becomes rancid by the action of lipase produced by bacteria that grow in the water droplets. The lipase acts on the fat surrounding the water droplets. Lipase rancidity in butter, therefore, is really a type of deterioration caused by bacteria.

Phospholipases, enzymes similar to lipase, are split phospholipids. Most phospholipids are similar to fats in that they contain glycerine, two alcohol groups of which are combined with fatty acids. The third alcohol group in this case is combined with a molecule containing phosphoric acid, a short chain of carbons, and a nitrogen group with carbons attached to it. Lecithin is a typical phospholipid. Phospholipase splits off fatty acids from phospholipids. Such action may cause deterioration in foods in that it results in a destabilization of proteins that causes a toughening of the tissues and a loss of succulence (juiciness).

ENZYMES THAT DECOMPOSE CARBOHYDRATES (CARBOHYDRASES)

Fruits contain pectin which supports the particular structure of the product. In processed fruit juices (for instance, tomato or orange juice), if the pectin is broken down, the solids tend to settle to the bottom, leaving a clear serum on top. Pectin consists of a long chain of galacturonic acid molecules with the carboxyl groups partially esterified with methyl alcohol. It has a high water-holding capacity. There are pectic enzymes that will either break down the pectin molecule to smaller units or completely decompose the molecule to its primary unit, galacturonic acid. The emulsifying properties of pectin may be lost, causing settling in fruit juices and softening in fruit. When the pectin in whole fruit breaks down, it may result in deterioration of the fruit as a result of the action of other enzymes or invasion of the tissues by microorganisms.

In the sugar cane plant, there is an enzyme, invertase, that breaks down cane sugar (sucrose), which has 12 carbon atoms, into glucose and fructose, each having six carbon atoms. Before sugar cane is harvested, therefore, a part of the plant must be removed to eliminate the source of the enzyme. Were this not done, there would be a loss of sucrose during the processing of the cane. Many other carbohydrases, which break down cellulose or starch or break down more complex sugars to smaller units, exist.

APPLICATIONS

Whereas enzymes may cause a deterioration of foods, they may also be used in the processing of foods to produce particular products or to modify the characteristics of particular products. Proteolytic enzymes obtained from plants may be used for tenderizing meat either by injecting animals with a solution of the enzyme prior to slaughter or by sprinkling the powdered enzyme on meat surfaces and allowing it to react, prior to cooking. In the manufacture of certain kinds of milk powder (e.g., to be used in chocolate), the lipases may be allowed to act on the milk fat prior to drying to obtain a particular flavor in the finished product. Proteases may also be used to chill-proof beers, a technique used to remove proteins that would cause clouding during cool storage. Protein hydrolysates can be produced from the use of proteases from both plant and animal sources. They have many uses, for example, as flavorings, nutrients, and stabilizers to improve texture.

The characteristic flavor of certain cheeses is due to the action of lipases on the milk fat contained therein. In order to obtain the particular flavors of Roquefort, Gorgonzola, or blue cheese, the milk fat must first be broken down to fatty acids that can then be oxidized. The lipases that decompose the fats to provide fatty acids are produced by the molds allowed to grow in these cheeses. Enzymes produced by molds then oxidize the specific fatty acids that ultimately produce the unique characteristic flavors. Lipases also can be added to dried egg whites to improve the whipping quality.

There are many applications of enzyme technology that involve the use of carbohydrate-splitting enzymes. In making malt, barley is germinated to obtain an enzyme that will convert starch to a sugar (maltose), which can be converted by yeasts to ethyl alcohol and carbon dixoide. By this means, various grains can be used as the source of sugar for fermentation. Carbohydrate-splitting enzymes are also used to modify starches used in foods and to modify starches used in sizing and laundering clothes. Others are used to form corn syrup from corn starch and then isomerize glucose to fructose in the formation of high-fructose corn syrup which is used extensively in the manufacture of soft drinks. Invertase (sucrase) is used in the production of chocolate-covered cherries. The cherries are rolled in a mixture of crystalline sucrose and the enzyme before they are covered in chocolate. During a short holding period, the enzyme splits the sucrose into a mixture of glucose and fructose, which has a higher water solubility and results in the sweet taste and creamy texture that is characteristic of the product. The presence of pectins in fruit juices may cause clouding. Pectinases can be added to remove the pectin and clarify the juice.

There are many other applications of enzyme technology in the food and other industries and it is expected that the number of applications for enzymes will continue to increase. One of the factors that will serve to widen the use of enzyme technology is the development of immobilized enzymes. It has been found that enzymes can be fixed chemically to the surface of inert substances, such as glass beads. In this form, they can be packed into a column through which a solution or suspension of the material to be acted on (called the substrate) is allowed to pass. In this manner, the enzyme responsible for the conversion or change in the substrate is not lost or washed out with the substrate. Thus, the enzyme can be used for a number of substrate conversions. Moreover, in this form (immobilized enzyme), the active agent is much less subject to inactivation such as, for instance, by high temperature.

Restoration and enrichment of flavor in some processed foods can be accomplished by addition of enzymes. Examples of this include the conversion of allin in garlic into garlic oil by allinase and the addition of an enzyme preparation from mustard seeds to dehydrated cabbage which restores flavor by converting flavor precursors into the volatile sulfur-containing compounds responsible for the familiar flavors.

9

Chemical Reactions

Foods sometimes deteriorate because of chemical changes not associated with microbial growth or enzyme-induced chemical reactions. Actually, certain components in foods are subject to reactions that involve either a combination with naturally occurring elements (such as oxygen) or with compounds present in the foods themselves.

OXIDATION

One of the changes that occur in foods is the oxidation of fats and oils. Fats and oils are chemically similar and are classified as lipids. Generally, fats are those lipids that are solid at room temperature (e.g., lard, suet), and oils are those lipids that are liquid at room temperature (e.g., olive oil, corn oil). Fats and oils are glycerol esters of fatty acids, such that each molecule of fat or oil contains three fatty acids that may be the same as or different from each other. Generally, they are different from each other. The glycerol esters are formed by a condensation reaction that results in the formation of one triglyceride molecule and three molecules of water:

$$\begin{array}{l} CH_2OH \\ | \\ CHOH \\ | \\ CH_2OH \end{array} \quad + \quad 3\ RCOOH \longrightarrow \quad \begin{array}{l} CH_2COOR_1 \\ | \\ CHCOOR_2 \\ | \\ CH_2COOR_3 \end{array} \quad + \quad 3\ H_20$$

glycerol — fatty acids — triglyceride — water

In the preceding formula, R is the symbol that represents one of many different hydrocarbon chains. Usually, the R groups in the fatty acids will differ from each other, although they could be the same. However, each of the R groups in the triglyceride must correspond to one of the R groups in the fatty acids from which the triglyceride was formed. A hydrocarbon chain simply means that it is made up of a number of carbon atoms bonded together, like a string of beads, with one or two hydrogen atoms linked to each of the carbon atoms. The carbon atoms on each end of the string may sometimes have three hydrogen atoms attached to them.

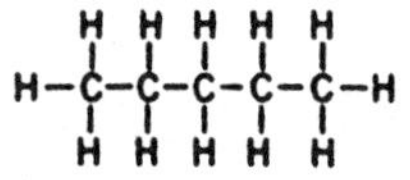

a hydrocarbon chain

Fats in nature usually contain long carbon chains with an acid (carboxyl) group at the end. The carbons in the chain are mostly saturated with hydrogen (e.g., butyric acid).

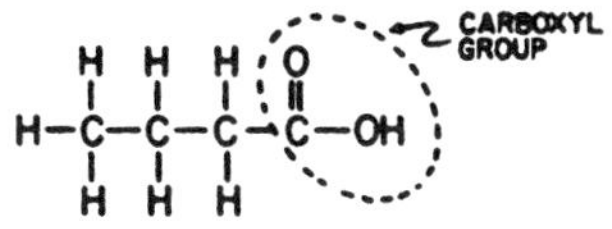

butyric acid

The formula for butyric acid is usually written $CH_3(CH_2)_2$ COOH.

Fats that are completely saturated with hydrogen atoms are called saturated fatty acids. However, some fatty acids contain one or several groups of two adjacent carbon atoms in which the carbons are not fully saturated with hydrogen (see the following formula).

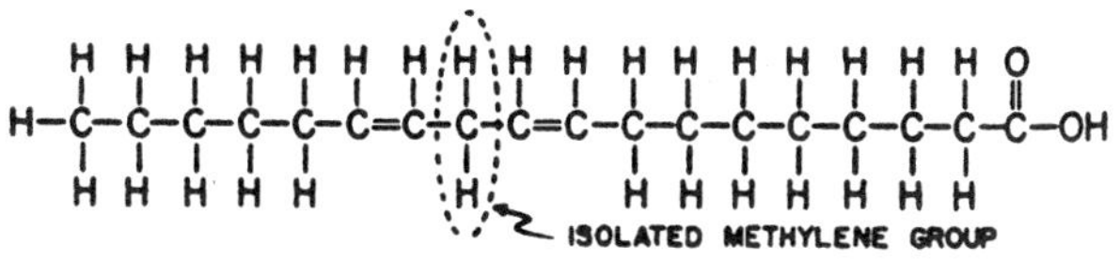

linoleic acid

As can be seen, there are two sites of incomplete saturation indicated by the double bonds (note that each of the unsaturated carbons has only one hydrogen atom attached to it). Fatty acids having one or more sites of incomplete saturation are called unsaturated fatty acids. Unsaturated fatty acids having a given number of carbon atoms have a lower melting temperature than saturated fatty acids having an equivalent number of carbon atoms. Thus, linoleic acid (unsaturated, with 18 carbon atoms) melts at 23°F (–5°C) whereas stearic acid (saturated, with 18 carbon atoms) melts at 157.3°F (69.6°C). When a lipid is made up of many unsaturated fatty acids, it will be liquid at room temperature. On the other hand, lipids, such as suet or beef fats, containing comparatively few fatty acids with unsaturated groups are solid at room temperature. Fatty acids that contain none or only one unsaturated group are not especially subject to oxidation, but those with more than one unsaturated group oxidize readily. However, oxidation proceeds much faster in fatty acids containing one or more sites in the carbon chain in which there is first a group of two carbons not fully saturated with hydrogen followed by a carbon saturated with hydrogen (an isolated methylene group) followed by another group of two carbons not fully saturated with hydrogen. In linoleic acid (see preceding chemical formula), the isolated methylene group is identified by the dotted enclosure. Notice the pair of incompletely saturated carbons on each side of the isolated methylene group. The greater the number of sequences of this particular configuration in the fatty acid molecule, the faster the rate of oxidation. Thus, when linoleic acid has an isolated methylene group (a sequence of two carbons not saturated with hydrogen, one carbon saturated with hydrogen, and two carbons not saturated with hydrogen), it is said to oxidize 10 times as fast as when it does not have an isolated methylene group.

Oxygen, or a source of it, must, of course, be present for oxidation of fats occur. However, a large amount of oxygen is not required for this reaction, and in foods, it is very difficult and commercially impractical to package them (under a vacuum or in an inert gas such as nitrogen) so that sufficient oxygen is absent to prevent changes in fats. The reason for this is that in foods oxygen dissolves to some extent in the water present and also become trapped or occluded in the tissues.

Certain sources of energy accelerate the oxidation of fats. One of these is heat, and generally, the higher the temperature at which a fat is held, the faster the rate of oxidation. Light of certain wavelengths, especially in the ultraviolet or near-ultraviolet regions, accelerates the oxidation of fats. High-energy radiation, such as cathode, beta, and gamma rays, also greatly accelerate the rate at which fats oxidize.

Metals and their compounds accelerate the oxidation of fats. Much less than one part per million to only a few parts per million of metals is required to accelerate the oxidation of fats. It is considered that the following is the order of decreasing activity as prooxidants for fats: copper, manganese, iron, nickel, zinc, and aluminum.

When fats oxidize, short-chain carbon compounds containing one or more groups in which two adjacent carbons are not saturated with hydrogen may be formed:

$$R-\underset{}{C}=\underset{H}{C}-\underset{H}{\underset{|}{C}}-\underset{H}{C}=\underset{H}{C}-R \quad \xrightarrow{-H^+} \quad R-\underset{H}{C}=\underset{H}{C}-\underset{H}{C}=\underset{H}{C}-\underset{H}{C}-R$$

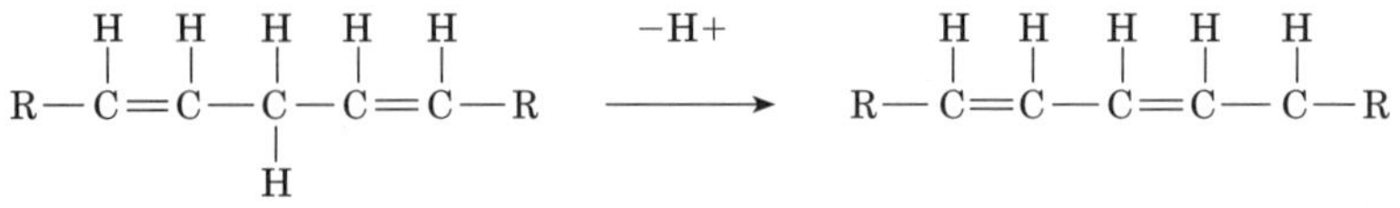

This reaction proceeds further in the presence of oxygen, yielding peroxide radicals and finally hydroperoxides.

$$R-\underset{H}{C}=\underset{H}{C}-\underset{H}{C}=\underset{H}{C}-\underset{H}{C}-R \xrightarrow{+O_2} R-\underset{H}{C}=\underset{H}{C}-\underset{H}{C}=\underset{H}{C}-\underset{OO}{\underset{|}{C}}-R \xrightarrow{+H} R-\underset{H}{C}=\underset{H}{C}-\underset{H}{C}=\underset{H}{C}-\underset{OOH}{\underset{|}{C}}-R$$

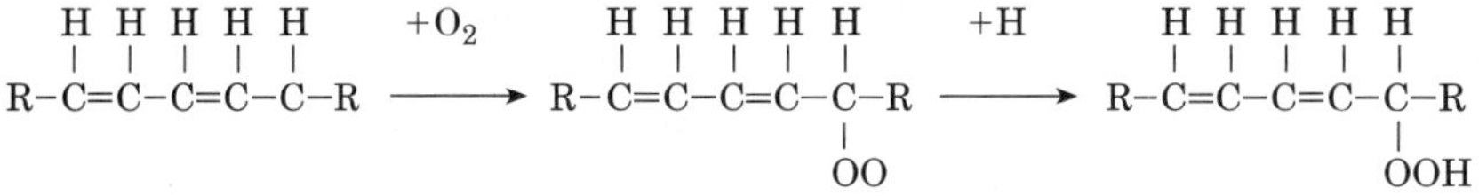

$$\text{peroxide radical} \qquad\qquad \text{hydroperoxide}$$

These compounds have odors and flavors that are generally undesirable and unacceptable. This change in fats or in the fats present in foods is called rancidity. This should be distinguished from hydrolytic rancidity, in which lipase hydrolyzes fat, splitting off short-chain fatty acids such as butyric acid. Butyric acid has an undesirable taste and odor.

Oxidative-rancidity odors are generally sharp and acrid and usually may be described as linseed-oil-like, tallowy, fishy, or perfumelike. The more solid fats, such as beef and mutton fat, are more apt to be tallowy when rancid, while pork fat, vegetable oils (soybean oil, cottonseed oil, corn oil, etc.), and fish and whale oils usually have the linseed oil flavor and odor when rancid. Under some conditions, vegetable or marine oils may be fishy or have a perfumelike odor when rancid. This appears to be a preliminary stage that eventually develops into the linseed-oil-like odor and flavor.

Oxidative rancidity may be a cause of deterioration in many forms of fish, meat, and poultry; there are instances in which oxidation of fats may be responsible for undesirable changes in shellfish and even vegetables.

When food products are dried, their fats may soon oxidize, even those that contain small amounts of fat, such as vegetables, shrimp, and some fish, because whatever fat or oil is present is now exposed to oxygen over a large area. The colors of certain vegetables (e.g., carrots), spices (e.g., red pepper and paprika), and crustaceans (e.g., shrimp) are due to chemical compounds known as carotenoids. These compounds contain many groups in which adjacent carbon atoms are not fully saturated with hydrogen. If the small amounts of fat present undergo the first stages of oxidation (peroxide formation), a fading of color may occur in these compounds, such as from deep yellow or red to light yellow or the color may be bleached out entirely. Off-flavors and odors may or may not be associated with such changes in these foods, but the color change represents a deterioration of the product.

NONENZYMATIC BROWNING

There are several types of nonenzymatic browning. One type is caused by a chemical reaction known as the Maillard reaction. This chemical change is initiated by a combination of an amino acid and a sugar. The amino acid may be present in the food as a separate entity (free amino acid) or it may be an amino acid present in the food as part of a protein. The sugar must be of a reactive type, containing a reactive portion known as a carbonyl group.

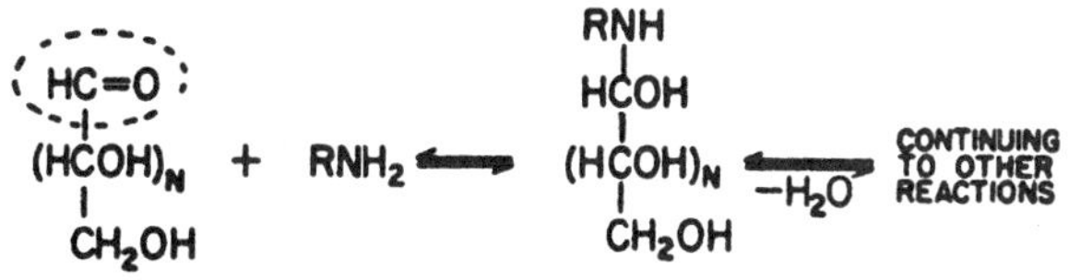

sugar + amino acid addition compound

The carbonyl group in the sugar is identified by the broken-line circle. A carbonyl group also occurs in other chemical compounds, the aldehydes and ketones. Once the reaction is initiated, it proceeds through a long series of chemical changes resulting in complex compounds that are flavorful and brown or black in color, causing changes in the flavor and color of the food.

When sugars are heated at high temperatures, they turn brown and then black. This reaction involves the dehydration or removal of water from the sugar, resulting, through a series of reactions, in the formation of furfurals, ring compounds of four carbons. These ring compounds further combine to form complex chemical compounds that are brown or black in color and that have an odor and flavor entirely different from that of the sugar. This is called caramelization, although the flavor produced may be different from that which we know as caramel candy where the components of milk are involved in the typical flavor of the confectionery. Caramelization proceeds very fast at high temperatures such as are attained when sugars are heated directly. However, at lower temperatures, such as those encountered in the normal handling of many foods (usually above 50°F [10°C]), when the right conditions are present, caramelization will proceed fast enough to cause a deterioration of foods.

A third type of browning may be caused by the oxidation of ascorbic acid (vitamin C). Once oxidized, the same type of compounds may be formed as in the case of caramelization of sugars.

Browning may take place in foods of all moisture contents. Hence, in canning, at least part of the difference in flavor between that of the fresh food and that of the canned product is due to reactions that are manifest by browning, and some types of food cannot be canned successfully because of browning (e.g., scallops and cauliflower). However, browning proceeds fastest in foods of low to medium moisture content. Thus, some dried foods are especially subject to changes caused by nonenzymatic browning.

As has been pointed out, the color difference, and in some cases, flavor differences, between fresh and heat-processed foods is probably due to browning reactions. The browning reactions can be minimized if canned foods are given a high-temperature–short-time process (HTST process) as compared with regular processing temperatures and times. However, browning reactions may proceed in HTST-processed foods during storage if they are held at temperatures above 50°F (10°C). Apparently, browning occurs slowly, if at all, when the moisture content is less than 2%. Only by freeze-drying can the moisture in foods be reduced to this low level. Therefore, these products are freeze-dried to a low moisture content and packaged so that moisture is not reabsorbed from the atmosphere. For this reason, fruit juices are generally not dried except by freeze-drying. The moisture content must be kept to a low level throughout storage if the freeze-dried product is to be stable. Dried egg products, whites, and whole egg mixtures (usually spray-dried to a moisture content of about 5%) are quite subject to nonenzymatic browning reactions, resulting in a low acceptance of these products. For some time now, methods have been available to remove the sugar (glucose) from these products prior to drying, resulting in much improved products because adverse nonenzymatic reactions are prevented.

Sugar can be removed from food by subjecting it to fermentation by bacteria or yeast before drying. The microorganisms consume the sugar during the fermentation process, thereby eliminating one of the necessary components required in the Maillard reaction. Another process for preventing the Maillard reaction involves the addition of two enzymes (glucose oxidase and catalase). The oxidase converts glucose to gluconic acid, which does not combine with amino groups, but one of the products of the conversion is hydrogen peroxide, an undesirable compound, and the role of the catalase is to break down the peroxide.

Although many processed foods may be adversely affected by nonenzymatic browning, there are other foods dependent on this type of reaction for their typical odor, flavor, and color. Maple syrup develops its typical color and flavor because the Maillard reaction takes place between sugar and an amino acid normally present in the sap of the maple tree. The typical color and taste of prunes is due to nonenzymatic browning that occurs when prune plums are dried and stored. The flavor and color of coffee are due to nonenzymatic browning that occurs in the components of the coffee beans when they are roasted (comparatively high temperatures are used for roasting). In the cooking of many foods, browning flavors and odors are produced. Roasted, broiled, or fried meats, or broiled or fried fish are examples of this type of change. The flavors and colors produced in the crusts of breads during baking are another example of the importance of the Maillard reaction resulting in desirable changes.

THE STRECKER DEGRADATION

The Strecker degradation is a reaction that takes place between an amino acid and certain sugar fragments or compounds produced by bacteria that contain reactive groups known as dicarbonyls, as follows:

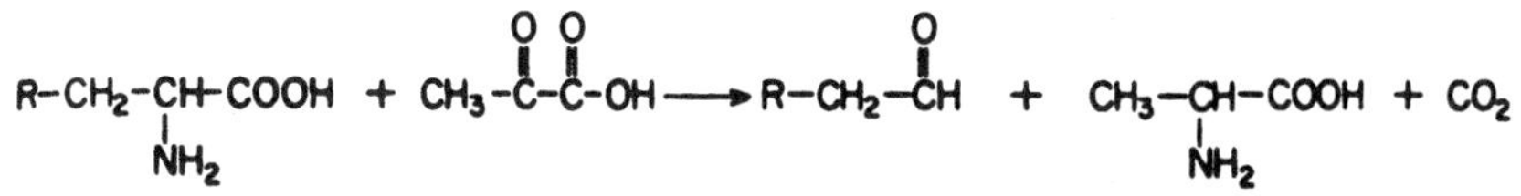

$$\text{R-CH}_2\text{-CH(NH}_2\text{)-COOH} + \text{CH}_3\text{-C(O)-C(O)-OH} \longrightarrow \text{R-CH}_2\text{-CH(O)} + \text{CH}_3\text{-CH(NH}_2\text{)-COOH} + \text{CO}_2$$

amino acid + pyruvic acid → aldehyde + amino acid
(a dicarbonyl) (different from the original amino acid)

The reaction results in the formation of an aldehyde, a different amino acid, and the splitting off of carbon dioxide. The new compound containing the aldehyde group is flavorful, which may or may not be desirable.

The Strecker degradation takes place when milk is heated to high temperatures for long periods or during storage after milk is heated to high temperatures for short periods and then stored at room temperature. This is probably the reason that fluid milk cannot be canned or heat-processed by ordinary methods. If fluid milk is given a high-temperature–short-time type of heating processing (HTST), it will not develop the off-flavor, provided the canned product is stored at temperatures below 50°F (10°C). Canned fluid milk, given a HTST treatment and held at room temperature, soon develops a caramelized flavor. In the manufacture of caramel confectioneries, condensed milk, cream, or some combination of the two, together with corn sugar, cane sugar, or both, is heated to high temperatures. The caramel flavor that develops is probably due to the Strecker degradation. This type of change in foods, therefore, is sometimes desirable and sometimes not desirable.

THE AGGREGATION OF PROTEINS

The aggregation is the bonding of protein chains to form a more closely knit network of proteins. This causes deterioration of some foods and it would appear that in this process some of the water loosely held by the protein is squeezed out, causing drip loss from frozen food on defrosting. Protein aggregation occurs mainly in low-fat fish, such as cod, during frozen storage and is correlated with the liberation of free fatty acids from phospholipids by the phospholipases in the muscle. Some studies indicate that the protein aggregation and toughening that occur in some fish during frozen storage are due to enzymes that cause a breakdown of trimethyl amine oxide (present in the flesh of some fish) to dimethyl amine and formaldehyde.

$$(\text{CH}_3)_3 \equiv \overset{+}{\text{N}}\text{—}\overset{-}{\text{O}} \xrightarrow{\text{enzyme}} (\text{CH}_3)_2=\text{NH} + \text{H-CH(=O)}$$

trimethylamine oxide   dimethyl amine   formaldehyde

It is well known that formaldehyde causes an aggregation or denaturation of proteins.

It has been theorized that in the presence of the free fatty acids, there is a reaction between free fatty acids and proteins to form a crosslinked network within the muscle, resulting in a close association of protein fibers.

Another theory of what happens during protein aggregation is that the splitting off of fatty acids destabilizes the protein molecules, causing them to form a closely bonded mass, as in normal muscle, the phospholipid itself, from which the fatty acids are split off, is bonded or conjugated with the protein.

Protein aggregation does not take place in high-fat fish, such as salmon. It is theorized that the free fatty acids split off by phospholipases dissolve in the fat present in the tissue of high-fat fish, and as a result of dilution, become unavailable for bonding with the protein. There is also the possibility that the proteins in high-fat species are different from those in species containing little fat, hence are not destabilized by a splitting of fatty acids from phospholipids. Another possible explanation is the absence of the enzyme that splits trimethyl amine oxide, which is believed to cause toughening in the lean species of fish in which this enzyme has been found.

When protein aggregation occurs in frozen products, the tissues become tough and dry and lose succulence when cooked. This change is temperature-dependent and does not occur to any extent in marine products held at temperatures below –22°F (–30°C). At –40°F (–40°C), cod held for 1 year may not be distinguished from the fresh product. The reason why the protein aggregation change does not occur at very low temperatures may be that water is not available to provide for the hydrolysis or splitting off of free fatty acids from phospholipids as brought about by the phospholipases, or that enzymes that decompose trimethyl amine oxide are not active.

Part II

Food Processing Methods

10

Thermal Processing

The development of the modern heating process started in France during the first decade of the 1800s by Nicholas Appert, who preserved foods in sealed glass jars in boiling water. In 1810, Peter Durand of England developed the metal can, which was fabricated and sealed by hand soldering. In 1819, William Underwood of the United States started the first canning factory in Baltimore. But preserving foods in boiling water took too long, requiring about 6 hours, so salt was added to the water bath, which increased the boiling temperature, thereby shortening the processing time. However, the salt corroded the cans, so the next innovation was to heat in steam under pressure. The higher the pressure, the higher the temperature at which water boils (see Table 10.1) and the shorter the processing time. These early pressure chambers evolved into the modern retort.

Table 10.1. Boiling Points of Water at 0–20 PSIG (0–1408 g/²CM) (at Sea Level)

Pressure (at Sea Level)		Temperature at Which Water Boils	
(psig)	(g/cm²)	(°F)	(°C)
0	0	212.0	100.0
1	70.4	215.4	101.9
2	140.8	218.5	103.6
3	211.2	221.5	105.3
4	281.6	224.4	106.9
5	352.0	227.1	108.4
6	422.4	229.6	109.8
7	492.8	232.3	111.2
8	563.2	234.7	112.6
9	633.6	237.0	113.8
10	704.0	239.4	115.2
11	774.4	241.5	116.4
12	844.8	243.7	117.6
13	915.2	245.8	118.8
14	985.6	247.8	119.9
15	1056.0	249.8	121.0
16	1126.4	251.6	122.0
17	1196.8	253.4	123.0
18	1267.2	255.4	124.1
19	1337.6	257.0	125.0
20	1408.0	258.8	126.0

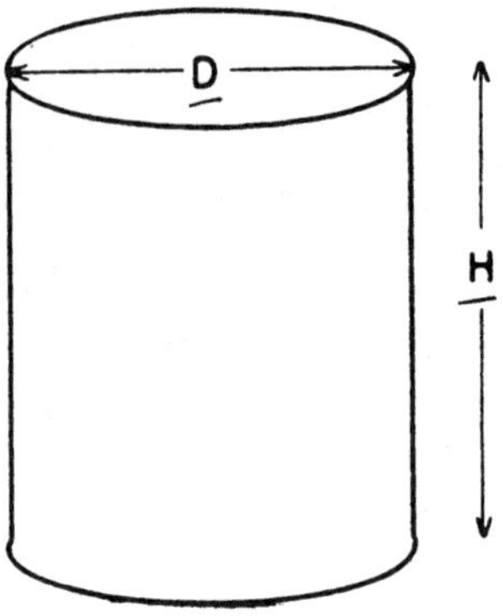

Figure 10.1. Diameter (*D*) and height (*H*) of a metal container.

By the early 1900s the manufacturing and sealing of cans were done by machines. The lid used for the cans contained a rim to which a plastic gasket could be added, and the rim could be sealed tightly by machine by first crimping it over and under the flanged top of the can body and then pressing the two together by a second roller operation. The plastic gasket made the can airtight by filling the tiny voids produced between the rim of the can cover and the flange of the can as they were brought together. This is essentially the method used today for hermetically (airtight) sealed metal containers. Normally, cans are described in terms of the dimensions of the diameter (*D*) and the height (*H*) (see Fig. 10.1) and each of the dimensions is given in numbers having three digits each. The first digit is in inches (1 in. = 2.54 cm), and the next two digits are in $\frac{1}{16}$ of an inch (0.16 cm). Thus, when a can is described as a 202 × 214 can, this means that its diameter is $2\frac{2}{16}$ in. (5.40 cm) and its height is $2\frac{14}{16}$ in. (7.30 cm). Not all cans are round, however, and, as in the case of sardine cans, three dimensions are given: the length, the width, and the height.

Various lacquers have been developed for lining cans, especially to prevent food discoloration that could occur as a result of interactions between the food and the can. Other innovations include quick-opening cans with pull tabs, as are used for soft drinks, sardines, and nuts; cans that are opened with a slotted key that comes with the can, as are used for sardines and cured hams; pressurized cans that use a propellant to dispense whipped cream and cheese spreads; and extruded or drawn cans, as are used for beverages, tuna, and sardines. The extruded can has several advantages over the conventional can, a major one being the elimination of the bottom and side seams, which reduces the probability of seam failures, eliminates the use of lead (used in the side seam), and permits more stable stacking on shelves, requiring somewhat less vertical space.

PRETREATMENT OF FOODS

It is generally necessary for foods to undergo a treatment prior to the canning process, but these pretreatments differ depending on the foods, and no attempt is made to cover all of them here. Some pretreatments are applied to many different foods. One of these, usually applied to vegetables, is blanching. Vegetables are first washed, usually in water and detergent, then rinsed. They are then passed along on belts, where any

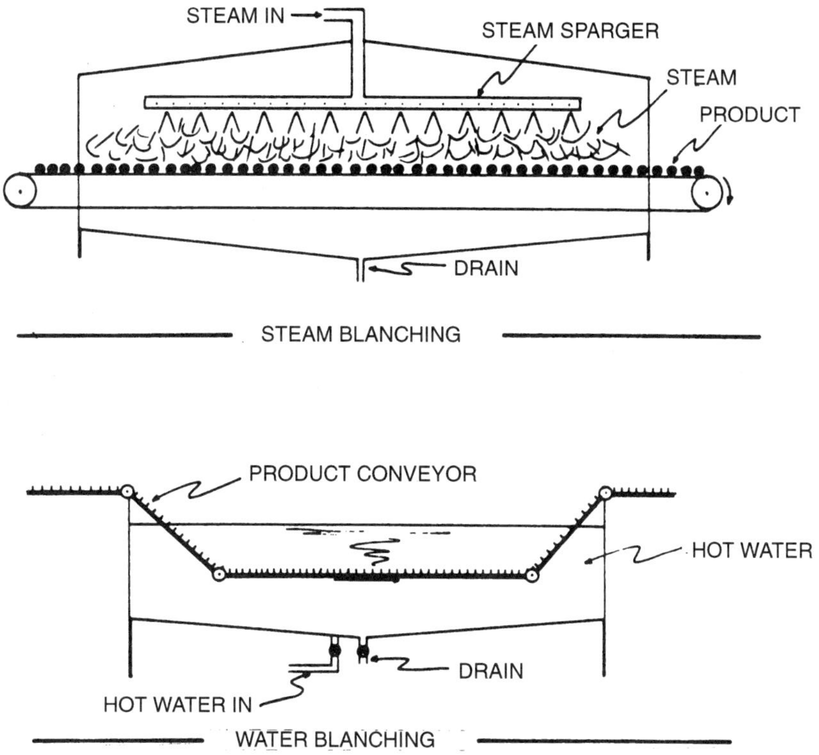

Figure 10.2. Continuous blanching.

remaining foreign matter, such as weeds or stalks, can be removed by hand. Blanching consists of heating in steam (no pressure) or hot water (usually about 210°F [98.9°C]) until the temperature of the food is brought up to about 180 to 190°F (82.2 to 87.8°C) in all parts (see Fig. 10.2), then cooling in water.

Blanching shrinks the product, providing for a better fill of the container, and it removes gases, allowing better vacuum after sealing. The blanching process also destroys enzymes in the food that otherwise might react during the initial heating in the retort and cause discoloration of off-flavors in the product. When very-high-temperature–very-short-time methods of heat processing are used, enzymes might not be inactivated, which could cause development of off-flavors in the food. However, this can be prevented by blanching. Finally, blanching tends to fix the natural color of vegetables and it provides a clearer brine in the canned products.

VACUUM IN CANS

The Need for Vacuum

Foods are packed under vacuum for several reasons. If canned foods were not under vacuum, the cans would swell should they be stored at higher temperatures or lower pressures than those at which they were packed. Thus, cans packed at sea level at ordinary temperatures would swell due to expansion of gas within the can if the cans were shipped to Denver, Colorado, which because of its high altitude (about 1 mi [1.61 km] above sea level), is under reduced pressure, or if the cans were shipped to tropical areas or held in unusually hot places. When cans are in a swelled state, they are normally suspected of containing food that has spoiled, because when bacteria grow, many of them produce gases. Thus a perfectly good canned food might be discarded because it could be suspected of being spoiled. Another reason for canning foods under vacuum is to remove oxygen. During heat processing, oxygen reacts with the food, causing undesirable changes in color and flavor of the product, as described in Chapter 9.

Finally, if at least some of the air is not removed from the container prior to sealing, then the product may have to be cooled under pressure (usually air pressure). Otherwise, the can will buckle (a permanent distortion along the cover seam), and such cans will be discarded. The reason for the buckling can be explained. During the heating process, the retort is under pressure, for example, 20 psi (1.41 kg/cm^2). As the temperature inside the can reaches the high temperature in the retort, moisture in the food or brine vaporizes, exerting pressure against the inside of the can. In addition, the residual air in the can tends to expand as the internal temperature increases, and the expanding air also exerts a pressure against the inside of the can. By the end of the process, the can is under equilibrated pressures from both the inside and the outside (see Fig. 10.3). In this situation, the can is not unduly stressed, because the internal pressure is counterbalanced by the outside pressure and vice versa.

At the end of the process, the external pressure, relative to the can, may be immediately released by shutting off the steam supply to the retort and opening the retort valve to the outside, allowing all its internal steam to escape. However, the internal pressure in the can cannot be released immediately because the can cannot be vented to the outside. Thus, the wall and lids of the can are under internal pressure, causing distortion, possibly even damage, to the seals (formed by the lids and the can body). This problem can be eliminated by replacing the hot steam in the retort with air under pressure. In this way, the heat source is removed without altering the pressure equilibrium. With the removal of the heat source, the contents of the can will cool down, condensing water vapor and cooling the residual air, returning it to its original

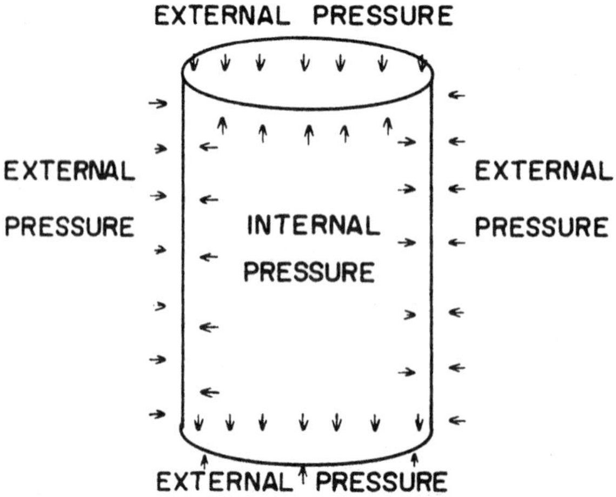

Figure 10.3. Pressure equilibrium during thermal processing.

volume, thus reducing the internal pressure until it no longer pushes against the inside of the can. Removing as much air as possible from the container prior to processing tends to minimize buckling. However, even with vacuum packing, cans of large diameter must be cooled under pressure to prevent distortion of the container.

Some foods, such as whole kernel corn, may be packed under a very high vacuum in order to provide for good heat penetration, when only a small amount of liquid is used with the foods. When this is done, a special beaded can with protruding ridges around the central part of the can body to strengthen the can may be needed. Otherwise, the container body may panel (become flattened) as the result of the difference in pressure between the inside of the can and atmospheric pressure on the outside.

Obtaining the Vacuum

Vacuum in canned foods may be obtained in several ways. One of the most common methods is to add hot food to the container. In this way, the residual air is removed, resulting in a partial vacuum. A second method is to add cold food to the container and to preheat it by passing it through a steam box uncovered, or only partially sealed, prior to sealing. In either of these techniques, the heat causes the product and the air in the headspace to expand, pushing air out of the container. In addition, the water vapor in the headspace displaces air, and trapped air in the food is driven out. In this condition, the can is sealed, and when the product is cooled it will be under vacuum, as much of the air has been removed from the container.

A vacuum in canned foods can also be obtained by subjecting the container to a mechanical vacuum just prior to sealing it in the chamber. A good vacuum can be obtained in this manner, but there are some limitations. For instance, if the product is packed in liquid, such as vegetables packed in canner's brine or fruit packed in syrup, when the vacuum is applied, much of the liquid may be flashed out. This is caused by the dissolved and occluded (trapped bubbles) air in the liquid, which comes out as a gas when a sudden vacuum of high intensity is applied. The sudden release of air causes some of the liquid to spatter out of the can. To avoid this, liquids must be subjected to vacuum treatment prior to filling into the can in order to remove dissolved and occluded air.

The third type of vacuum used in canning foods is called a steam jet vacuum. Just before the cover is placed on the can to be sealed, a jet of steam is forced over the contents of the can. This does not provide a high vacuum and removes only air from the headspace of the food in the can. It is used mainly for materials packed without liquid.

LIQUID IN CANS

Vegetables and fruits are usually packed in liquid. Canner's brine, a weak solution of sugar and salt, is ordinarily used for vegetables, and sugar solutions that may be as concentrated as 55% sugar or as dilute as 25% sugar are used for fruits. These liquids afford some protection against heat damage because they permit convection heating, which occurs at a faster rate than does conduction heating. With solid packs, such as tuna fish, corned beef hash, and even concentrated soups, such as pea or mushroom, heating takes place in the container through conduction. In convection heating which occurs when liquids are present, a hot layer of liquid rises along the

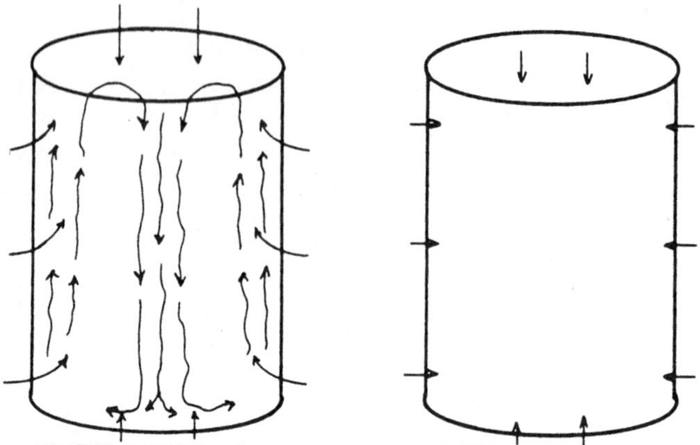

Figure 10.4. Convection and conduction heating patterns in cans of foods.

sides of the can body, travels over the top to the center, and flows down the central axis as more hot liquid moves up the sides. This mixing serves to speed up the transfer of heat. Thus, in Figure 10.4 the heat transfer pattern in convection can be seen to more widespread.

In the conduction heating of solid packs, heat penetrates from all sides of the container, but because solid foods are poor conductors of heat, heat travels toward the center of the container only relatively slowly.

In convection heating, the slowest heating point in the container is along the central axis ¾ to 1½ in. (1.91 to 3.81 cm) from the bottom (can in the upright position). In conduction heating, the slowest heating point in the container is along the central axis at the geometric center of the container.

FILLING THE CANS

Food must, of course, be filled into cans before they are sealed, and this is usually done by machine. For instance, with peas, a central hopper, which is kept filled with peas, is located above a rotating plate with openings through which the peas fall. When the plate rotates to the point at which the plate opening is just below the hopper, the opening is just above a container (below the plate), which when filled will hold a certain volume or weight of peas. The peas thus fall down and fill the bottom container, which then rotates to a point where the peas are released into a can. In the case of vegetables, after the food is added to the container, canner's brine is added automatically to fill the can, filling all voids and covering the vegetables. This liquid is usually added hot, and because such foods are not packed under vacuum, the hot liquid provides whatever vacuum will be present in the container after processing and cooling. Canner's brine is held in a reservoir on the filling machine and is heated in this reservoir. As explained later, the temperature of the brine should be held at 170°F (76.7°C) or higher.

Semiliquid packs that heat by conduction, such as concentrated pea soup or mushroom soup, may be filled hot and can be filled volumetrically by automatic means.

Solid packs such as tuna fish and corned beef may be filled by hand. A small amount of hot liquid (brine or oil) may be added to provide some vacuum, or the product may be passed through a steam (exhaust) box to be heated and provided with some vacuum before the can is covered and sealed.

SEALING THE CANS

Cans are sealed automatically by machine. In sealing, the cover falls onto the top of the can automatically; the base plate of the sealer (upon which the can rests) raises the can with cover up tightly against the chuck. The edges of the can cover and the flanged body top are subjected to the action of two different rotating rollers. The first roller crimps the cover and body flange so that the edge of the cover is bent around and under the edge of the body flange; the second roller flattens and presses the top seam together so that it forms a tight seal with the help of the plastic gasket located in the outer rim of the cover (see Fig. 10.5).

In the canning operation, it is imperative that the sealing machines provide a tight

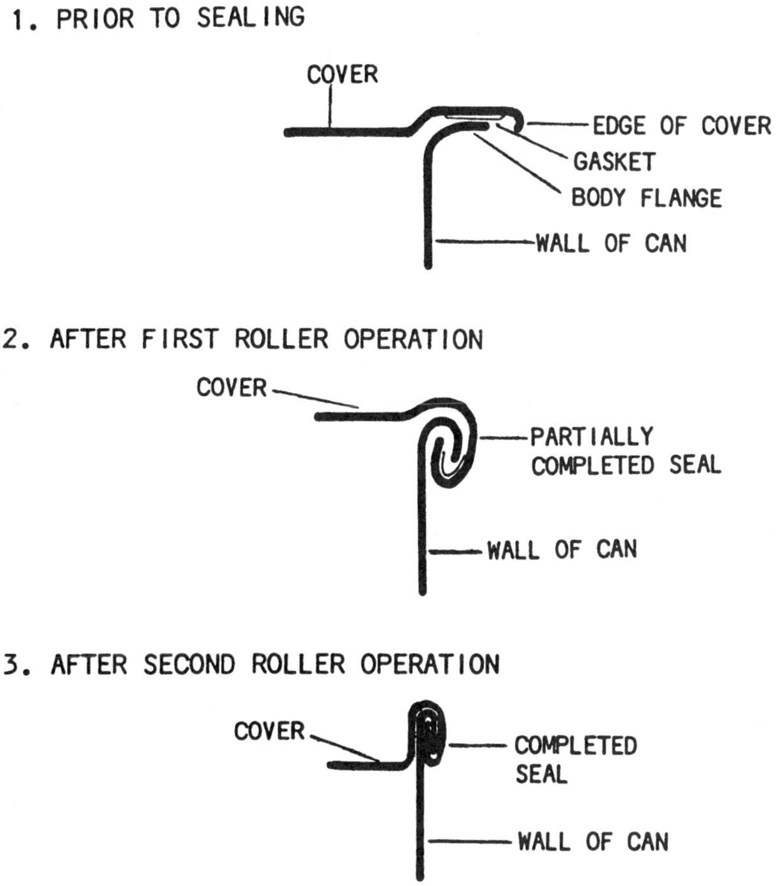

Figure 10.5. Cross-sections of cover and body of can at three stages of sealing.

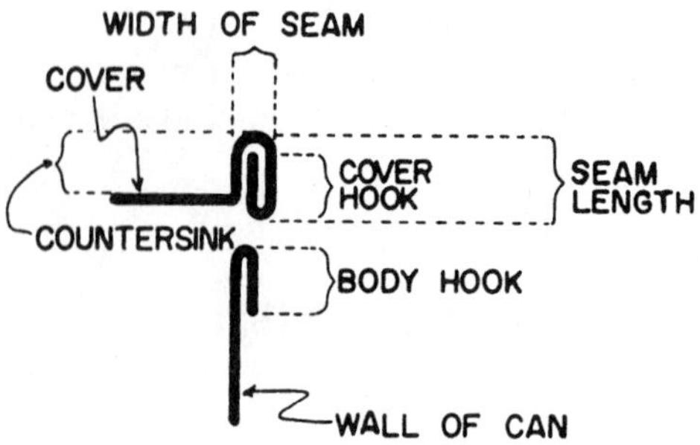

Figure 10.6. Components of can seam in exploded view.

seal. This can be checked by removing the cover to observe the configurations of the cover hook and body hook (see Fig. 10.6). The dimensions of the seal components may then be measured. It is known that such dimensions should fall within certain limits. Tools for exposing the seam components and special micrometers for measuring their dimensions are available. A simple way to determine that the base plate, first operation (first roller), and second operation (second roller) of the sealing machine have been adjusted properly is to fill a can with boiling water, seal, cool, and then measure the vacuum in the can by means of a gauge that reads in inches of vacuum (1 in. = 2.54 cm). The vacuum gauge has a sharp shaft that is pushed through the cover of the can, and a rubber gasket prevents air from entering the can. If a vacuum of 10 in. (25.4 cm) or more is obtained, it can be assumed that the can is effectively sealed.

The sealing operation should be checked at the start of the day's operation and periodically throughout the processing period. The importance of an effective seal cannot be overemphasized owing to the unfavorable economic impact that will certainly result from FDA seizure of product and product recalls from the market or from lawsuits if food poisonings should occur.

THE HEAT PROCESS

The next operation in food canning is heat processing. Heating times and temperatures for canned foods are based on the destruction of the spores of *Clostridium botulinum*, as this organism has been known to survive an inadequate process and thereafter grow out, producing toxins that can cause sickness and, in some instances, death. To obviate this possibility, a system was set up to ensure the destruction of any spores of *Cl. botulinum* that might be present in the food.

The first step in developing the new process was to find the most resistant spores of *Cl. botulinum*, which turned out to be those of certain type A strains. (There are at least seven known types of *Cl. botulinum*, each having a number of strains. More about this organism is found in Chapter 3.) The reason that the minimum process was necessarily based on *Cl. botulinum* is that the toxin produced by the various strains of this particular organism is the most powerful toxin known. In one experiment performed by Nickerson and Ronsivalli, the toxin produced by one strain of type B was so powerful that when a bite-size piece of beef containing it was diluted about 10^8 (100,000,000) times, all the mice into which it was injected died. It is considered from these results that if a human had eaten a small amount of that meat, he would surely have died.

The *D*-Value

To be sure that the spores of *Cl. botulinum* are destroyed, a more heat-resistant, nonpathogenic organism is used in laboratory experiments done to determine a safe heat process. An organism known as Putrefactive Anaerobe 3679 (PA 3679) or *Bacillus stearothermophilus* (FS 1518) is usually chosen.

Bacteria are killed by heat at a rate that is referred to as a logarithmic order of death. This means that at a constant temperature, the same percentage of bacteria will be killed in a certain time interval. The number of bacteria in the solution will not affect this time–temperature–destruction relationship. The value that describes this is known as the *D*-value and it is defined as the time needed at a given temperature to destroy 90% of a microbial population. For example, if the *D*-value for a particular bacteria is 2.5 minutes at 250°F (121°C) and there were 1,000,000 (10^6) bacteria in the initial population, after 2.5 minutes, 100,000 (10^5) would remain (a reduction of 900,000 or 90%). After the next 2.5 minutes, 10,000 (10^4) would remain. Each 90% reduction can be described as a reduction of one "log cycle." The time of heating at a constant temperature is determined by how many of these "log cycles" one wishes to reduce the population of bacteria. The generally accepted heating time to ensure safety in canned foods is that equal to the time needed to decrease the bacterial population by 12 *D*-values. Because even spoiled foods rarely support more than a billion bacteria per can, a heat process which is equal to the reduction of 12 *D*-values should destroy all pathogens and commercially sterilize the food. This would reduce the number from 1,000,000,000,000 to 1!

To calculate the *D*-value for a particular organism, the number of living organisms is determined, the temperatures is chosen, and the time the food is heated at that temperature is recorded. Samples of the food are then removed after chosen time intervals and a graph is constructed (see Fig. 10.7).

To calculate the processing time for a particular food, more than just the *D*-value is needed. Depending on the food being processed, and the size, shape, and material the container is made from, different times will be needed to raise the temperature of the food to that used in finding the *D*-value.

Estimating Processing Times

Once the thermal death rate curve (*D*-value) at a constant temperature has been determined for a specific organism in a given food, the length of time it takes to reach that temperature at the coolest point in the can must be found. As mentioned, depending on the type of food being processed, the heating pattern will differ. In foods that are heated through convection, the "cold point" location will differ from those heated through conduction. Figure 10.8 shows different locations of the cold point that can occur in foods of different consistencies and moisture levels.

To be sure a can is processed a sufficient time at a specific temperature to ensure the reduction of 10^{12} heat-resistant *Cl. botulinum* spores to 1 (a reduction of 12 *D*-values), the time it takes to heat the cold point to the tested temperature must be determined and then added to a time that is equivalent to 12 times the *D*-value. For example, if it takes 25 minutes to heat the "cold point" of a canned food (that has been filled at a constant temperature) to 250°F (121°C) and 2.78 minutes to reduce the numbers of *Cl. botulinum* spores in this food one *D*-value, the food must be processed for a total

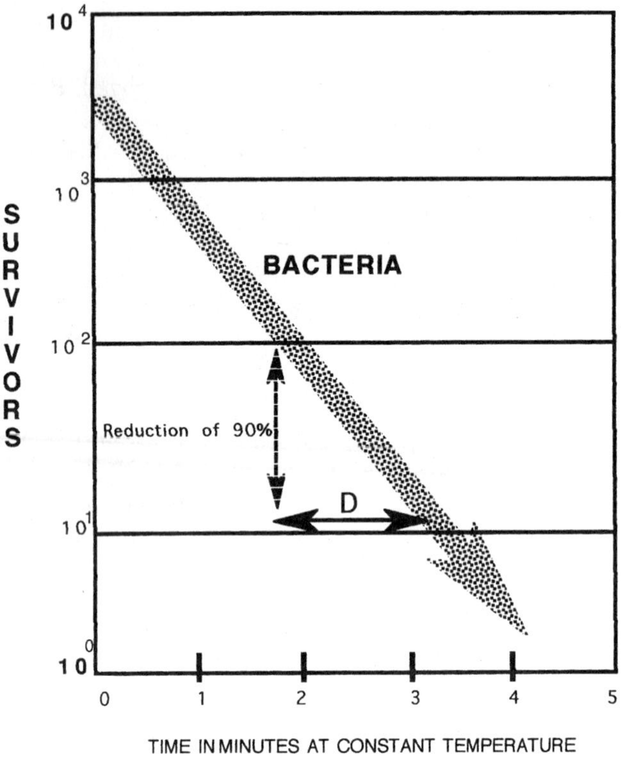

Figure 10.7. *D*-value of 1.25 minutes.

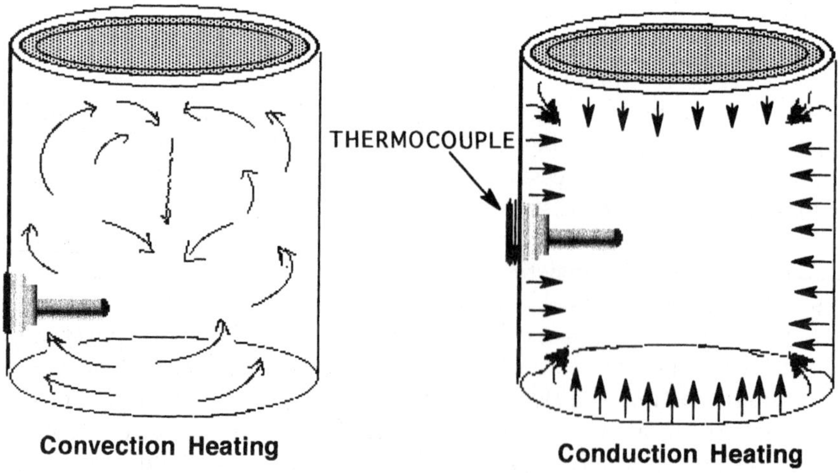

Figure 10.8. Cold Point in cans with convection and conduction heating.

of: 25 + (2.78 × 12) or 58.36 minutes at 250°F (121°C). This will ensure the reduction of 12 *D*-values and essentially sterilize the food.

In the thermal process, the degree of lethality at a given temperature can be reached at different time–temperature exposures. The times needed to raise the temperature to the processing level also accomplish some microbial destruction. This is commonly accounted for in commercial processes by decreasing the time at the processing temperature accordingly. After the process time and temperatures have been achieved, more microbial destruction takes place in the cooling period.

The term "unit of lethality" is important in understanding how a process is calculated. A unit of lethality is defined as the heat kill equivalent to 1 minute at 249.8°F (121°C) against an organism of a given *z*-value (see definition later in the chapter). Once the time and temperature needed at 249.8°F (121°C) for safe processing are established, the different fractions of a unit of lethality or times at different temperatures that correspond to killing times equal to part of a unit of lethality are determined. These fractions are "lethality rates." To calculate these lethality rates and incorporate them into the final heat process, the industry has developed mathematical formulas that determine the safe processing times that will not over process the food but yield a high-quality, *wholesome* product. These formulas and others can be found in more advanced texts on thermal processing of foods.

Processing times can be found for other processing temperatures as long as the heating time will reduce the number of *Cl. botulinum* spores 12 *D*-values for that particular temperature. Table 10.2 shows a comparison of different *D*-values for *Cl. botulinum* at different temperatures.

Table 10.2. *D*-Values at Different Temperatures for *Cl. botulinum*

2.78 min at 250°F (121°C)	10 min at 240.8°F (116°C)
0.78 min at 260.6°F (127°C)	36 min at 230°F (110°C)
1.45 min at 255.2°F (124°C)	150 min at 219.2°F (104°C)
5.27 min at 244.4°F (118°C)	330 min at 212°F (100°C)

The *z*- And F_0-Values

As mentioned, the *D*-value is a thermal death *rate* value. Thermal death *time* values can also be constructed for bacteria. To do this, a graph is constructed from different *D*-values at different temperatures. The *z*-value is a thermal death time value that measures the number of degrees required for a specific death time curve to pass through one log cycle of time. The *z*-value for Figure 10.9 is 18°F. The F_0-value is a reference that equals the number of minutes at 250 F° (121°C) required to destroy a specific number of organisms whose *z*-value is 18°F (10°C). This value is known as the "sterilization value" of heat treatment.

D-values for other organisms such as *Listeria* and *Salmonella* are also of significant importance to the food scientists. These organisms can be very dangerous in foods such as dairy products and liquid eggs that are heat processed at pasteurization temperatures that are much lower than those used for low-acid canned foods. Similar calculations finding *D*- and *z*-values for these organisms at pasteurizing temperatures are found and process controls are used to ensure adequate lethality rates are used here as with canned products.

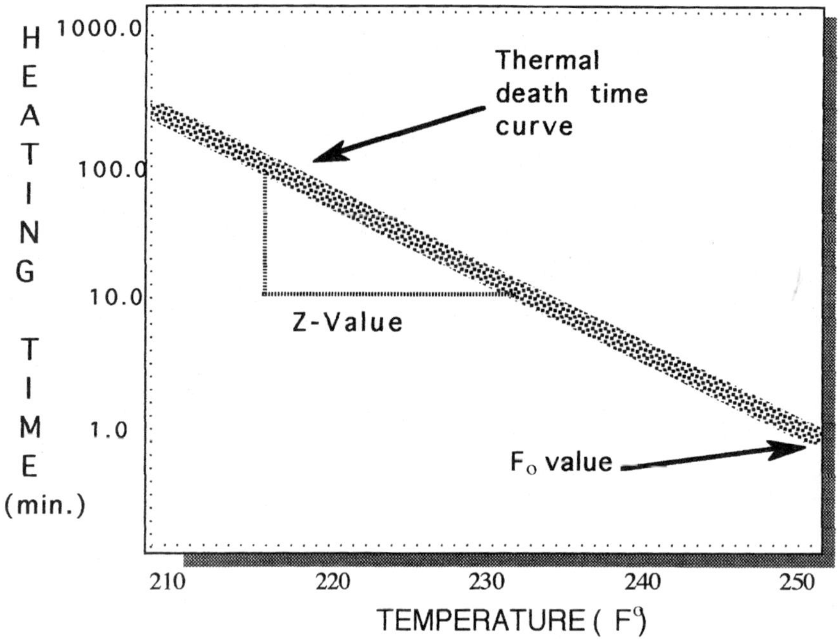

Figure 10.9. *z*-Value of 18°F and F_0-value.

THE CONVENTIONAL HEAT-PROCESSING CHAMBER (RETORT)

The ordinary retort (see Fig. 10.10) is the three-crate retort in which one crate of cans is placed on top of another in the retort. When cans are placed in these crates, they may be allowed to fall into the crate haphazardly or they may be placed in the crate on end in an orderly manner. If the latter system is used, perforated metal separators must be placed between the layers of cans. This provides for an adequate circulation of steam around the cans necessary for optimum heat transfer to the product.

All retorts should be fitted with automatic steam valves that can be set to allow steam to flow into them to raise the processing temperatures only to the desired point. Retorts should also have a temperature sensor (inside of the retort) attached to a recorder on the outside and a chart showing temperature and time of processing for each batch of product. Processing charts should be identified with the code on the cans of each batch and should be kept on file. If the product is to be cooled in the retort after processing, an air pressure system with automatic inlet valve that can be set for a definite air pressure should be a part of the installation. In such installations, there should be an automatic pressure outlet valve attached. Otherwise, when cans with high vacuum are cooled, the outside pressure may become high enough to cause paneling of the cans.

All retorts must be fitted with fast-opening valves to allow venting. When the steam is first turned on, the vent valve should be opened wide and left open until the temperature is raised to 220°F (104°C). This removes, from the retort, air that would otherwise cause cold spots around some of the cans and thus prevent adequate heating. The

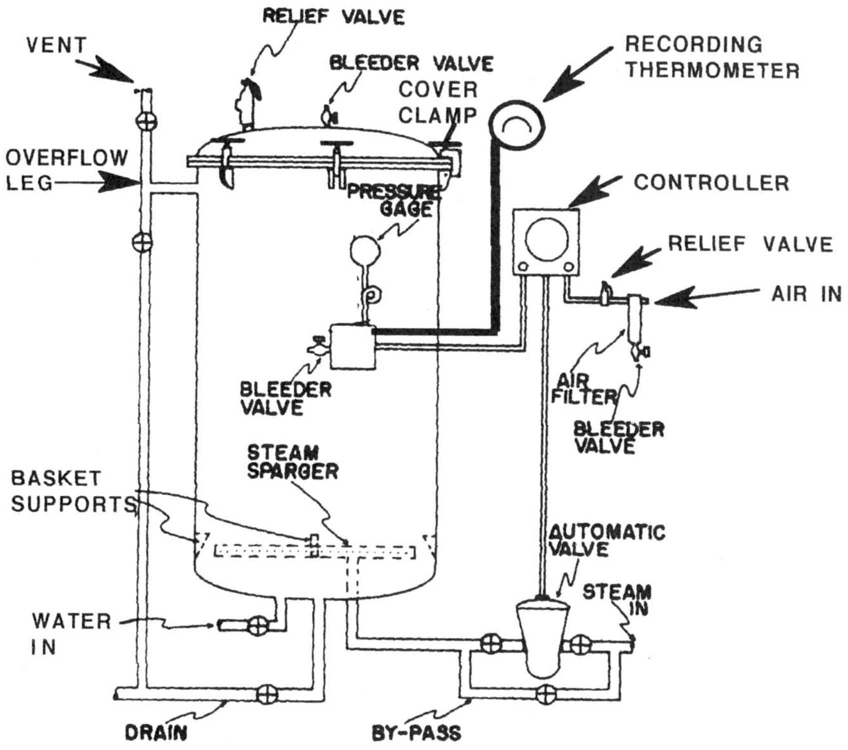

Figure 10.10. Conventional retort with recording thermometer.

reason for this is that air is a poor heat conductor. Also, during retorting, a small bleeder petcock valve at the top of the retort should be kept in the wide open position. This removes air that may come in with the steam and also provides for good circulation of the steam.

COOLING HEAT-PROCESSED FOODS

Cans of food that have been heat processed may be cooled in the retort by allowing water to flow in after the steam has been turned off, in which case they may be cooled under a pressure of air or steam exerted over the water level in the retort. In other operations, the retort may be blown down and the crates of cans removed and moved slowly through a cooling canal. In either case, the cooling water should be potable (drinkable) and should not have a high bacterial count. The reason for this is that as the vacuum is forming in the can as a result of the cooling, the gasket in some cans may be soft enough to permit microscopic amounts of the cooling water to be sucked into the cans. If the cooling water is high in bacterial count, enough bacteria, which may cause spoilage or even disease, may thus enter the can and contaminate the product. Thus, it is desirable to chlorinate cooling water so that it contains a residual of 5 ppm available chlorine. This level of chlorine is enough to keep the water relatively free from bacteria.

During cooling, the average temperature of the product in the can should be brought to 95 to 100°F (35 to 43.3°C) as quickly as possible. The reason for this is twofold. First, if this temperature is maintained, the surface of the can will be warm enough to evaporate moisture remaining from the cold water bath. Should this water not be evaporated, it would cause rusting of the outside of the can, possibly spoiling the label and making the container unsightly, so that it would have to be rejected. Second, if the temperature is not lowered to the above-stated range, it may be sufficiently high to promote the growth of residual thermophilic bacteria, which would cause spoilage of the food within the can.

Thermophilic bacteria form spores and many types of thermophilic bacteria are unusually heat-resistant. Therefore, in canning operations, an attempt is made to keep as many thermophilic bacteria out of the product as possible. This can be accomplished by preventing the buildup of these bacteria at various points in the canning operation in the plant. The commercial process used today will reduce the number of thermophiles in foods to a minimum so that only a few thermophilic spores will be present in the product. The few remaining spores will cause no problems, provided the product is held at temperatures below 110°F (43.3°C). Therefore, in cooling freshly processed cans, the average product temperature is brought to 110°F (43.3°C) or below, as thermophilic bacteria require only a few hours of growth to cause spoilage.

After or before cans are heat-processed, they must be washed to remove grease and food particles that get onto the outside of the can during the various procedures involved in the operation. This is done by passing the can through alkaline or detergent solutions, then through rinse water. If done after heat processing, the temperature of the rinsing solutions should be high enough to provide for evaporation of the water, thus preventing corrosion of the outside of the container.

After cooling, cans of food are usually packed into cases or stored in large masses in a warehouse under conditions in which they cool very slowly to room temperature. To prevent growth of thermophiles in the product prior to canning, canner's brine in the reservoir on the filling machine should be held at 170°F (76.7°C) or higher, a temperature at which no growth can take place.

OTHER METHODS OF HEAT PROCESSING

There are a number of types of processing retorts for canned foods that have come into use in recent years and are quite different from the conventional retort.

CONTINUOUS AGITATING RETORT

In the agitating retort, cans enter the retort continuously on a conveyor through a special inlet that prevents loss of steam. In the retort, the cans are conveyed back and forth for whatever period of time is necessary for sterilization. Also, while being conveyed, they are rotated around their long axes. Except for solid-packed foods, this action causes some agitation of the product within the can, speeding up heat penetration and shortening the processing time. Cans exit from the continuous agitating retort through a special valve and enter the cooling system, which is set up much in the manner of the retort except that it is filled with cooling water.

In another type, the agitort, the cans or containers of cans are attached to a wheel that rotates during processing. In this system, the cans are rotated end over end, so

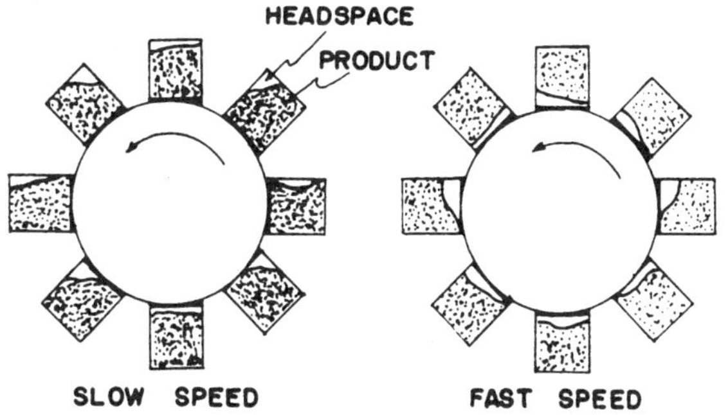

Figure 10.11. Headspace patterns in fast-speed and slow-speed agitorts.

the air in the headspace will travel along the sides of the can to mix the food, and back again (see Fig. 10.11). The agitation of the contents hastens heat transfer even in semisolid packs. In this system, the cans are cooled within the agitort. Because the process depends on the rate at which the cans rotate, special equipment has been devised to count the rotations the cans undergo during the process.

The Hydrostatic Cooker

Hydrostatic cookers (see Fig. 10.12) are used in Europe and to some extent in the United States. In this system, there is a central chamber with a narrow entrance on one side and an exit chamber on the other side. The system is partially filled with water, and when steam is turned on in the central chamber, the water is forced up to higher levels in the entrance and exit legs (chambers).

Cans enter through the warm water of the entrance chamber on a conveyor, gradually entering warmer water as they approach the steam chamber. They are conveyed through the steam chamber at temperatures and for times that provide for commercial sterilization of the product. The cans are then conveyed out of the steam chamber, first through the warm water and eventually through the cold water of the exit chamber. The cans may be rotated around the long axis as they are carried along. Because the pressure in the hydrostatic cooker depends on the water head (height of water), this type of cooker is quite tall (about 40 ft [12.2 m]).

The agitort and the hydrostatic cooker are somewhat faster than other heat-processing methods, both because of the agitation that speeds up heat penetration and because in such conditions somewhat higher temperatures (up to 270°F [132.2°C]) may be used without causing excessive heat damage to the product. This allows for the canning of some semiliquid foods, such as mushroom soup or cream-style corn, in large cans. Because the labor involved and the cost of the container per unit weight of food are smaller as the can size is increased, it is economical to can foods in larger containers. This is desirable for institutions feeding large numbers of people, because it costs them less to buy and handle large cans.

The High-Temperature–Short-Time (HTST) Process

There are systems for processing canned foods at high temperatures for short time. These are referred to as HTST processes. In such systems, commercial sterilization is

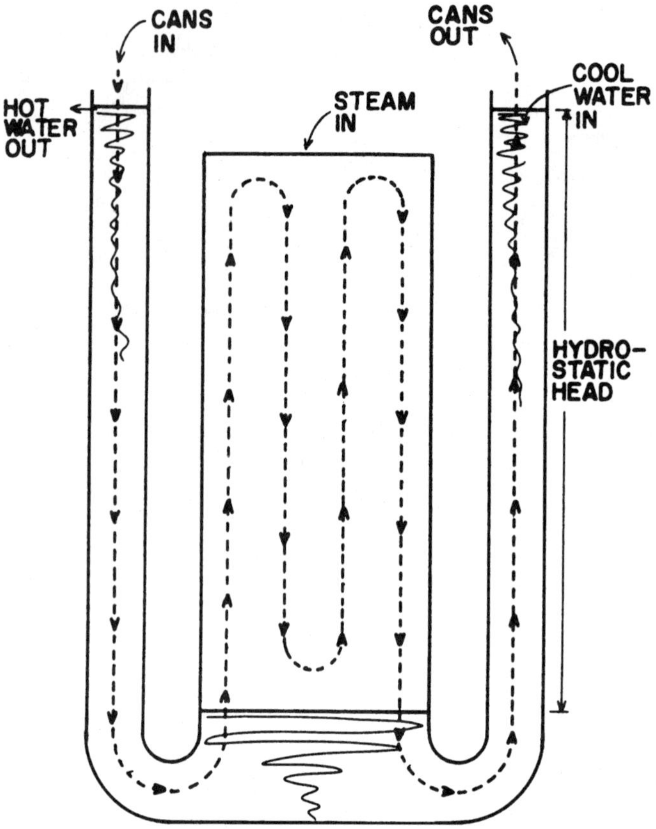

Figure 10.12. Hydrostatic retort.

achieved at temperatures of 280 to 300 °F (137.8 to 148.9° C) in 15 to 45 seconds. Large discrete particles cannot be processed by HTST methods because they require some time for heat to penetrate their centers. HTST methods are applied only to liquids and to foods that have been pureed (mashed bananas, concentrated pea soup, etc.)

Aseptic Fill Method

The aseptic process was deemed by the Institute of Food Technologists to be the most important food science innovation during the 50-year period 1939 to 1989. In one system, the pureed material or liquid is passed in thin layers through a heating system wherein the temperature of the product is raised quickly. The product is then pumped to a holding chamber where it is held at high temperatures for 10 to 20 seconds, then pumped in thin layers through a cooling system wherein the temperature of the product is lowered quickly. It is next pumped to a filling system where it is forced into presterilized containers, and the cans are sealed with presterilized covers. Such systems can be presterilized with very high temperature gas or steam before the operation is started. Flexible or plastic packages that will not stand the high temperatures of sterilization may be presterilized by the use of lower temperatures in combination with chemicals or ultraviolet light. Hydrogen peroxide is often used in this process. The filling and sealing of the containers must be carried out under sterile conditions. More on this system is discussed in Chapters 15 and 20.

Cooking Under Pressure

Another HTST method of sterilizing canned food is carried out in pressure chambers. As the pressure of the air in a chamber is raised above atmospheric pressure, the temperature at which water boils is raised; and if suitable pressures are used, temperatures of 280°F (137.8°C) or higher may be attained. In these conditions, foods that contain discrete particles may be heated in open steam kettles in the pressure chamber for sufficient time to allow sterilization at the center of the food particles, then filled into cans and sealed. The cans are then inverted, allowed to stand for a few minutes, then cooled. It is not necessary to presterilize the cans in this case because the high temperature of the product, when filled, will destroy whatever bacteria may be adhering to the inside. The containers are inverted to ensure contact between the bacteria adhering to the inside of the cover and the very hot product that would destroy them. Personnel working in chambers under pressure may need to be pressurized slowly in locks when entering and depressurized slowly in locks when leaving. Otherwise, they may be subject to the bends, a condition caused by nitrogen dissolving in the blood when humans are subjected to high pressures, and coming out as gas bubbles when the pressure is suddenly released. A system employing both cooking under pressure and aseptic filling is now being used to can products such as chicken salad.

The Sous-Vide Process

This process was developed in France and used in various European countries. *Sous vide* is the French term for *under vacuum* and whenever food is packaged under vacuum, the potential for the growth of *Cl. botulinum* must be of primary concern. By this process, foods are packed in vacuum and heat-processed but not enough to provide for the destruction of botulinum bacteria. The products, reputed to be of superior quality, must be held under refrigeration until prepared for consumption; otherwise surviving bacterial spores can vegetate and produce toxin. The U.S. Food and Drug Administration (FDA) states that only licensed food processing operations are allowed to use the sous-vide process on site. Food service operations are allowed to purchase these products only from reputable suppliers that operate under a verified HACCP system. Foodservice operations are not allowed to prepare sous-vide on site. Storage of these products should be at temperatures from 32 to 38°F (0 to 3.3°C) to ensure no growth of *Cl. botulinum*.

Modified-Atmosphere and Controlled Atmosphere Packaging (MAP)

This technology involves removing some oxygen from a package and replacing it with a more inert gas such as carbon dioxide. Packaging here is very important as it must prevent the entrance of oxygen and the escape of the modified atmosphere gasses. This type of packaging will enhance the growth of anaerobic organisms such as *Cl. botulinum*, so extreme care must be taken in the handling and storage of such products.

Temperatures must be kept as low as possible and never be allowed to exceed 40°F (4.4°C).

Microwave Processing

Microwave energy, usually generated by a magnetron, occurs as an alternating current at either 915 or 2450 megahertz. Both frequencies are authorized by the federal government for use with microwave ovens. Because water molecules are polar (have positive and negative ends), they tend to oscillate as they try to align themselves alternately between the positive and negative charges of the microwave energy. When frozen foods are exposed to microwave energy, the liquid water in them (all frozen foods contain some water in the liquid state) begins to heat due to the friction created by high-speed oscillations of the water molecules. Most of the water in frozen foods is in the ice state and is not affected by the microwave energy, until heat generated by the liquid water molecules is conducted to the ice, melting it, and thereby producing more liquid water to become involved in the heating process.

The first application of microwave energy in industry occurred during the 1970s when cooperative experiments between the Gloucester Laboratory, a USDC facility, and the Raytheon Co. resulted in the introduction of a continuous microwave tunnel in the seafood industry for thawing or tempering of shrimp and fish blocks. In tempering, the product is not thawed, yet lends itself to further processing such as cutting or coating with a breading.

Those early experiments also tested the feasibility of using microwave energy for shucking oysters. The innovation soon spread to the meat industry, and now microwave processing is extended to other applications including baking, pasteurization, and sterilization. Because the highest temperature attainable in conventional microwave units is 212°F (100°C), it is necessary to pressurize the processing chamber of the microwave unit in order to achieve the high temperature required to effect sterilization in a reasonable time period.

Microwaves have many current uses in the food industry and more are possible in the future. Examples exist in processes used in production of concentrates, blanching, and puffing (for snack foods or cereals).

CONTAINERS

Cans

For general commercial applications, the speed with which cans are handled during processing and the structural protection they lend to the contents may make them still the most desirable overall container for heat-sterilized foods.

Flexible Pouches

Flexible pouches have several advantages over metal and glass containers: (1) Their shape adapts to available space. (2) They are lighter in weight. (3) When empty, they require less space. (4) When filled, their thickness dimension is small, permitting faster processing times and mitigating heat damage to foods. (5) Heat penetration is greater. (6) They are not subject to corrosion or breakage. (7) They are easier to open.

Flexible pouches have a few disadvantages: (1) Pouches cannot be filled as rapidly. (2) They are awkward to handle (e.g., stacking), especially the larger ones. (3) They give no structural support to fragile contents. (4) They are somewhat vulnerable to tearing and cutting.

The materials used in pouches are generally composite laminates and quite strong. They include an outer layer of polyester, which has nearly the tensile strength of steel and resists wear and tear, an inner layer of nylon-11 or polypropylene that resists wear and tear and seals well, and aluminum foil as a middle layer for eliminating light from the contents as well as for making the pouch material impermeable to gases. The layers are bonded together with an adhesive. While flexible pouches can be processed in conventional retorts, special systems have been developed; in most cases, the systems are continuous.

Glass Containers

As a container material, glass has several desirable properties. It permits visibility of the product, imparts no taste, and is noncorrosive. The use of glass containers precedes by centuries the use of metal cans and, of course, flexible pouches. The advances in closures of glass containers are probably more dramatic than the advances in the design of the containers and, in fact, have forced the changes in container design. Cork closures have been perhaps the most important. Cork is light, compressible, hermetically sealable, inexpensive, plentiful, and stable. Corks have long been in use for closing glass bottles, and still are. Ordinary cork closures (stoppers), however, are unable to withstand the buildup of internal pressure, except where the pressure is not too great and when they are used with bottles having small openings, such as in wine and champagne bottles. The diameters of the openings of glass jars used for preserving fruit and vegetables are so large that a fitted cork would be easily pushed out (for any given pressure, the total force increases as the area of the opening of the glass container increases). Even for small openings, it was necessary to develop holddown devices, such as clamps and wire fasteners, when high pressures were expected, as in the bottling of beer. The crimped metal crown with cork liner was an effective closure for beer and soda bottles. However, aluminum screw-caps are now used in nearly all cases.

Variations of screw-top, wide-mouth jars evolved to the modern Mason jar (named after its inventor). Tight-fitting, screw-type closures find a wide variety of uses in modern glass containers for commercially pasteurized and heat-processed foods as well as in home-canning operations.

However, it was not until more advanced jar closures were developed that canning in glass became significant. These closures include the Phoenix cap, the Sure Seal cap, and the Vacuum Side Seal cap, types found in many tumbler-sized tapered jars such as those used for cheese spreads and jellies; and the Amerseal cap, a modified screw-type cap such as found in capping jars for apple sauce, jellies, and wide-mouth fruit juice bottles.

Screw-type and pressed/pry-off-type lids and caps are currently in widespread use. Also, most containers are now tamperproof or to show when they have undergone tampering.

Microwavable Containers

The ease and speed with which microwavable entrees can be heated and served will accelerate an already growing demand for them. Not only are microwave ovens found in nearly every home, but they are also available in nearly every work place for use by employees. Packaging for microwavable meals includes trays made of high-gas-barrier materials (such as polypropylene bonded to an ethylene vinyl alcohol polymer) that satisfies the criteria for such containers: (1) microwavable, (2) lightweight, (3) unbreakable, (4) easy to open, and (5) reclosable.

Plastic, Flexible Containers

The advances in aseptic packaging have allowed packing of heat-processed foods such as ketchup in "squeezable" bottles that are lighter, unbreakable, and very convenient in dispensing the product.

CANNING OF ACID FOODS

Foods that have a pH of 4.5 or lower are considered to be acid foods. Acid foods that are canned need not be heated at high temperatures to attain commercial sterilization. The reason for this is that bacteria, including those that form heat-resistant spores, are more easily destroyed by heat when present in acid solutions. Moreover, spore-forming bacteria generally will not grow in foods having pH values of 4.5 or less. There are some exceptions to this; for instance, *Bacillus thermoacidurans* may grow in tomato juice (maximum pH 4.5) and cause spoilage.

Acid foods are ordinarily processed by heating the cans in boiling water until all parts of the product have reached 180 to 210°F (82.2 to 98.9°C), and then cooling the cans. An exception is tomato juice, which is now often processed by flash heating to 250°F (121.1°C), holding at this temperature for 0.7 min, cooling to 200 to 210°F (93.3 to 98.9 C), filling into presterilized can, sealing, and inverting the can so that the sterilizing effect of the heat (200 to 210°F [93.3 to 98.9°C]) at that pH will act on the can cover.

Those foods that have a pH of 4.5 or less include apples and apple juice, apricots, blackberries, blueberries, boysenberries, cherries, cherry juice, all citrus fruits and their juices, currants, gooseberries, loganberries, papaya juice, peaches, pears, pickles, pineapple in various forms and pineapple juice, plums, prune juice, raspberries, rhubarb, sauerkraut and sauerkraut juice, strawberries, tomatoes, tomato juice, and youngberries.

WAREHOUSE STORAGE OF CANNED FOODS

When canned foods have been heat-processed and cooled and the cans have been cleaned and dried, they are either stored in warehouses in bulk until labeled, cased, and shipped out, or they are labeled, cased, and stored in a warehouse until shipped out. Lithographed cans may be used, in which case labeling is not necessary.

Warehouses should be maintained so that the temperature does not rise much above

85°F (29.4°C) or fall below 50°F (10°C). Very high temperatures may promote the growth of thermophilic bacterial spores present in small numbers in the food. Very low temperatures may lower the temperature of the cans to the point that, should a sudden hot spell occur, cans will sweat (condense moisture), eventually causing external corrosion of the cans.

FDA REGULATIONS AND SAFETY

The Good Manufacturing Practices (GMPs) are procedures approved by the FDA for use in both low-acid foods [those foods with a pH greater than 4.6 and a water activity (a_w) of greater than 0.85] and acidified foods. Use of the GMPs will help assure food safety and wholesomeness.

The Emergency Permit Control regulations also help ensure safety by requiring all processors to register their production plants and methods with the FDA. The manufacturers must also keep detailed records of all production and have them available for FDA inspection. These records are so very important if a problem is discovered and a recall of all potentially dangerous products must be made. The purpose of all these regulations is to ensure the destruction of all *Cl. botulinum* organisms and ensure that all canned food is safe for human consumption.

In foods for which a vacuum is not achieved and *Cl. botulinum* is not the target organism, pasteurization is used to eliminate pathogens or heat-resistant spoilage organisms. Some species of *Listeria* and *Salmonella* have been found to be more heat-resistant than once thought. Use of the *D*-values for these organisms is crucial in developing new pasteurization times and temperatures that are effective in their destruction.

11
Drying

Drying is probably the oldest food preservation process practiced. It is believed that many foods, especially grains and fruits of high sugar content, were preserved by primitive peoples by allowing the foods to dry in the sun. Spices and fish, cut into thin strips, were also preserved in this manner.

There are a number of different methods of drying food for preservation. The most important ones include sun drying, tunnel or cabinet drying, drum drying, spray drying, and freeze-drying.

PRETREATMENT

Foods to be dried must be washed, and some peeled and cut. Others may be precooked. Cut fruits are subject to darkening through enzyme action and must be either blanched or treated with salts or sulfur dioxide. However, if treated with sulfur dioxide, the product must be so labeled, because of known allergic reactions in some consumers. Certain vegetables may be pretreated in the same manner. Sulfuring may also be required to limit nonenzymatic browning (the Maillard reaction).

Various dried egg products (egg white, egg yolk, and whole egg products) are also subject to browning and are susceptible to the development of off-flavors. In this case, the reaction involves a combination of a small amount of glucose, which is naturally present in eggs, with the proteins. Because of this, dried egg products, especially egg whites, may be treated with glucose oxidase and catalase. The glucose oxidase converts glucose to gluconic acid (which does not combine with amino groups) and hydrogen peroxide. The purpose of the catalase is to convert the undesirable peroxide to water and oxygen. The elimination of glucose may also be done by natural fermentation, using microbes. However, this process can be considered unsanitary, and to avoid product spoilage and even food poisoning, it is necessary to hold the product at 130°F (54.4°C) for several hours after drying.

Because plums are naturally coated with a thin layer of wax, drying them to prunes is greatly speeded up by predipping the fruit in dilute lye solution, then in hot water, prior to drying.

METHODS OF DRYING

Sun Drying

Sun or natural drying is still used in hot climates for the production of dried fruits or nuts. This may be done in direct sunlight or in shaded areas where the drying is accomplished by the hot dry air. It should be apparent that sun-dried fruit is produced only in areas where the climate provides periods of relatively high temperatures, relatively low humidities, and little or no rainfall. Plums, grapes, apricots, peaches, and pears are dried in this manner. Some of these fruits are also dried in tunnel or cabinet dehydrators.

In sun drying, small fruits are prepared and spread on trays to dry in the sun for several days, then stacked to complete the drying cycle in shaded areas. Larger fruits, such as apricots, peaches, and pears, are halved and pitted, and apples are peeled, cored, and sliced prior to drying. Such fruits are sulfured to prevent enzymatic browning. Sun drying times vary between 4 and 25 days, depending on the size of the product, the type of pretreatment, and so on. During sun drying, precautions must be taken to prevent contamination from wind-blown dust and dirt. Moisture contents in the sun-dried products vary between 10% and 35% depending on the tendency of the dried product to absorb moisture. After drying, some fruit may require moisture-vapor-proof containers.

Hot-Air Drying

When mechanical dehydrators are used, the product is placed on metal mesh belts in a tunnel, or in a cabinet on trays where controlled, elevated temperatures are used. Heated air is circulated by blowers and the air temperature, relative humidity, and air velocity are controlled. Hot-air driers of this type are classified as parallel flow, counter flow, direct flow, or cross flow, depending on the direction in which the product moves in relation to the direction of the flow of the heated air. In bin, loft, and fluidized-bed driers the heated air is blown upward through the product.

The hot air used in tunnel or cabinet driers may or may not be recirculated. If it is recirculated the relative humidity must be carefully regulated because during each passage over the food, the air takes up moisture, raising its relative humidity.

The air in dehydrators is heated either by steam tubes or coils, or by being mixed directly with the combustion gasses of gas or oil. Electric resistance heaters are used in rare instances. In all cases, except natural draft drivers, the hot air or gas is circulated by blowers or fans of different designs and is discharged from the drier through a ventilator that may be equipped with a fan to increase its capacity, thus increasing the amount of air that can be circulated through the drier (Fig. 11.1).

The time required for drying of a particular product depends on the characteristics of the raw material (moisture content, composition, shape, and size), the temperature and humidity of the air in the drier, and the rate of air circulation in the drier.

Initially, drying of food occurs through evaporation of moisture from the food surface. Later, in the drying cycle, drying involves the diffusion of water, water vapor, or both, to the surface of the food.

In the initial stages of drying, air velocities are usually regulated at about 1000 ft/min (304.8 m/min), but in the later stages, the air velocity is usually lowered to about 500 ft/min (152.4 m/min), as this rate will remove all moisture available at surfaces at this point of the drying cycle. High initial rates of drying are said to prevent adherence of the dried particles to drying trays and belts, thus facilitating unloading. Water-repellent plastics, such as polyethylene or Teflon, may be used for coating trays and belts to prevent sticking of the dried food.

Due to evaporation of moisture, which lowers the temperature, product temperature is below that of the air in the drier during the initial stages of drying and up to a

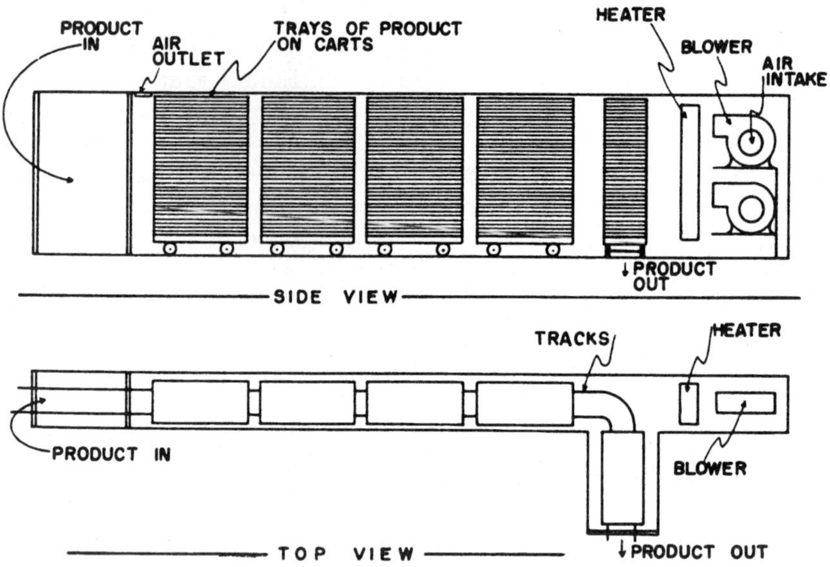

Figure 11.1. Continuous tunnel drier.

point where approximately one-half of the moisture has been evaporated. The product temperature then starts to rise, and at the end of the drying cycle approaches that of the air.

Foods undergo some form of breakdown, or loss in quality, when they are exposed to heat. The amount of damage they undergo increases as the temperature to which they are exposed increases, and it also increases as the time of exposure is increased. Therefore, it is important to control both time and temperature during the drying cycle. In the beginning of the drying cycle, higher temperatures can be used for two reasons: (1) The amount of surface water in the foods is highest in the beginning of the drying cycle, so most of the energy of the hot air is expended in vaporizing surface water, and (2) the tendency of the food to be heated by the hot air is partially counterbalanced by the cooling effect of the evaporation of the surface water.

In the later stages of drying, the amount of surface water in the product is relatively low; therefore, it is prudent to lower the temperature of the air to a point where the energy is just sufficient to vaporize the surface water. If the air temperature is not lowered, the excess energy will go into raising the temperature of the product, which by this stage is not being cooled sufficiently by evaporative cooling. For the drying of vegetables, the initial temperature of the air is 180 to 200°F (82.2 to 93.3°C). In the later stages of drying, the temperature is reduced to 130 to 160°F (54.4 to 71.1°C).

Fluidized-Bed Drying

In fluidized-bed drying (a special type of hot-air drying), the product is fed in at one end of the dryer to lie on a porous plate and is agitated and moved along toward the exit at the other end by hot air that is blown up through the product. The air that has passed through the product, picking up moisture, exits through an outlet at the top of the drier. More about fluidized-bed processing is found in Chapter 12.

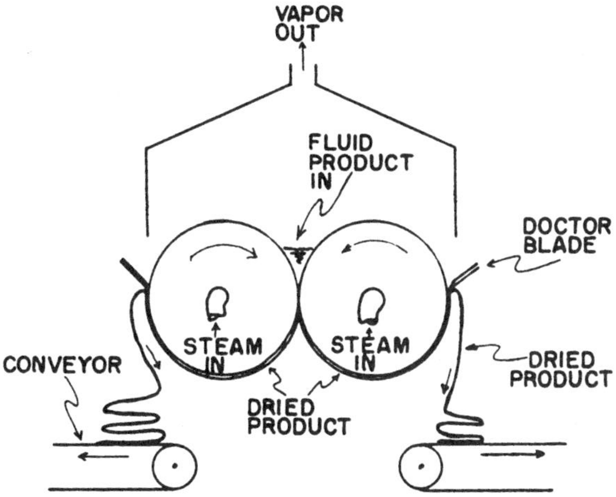

Figure 11.2. Drum drier.

Drum Drying

Milk, fruit, vegetable juices, purees, and cereals may be dried with drum driers. These products are allowed to flow onto the surface of two heated stainless steel drums located side by side and rotating in opposite directions with little clearance between them. The product dries on the drums and is scraped off by stationary blades fixed along the surface of the drum (see Fig. 11.2). Refrigeration may be used to lower the temperature of the dried product quickly. Drum drying can also be carried out under vacuum, in which case the drying is accomplished at lower temperatures and the product is protected from oxidation. In ordinary drum drying, the process is controlled by varying the moisture content of raw material (preconcentration), the temperature of drum surfaces, the space between the drums, the speed of rotation of the drums, and the amount of vacuum applied.

Spray Drying

Milk, eggs, soluble or instant coffee, syrups, and other liquid or semiliquid foods are spray-dried (see Fig. 11.3). The liquid material is sprayed into the top of a chamber simultaneously with hot air, which is also blown in at the top. The cool, moist air exits near the bottom, and the dried particles fall to the bottom and are collected by gravity flow, or with the aid of scrapers that may also be used to remove dried material from the walls or bottom. Cyclones (conical-shaped collectors) may be used to collect particles escaping with the exit air.

Particle size is an important factor in spray drying, and the liquid is, therefore, dispersed into the drying chamber through a pressure nozzle or by centrifugal force generated by a disc rotating at high speed. Both methods atomize the liquid, producing droplets of a size that will be dried by the heated air through which they must fall on their way to the bottom of the drying chamber. The material may be preconcentrated in some instances, or aids, such as gums, pectin, or milk solids, may be added prior to spray drying. In any case, it is necessary to lower the moisture content and tempera-

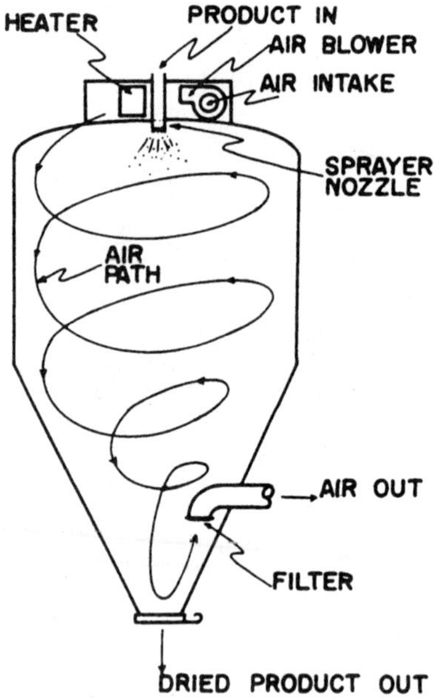

Figure 11.3. Spray drier.

ture of the product to the point where particles will neither stick together nor stick to the wall of the spray drier.

In spray drying, the temperature of the food reaches a maximum of about 165°F (73.9°C) and remains there for only a short period of time. However, in spray drying the moisture content can be lowered only to about 5%. Thus, spray drying does not cause much heat damage to foods, but because of their relatively high moisture contents, spray-dried foods may undergo some spoilage during storage. The shearing action of the atomization step of spray drying may affect the functional characteristics of some proteins, such as those of egg products (whipping qualities, etc.).

Spray drying tends to produce fine powders that neither wet nor disperse readily when reconstituted. The material thus tends to clump and form a mass through which water does not penetrate. To avoid this, a process known as agglomeration may be used. In this case, the dried material is slightly rehumidified; for instance, dried milk powder may be heated with steam under controlled conditions that surface-moisten the product. By other methods, the preconcentrated product is injected into the drying chamber with a steam injector. The latter procedure causes the dried product to form in the shape of beads or bubbles, which break into flakes that disperse readily in water. Vacuum puff drying may also be used to produce a product that disperses in water without forming clumps.

Freeze-Drying

The freeze-drying of foods is carried out by first freezing the product and then subjecting it to a very high vacuum wherein temperatures are high enough to assist in the evaporation of moisture but low enough to prevent melting of the ice in the product (see Fig. 11.4). In this method, the water, existing as ice in the food, is evaporated directly as a vapor without passing through the liquid phase (sublimation). The

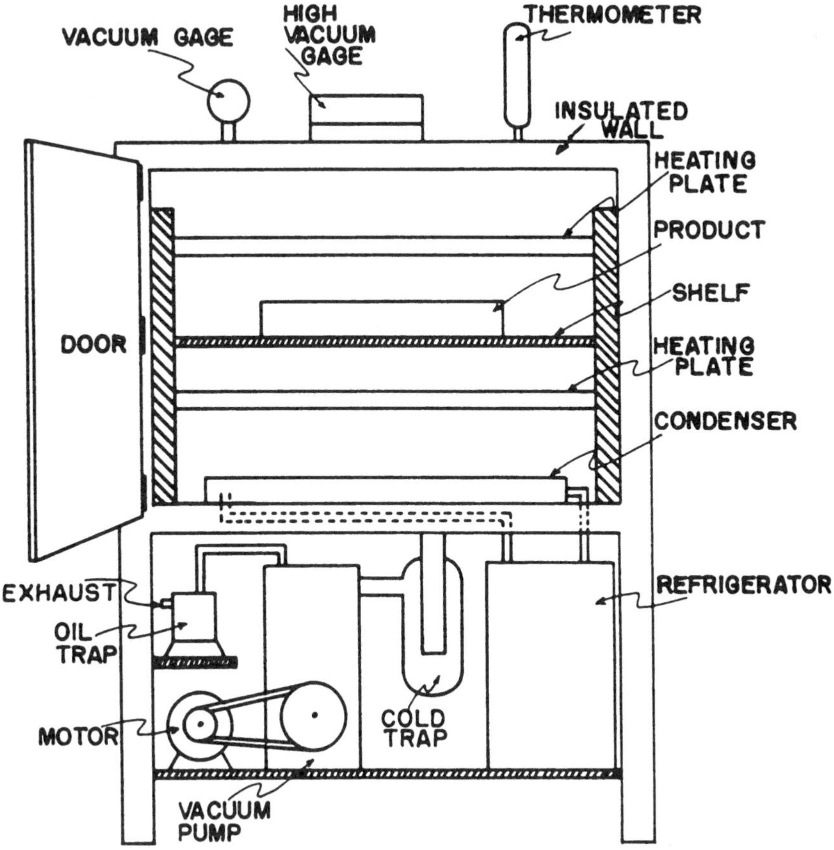

Figure 11.4. Freeze drier.

vapor is condensed outside the evaporation chamber. The resulting product has a honeycombed structure containing much surface area, and because it maintains its original shape, its specific gravity is reduced considerably, as is the moisture content.

In freeze-drying, the maximum surface temperatures used depend on the composition of the food. Some foods, such as vegetables and mushrooms, can withstand 180°F (82.2°C); others, such as fatty fish, require temperatures as low as 100°F (37.8°C). The vacuum applied must be very high (less than 0.02 in [0.05 cm] mercury), and most foods require even higher vacuums for good results.

As in the case of foods dried by other methods, it may be necessary to blanch or to treat the foods prior to freezing them, or to treat them with chemicals to provide a source of sulfurous acid to inactivate or inhibit enzymes. This is especially the case with vegetables (garlic and onions excepted) and mushrooms. Meats are sliced or cut into small enough portions to permit reasonably fast drying. Chicken is usually precooked, boned, and diced prior to freezing and drying.

Although it has been stated that faster freezing rates with the resulting uniformly distributed small ice crystals are preferable for foods that are to be freeze-dried, some investigators have found that slower rates of freezing result in better rehydration (reabsorption of water) of the dried food, hence better quality. The temperature to which the frozen food is brought prior to the drying process will depend on the product itself and should be low enough to rule out significant amounts of melted material. For most foods this temperature is around –7°F (–21.7°C), but for some fruit juices it may be as low as –26 to –30°F (–32.2 to –34.4°C).

During freeze-drying, by most commercial methods, the highest food temperature is that of the dried surface layers. The temperature of the ice in the food is determined by the degree of vacuum present in the evaporation chamber. The higher the vacuum, the lower the ice temperature.

Freeze-dried foods are dehydrated to lower moisture contents than those dried by other methods. Usually, the moisture content of the freeze-dried food is below 3%. Because large surface areas are present in freeze-dried foods, it may be necessary to use an inert gas, such as nitrogen, in place of air, to break the vacuum and so prevent chemical changes, such as oxidation of fats, which may occur shortly after contact with air is made.

Puff Drying

Puff-drying may be used for some foods that are temperature-sensitive, such as fruit or vegetable juice concentrates. With this method, the product may be heated in an oven and suddenly subjected to a high vacuum. In other applications, the product is evacuated to remove air, then heated with steam, then puffed by reapplying the vacuum. Puffing may also be attained by raising the temperature of the food under conditions that raise the temperature of the water in the food above 212°F (100°C) to provide pressure. The product expands or puffs when the external pressure is released. This is done with some cereals, providing a porous, puffed structure.

Microwave Drying

Microwaves are used in different drying applications including dehydration, concentration, freeze drying, and finish drying where traces of moisture remaining after conventional drying can quickly be removed from the inner portions of the product without overheating the already dried product.

RECONSTITUTION OF DRIED FOODS

Dried foods ordinarily must be reconstituted (water must be added back to them) before they can be eaten. During drying and during storage, changes take place in the food that affect the rate and extent to which water will be taken up by the dried product. There is also a redistribution of soluble components during rehydration.

It has been observed that some dried foods rehydrate better when the water is held at low temperatures (below 40°F [4.4°C]). Other foods rehydrate better when the water used is at higher temperatures, with some foods rehydrating satisfactorily in boiling water.

PACKAGING OF DRIED FOODS

In the packaging of dried foods, some products require only minimal specifications (cereals, some vegetables, etc.). Others require packages that are essentially moisture-vapor-proof. Hydroscopic materials, especially some dried fruit juices, which readily take up moisture, must be packaged to prevent moisture from entering.

Freeze-dried foods have a very low moisture content. If their low moisture content is not retained, they lose their desirable characteristics because of nonenzymatic browning. Moreover, because the water has been removed, components such as fats are exposed to the oxygen of the air over a large surface area, subjecting them to an accelerated rate of oxidation that eventually leads to rancidification and off-flavor development.

Certain freeze-dried foods, such as fatty meats (especially pork), lobster meat, crab meat, and shrimp, require protection against both oxidation and moisture absorption. They must, therefore, be packaged under vacuum or in an inert atmosphere in packages that are impervious to both moisture and oxygen, such as hermetically sealed metal containers and flexible pouches made of laminated aluminum foil and plastic. Fruit juices, dehydrated whole milk, and certain freeze-dried egg products must also be protected against both oxygen and moisture.

THE EFFECT OF DRYING ON MICROORGANISMS

The main purpose of drying foods is to lower their moisture content to a particular level that will exclude the growth of microorganisms (bacteria, molds, and yeasts). For any given moisture content, one food may support the growth of microorganisms while another will not. Whether or not microorganisms will grow in a dried food depends on the water activity.

The lower the water activity of a food, the less probable it is that microorganisms will grow. Generally, molds will grow at lower water activities than bacteria. For this reason, molds are more apt to grow in dried foods than are yeasts or bacteria. In drying foods, therefore, the moisture content is lowered to the point where microorganisms will not grow and it is kept that way through packaging, which excludes moisture.

Water activity can also be lowered by soluble components, such as sugar or salt. Thus, certain syrups and salted, partially dried foods (e.g., fish) are relatively stable as far as the growth of microorganisms is concerned, although there may be conditions in which they become subject to the growth of yeasts or molds because of their ability to grow at lower water activities.

DETERIORATION OF DRIED FOODS

Oxidative Spoilage of Dried Foods

Regarding the chemical changes that may take place in dried foods, oxidation of fats is one of the chief causes of deterioration. This is especially the case with fish, shrimp, crab meat, lobster, and other seafoods, and also a factor of some concern with meats such as pork. The pigment that provides the red color of cooked crustacean foods may also be changed or entirely bleached through the oxidation. Packaging under vacuum or in an inert atmosphere, such as nitrogen, and in such a manner as to exclude oxygen, may be used to protect dried foods against oxidation of fats. There may be instances in which antioxidants can also be added to the food to inhibit oxidation of fats. Antioxidants are chemicals that tend to interfere with the type of oxidation to which unsaturated (adjacent carbons of the fatty acids not fully taken up with hydrogen atoms) fats are subject. For instance, the tocopherols (vitamin E) are antioxidants.

Very small amounts of chemical antioxidants are required to provide a considerable protection against the oxidation of fats. For instance, only about 0.02% of a chemical antioxidant may be added to fats or oils subject to oxidation. However, antioxidants are fat-soluble and are not generally soluble in water. For many foods, therefore, in which the fat is solid or generally distributed throughout the food, there is no known method of getting antioxidants into the fat itself. Hence, there is no good method of preventing rancidification of the fat of some foods through the use of antioxidants. The packaging of dried products in such a manner as to exclude light is another procedure that may be used to assist in the prevention of fat oxidation because light energy accelerates fat oxidation and rancidification.

Nonenzymatic Browning in Dried Foods

Nonenzymatic browning is another cause of deterioration in dried foods. This may be due to caramelization (a dehydration of sugars) or to the combination of certain sugars and proteins, either process leading eventually to the formation of brown- or black-colored compounds that not only cause off-flavors but also discolor the food. The best protection against such changes is to lower the moisture content to the point where the rate of nonenzymatic browning is greatly reduced. In some foods this may mean lowering the moisture to less than 2%. Sulfurous acid or a source of this compound may also be used to inhibit nonenzymatic browning. Browning may also change proteins, lowering their nutrient quality and affecting their rehydration properties.

Enzymatic Changes in Dried Foods

Enzymatic changes may take place in dried foods during drying, storage, or rehydration. Generally, such changes are prevented by preblanching of the food, or by using sulfurous acid to inhibit enzyme action. Lowering and retaining the moisture content to 2% or less will inhibit enzyme changes during storage, but this process will not prevent such changes from taking place during rehydration of the food.

12

Low-Temperature Preservation

Refrigeration at Temperatures Above Freezing

Unlike drying or freezing, both methods of preserving foods that have been used in certain areas of the world for centuries, the refrigeration of foods at temperatures above freezing is of comparatively recent origin, and the development of this method of food processing is well documented. Ice and snow had been used by the Romans and the early French for the preparation of iced drinks, but the application of refrigeration at temperatures above freezing as a means of extending the storage life of foods was started in the United States in the middle 1800s. It is now the most popular method of food preservation, and it is estimated that more than 85% of all of our foods are refrigerated (held at temperatures above freezing) at some point in the chain of food handling from harvesting to consumption.

At the start, ice, harvested from lakes and ponds that had frozen over in the winter, was put in ice-warehouses and used during warm weather to keep foods cool in ice boxes, the forerunners of the modern refrigerator.

The next step was to develop mechanical refrigeration systems that would produce ice to replace natural ice, and, eventually, mechanical refrigeration systems were used to cool rooms, trucks, and boxes where food could be held without the use of ice. The use of ice was at its peak in the United States as late as the early 1930s.

The first household refrigerators, of course, were ice boxes in which ice was placed in a chamber at the top in order to cool a lower chamber where food was kept. Water from melting ice was allowed to run down to a container beneath the ice box and this had to be emptied periodically.

At this point, the refrigeration capacity of ice deserves mention. The cooling capacity of ice is such that if one should supercool 1 lb (0.45 kg) of it to 20° F (–6.7°C), then this ice would have the capacity to lower the temperature of 1 lb (0.45 kg) of food about 6°F (2.83°C) before any of the ice would melt. At that point, the temperature of the ice would have risen to 32°F (0°C). However, when heat is added to ice that is at 32°F (0°C), the ice temperature remains at 32°F (0°C). It so happens that when 1 lb (0.45 kg) of ice melts, it lowers the temperature of 24 lb (10.9 kg) of food about 6°F (2.83°C). It can be seen from this that the cooling effect of ice is greatest at the point when it is melting. This principle can be used in ice-cooled, self-serve salad bars and in coolers used with mobile units. A small amount of water added to the ice will start the melting process and increase the cooling effect of the ice.

Because little was known about the theory and principles of refrigeration in the early days, much research was required to develop effective systems for circulating cold air in cold storage rooms. As a matter of fact, in fishing boats, a pile of ice in one corner of the pen was used to refrigerate fish instead of being applied directly to the fish as it is today.

Mechanical Refrigeration

While ice is still used for refrigerating fish and some shellfish, and in some cases, fresh produce during shipment and during holding or display at the retail level, most foods are refrigerated by mechanical systems.

A mechanical refrigeration system consists of an insulated area or room (the refrigerator) and a continuous, closed system consisting of a refrigerant, expansion pipes or radiator-type evaporator located in the refrigerator, pump or compressor, and a condenser (see Fig. 12.1). The compressor and condenser are located outside the refrigerator. The refrigerant, usually ammonia, or one of the freons, flows into the expansion pipes as a liquid. Here, it evaporates to a gas, and in changing from the liquid to the gaseous state, it absorbs heat. The gas is pulled into the compressor by the suction action of the pump and is then compressed into a smaller volume. The latter action causes the gas to heat up. This heat must be removed, which is accomplished by passing the condensed gas through a system of pipes or radiators usually cooled by water, sometimes by forced air. Cooling the condensed gas liquefies it, whereupon it is returned to the evaporator in the refrigerator. The conversion of the gas to a liquid also produces heat that is transferred to the water or air of the condenser. Special valves at both ends of the evaporator allow the required flow of liquid refrigerant in, and of gas out, of the expansion system in the refrigerator.

There are a number of ways in which refrigeration may be applied to the insulated area to be cooled. Expansion pipes where the refrigerant is evaporated may be located along the walls of the refrigerator. In this case, natural circulation of air (the cold air

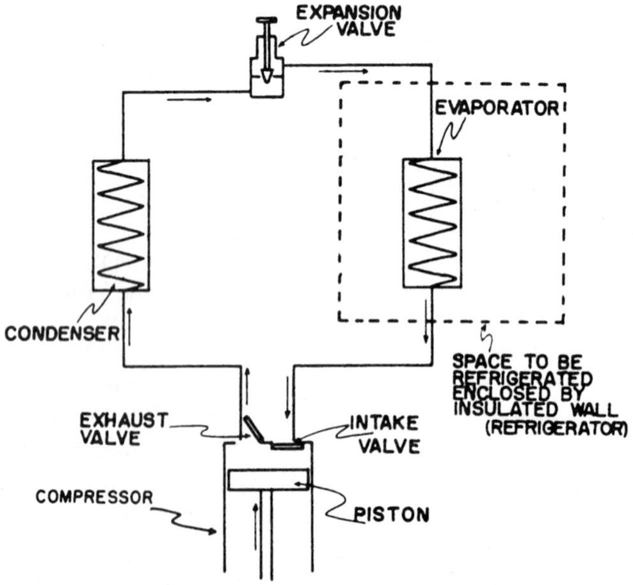

Figure 12.1. Principle of mechanical refrigeration.

being heavier) may be depended on to refrigerate areas within the room away from the expansion pipes, or some type of forced-air circulation may be used. In some instances, radiation-type evaporation units are used. A fan that blows air through the radiator fins provides circulation of cold air throughout the refrigerator. An indirect method of cooling refrigerators depends on the use of refrigerated brine in an outside container, cooled (usually below the freezing point of water), then sprayed into a chamber within the refrigerator and returned to the outside cooler. The cold brine spray refrigerates and humidifies the air within the chamber and causes it to circulate outside of the chamber throughout the refrigerator. Refrigerated brine circulating through pipes may also be used.

Refrigeration Practices

Raw materials, such as fish, meats, poultry, vegetables, and fruits, should be held in a different refrigerator from cooked foods. If cooked and raw foods must be stored together in the same refrigerator, the cooked foods should be placed in covered containers and stored above and separate from the raw foods. This precaution is necessary to prevent bacteria on raw materials from contaminating the cooked foods.

The available refrigeration capacity for all refrigerators used to hold foods at temperatures above freezing should be sufficient to take care of peak loads. That is, when comparatively large amounts of material are to be placed in the refrigerator within a short period of time, the refrigerator capacity must be adequate to cool all parts of all products down to the desired holding temperature within 1 to 3 hours. Often this is not possible and some alternative must be used. The usual alternative is to precool the product to or near to the desirable holding temperature before it is placed in the refrigerator. The product can be precooled in an insulated area where refrigerated air can be blown through the product, or in a chamber where the product can be placed under vacuum. (The basis of vacuum cooling is the evaporation of water from the product under low pressure.) Heat exchangers and thin film evaporators may be used for liquid or semiliquid materials.

Products should not be placed in the refrigerator in large bulk form unless they have been precooled. Individual units may be placed on trays and the trays placed on racks to facilitate cooling of the product within the refrigerator. Once cooled, the individual units may be stacked closer together to conserve space. Liquids to be held in large quantities must be precooled before placing in the refrigerator, or they can be poured into small containers to facilitate cooling.

Large commercial refrigerators should not open directly to the outside but rather to an anteroom that in turn opens to the outside, to reduce the amount of heat entering the refrigerator from the outside. The doors should be of the swinging, self-closing type.

Fish, meats, and poultry should be held at temperatures above, but as near to 32°F (0°C) as possible, to maintain good edible quality and prevent spoilage because of excessive enzyme action or excessive bacterial growth. Because these products generally freeze below 31°F (–0.56°C), a good temperature range for them is 31 to 35°F (–0.56 to 1.67°C). Such a temperature range is attainable in the well-regulated refrigerator.

Other Refrigeration Methods

An early development for chilling some foods, especially seafoods, was the use of brine (salt solutions of varying concentrations) as a cold dip or spray. Chilling with brine is faster and more uniform than chilling in ice, and the practice is still used today. Because the temperature of the brine can be reduced to well below freezing (down to –6°F [–21.1°C] when sodium chloride is used and down to –67°F [–55°C] when calcium chloride is used), brines have also been used for freezing and, to a degree, are still used today.

Chilled seawater (CSW) is used to chill fish at sea. CSW consists of seawater and a predetermined amount of ice (usually in crushed form) sufficient to cool the catch and to hold it at the temperature of melting ice, which in seawater is somewhat below 32°F (0°C). As with brine, CSW cools more rapidly and more uniformly than does ice alone.

Refrigerated seawater (RSW), also used to chill fish at sea, is similar to CSW except that cooling is accomplished with mechanical refrigeration, which permits control of the RSW temperature. Actually, the salt concentration of RSW can be increased if desired, to operate at temperatures below the freezing point of seawater.

In dehydrocooling, foods are subjected to evaporative cooling under vacuum. Heat loss is at the rate of 970 Btu/lb water (539 kcal/kg). If dehydration may be a problem, the product may be sprayed with cold water.

There are a number of new processed foods that require refrigeration and must be handled with extreme care. These include the new-generation refrigerated foods and the modified atmosphere packaged (MAP) foods.

The new-generation foods include fresh pasta, meat, and vegetable salads. Some of these products have been precooked or processed in some manner that may have reduced the total number of bacteria but not destroyed some potential pathogens such as *Cl. botulinum* or *Listeria monocytogenes*. These foods may also not show the normal spoilage warnings, so the recommended shelf-life storage times must be closely followed and they should also be stored at temperatures below the recommended upper limit of 40°F for most refrigerated foods.

The MAP foods, including those produced through the sous-vide process, are discussed in the chapter on thermal processing (Chapter 10) because they are sometimes heat processed prior to packaging. As mentioned, this process can be quite dangerous if the food is not handled correctly after packaging is complete. *Cl. botulinum* can grow in these newly created, oxygen-free atmospheres and extreme care must be taken when storing them. Any sous-vide product must be stored at temperatures between 32 and 38°F (0 to 3.3°C) or frozen. When the frozen products are thawed, they must never get above the temperatures mentioned previously unless they are cooked from the frozen state.

In storing of fruits and vegetables, controlled atmosphere storage can also be employed. This involves replacing oxygen with different percentages of more inert gasses such as nitrogen or carbon dioxide. This storage is used to slow down the ripening process in products such as apples which are harvested in a seasonal, short period of time but are distributed as fresh product throughout the year. Controlled atmosphere storage can also prevent financial losses if severe winter storms stall deliveries, which will substantially affect shelf-life.

FREEZING

The preservation of foods by freezing dates from antiquity, having been used by such ethnic groups as Eskimos, and Native Americans of certain cold areas, among others.

Fish caught in the winter months in cold climates were frozen and held frozen in cold ambient air (the air outside). Red meats were also frozen and held in natural, freezing, ambient conditions.

However, so-called "quick freezing" as we know it today was started in the United States in the early 1920s. There is no universally accepted definition of quick freezing, but one description states that the fall in temperature from 32 to 25°F (0 to –3.9°C) must occur within 30 min or less. Clarence Birdseye is credited with starting the quick freezing of foods, and he must be given credit for promoting the development of frozen foods.

The Preservation Effect of Freezing

The spoilage of foods by microbes and by chemical reactions is possible because of a sufficiently high water activity in foods due to their high water content. Foods can be preserved by freezing, because the water activity can be lowered to levels that prevent the functioning of microbes and significantly reduce the rates of chemical reactions. The water in foods does not start to freeze at 32°F (0°C) because of dissolved substances in it. It starts to freeze at a slightly lower temperature. As the temperature is lowered, in the freezing cycle, water molecules in the foods begin to form ice (crystallize); there is a tendency for randomly distributed water to form the orderly network pattern of ice crystals and, as the water freezes, molecular freedom of movement becomes restricted.

When foods are allowed to freeze slowly, water molecules, even though they are slow moving, have time to migrate to ice crystals, resulting in the formation of large ice crystals. When foods are made to freeze rapidly, the sluggish water molecules do not have enough time to migrate to ice crystals at any distance and instead are "frozen in their tracks," so to speak, forming relatively small ice crystals made up of local water molecules. In most cases, to maintain top quality, small ice crystals are more desirable than large ice crystals, and rapid freezing methods are employed.

Even when food is rigid from exposure to temperatures of less than 28°F (–2.2°C), some of the water remains in liquid form, and it is not until the temperature of the food is lowered to about –76°F (–60°C) that all detectable water is converted to ice. Figure 12.2 shows the temperatures at which various percentages of the water are frozen.

FREEZING METHODS

According to some records, the earliest use of artificial freezing was in the mid-1800s, when fish were frozen in pans surrounded by ice and salt. In the late 1800s, fish, meats, and poultry were frozen by ammonia refrigeration equipment, with fish being the most important in terms of volume. The commercial freezing of fruits and vegetables started in the early 1900s, the former preceding the latter. Today, there are a number of methods that were developed for speed, quality, or for specific types of food.

Air-Blast Freezing

Air-blast freezing is one of the most commonly used procedures for freezing foods. The foods are packaged and placed on racks, and the racks are wheeled into insulated

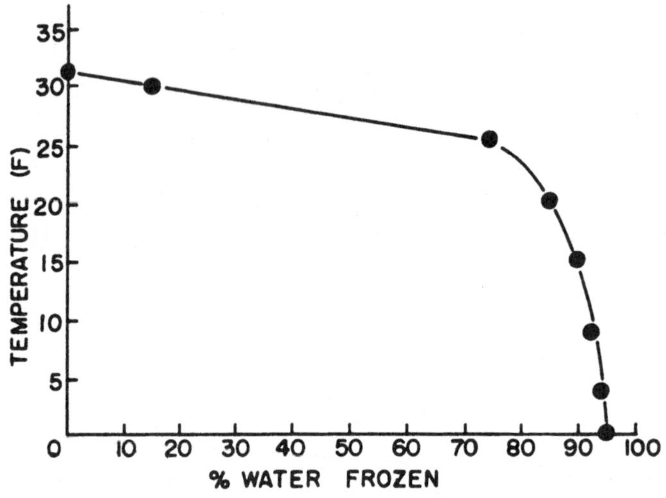

Figure 12.2. Freezing of water in foods at 0° to 32°F (–17.8° to 0°C).

tunnels (see Fig. 12.3) where air at –20 to –40°F (–28.9 to –40°C) is blown over the product at a speed of 500 to 1500 ft/min (152.4 to 457.2 m/min). When the temperature of the product reaches 0°F (–17.8°C) in all parts, the packages are put into cases and the cases are placed in storage at 0°F (–17.8°C) or below. Air-blast freezing may also be applied to packaged products placed on a belt and carried through cold air tunnels.

A modification of air-blast freezing is used to obtain free-flowing frozen products, such as peas. In this case, the product is frozen, prior to packaging, on a belt operating in a refrigerated, insulated tunnel. Air at –20°F (–28.9°C) or below is blown over the belt as it moves along. The frozen product empties from the belt into a hopper from which it is promptly removed, then packaged, cased, and stored. One drawback of the cold air-blast method is that moisture is lost to the cold air because the product is not packaged. Peas, for example, lose about 5% moisture when frozen by this method.

High-altitude freezing, a form of blast freezing, appears feasible but has not been tested and will therefore be given only minor mention. By this process, foods are frozen,

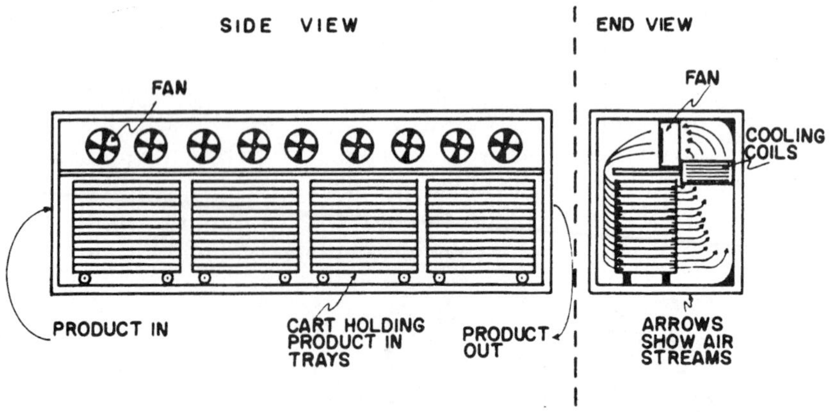

Figure 12.3. Tunnel blast freezer.

during transport, in a modified jet aircraft (Boeing 727) at the high altitudes at which these aircraft normally fly, where the ambient temperatures are as low as –60°F (–51.5°C).

Fluidized-Bed Freezing

The fluidized-bed process was described in Chapter 11 as a means of drying food particles, using hot air as the medium. The same process can be used to freeze foods when cold air is used as the medium. The use of the word *fluidized* derives from the fact that the food particles in air-suspended motion undergo a flow from the starting edge of the bed or perforated plate to the opposite edge where the process ends, thus simulating fluidity. A major advantage of this process is that the food particles are individually quick-frozen (IQF). Even particles that normally tend to agglomerate may be IQF.

Dehydrofreezing

The removal of heat from foods can be accomplished by evaporative cooling, which can be accelerated by subjecting the foods to reduced pressure (as in a vacuum). However, this process entails the loss of water, which can be prevented by spraying the product with cold water.

Plate Freezing

In plate freezing layers of the packaged product are sandwiched between metal plates. The refrigerant is allowed to expand within the plates to provide temperatures of –28°F (–33.3°C) or below, and the plates are brought closer together mechanically so that full contact is made with the packaged product. In this manner, the temperature of all parts of the product is brought to 0°F (–17.8°C) or below within a period of 1.5 to 4 hours (depending on the thickness of the product). The packages are then removed, put into cases, and stored. This method can be used for meat, fish, and dairy, and other products where agglomeration of the food particles is not a concern.

Continuous-operating plate freezers are used in commercial plate freezing. In one such system, the freezer is loaded at the front and unloaded at the rear after completion of the freezing cycle. This is done automatically and continually. In another continuous system, the packages are fed automatically on belts that place them in front of eight levels of refrigerated plates. The packages are forced into the spaces between the plates and the plates closed to provide contact. As freezing proceeds, the packages are advanced so that with each opening of the plates the packages are advanced by one row with a new set of packages entering the front row. By the time the packages reach the far side of the plates, the foods are completely frozen, and they are pushed or unloaded out of the freezer to be cased and stored.

Liquid Freezers

Liquid refrigerants such as liquid nitrogen and dry ice (carbon dioxide) may be used for the quick freezing of foods. Liquid nitrogen has a temperature of –320°F (–195.6°C), and CO_2 is at –110.2°F (–79°C) in its solid state.

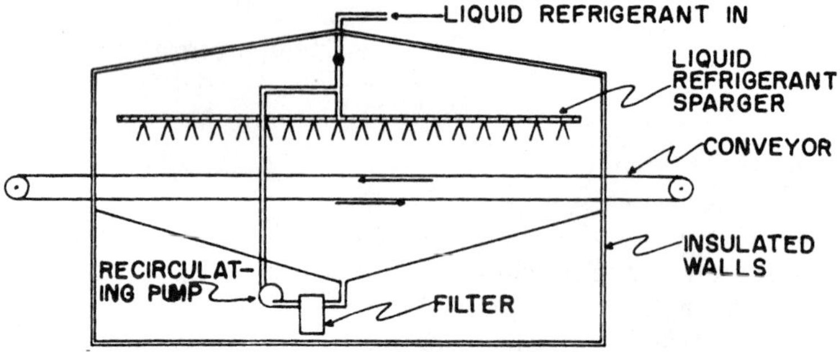

Figure 12.4. Continuous liquid-refrigerant freezer.

With liquid nitrogen, the individual food portion is placed on a moving stainless steel mesh belt in an insulated tunnel where it is sprayed with liquid refrigerant (Fig. 12.4). Excess refrigerant is recovered, filtered, and recycled. The food leaves the freezer in the frozen state and is thereafter packaged, cased, and stored. This method, often referred to as the IQF (individually quick frozen) method, provides very fast freezing and is used mostly for certain marine products, such as various forms of frozen shrimp.

When using carbon dioxide, two general methods are employed. In one, powdered CO_2 is physically mixed with the food to quickly freeze it. In the other, liquid CO_2 under pressure is sprayed onto the food surface. As the pressure is released during the spraying process, the liquid CO_2 becomes dry ice at –110.2°F (–79°C) and the food is quickly frozen.

Slow Freezing

Some foods, such as whole fish and fruit in barrels (used for the manufacture of jams and jellies), are frozen in bulk by placing them in a cold room on racks (individually or in pans) or standing them on cold room floors. The temperature of such rooms may be as low as –10 to –30°F (–23.3 to –34.4°C), and some air circulation may be used within the storage room. In such instances, freezing proceeds at a slow rate. Fish frozen in this manner are immersed in cold water or sprayed with cold water (in pans) to form a glaze coating of ice) that helps to protect it against dehydration during frozen storage. The glaze is built up in layers and, during long storage periods, must be replaced as it is lost by sublimation.

GENERAL CONSIDERATIONS OF FREEZING PRESERVATION OF FOODS

There are three recognized methods of freezing foods: fast or quick freezing, sharp freezing, and slow freezing. There is no generally accepted definition for differentiating among the various freezing rates. It is obvious that with some methods, such as for food frozen in bulk, the product is frozen slowly.

It is generally agreed that the quality of foods that are frozen quickly is better than the quality of foods that are frozen slowly, and that the lower the temperature to which the food is brought the better the retention of the characteristics of the fresh product. The reasons are that (1) rapid freezing results in the formation of a large number of very small ice crystals that are evenly distributed, and this causes less damage to the tissues of the food; (2) soluble components move about within the food to a lesser degree when the product is quick-frozen because the time required for solidification of the food is shortened; (3) the rates of chemical and biochemical changes are reduced or prevented by decreasing the temperatures rapidly to a certain point. "Quick freezing" as we know it today was started in the United States in the early 1920s.

Although the freezing rate is important to the quality of the frozen food, it should be noted that the temperature at which a frozen product is held after freezing is more important than the temperature to which it is brought during freezing. It is obvious that if a food is frozen to 0°F (–17.8°C), then stored at 10°F (–12.2°C), the same changes will take place and at the same rate as if the food were brought to only 10°F (–12.2°C), originally. In fact, additional damage is incurred when the product goes initially from 10 to 0°F (–12.2 to –17.8°C) and back to 10°F (–12.2°C), as any freezing cycle has an adverse effect on the quality of foods. Because frozen foods are stored for much longer periods than those required for freezing, storage changes are of much greater significance to the quality of the product than are the initial changes due to freezing. It has been shown by government research workers that changes causing deterioration of many frozen foods occur approximately twice as fast at 5°F (–15°C) as at 0°F (–17.8°C), twice as fast at 10°F (–12.2°C) as at 5°F (–15°C), and so on. Actually, frozen storage warehouses usually attempt to maintain temperatures of 0°F (17.8°C), an FDA requirement for all foods shipped interstate and a requirement for all food-handling establishments, including restaurants. It should be remembered that the lower the temperature, the longer the shelf-life and the better the quality.

Although changes often take place in frozen foods during storage after freezing, the major changes take place during distribution after freezing. Frozen foods may be carried to retail outlets in trucks that are not refrigerated or in which the refrigeration is not adequate to hold the temperature at 0°F (–17.8°C). Frozen foods may be delivered to the retail outlet and left on the unloading platform at ambient temperatures for several hours, sometimes even subjected to the direct rays of the sun. At the retail outlet, products are sometimes placed out of refrigeration, either in an open top display case above the load line, or out of the case entirely. Open-shelf-type frozen food display cases now used in many retail stores are usually incapable of maintaining temperatures of 0°F (–17.8°C), especially around the products displayed at the front of the shelf, because refrigerated air is blown from the back of the shelf to the front, and cold air, being heavier than warm air, tends to flow out and down to the floor area in front of the case. As a matter of fact, in stores using this type of display case, the cold air can be felt as one walks through the aisle where such storage cases are located. Because high storage temperatures greatly accelerate deteriorative changes in frozen foods, loss of their quality occurs more often during distribution and display at retail stores than during freezing and subsequent storage in frozen food warehouses.

PREPARATION OF FOODS FOR FREEZING

Many foods preserved by freezing require some type of pretreatment, but only a selection will be described here. Citrus juices are often concentrated to about one-fifth of their original volumes, then diluted back to one-fourth of the original volumes with fresh juice prior to freezing. This dilution with fresh juice is done to add back flavor components because most of the flavor compounds are removed during vacuum concentration of the juice. When this product is to be reconstituted it requires the addition of water equal to three times its volume.

With a few exceptions, such as onions, vegetables must be blanched before they are frozen and stored. Vegetables are blanched by heating in steam or hot water (about 210°F [98.9°C]), while being carried along a perforated metal belt or screw conveyor, until the temperature of all parts reaches 185 to 200°F (85 to 93.3°C), then cooling with sprays of water or in water flumes. Vegetables are blanched prior to freezing because enzymes are present, and if not inactivated by the heat, they would cause certain spoilage reactions to take place in the food during frozen storage. These reactions would produce off-flavors (haylike flavors) in the vegetable.

Blanching times vary with the type of food, the type of heat used (steam or hot water), and the bulk of material being heated. The heating must reach a high enough temperature in the product and remain at that temperature long enough to inactivate the particular enzymes that cause off-flavors. Very high temperatures, used for sterilizing certain types of canned foods, applied for a few seconds, will not destroy all the enzymes; the time is too short. Blanching times used for frozen vegetables may be relatively short (about 1 minute in water at 210 to 212°F [98.9 to 100°C] for peas) or relatively long (about 9 minutes in steam for corn on the cob).

Besides destroying enzymes that would cause off-flavors and texture changes, blanching has a cleaning effect and destroys contaminating bacteria. Moreover, blanching fixes the color of green vegetables by removing air from beneath surface tissue and possibly affecting the chlorophyll.

PACKAGING

Packaging is an important factor in the freezing of foods. It is not possible to maintain high humidities in frozen storage rooms because moisture present in the atmosphere of freezer storage rooms tends to condense and freeze on the expansion pipes or other apparatus used to cool the room, thereby reducing the cooling efficiency of the system. Packages used for frozen foods, therefore, must be reasonably moisture-vapor-proof. Otherwise, the product will dehydrate, causing undesirable changes in its appearance and accelerating loss in flavor and juiciness. Toughness and other manifestations of deterioration in quality will also be accelerated. Even when food is packaged in vapor-proof material, however, some moisture can be lost from the food. This may happen when there is too much space inside the package not occupied by the food itself. This results in the formation of what is called cavity ice or ice within the package. In such instances, the product, which is warmer than the package, loses water to the cavity space by evaporation, as any water vapor in the cavity space is condensed on the colder package. Under these conditions, there is a transfer of water from the food to the inside of that part of the package surrounding the cavity. Cavity ice formation may cause the same undesirable changes to take place as moisture loss to the atmosphere outside the package, although it usually occurs in localized areas.

The frozen-food package should have sufficient mechanical strength; that is, it should have adequate burst strength and tear strength at low temperatures and high wet strength against water exposure.

During freezing, the package will be subjected to stresses as a result of the expansion of the product. Because of this and possible cavity ice formation, the package should conform closely to the shape of the food. A suitable flexibility is, therefore, a desirable property of the packaging material.

Packages should be liquid-tight because some frozen foods (such as fruits packed in syrup) may have some free liquid that could leak out, and some are thawed in the package, which could also cause leakage.

Transfer of moisture-vapor through packaging may take place through pores or cracks in the container or by diffusion of moisture through the packaging material. The loss of moisture through seals, and package imperfections such as pores, may account for the greatest loss of moisture from frozen foods.

Tin or aluminum cans, waxed paper tubs or cylindrical containers, rectangular paper cartons treated with special waxes, plastic bags of polyvinylidene chloride or polyethylene, and aluminum foil wraps or aluminum dishes are all used for the packaging of frozen foods. Many of the paper containers are used in combination with cellophane, waxed paper liners, or overwraps.

PROBLEMS WITH FREEZING OR THAWING OF BULK FOODS

More of the changes occurring in frozen foods take place during frozen storage rather than during the freezing process itself. There are some instances, however, when foods are frozen very slowly, during which time undesirable changes occur. When strawberries in sugar are frozen in bulk form in cold rooms, the freezing time is long, and the deterioration during freezing is equivalent to that which would occur in 2 years of storage at 0°F (–17.8°C) after freezing.

The defrosting of bulk-packed foods, fruit in barrels, liquid egg frozen in 30-lb (13.6-kg) tins, and so forth may be responsible for a considerable loss of quality. When frozen, bulk-packed foods are allowed to defrost at room temperature, the outside layers of the food will have been held at room temperature for a long period of time before the inner layers become defrosted. This period of holding at room temperature may result in enzyme changes or in loss of quality caused by the growth of microorganisms.

QUALITY CHANGES DURING FROZEN STORAGE

A number of different changes may take place in frozen foods during storage. These changes may be physical, chemical, enzymatic, and, in rare cases, microbial. Microbial changes occur when refrigeration is inadequate.

Desiccation or drying out is a kind of change that takes place in frozen foods under conditions of poor packaging or varying storage temperatures. In poultry such loss of moisture occurs first around the area of the feather follicles, causing a speckled or "pock-marked" appearance. Protein and fat changes may also be accelerated when moisture is lost from the surface areas of frozen foods.

Crystallization is a physical change that may occur in some types of frozen foods. Certain types of dairy products, such as ice cream and concentrated milks or creams, sometimes undergo this type of change, which is due to the crystallization of lactose or other sugars that do not readily dissolve on defrosting, causing an undesirable texture called sandiness. A similar change may occur in some types of sweetened citrus juices during long storage.

Loss of volatile flavor components may occur in some frozen foods, such as fruit, during frozen storage because these compounds boil or evaporate at temperatures that are even lower than the temperature at which the foods are stored. This causes a loss of the typical flavor components of the foods.

The breaking of gels or emulsions during defrosting may take place with some foods. High-moisture fruits, such as tomatoes, the liquid of which is held by pectin gels, are subject to this type of change and have, therefore, not been frozen successfully. Foods packed with a white sauce or gravy are also subject to this type of physical change and it has been found that fluctuation of storage temperatures greatly accelerates the curdling and "weeping" of the gravy or white sauce.

Protein denaturation is a general term for the physiochemical change occurring especially in flesh-type foods during frozen storage. It results in a toughening of the tissues and dryness or loss of succulence. It is believed that this type of change in stored frozen fish is due to enzymes that break off fatty acids from phospholipids in tissues or to enzymes that decompose trimethylamine oxide and produce formaldehyde. The free fatty acids combine with protein chains, which causes them to aggregate or bind themselves into a closer matrix, resulting in the squeezing out of water (drip occurring during defrosting) and a firming or toughening of the tissues. Protein changes of this type occur especially in lean fish but also, to some extent, in poultry and meats. Grinding of tissues of this kind appears to accelerate protein changes during frozen storage.

As might be expected, the oxidation of some of the components of frozen foods is a cause of some of the major changes that may take place during storage. Oxidation of ascorbic acid or vitamin C is a change of this kind. In fruits in which enzymes have not been inactivated by heating, this may be accelerated by enzyme action, but it can also occur through straight atmospheric oxidation. Although this change may cause no loss of flavor or texture, it results in a loss of the nutritional quality of the food.

The fats of foods may oxidize during frozen storage, resulting in a change recognized as rancidity. The fats of meats may undergo this change but fish fats are especially subject to this kind of deterioration because the fatty acids in fish fats contain many adjacent carbons not fully saturated (combined) with hydrogen. They are, thus, very subject to reacting with oxygen, a process that eventually leads to the formation of compounds that cause the off-flavors recognized as rancid flavors. Fatty fish become rancid faster in frozen storage than do lean fish, but even lean fish are subject to rancidification during relatively short periods of frozen storage.

The colors of fruits and vegetables may deteriorate through oxidation during frozen storage, resulting in a less desirable appearance.

Enzymes may cause various changes in foods during frozen storage. It is not possible to freeze and store whole uncooked lobster, because, during storage, enzymes react with the proteins, causing the meat to become soft and crumbly after cooking. On the other hand, if whole lobsters are cooked to inactivate the enzymes and then frozen and stored, the oil in the digestive gland (tomalley) oxidizes and becomes rancid, causing off-flavors that spread into the meat itself.

In fruits, enzymes, called polyphenolases, accelerate the oxidation of certain chemicals leading to the formation of brown- or black-colored compounds. This is the reaction that one sees when an apple or peach is cut and allowed to stand at room temperature for a short period of time. In preparing apples for freezing, therefore, peeled apple slices are treated with salt that liberates sulfurous acid (sulfur dioxide) and are held under refrigeration for sufficient time to allow this compound to diffuse into the tissues.

Table 12.1. Approximate Time[a] (in months) of High-Quality Shelf-Life of Some Foods

Product	0°F (–17.8°C)	10°F (–12.2°C)	20°F (–6.7°C)
Orange juice (blanched)	27	4	1
Peaches	12	<2	0.2
Strawberries	12	2.4	10 days
Cauliflower	12	2.4	10 days
Green beans	11–12	3	1
Green peas	11–12	3	1
Spinach	6–7	<3	0.75
Raw chicken	27	15.5	<8
Fried chicken	<3	<1	<0.6
Turkey pies or dinners	<30	9.5	2.25
Beef (raw)	13–14	5	<2
Pork (raw)	10	<4	<1.5
Lean fish (raw)	3	<2.25	<1.5
Fat fish (raw)	2	1.5	0.8

[a]It should be noted that the above storage life times refer to those periods at which a quality difference between the frozen and the fresh product can first be detected by an expert panel and do not refer to spoilage or rejection times.

The sulfur dioxide inactivates or inhibits the enzymes that promote enzymatic browning. Also, in preparing sliced peaches for freezing, small amounts of ascorbic acid (vitamin C) are added to the syrup because this compound is a reducing agent (counteracts oxidation reactions).

In general, microorganisms will not grow in frozen foods unless they are held at temperatures above 15°F (–9.4°C). At this temperature there may be enough free liquid in the food to allow the growth of some microorganisms, especially molds. However, at such a rate the food would soon be spoiled because of changes other than those caused by the growth of microorganisms.

Frozen foods have not often been involved in the transmission of foodborne disease, although, if grossly mishandled, such as by defrosting and then holding out of refrigeration for some time, they may constitute a public health hazard.

SHELF-LIFE OF FROZEN FOODS

The storage life of frozen foods, at temperatures that today are considered to be economically practicable, is not without definite limits. Some idea of the length of time that frozen foods may be expected to retain high quality may be obtained from Table 12.1.

THAWING

Although freezing is one of the most effective means of preserving foods over long periods, the need to thaw them prior to reprocessing in food plants or for domestic use represents one of the undesirable aspects of freezing preservation. Thawing is time-consuming and, in some cases, is associated with loss of product quality. Evidence shows that it normally takes food longer to thaw than to freeze under similar heat transfer conditions. The Educational Foundation of the National Restaurant Association and the FDA recognize three accepted methods for thawing foods:

1. Under refrigeration below 40°F (below 4.4°C).
2. Under running water below 70°F (below 21.1°C).
3. As part of the cooking process. This includes microwave thawing, which should be used only on foods to be prepared for service immediately after thawing.

Foods frozen in bulk, such as barrels of fruits, 30-lb (13.6-kg) containers of egg mixtures, fruit, large fish, or blocks of meat used for sausage products, may present a defrosting problem. Because bulk-frozen foods take long to defrost and because it is well known that the rate at which the food defrosts is dependent on the temperature to which it is exposed, there may be a tendency to defrost the food at relatively warm temperatures. When this is done, the outside portion of the food is subject to bacterial decomposition or the growth of yeasts and molds before the inner portions defrost. Some methods of alleviating the problems associated with the defrosting of bulk-frozen foods have been developed.

Refrigerator defrosting (holding at temperatures of 35 to 40°F [1.7 to 4.4°C]) is probably the best method of defrosting bulk-frozen foods when no fast method is available. This would apply to large whole fish, fruit in barrels, and apples in 30-lb (13.6-kg) containers, as bacterial or mold growth would be limited under these conditions. However, in industrial processing, where bulk-frozen products are thawed as an intermediate step in the manufacture of the company's line of products, the refrigeration space required may be so large as to discourage this practice.

In defrosting eggs in 30-lb (13.6-kg) containers, it maybe possible to use a machine that breaks up or grinds the product into a kind of slush that can be used in that form in preparing baked foods.

Large blocks of meat used for manufacturing sausage products can be ground in the frozen state and used as such.

By the use of microwave energy, food can be thawed rapidly and with virtually no quality loss. If microwave thawing is used, it is recommended that the food be cooked or processed immediately.

Because the heat generated in foods by microwaves is quite rapid (about 10 times more rapid than by baking), when uneven heating in a frozen product does occur, the temperature differences within a food can become great. This, however, happens only under certain conditions, and it can be dealt with quite easily. For this condition, and also when one wants to ensure uniform temperature control, one solution is to apply the microwave energy in intermittent bursts. By this technique, the absorbed thermal energy generated during a burst of microwaves is allowed to be distributed by conduction during the intervals between the bursts, thereby permitting the temperature of the food to increase more uniformly, albeit more slowly. Developments such as wave guides and turn tables have improved the distribution of microwave energy significantly.

The particular advantage of using microwave energy for thawing foods is that deterioration by microorganisms is not a factor. The feasibility and benefits of microwave thawing of frozen meats and fish have been adequately demonstrated, especially for thawing frozen shrimp blocks.

13
Food Additives

Up until about 1906, food handled in the United States was often produced and processed under unsafe and unsanitary conditions and there was little control over chemical additives used as preservatives or colorings. In 1906, Upton Sinclair published his book entitled *The Jungle*, which was based upon the meat-packing industry in Chicago. In the book, Sinclair intended to focus on poor working conditions and exploitation of workers, but his description of how meat products were handled led to the passage of the Meat Inspection Act of 1907 and the Federal Food and Drug Act of 1906. At that time, it was quite common to have floor sweepings added to pepper, ash leaves to tea, brick dust to cocoa, copper salts to pickles and peas, and lead salts to candy. Approximately 80 different dyes were used in foods, and sometimes the same batch used for coloring textiles was used for coloring food.

In 1903, Dr. Harvey Washington Wiley, then the Chief of Bureau of Chemistry of the U.S. Department of Agriculture, established a "poison squad" that consisted of young men who consumed foods treated with known amounts of chemicals commonly used in foods. The goal of the project was to determine whether these compounds were deleterious to health. The result of the efforts of Dr. Wiley and the "squad" was the passage of the Food and Drug Act of 1906, which is also referred to as "The Pure Food Act."

The Bureau of Chemistry was the enforcement agency of the 1906 Act until 1927, when research and enforcement functions were separated and the Food, Drug and Insecticide Administration was established. The name of this organization was changed to the Food and Drug Administration (FDA) in 1931, and in 1940 it became a unit of the Department of Health and Human Services. In 1938, the Food, Drug and Cosmetic Act (FD&C Act) was passed into law. This Act gave the government authority to conduct on-site inspections and provide for the establishment of standards of identity for individual food products. The Act also allowed the government to obtain federal court injunctions against violators.

Food additives have been defined as chemical substances deliberately added to foods, directly or indirectly, in known and regulated quantities, for purposes of assisting in the processing of foods; preservation of foods; or in improving the flavor, texture, or appearance of foods.

In September of 1958, the FD&C Act was amended to prohibit the use of food additives that had not been adequately tested to establish their safety. The term food additive was defined as follows:

> . . . any substance the intended use of which results or may reasonably be expected to result, directly or indirectly, in its becoming a component or otherwise affecting the characteristics of any food (including any substance intended for use in

> producing, preparing, treating, packaging, transporting or holding food; and including any source of radiation intended for any such use), if such substance is not generally recognized among experts qualified by scientific training and experience to evaluate its safety, as having been adequately shown through scientific procedures (or in the case of a substance used in food prior to January 1, 1958, through either scientific procedures, or experience based on common use in food) to be safe under the conditions of its intended use . . .

An additive may be reactive or inactive; it may be nutritive or nonnutritive; it should be neither toxic nor hazardous. Some substances, such as pesticides, are added to foods unintentionally, and these are, of course, undesirable, and may be hazardous to health. Because of their toxicity, their presence is closely regulated by strict government tolerance guidelines. The Environmental Protection Agency (EPA), established in 1970, is responsible for establishing tolerances for pesticides and the FDA is responsible for monitoring and ensuring compliance to these tolerances for agricultural commodities.

As a result of the Food Additives Amendment of 1958, the term "Generally Recognized as Safe" (GRAS) evolved. Additives are classified as GRAS when they have been used without apparent harm for long periods of time, long before regulations had been put into effect. These include substances such as baking powder chemicals (e.g., sodium bicarbonate), fruit acids such as citric and malic, and gums such as agar-agar. The purpose of this list was to recognize the safety of basic substances without complicated safety testing.

To get new food additives approved, a petition must be submitted to FDA that contains scientific data that clearly show that the intended chemical is harmless in the intended food application at the intended use level. The burden of proving the safety of the additive lies with the company that wishes to use or sell the chemical. This testing can often require several years because the FDA often requires that the additive undergo at least a 2-year feeding study in two species of animals. These studies must reveal both long-term and short-term effects.

A very controversial clause was included in the Food, Drug and Cosmetic Act. A portion of it stated that:

> ". . . no additive shall be deemed to be safe if it is found to induce cancer when ingested by man or animal, or if it is found, after tests which are appropriate for the evaluation of the safety of food additives, to induce cancer in man or animal."

This clause, known as the Delaney Clause or Amendment, is named after Representative James Delaney of New York, who introduced it into the legislation. It essentially states that if the additive, at any level, can produce cancer in humans or animals or can be shown to be carcinogenic by any other appropriate test, it will not be allowed. This has been a very controversial clause because, although it was added to ensure safety, some additives may cause cancer in some species of animal when fed at levels that would not be possible for humans to ingest under normal food consumption. The use of informed scientific judgment in regulatory decisions may result in a modification of the Delaney Clause in the future.

Ionization radiation is considered an additive because the treatment may induce changes in food. In this case, the FDA tests foods that have been irradiated and approves irradiation sources and maximum dosages.

Besides the requirements mentioned, food additives must also meet five general criteria:

1. Intentional additives must perform their intended function.
2. Additives must not deceive the consumer or conceal faulty ingredients or defects in manufacturing practices.
3. Additives cannot considerably reduce the food's nutritional value.
4. An additive cannot be used to achieve an effect that could be gained by good manufacturing practices.
5. A method of analysis must exist to monitor the use of the additive in foods or its incidental occurrence in foods (such as migration from a packaging material).

PHILOSOPHY OF FOOD ADDITIVES

Foods are made entirely of substances that, in the pure form, can be described as chemicals or chemical compounds. It is important to note that our knowledge of the composition of foods, because of its complexity, is by no means complete. For instance, it is reported that one of the most important of our natural foods, human milk, contains several hundred chemical compounds.

Unfortunately, the interpretation of the word *chemical* is too often inaccurate. Thus, some consumers are apprehensive about purchasing a food that is preserved by treating it with a chemical with which they are unfamiliar. However, a number of foods may be preserved with table salt, which is a chemical. Consumers are not apprehensive about using salt as a preservative, because they are familiar with it, at least for adding taste and sometimes for bringing out the flavor in foods, yet table salt is definitely a chemical, with the name sodium chloride, and the formula NaCl. Refined sugar, vinegar, spices, and other substances that are routinely added to foods are also chemicals or mixtures of chemicals, and the use of these is not questioned. The characteristics of chemicals that we use with confidence are familiarity and frequent use. The characteristics of chemicals that arouse skepticism in consumers are that they are uncommon and unfamiliar.

A large number of chemical additives are unfamiliar, and there is a need for regulatory agencies to question their use from the standpoint of safety. Obviously then, we should not fear the use of chemicals, but we do need to screen them for safety when their effects on human health are not known. Some lessons have been learned along these lines. For example, indiscreet use of certain additives for coloring candy and popcorn was reported to have caused diarrhea in children, resulting in the removal of these dyes from the FDA approved list of additives. There are a number of related concepts that must be remembered when dealing with food additives:

- All foods are composed of chemical compounds, many of which can be extracted and added to other foods, in which case they are classified as additives.
- Any additive or chemical compound can be injurious to health when particularly high levels of that compound are added to foods.
- Any additive or chemical compound can be safe to use when particularly low levels of that compound are added to foods.
- It is necessary to evaluate each additive for usefulness and toxicity in a sensible, scientific way, regardless of how safe its proponents say it is and how toxic its opponents say it is.

The use of radiation for preserving foods has been declared an additive, and whether or not it should be approved by the FDA makes it the prime example of extreme opposition and extreme favor. Quite often, the tendency to take a strong position for the use of an additive might make a proponent rationalize or overlook undesirable investigative scientific data concerning the additive. On the other hand, opponents tend to make irrational demands of investigators to prove the safety of an additive; for example, opponents of the use of radiation for preserving foods have suggested that radiation should not be approved for preserving foods until all possible chemical effects of the process have been identified. This, without going into detail, is an impossible task. It would be just as impossible to identify all the chemical effects of frying foods and of baking food.

Given present capabilities, our most reasonable evaluation of an additive for safety can be made through conventional animal feeding studies. The overall physiological effects that an additive may have on animals of two or three different species over a specified number of generations is the most comprehensive, as well as the most reliable, way to evaluate the safety of a food additive.

It should be remembered that chemical materials cannot be added to foods unless their use, in the quantities added, has been approved by the FDA. Moreover, additives are tested for toxicity in concentrations much greater than those allowed in foods. It should also be remembered that most food additives are components of natural foods and that without these additives the quality of many foods would be greatly inferior to that to which we have become accustomed. The shelf-life or availability of many foods would also be greatly limited were all additives to be eliminated from foods. Food additives are difficult to classify mainly because they overlap each other in numerous combinations of effects. It should be remembered, therefore, that the following classifications are not precise.

Food additives may be used for a number of reasons. At present, over 3000 intentional additives are allowed and they can be divided into several major groups. In this chapter, the major groups are covered and representative additives from each group are mentioned. No attempt is made to cover every food additive that exists.

ANTIOXIDANTS

Antioxidants are food additives used, since about 1947, to stabilize foods that by their composition would otherwise undergo significant loss in quality in the presence of oxygen. Oxidative quality changes in foods include: (1) the development of rancidity from the oxidation of unsaturated fats resulting in off-odors and off-flavors and (2) discoloration from oxidation of pigments or other components of the food.

There is a large number of antioxidants, and although they may function in different ways, the purpose of each is to prevent, delay, or minimize the oxidation of the food to which they are added. One of the ways by which some antioxidants function involves their combination with oxygen. Others prevent oxygen from reacting with components of the food. When only a limited amount of oxygen is present, as in a hermetically sealed container, it is possible for some antioxidants to use up all of the available free oxygen, because they have a relatively great affinity for it. Some antioxidants lose their effectiveness when they combine with oxygen; therefore, there is no advantage to using this type of antioxidant unless the food is enclosed in a system from which oxygen or air can be excluded. With the use of antioxidants, it should be noted that other precautions are necessary to minimize oxidation, because heat, light, and metals are prooxidants, that is, their presence favors oxidative reactions. Many of the antioxidants used commercially occur naturally in foods (e.g., vitamin C, vitamin E, citric acid, amines, and certain phenolic compounds). However, the amines and the phenolic compounds can be toxic to humans in low concentrations; therefore their use and that of synthetic antioxidants require strict regulation. It should be pointed out that the potency of the naturally occurring antioxidants is not as great as that of the commonly used synthetic antioxidants. The antioxidants that are considered to be the most effective and therefore are most widely used are butylated hydroxyanisole (BHA), butylated hydroxytoluene (BHT), and propylgallate. These are generally used in formulations that contain combinations of two or all three of them, and often in combination with a fourth component, frequently citric acid. The main purpose of adding citric acid is that it serves as a chelator or sequestrant (a chelator ties up metals, thereby preventing metal catalysis of oxidative reactions).

Fats and shortenings, especially those used in bakery goods and fried foods, are subject to oxidation and the development of rancidity after cooking. To prevent this, chemical antioxidants in concentrations up to 0.02% of the weight of the fat component may be added.

The use of antioxidants is regulated by the FDA and is subject to other regulations, such as the Meat Inspection Act and the Poultry Inspection Act. Their use is limited so that the maximum amount that can be added is generally 0.02% of the fat content of the food; however, there are some exceptions to, and variations of, that rule.

NUTRIENT ADDITIVES

The need for a balanced and ample nutrient intake by the human body is well known. Although nutrients are available in foods, losses of fractional amounts of some of them through processing and increasing frequency of improper dieting have led to the practice of adding minimum daily requirements or sizable fractions of minimum daily requirements of a number of nutrients to popular foods, such as breakfast cereals, baked goods, pasta products, and low-calorie breakfast drinks. Nutrient additives include mainly vitamins, proteins, and minerals.

Vitamin D is an exceptional example of the value of the food additive concept. The major source of vitamin D for humans is a precursor compound called 7-dehydrocholesterol which is produced in the liver. It circulates to an area just under the skin and is converted to previtamin D_3 by the ultraviolet rays of sunlight. Previtamin D_3 then goes through a number of steps and is converted to vitamin D_3 and finally to active vitamin D. However, in many cases, exposure to the sun is sporadic and insufficient, especially in areas where there is normally insufficient sunshine or in cases where sunlight exposure is of insufficient duration. Thus, vitamin D is added to nearly all commercial milk in a ration of 10 micrograms of vitamin D (as cholecalciferol). This is equivalent to the old 400 IU per quart (0.95 liter). Vitamins A and C and some of the B vitamins are also added to some foods.

The addition of protein concentrate (produced from fish or soybeans) to components of the diet of inhabitants of underdeveloped countries has been used successfully to remedy the high incidence of protein malnutrition. It should be noted that soybean protein is incomplete and requires the addition of some amino acids in which it is deficient. Children, especially, succumb in large numbers to the disease, kwashiorkor, that results from insufficient protein intake.

Among minerals, iron has received major attention as a food additive, mainly because of its role in preventing certain anemias.

FLAVORINGS

Flavorings are compounds, natural or synthetic, that are added to foods to produce flavors or to modify existing flavors. In the early days of human existence, salt, sugar, vinegar, herbs, spices, smoke, honey, and berries were added to foods to improve their taste or to produce a special, desirable taste. The variety of natural and synthetic flavorings available to the modern food technologist is very large. Essential oils provide a major source of flavorings. Essential oils are odorous components of plants and plant materials that give the characteristic odors of the materials from which they are extracted.

Because of the large production of orange juice, quantities of essential oil of orange are produced as byproducts. For this reason, there is little need for the production of synthetic orange flavoring.

Fruit extracts have been used as flavorings, but these are relatively weak when compared to essential oils and oleoresins. An oleoresin is a solvent extract of spices from which the solvent, usually a hydrocarbon, has been removed by distillation. Because of their weak effects, fruit extracts may be intensified by combining them with other flavorings.

Synthetic flavorings are usually less expensive and more plentiful than natural flavorings. On the other hand, natural flavorings are often more acceptable, but they are quite complex and difficult to reproduce synthetically. In fact, one of the problems with natural flavorings is that they may vary according to season and other uncontrollable factors. Synthetic flavorings, however, can be reproduced quite accurately. Many artificial flavors, such as amyl acetate (artificial banana flavor), benzaldehyde (artificial cherry flavor), and ethyl caproate (artificial pineapple flavor), are added to confectioneries, baked products, soft drinks, and ice cream. These flavorings are added in very small amounts, often 0.03% or less.

FLAVOR ENHANCERS

Flavorings either impart a particular flavor to food or modify flavors already present. Flavor enhancers, on the other hand, intensify flavors already present, especially when the desirable flavors are relatively weak. Monosodium glutamate (MSG) is one of the best known and most widely used flavor enhancers. This compound occurs naturally in many foods and in a certain seaweed that was used for centuries as a flavor enhancer in soups and other foods. It is only within the last hundred years that the reason for the effectiveness of the seaweed was discovered to be MSG. While it is effective at relatively low levels (parts per thousand), there are other compounds called flavor potentiators that also enhance flavors but are extremely powerful, effective in parts per million and even per billion. These compounds have been identified as nucleotides, and their effect is attributed to their synergistic properties (properties that intensify the effect of natural flavor components). Two in this group are disodium inosinate and disodium GMP.

Several theories attempt to explain how MSG and other flavor enhancers and potentiators work. One theory is that they increase the sensitivity of the taste buds, thus increasing flavor. A second suggests that an increase in salivation as a result of the flavor enhancers will increase flavor perception. A third theory of intensified flavor perception is based on the observation that flavor enhancers produce certain physical sensations in the mouth such as coolness and heat.

ACIDULANTS

From the root word, *acid,* in acidulants, one can conclude that this class of compounds tends to lower the pH of any food into which the compounds are incorporated. Acidulants also enhance desirable flavors, and in many cases, such as in pickled products, are the major taste component. Vinegar (acetic acid, CH_3COOH) is added to relishes, chili sauce, ketchup, and condiments as a flavor component and to aid in the preservation of these products. Because the microbial spoilage of food is inhibited as the pH is lowered, acidulants are used for that purpose in many cases. Many acidulants occur naturally in foods (e.g., citric acid in citrus fruits, malic acid in apples, acetic acid in vinegars; all three are contained in figs). Tartaric acid is widely used to lend tartness and enhance flavor. Citric acid is widely used in carbonated soft drinks. Phosphoric acid is one of the very few inorganic acids used as an acidulant in foods. It is widely used, comprising 25% of all the acidulants in foods. Citric acid accounts for 60% of all acidulants used in foods.

In addition to their preservative and flavor enhancing effects, acidulants are used to improve gelling properties and texture. Acidulants are also used as cleaners of dairy equipment.

Acidulants may be used in the manufacture of processed cheese and cheese spreads for the purpose of emulsification as well as to provide a desirable tartness.

Acid salts may be added to soft drinks to provide a buffering action (buffers tend to prevent changes in pH) which will prevent excess tartness. In some cases, acid salts are used to inhibit mold growth (e.g., calcium propionate added to bread).

As has been pointed out, all microorganisms have a pH at which they grow best (see Chapter 3), and a range of pH above or below which they will not grow. Generally, it is not possible to preserve all foods by adding acid to the point where microorganisms will not grow. Most foods would be too acid to be palatable. The amount of acid may be enough to inhibit the growth of microorganisms provided that such treatment is combined with some other method of preservation. Certain dairy products, such as sour cream, and fermented vegetables, such as sauerkraut, are preserved with lactic acid produced by the growth of bacteria. Addition of the acid, along with holding at refrigerator temperatures above freezing, in combination will prevent growth of pathogenic and spoilage organisms. When sauerkraut is canned, it is given a heat process sufficient to destroy all spoilage and disease microorganisms.

Pickles are preserved by the addition of some salt, some acid, and a heat process sufficient to raise the temperature of all parts of the food to or near 212°F (100°C).

Pickled herring are preserved by the addition of some salt, some acetic acid (vinegar), and holding at refrigerator temperatures above freezing. In this case, the nonacid part of the acetic acid molecule has an inhibiting effect on the growth of microorganisms.

ALKALINE COMPOUNDS

Alkaline compounds are compounds that raise the pH. Alkaline compounds, such as sodium hydroxide or potassium hydroxide, may be used to neutralize excess acid that can develop in natural or cultured fermented foods. Thus, the acid in cream may be partially neutralized prior to churning in the manufacture of butter. If this were not done, the excess acid would result in the development of undesirable flavors. Sodium carbonate and sodium bicarbonate are used to refine rendered fats. Alkaline compounds are also added to chlorinated drinking water to adjust the pH to high enough levels to control the corrosive effects of chlorine on pipes and equipment. Sodium carbonate is also used in conjunction with other compounds to reduce the amount of hardness in drinking water. Sodium hydroxide is used to modify starches and in the production of caramel. Sodium bicarbonate is used as an ingredient of baking powder, which is used in baked products. (It is also a common household item used in a variety of cooking recipes.) Its action is described in the "Leavening Agents" section of this chapter. Alkaline compounds are used in the production of chocolate and to adjust the acidity level in grape juice and other fruit juices that are to be fermented in the production of wine.

It is important to note that some alkaline compounds, such as sodium bicarbonate, are relatively mild and safe to use, while others, such as sodium hydroxide and potassium hydroxide, are relatively powerful reagents and should not be handled by inexperienced people.

SWEETENERS

Sweetening agents are added to a large number of foods and beverages. Table sugar (sucrose), the most commonly used sweetener in the country, and corn syrup, are covered in Chapter 22 and therefore are not described in any detail here. Sweeteners include other sugars, as well as an abundance of natural and synthetic agents of varying strengths and caloric values.

Many sweeteners are classified as nonnutritive sweeteners. Although this classification might imply a lack of nutritional value, the implication is correct only in a relative sense. That is, the caloric value of a nonnutritive sweetener, such as aspartame, is about 4 cal/g, the same as that for sugar. However, because it takes only 1 g of aspartame to provide the same sweetness level as about 180 g of sugar (sucrose), it can be seen that the caloric contribution of aspartame is only about 0.5% that of sucrose. It is on this basis that a nonnutritive sweetener is classified as such.

The sweeteners described in this chapter are fructose, molasses, honey, maple sugar, lactose, maltose, some polyhydric alcohols (xylitol, sorbitol, mannitol), aspartame, saccharin, glycyrrhizin, and acesulfame K.

Fructose

Of the natural sugars other than sucrose used by humans, fructose (also known as levulose), a monosaccharide ($C_6H_{12}O_6$), is the sweetest (nearly twice as sweet as table sugar, sucrose) and it is the most water-soluble. It is hygroscopic, making it an excellent humectant when used in baked goods. The value of a humectant in baked goods is that it retards dehydration. Solutions of fructose have a low viscosity that results in lower "body" feel than sucrose but have greater flexibility of use over a wide range of temperatures. Because of its greater solubility and more effective sweetness than sucrose, fructose is a better choice than sucrose when very sweet solutions are required, as fructose will not crystallize out of solution, whereas sucrose will. Fructose has sometimes been called fruit sugar, since it occurs in many fruits and berries. It also occurs as a major component in honey, corn syrup, cane sugar, and beet sugar. In fact, sucrose, a disaccharide, is composed of glucose and fructose. Of these two components, the glucose moiety, or portion, cannot be metabolized by people with diabetes, and it is for this reason that the ingestion of sucrose cannot be tolerated by them. Fructose, on the other hand, does not require insulin for its metabolism and can, therefore, be used by diabetic individuals. Its use also appears to reduce the incidence of dental caries. When used with saccharin, it tends to mask the bitter aftertaste of saccharin. As it apparently accelerates the metabolism of alcohol, it has been used to treat those suffering from overdoses of alcohol. It has been recommended as a rapid source of energy for athletes and, in combination with gluconate and saccharin, as an economic, effective, safe, low-calorie sweetener for beverages.

Molasses

Molasses can be considered a byproduct of sugar production (see Chapter 22). The use of molasses as a sweetener in human foods is largely in baked goods that include bread, cookies, and cakes. In addition to sweetening, molasses adds flavor and acts as a humectant. It is also used in baked beans and in the production of rum and molasses alcohol. (The greatest use of molasses, however, is in the production of animal feed). Molasses comprises about 60% sucrose, but the sucrose content can be lower, depending on the grade of the molasses and on the raw material from which it was produced. Thus, the sucrose content of cane blackstrap (the final fraction of cane molasses) is only about one-half that of beet blackstrap (the final fraction of beet molasses). The fractions produced before the blackstrap are of higher grades and are those usually used for human consumption. Blackstrap generally is used for industrial purposes.

Honey

Honey, a natural viscous syrup, comprises mainly invert sugar. It is produced from the nectar of flowers, which is mainly sucrose, by the action of an invertase enzyme that is secreted by the honey bee. Honey is used as a direct sweetener, as an additive in a number of products, including baked goods, as well as in other ways. It is relatively expensive.

Invert sugar, corn sugar, and corn syrup are covered in Chapter 22 and are not covered here.

Maple Sugar

Maple sugar is produced from the sap of the sugar maple tree. It is comprised mainly of sucrose and small amounts of other sugars, including invert sugar. Maple sugar is used in the manufacture of candies, fudge, baked goods, and toppings. It is among the most expensive of sweeteners.

Lactose

Lactose ($C_{12}H_{22}O_{11}$), the sugar component of mammalian milk, is less sweet and less water-soluble than sucrose. Although most babies and young children generally are able to metabolize this sugar, some are unable to do so. The ability to metabolize the sugar appears to decrease with age. When a person is unable to metabolize lactose, the ingestion of milk may cause intestinal discomfort, cramps, and diarrhea. The major source of lactose is whey, a cheese byproduct. Because lactose is not as sweet as sucrose, larger amounts can be used in those foods in which the texture benefits from a high solids content.

Maltose

Maltose ($C_{12}H_{22}O_{11}$), or malt sugar, is produced during the malting process in brewing (enzyme conversion of starch). It is converted to alcohol by the action of yeasts through an intermediate conversion to dextrose. This sugar is much less sweet than sucrose, and it is used mainly in the manufacture of baked goods and infant foods.

Xylitol

Xylitol is a polyhydric alcohol having the formula ($C_5H_7(OH)_5$). At present it is used as a sweetener in chewing gum, mainly because of its noncariogenic property (it has not been found to cause tooth decay). It occurs naturally as a constituent of many fruits and vegetables, and is a normal intermediary product of carbohydrate metabolism in humans and in animals. Commercially, it is produced by the hydrolysis of xylan (which is present in many plants) to xylose, which is then hydrogenated to produce xylitol. The xylitol is then purified and crystallized. Xylitol imparts a sweet taste, which also appears to have a cooling effect. As it is not metabolized by many microorganisms, it is quite stable.

Sorbitol

Sorbitol is a polyhydric alcohol ($C_6H_8(OH)_5$) that is found in red seaweed and in fruits (apples, cherries, peaches, pears, and prunes). It was first isolated from the sorb berries of the mountain ash, hence its name. It is used as an additive because of its humectant property as well as its sweetening effect. It is used in cough syrup, mouthwashes, and toothpastes. Another of its desirable properties is that it is not easily fermented by microorganisms. Because sorbitol is largely transformed to fructose by liver enzymes in the body, it is tolerated by diabetic individuals, as fructose is not dependent on the availability of insulin for its metabolism. Sorbitol can be produced industrially by the electrochemical reduction or catalytic hydrogenation of glucose.

Mannitol

Mannitol is a polyhydric alcohol having the formula ($C_6H_8(OH)_6$). It is used in chewing gum, pharmaceuticals, and in some foods. It is a naturally occurring sweetener in many plants, algae, and mold. It occurs in the sap of the manna tree, an ash native to southern Italy, and can also be made by the reduction of either of the monosaccharides mannose or galactose. Industrially, it is produced by electrochemical reduction or catalytic hydrogenation methods. Although it is similar to sorbitol in many respects, it is less soluble than sorbitol.

Aspartame

Aspartame is the common name for aspartyl-phenylalanine. It is a combination of the two amino acids from which its name is derived. First produced in 1969, it is reputed to be about 180 times sweeter than sucrose. Like cyclamate, it was approved and later banned by the FDA. Exhaustive evidence of its safety has been presented by animal testing and by definition of its metabolic fate in animals and humans. It was subsequently reinstated as safe for use by the FDA.

Unlike saccharin and cyclamate, aspartame leaves no bitter aftertaste. It is quite expensive, about 200 times more so than sucrose, but as it is about 180 times sweeter than sucrose, its cost for obtaining a given unit of sweetness is not much more.

Saccharin

Saccharin, the imide of *o*-benzosulfonic acid, is used as a sodium or calcium salt. It is about 300 times sweeter than sucrose (table sugar). It may leave a bitter aftertaste, and its safety has been question as a result of some animal feeding tests. As an intense sweetener it is useful for diabetic individuals, and it reduces the incidence of dental caries.

STARCHES

Although starches differ from each other somewhat, depending on the plant from which they are extracted, they are sufficiently similar chemically to be often classified together as starch. The two basic starch polymers are amylose and amylopectin. Starch is used as a source of carbohydrate, and because it is relatively inexpensive, is often used as an extender. Its properties also make it useful as a thickening agent. The major source of starch is corn, but some starch is also produced from sorghum, potatoes, and wheat. More on starches is discussed in Chapters 2 and 18.

GUMS

Gums, a class of complex polysaccharides, are defined as materials that are dispersible in water and capable of making the water viscous. Many gums occur naturally in certain land and sea plants. Examples are gum arabic and agar. Many gums, such as the cellulose derivatives, are modified or semisynthetic, and some gums, such as the vinyl polymers, are synthetic. Gums are used to stabilize ice cream and desserts, thicken certain beverages and preserves, stabilize foam in beer, emulsify salad dressings, and form protective coatings for meat, fish, and other products. Gums add "body" and prevent settling of suspended particles in chocolate milk, ice cream, and desserts.

They may also prevent the formation of large ice crystals in frozen desserts. A significant potential for the use of gums lies in the production of certain low-calorie foods. For example, the oil in salad dressing can be replaced with gums to result in a product with the normal appearance, texture, and taste but without the calories normally associated with the product.

ENZYMES

Enzymes occur naturally in foods, and their presence may be either beneficial or detrimental, depending on the particular enzyme (see Chapter 8). When the presence of enzymes is undesirable, steps are taken to inactivate them. When their presence is desirable, either the enzymes or sources of the enzymes are intentionally added to foods. For example, the enzyme papain (from the papaya fruit) is added to steak to tenderize it. Many of the useful enzymes used in food processing are produced by microbes; consequently those microbes producing the desired enzyme may be added intentionally to food. For example, specific yeasts are intentionally added in the production of bread, beer, or cheese.

The use of enzymes as food additives presents no problem from the standpoint of safety, because enzymes occur naturally, are nontoxic, and are easily inactivated when desired reactions are completed. Enzymes called amylases are used together with acids to hydrolyze starch in the production of syrups, sugars, and other products.

Invertase

Certain enzymes, such as invertase, split disaccharides, such as sucrose (table sugar), to lower sugars (glucose and levulose). Invertase has many applications, and is used, for example, to prevent crystallization of the sucrose that is used in large amounts in the production of liqueurs. Without invertase, the liqueurs would appear cloudy.

Pectinase

Pectinases are enzymes that split pectin, a polysaccharide that occurs naturally in plant tissues, especially those of fruit. Pectin holds dispersed particles in suspension, as in tomato juice. Because it is desirable to keep the thick suspension in tomato juice, pectinases that occur naturally in it are inactivated by heat. On the other hand, products such as apple juice are customarily clear, and this is accomplished by adding commercial pectinase to the product, which degrades the pectin in the apple juice, resulting in the settling out of the suspended particles, which are then separated from the clear juice. In the manufacture of clear jellies from fruits, it is first necessary to add pectinase to destroy the naturally occurring pectin in order to clarify the juice. This pectinase must now be inactivated by heat. Then more pectin must be added to the clarified juice to produce the thick consistency of jelly. If the pectinase is not inactivated after clarification, the enzyme would also break down the newly added pectin required to produce the thick consistency.

Cellulases

Cellulases are enzymes that can break down cellulose, said to be the most abundant form of carbohydrate in nature. Cellulose, the principal structural material in plants, is insoluble in water and is indigestible by humans and many animals. Ruminants are able to digest cellulose because of a cellulase (produced by microorganisms in the large stomach) contained in their gastric juice. Commercial applications of cellulases are not widespread at present. Cellulases are used for tenderizing fibrous vegetables and other indigestible plant material for the production of foods or animal feed.

Proteases

Proteases are enzymes that break down proteins, polypeptides, and peptides. Peptides are the structural units of which polypeptides consist, and polypeptides are larger structural units that make up the protein. There is a large number of specific proteases, and each attacks protein molecules at different sites, producing a variety of end products. Proteases are used to produce soy sauce from roasted soybeans, cheese from milk, and bread dough from flour. They are also used to tenderize meat and chill-proof beer which, if untreated, develops an undesirable haze when chilled.

Lipases

Lipases, the lipid (fat or oil) splitting enzymes, have limited commercial application, with oral lipases having the widest. Lipases prepared from oral glands of lambs and calves are used in a controlled way in the production of certain cheeses and other dairy products, as well as lipase-treated butter fat used in the manufacture of candles, confections, and baked products. Lipases are also used to remove fat residuals from egg whites and in drain cleaner preparations.

Glucose Oxidase

Glucose oxidase is an enzyme that specifically catalyzes the oxidation of glucose to gluconic acid. This reaction is important in preventing nonenzymatic browning, because glucose is a reactant in the undesirable browning reaction. The most important application of this enzyme is in the treatment of egg products, especially egg whites, prior to drying. Eggs treated with this enzyme before they are dried do not undergo nonenzymatic browning during storage, because the sugar has been removed. In some cases, the enzyme is added to remove traces of oxygen to prevent oxidative degradation of quality. Examples of this type of application are mayonnaise and bottled and canned beverages (especially beer and citrus drinks).

Catalase

Catalases are used to break down hydrogen peroxide (H_2O_2) to water and oxygen. Therefore, catalases are used when the presence of hydrogen peroxide is undesirable or when hydrogen peroxide is used for specific purposes, such as in bleaching, but then must be removed from the system. Examples of the latter case are the uses of hydrogen peroxide for preserving milk in areas where heat pasteurization and refrigeration are unavailable and in the manufacture of cheese from unpasteurized milk. Hydrogen peroxide is produced during the spray-drying process. Catalase is used to convert the unwanted H_2O_2 to water and oxygen.

SEQUESTRANTS

The role of sequestrants is to combine with metals, forming complexes with them and making them unavailable for other reactions.

$$M + S \rightarrow MS$$

where

M = metal
S = sequestrant
MS = complex.

Sequestrants, like many other additives used for enhancing specific properties of foods, occur naturally in foods. Many sequestrants have other properties; for example, citric, malic, and tartaric acids are acidulants but they also have sequestering properties.

Because metals catalyze oxidative reactions, sequestrants can be considered to have antioxidant properties. Thus, they stabilize foods against oxidative rancidity and oxidative discoloration. One of the important uses of sequestrants as additives is to protect vitamins, as these important nutrients are especially unstable when exposed to metal-catalyzed oxidation. Sequestrants are used to stabilize the color of many canned products and they help stabilize antioxidants. Sequestrants are especially helpful in stabilizing color and lipids in canned fish and shellfish. Because fish and shellfish naturally contain relatively high concentrations of metal, these products normally have poor color stability, and the lipids tend to rancidify during storage.

Sequestrants are also used to stabilize the flavors and odors in dairy products and the color in meat products.

POLYHYDRIC ALCOHOLS

In addition to their use as sweeteners, many polyhydric alcohols (also called polyols) are used to improve texture and moisture retention because of their affinity for water. Many polyols are present in foods naturally, glycerine (glycerol) being the predominant one. However, only four of the many polyols are allowed as food additives. They are glycerine, sorbitol, mannitol, and propylene glycol. All but the last have a moderately sweet taste (see section on sweeteners), although none are as sweet as sugar. Propylene glycol has a somewhat undesirable bitter taste, but is acceptable in small amounts. Sorbitol imparts a cool sensation. Glycerine, on the other hand, imparts a hot sensation.

Polyols are used in the production of dietetic products including beverages, candy, gum, and ice cream to contribute to texture as well as to sweetness. These compounds have a less adverse effect on teeth than sugar, because they are not fermented as quickly as sugar and are usually washed away before they can be utilized by microorganisms.

SURFACE-ACTIVE AGENTS

Surface-active agents affect the physical force at the interface of surfaces. Commonly called surfactants, they are present in all natural foods, because by their nature they play a role in the growth process of plants and animals. They are defined as organic compounds that affect surface activities of certain materials. They act as wetting agents, lubricants, dispersing agents, detergents, emulsifiers, solubilizers, and so forth. One use for wetting agents is to reduce the surface tension of materials to permit absorption of water by the material. An example of their use is in powdered chocolate mixes used to prepare chocolate milk by addition of water.

Dispersions of materials depend on the reduction of interfacial energy, and this can be accomplished by certain surfactants.

Surfactants are used in the production of foods to prevent sticking, such as in untreated peanut butter. Surfactants are also used in cleaning detergents used on food equipment, and they can stabilize or break down foams.

Emulsifiers, such as lecithin, mono- and diglycerides, and wetting agents, such as a class of chemicals known as "tweens," may be added to bakery products (to improve volume and texture of the finished products and the working properties of the dough and to prevent staling of the crumb), cake mixes, ice cream, and frozen desserts (to improve whipping properties). Except for the tweens, the chemicals cited above are natural components of certain foods.

LEAVENING AGENTS

Leavening agents are used to enhance the rising of dough in the manufacture of baked products. Inorganic salts, especially ammonium and phosphate salts, favor the growth of yeasts, which produce the carbon dioxide gas that causes dough to rise. Chemical reagents that react to form carbon dioxide are also used in baked goods. When sodium bicarbonate, ammonium carbonate, or ammonium bicarbonate is reacted with potassium acid tartrate, sodium aluminum tartrate, sodium aluminum phosphate, or tartaric acid, carbon dioxide is produced. Baking powder is a common household leavening agent that contains a mixture of chemical compounds that react to form carbon dioxide, producing the leavening effect. Baking powder can be either single acting or double acting, giving the desired leavening effect in different products. The chemistry is shown below.

Single-Acting Baking Powder (Quick-Acting Baking Powder):

The CO_2 is liberated when sodium bicarbonate (the base) reacts with potassium acid tartrate (the potassium salt of tartaric acid).

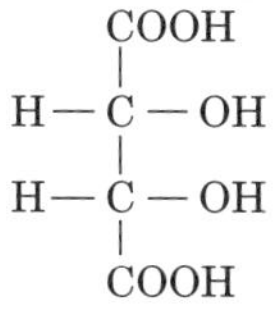

tartaric acid

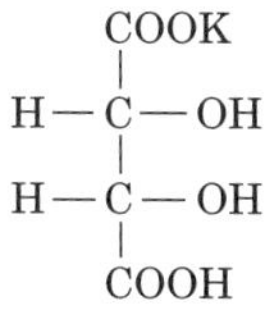

potassium acid tartrate

The reaction is as follows:

$$NaHCO_3 + KHC_4H_4O_6 \xrightarrow{H_2O} KNaC_4H_4O_6 + CO_2 + H_2O$$

sodium bicarbonate	potassium acid tartrate	sodium potassium tartrate	carbon dioxide	water

Double-Acting Baking Powder:

The double-acting type has two acid-reacting ingredients (monocalcium phosphate monohydrate and sodium aluminum sulfate). The hydrated form of monocalcium phosphate reacts with sodium bicarbonate to release a portion of CO_2 during mixing a batter or dough. The remaining sodium bicarbonate will react with sulfuric acid which is produced from the sodium aluminum sulfate in hot water.

First action:

$$3\ CaH_4(PO_4)_2 + 8\ NaHCO_3 \longrightarrow Ca_3(PO_4)_2 + 4\ Na_2HPO_4 + 8\ CO_2 + 8\ H_2O$$

monocalcium phosphate	sodium bicarbonate	tricalcium phosphate	disodium phosphate	carbon dioxide	water

Second action:

$$Na_2Al_2(SO_4)_4 + 6\ H_2O \xrightarrow{\text{heat}} 2\ Al(OH)_3 + Na_2(SO_4) + 3\ H_2SO_4$$

sodium aluminum sulfate	water	aluminum hydroxide	sodium sulfate	sulfuric acid

$$3\ H_2(SO_4) + 6\ NaHCO_3 \xrightarrow{H_2O} 3\ Na_2SO_4 + 6\ H_2CO_3$$

sulfuric acid	sodium bicarbonate	sodium sulfate	carbonic acid

In this reaction, the major portion of the CO_2 is released after the product is heated in the oven.

IONIZING RADIATION

Earlier in this chapter, the Food Additives Amendment of 1958 was discussed and it was shown that in the definition of a food additive, the statement: ". . . any source of radiation intended for such use . . ." is included. Thus, ionizing radiation is considered a food additive.

Irradiation does not leave a residue in food and it does not make it radioactive. The levels of irradiation allowed in food processing do not induce measurable radioactivity. Any radioactivity found in irradiated foods has been shown to be "background radiation" or that which is already present naturally. Irradiation does, however, cause small chemical changes in the food as do other methods of food processing.

Foods that have been irradiated must be labeled with the green international logo (see Fig. 13.1) to inform the consumer that the food has been processed by ionizing radiation. The words "Treated with Radiation" or "Treated by Irradiation" must also appear and must be in the same print style as the product name and be no smaller than one-third the size of the largest letter in the product name.

The effectiveness of this process is understood and agreed upon, but is it safe? The HACCP program discussed in Chapter 4 was developed to ensure the safety of the astronauts against food poisoning. They ate irradiated food. It has been established that ionization radiation of foods can destroy pathogenic bacteria but what about the

Figure 13.1. International Radiation Logo (green in color)

long-term effects of consumption of these products? The safety of irradiated foods has been tested in feeding studies for over 40 years. The studies include both animal and human subjects. Chemistry studies, feeding studies, and mutagenicity and teratogenicity studies have not revealed any confirmable negative evidence as to the wholesomeness of foods preserved by ionization radiation. Nutrient retention of irradiated foods is comparable to that of heat-processed foods. Irradiated foods may be more susceptible to oxidation but this can be controlled by use of low temperatures and elimination of oxygen.

The future of irradiation of foods is uncertain but it seems that scientific evidence and logic will have more effect on legislation in this area. There will always be risks, so food scientists must decide which will be the least likely cause of danger and proceed accordingly.

CHEMICAL PRESERVATIVES

The practice of preserving food by the addition of chemical is quite old, ordinary table salt (sodium chloride) having been used as a preservative for centuries. It might be surprising to think of a naturally occurring substance as a chemical preservative, but many chemical substances used in the preservation of foods occur naturally. When they are used with the proper intent, they can be used to preserve foods that cannot be easily preserved by other means. They should not be used as a substitute for sanitation and proper handling procedures. Sometimes chemicals are used together with other processes, such as holding at refrigerator temperatures above freezing.

To preserve food, it is necessary either to destroy all of the spoilage microorganisms that contaminate it or to create and maintain conditions that prevent the microbes from carrying out their ordinary life processes. Although preservation is aimed mainly at microbial spoilage, it must be remembered that there are other types of spoilage factors, such as oxidation.

Although foods can be sterilized (such as by heat processing) and contained in such a way as to prevent contamination by microbes during storage, it still is often necessary in some cases to forego sterilization, thus making it necessary to take other steps to prevent microbial degradation of the food. Foods can be protected against microbial attack for long periods (months to years) by holding them at temperatures below freezing (see chapter 12). They can be preserved for shorter periods (several days) by holding them in ice or in a refrigerator at temperatures in the range 32 to 40°F (0 to 7.8°C) (see Chapter 12). Foods can also be preserved by altering them to make them incapable of supporting microbial growth. Drying is an example of this type of preservation. Foods must also be preserved against color and texture changes.

Quite often it is either impossible or undesirable to employ conventional preservation methods, and a large variety of food additives is available for use, alone or in combination with other additives or with mild forms of conventional processes, to preserve foods. Usually, chemical preservatives are used in concentrations of 0.1% or less. Sodium diacetate and sodium or calcium propionate are used in breads to prevent mold growth and the development of bacteria that may produce a slimy material known as rope. Sorbic acid and its salts may be used in bakery products, cheeses, syrups, and pie fillings to prevent mold growth. Sulfur dioxide is used to prevent browning in certain dried fruits and to prevent wild yeast growth in wines used to make vinegar. Benzoic acid and sodium benzoate may be used to inhibit mold and bacterial growth in some fruit juices, oleomargarines, pickles, and condiments. It should also be noted that benzoic acid is a natural component of cranberries.

Salt is an excellent microbe inhibitor, mainly as a result of its suppression of the water activity of the material to which it is added. Its effectiveness is enhanced when the food is also dried or smoked or both. Smoking also imparts a partial preservative effect.

Weak acids, such as sorbic acid, or salts of weak acids, benzoates, propionates, nitrites, certain chelating agents (chemicals that tie up metals and prevent the catalytic action of metals), and other chemical additives are effective preservatives. Natural spices also have antimicrobial properties. Antibiotics have been used as food additives and are still used to preserve animal feeds and human foods in some countries. Their use in human foods is banned in the United States and in some other countries.

Because many antimicrobial agents are generally toxic to humans, their use must be regulated not to exceed established levels beyond which they are hazardous to human health.

Nitrites, proven inhibitors of *Cl. botulinum,* and nitrates are added to cured meats, not only to prevent botulism, but also to conserve the desirable color as well as add to the flavor of the products. Some of these preservatives are discussed further in this chapter.

Sodium Chloride

When sufficient salt is added to food, it makes water unavailable to microorganisms. Because microorganisms require water to survive, they cannot exist when their water requirement is diminished by the addition of salt. We can reduce the amount of water available to microorganisms by lowering the water activity (a_w). (For more on a_w see Chapter 3.) Microorganisms require high levels of a_w. Most bacteria require a minimum a_w level of 0.96, although halophilic bacteria can grow at a_w of 0.75. Most yeasts grow at a_w levels of 0.90 and above, although a few can grow at an a_w level of 0.81. Molds can grow at lower a_w levels, with some able to grow at an a_w level of 0.62. While salt preserves foods mainly by lowering the a_w, the chloride ion is believed to inhibit bacterial growth, independently.

Some precautions must be observed in the salting preservation of flesh-type foods, such as fish or meats. When these products are salted, several days will be required before enough salt has diffused into all parts of the product to inhibit the growth of microorganisms. If, therefore, precautions are not observed, the growth of spoilage or even disease-causing bacteria may occur in some parts of the food before enough salt has diffused into the product to inhibit growth. The usual procedure is to hold products under refrigeration during salting until there has been an adequate "take-up" of salt throughout the food. Fish and meats should never be held at temperatures above 60°F (15.6°C) during salting. Preferably, holding temperatures during such procedures should be at 40°F (4.4°C) or slightly below.

Salted, undried meats, such as corned beef, should be held at 40°F (4.4°C) or below at all times after curing because there are some microorganisms that may still grow at salt concentrations present in such products. Chipped beef, which is dried as well as salted, has a low enough moisture content to prevent the growth of all microorganisms and may be held at room temperature.

Salt cod, which has a moisture content of 40% or higher, should be held at temperatures of 40°F (4.4°C) or slightly below because it is subject to spoilage through bacterial growth. On the other hand, well-dried salt cod and certain types of salted and smoked herring that have dried during smoking may be held at room temperature without spoilage.

Fatty Acids

The salts of certain fatty acids have an inhibitory effect on the growth of microorganisms. Thus, sodium diacetate (a mixture of sodium acetate and acetic acid) and sodium or calcium propionate

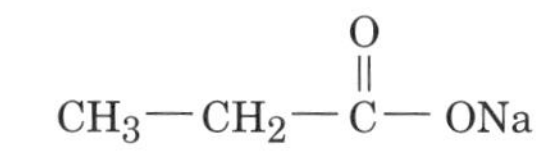

sodium propionate

are added to bread and other bakery products to prevent mold growth, as well as the development of a slimy condition known as "ropiness," which results from the growth of certain aerobic, spore-forming bacteria (see Chapter 3 for the definition of spore-forming bacteria). Caprylic acid, $CH_3CH_2—CH_2—CH_2—COOH$, or its salts or the salts of other fatty acids may be used in cheese to prevent the growth of mold.

As pointed out previously, it is the nonacid part of the molecule of fatty acids or their salts that inhibits the growth of microorganisms. It is believed that the effect of these compounds is the destruction of the cell membrane of microorganisms.

Sulfur Dioxide

Sulfur dioxide (SO_2) is used in some foods to inhibit the growth of microorganisms. Sulfur dioxide may be used as such, or a source of this compound such as sodium bisulfite ($NaHSO_3$) may be added to the foods. Sulfur dioxide inhibits a rather narrow range of microorganisms and is usually applied together with another chemical inhibitor to prevent the growth of undesirable yeasts or bacteria in fruit juices, which are stored prior to fermentation, in the production of wine or vinegar. Sulfur dioxide may inhibit microbial growth by preventing the utilization of certain carbohydrates as a source of energy or by tying up certain compounds concerned with the metabolism of some microorganisms.

For many years, sulfiting agents have been classified as GRAS (generally regarded as safe) substances by the FDA for use as food preservatives when used in accordance with GMP (good manufacturing practice). But in 1986, following several deaths from the consumption of fresh fruits or vegetables that had been treated with sulfites, the FDA withdrew the GRAS status of sulfites for this use. It was found that asthmatic individuals react, sometimes severely, when exposed to sulfites. Also, because sulfites have been found to destroy thiamin, these agents may not be used in meats and other foods containing thiamin. Although sulfites may still be used in foods that have not been excluded by the FDA, their presence must be declared on the label when their concentrations exceed 10 ppm (parts per million). Research has shown that in concentrations of 10 ppm or less, these agents should not cause adverse reactions in humans.

Sorbic Acid

Sorbic acid, CH_3—CH=CH—CH=CH—COOH, inhibits the growth of both yeasts and molds. This compound is most effective at pH 5.0 or below. This compound can be metabolized by humans, as can fatty acids, and hence is generally recognized as safe. Sorbic acid is used in certain bakery products (not yeast-leavened products, because it inhibits yeast growth), in cheeses, and in some fruit drinks, especially for the purpose of preventing molding. It is believed to inhibit the metabolic enzymes required by certain microorganisms for growth and multiplication.

Sodium Nitrite

Sodium nitrite ($NaNO_2$) is added to some food products to inhibit bacterial growth (especially *Cl. botulinum*) and to enhance color. It is added to most cured meats, including hams, bacon, cooked sausage (such as frankfurters, bologna, salami), and to some kinds of corned beef. Nitrite provides for the red or pink color of the cured and cooked sausages and of the other cured products after cooking. The nitrite combines with the reddish pigment of meat, the myoglobin, and prevents its oxidation. If the meat were not treated with nitrite, it would discolor to a brown color during cooking or during storage. When red meat is heated, as in cooking, the color turns from red to gray or brown as a result of the conversion of myoglobin to the oxidized form, metmyoglobin. On extremely long or extremely high heating or on exposure to light and oxygen, even the nitrited myoglobin may be oxidized to metmyoglobin, with the result that the red or pink color is lost.

In addition to stabilizing the color of cured or cured and cooked meats, the industry claims that nitrite acts as a preservative in that it tends to prevent the growth of spores of *Cl. botulinum* that may be present. *Cl. botulinum*, of course, is the most dangerous disease-causing organism.

Nitrite is also used in some fish products, such as smoked whitefish and chubs, for the specific purpose of preventing the growth of *Cl. botulinum*.

Sodium or potassium nitrite may not be used in meats or on fish that are to be sold as fresh. In cured products, it is allowed in concentrations not to exceed 0.0156% (156 parts of nitrite per million parts of the food). The amounts to ¼ oz or 7.09 g per 100 lbs of meat. In practice, however, it is closer to 120 ppm or less. The residual nitrite or that which actually appears in the final product is much less. A study done by Robert Cassens of the University of Wisconsin in 1995 showed residual levels of nitrite as low as 7 ppm for bacon, 6 ppm for sliced ham, and 4 ppm for hot dogs (expressed as nitrite ion). Numerous factors affect residual nitrite levels but it is accurate to say that the amounts are much lower than that added at the time of processing. There is some question as to whether or not nitrites should be allowed in food in any concentration. It has been found that nitrite-cured products, especially those cooked at high temperatures, such as bacon, may develop nitrosamines, compounds formed by the reaction of nitrites with amines, and nitrosamines are known to be extremely carcinogenic or cancer-promoting. It is clear that the amount of nitrites used must be minimized but the level must remain high enough to protect against botulism and other microbiologically based foodborne diseases.

Cured meats are not the only sources of nitrites in the diet. Vegetables, baked goods, cereals, and water are also sources. Nitrites can also be made in the body from intakes of nitrates (NO_3) by a simple reduction. This can happen in the oral cavity as a result of action of bacteria in the saliva. Vegetables account for about 85% of our dietary intake of nitrates.

Processing methods and formulations have changed dramatically. The result is lowered residual nitrites and decreased levels of volatile nitrosamines. Some estimates are that the levels are one-third of those of a decade ago. This must be placed in proper perspective, however, because it is minor in relation to nitrite synthesized by the body. Steps can be taken to reduce amounts of nitrites ingested but, to ensure microbiological safety, their continued use in cured meats seems inevitable.

Oxidizing Agents

Oxidizing agents, such as chlorine, iodine, and hydrogen peroxide, are not ordinarily used in food, but they are used to sanitize food-processing equipment and apparatus and even the walls and floors of areas where food is processed. Thus, there is no doubt that small residuals, especially of chlorine or iodine, can get into food.

Hydrogen peroxide may be used to destroy the natural bacterial flora of milk, prior to inoculation with cultures of known bacterial species, for producing specific dairy products. In such cases, all of the residual hydrogen peroxide must be removed by treatment with the enzyme catalase (see Chapter 8). This treatment with catalase must be carried out prior to the inoculation of milk with cultures of desirable bacteria; otherwise the hydrogen peroxide will destroy the added bacterial culture, the growth of which is the objective of culturing milk.

Oxidizing agents are believed to inhibit and destroy the growth of microorganisms by destroying certain parts of the enzymes essential to the metabolic processes of these organisms.

Benzoates

The benzoates and parabenzoates have been used as preservatives mainly in fruit juices, syrups (especially chocolate syrup), candied fruit peel, pie fillings, pickled vegetables, relishes, horseradish, and some cheeses. The probable reason that the benzoates and the related parabenzoates have been allowed as additives to food is that benzoic acid is present in cranberries as a natural component in concentrations that are higher than 0.1%. Benzoic acid or its sodium salt is allowed in food in concentrations up to

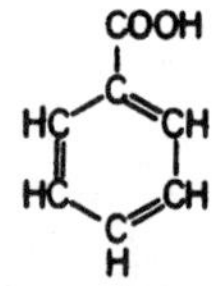

benzoic acid

0.1%. Parahydroxybenzoic acid or its esters, for instance, propyl para-hydroxybenzoic acid, may also be used.

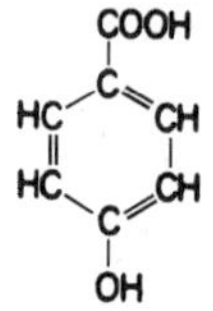

para-hydroxybenzoic acid

The benzoates are most effective in acid foods in which the pH is as low as 4.0 or below. The parabenzoates are said to be more effective than the benzoates over a wider range of pH and against wider groups of microorganisms.

Investigations have indicated that benzoates prevent the utilization of energy-rich compounds by microorganisms. It has also been found that when bacteria form spores in the presence of benzoate, the spore may take up water and germinate to the point of bursting and shedding the spore wall, but enlargement and outgrowth into the vegetable cell with subsequent cell division and multiplication does not occur.

COLORANTS

We are accustomed to specific colors in certain food, and colors often provide a clue to the quality of the foods. Color additives can be categorized into three major types: natural, nature-identical, or synthetic.

Many colorants (compounds that add colors to foods) are natural, and these include the yellow from the annatto seed; green from chlorophyll; orange from carotene; brown from burnt sugar; and red from beets, tomatoes, and the cochineal insect. Natural colors are simply pigments obtained from animal, vegetable, or mineral sources.

If synthetic counterparts of colors and pigments are derived from natural sources, the term "nature-identical" applies. These include the pure carotenoids such as canthaxanthin (red), apocarotenal (orange-red), and beta-carotene (yellow-orange). These have all gone through toxicological studies and are approved by the FDA. Canthaxanthin and apocarotenal have maximum addition limits but beta-carotene can be added at the necessary level to accomplish its intended purpose.

Some colorants, however, are derived from synthetic dyes. The synthetic dyes in use have been approved and certified by the FDA. These certified color additives are divided into two groups: FD&C dyes and FD&C lakes. Dyes are water-soluble and are available in powders, granules, liquids, blends, and pastes. GMPs suggest that they not be used in amounts exceeding 300 ppm. The lakes are water-soluble FD&C certified dyes on a substratum of aluminum hydrate or aluminum hydroxide. The lakes must also be certified by the FDA. They are useful in foods that have very little water such as coloring oils. They are used in icings, fondant coatings, cake and doughnut mixes, hard candy, and gum products. They do not solubilize as do dyes but color by dispersion rather than solution.

In 1960 the Color Additive Amendments separated "color additives" from "food additives." Colors (which include black, white, and intermediate grays) no longer were to be classified as food additives. In determining whether a color additive is safe, the FDA must take into account the probable consumption of the additive and of any substance formed as a result of its use. There is also a cancer clause similar to the Delaney Clause in the amendments.

Some compounds are not color additives but are used to produce a white color. Thus, oxidizing agents including benzoyl peroxide, chlorine dioxide, nitrosyl chloride, and chlorine are used at the end of the production cycle to whiten wheat flour, which is pale yellow in color if untreated. Titanium dioxide, on the other hand, is considered a color additive and may be added to some foods, such as artificial cream or coffee whiteners to add a white color.

Part III

Handling and Processing of Foods

14

Meat

Of the many foods obtained from the land, humans tend to prefer animal foods, mainly beef, pork, poultry, and lamb as well as their byproducts (e.g., cheeses, milk, and eggs). The most important source of meat in the United States is cattle. Horses have not been an important source of meat, except during wars when meat shortages have occurred.

A market exists for domestically produced game-type meats. These include buffalo, bear, elk, kangaroo, and rabbit. They are usually expensive and selected mainly as gourmet items. The meat is generally tough and very "gamey." In general, this type of meat is prepared in moist heat.

The proportion of meat-derived foods that humans consume is related to the general affluence of the society in which they live. The reason for this is that the use of animal flesh as food tends to use up a much greater amount of calories, proteins, and other nutrients that might be available directly from plants. It is reported that it takes as much as 10 lb (4.5 kg) of plant food to produce about 1 lb (0.45 kg) of beef. Thus, in deprived societies, the population is compelled to derive most of its nutrition from plants rather than from animals.

Animal foods offer more than just palatability. Because animals are biologically similar to humans, it should be obvious that animal foods contain many of the nutrients that are required by humans to carry out their own body functions. For example, in most cases, the animals are good sources of the essential amino acids, as well as the vitamins and minerals required by humans.

Meat and meat products include the muscle tissues of cattle, hogs, sheep, and other animals, as well as the organs of these animals. Among the organs used are the tongue, heart, brain, liver, and kidneys. Some of the byproducts derived from the animals that are used in the meat industry include the intestinal walls, used as casings for making different types of sausages. Fat is used for the production of lard and tallow; the tallow is used as a source of raw material for making soap and candles. Other byproducts include leather for shoes and a variety of other commodities; wool for textiles; gelatin, which is used in the production of gelatin desserts and other similar food products; blood, which is used in some sausages and in feeds; bone, which is used mainly in fertilizers and feeds; animal scraps from the slaughtering plants, which are used in the manufacture of feeds; and a variety of enzymes and other chemicals used in different industries, especially in the food and pharmaceutical industries.

Except for tongues, head meats are generally used in the production of sausages. Tripe, prepared from the first and second stomachs of cattle, is cooked for about 3 hours and sold as is, or it may be cured. The sweetbreads (pancreas and thymus) are marketed either fresh or frozen. Both gelatin and animal glue are prepared from collagen derived mainly from the bones and the hide, but also from other parts, such as the sinews, ears, snouts, and trimmings. Oxtails, although quite tough, have a distinctly desirable flavor, and are used in making oxtail soup.

When livestock animals are grown for meat production, the males are castrated. There are two reasons for this. On castration, the hormone balance in the animal is affected so there is an increased amount of fat deposited throughout the musculature. Because of this, the meat tends to be more succulent and more tender. The second reason for castration is that the strong odor usually associated with viable males tends to disappear.

As explained in Chapter 24, the tenderness of beef is inversely related to the age of the animal, and marbling, an effect describing a uniform distribution of relatively high amounts of fat in the muscle, is not always a reliable indicator. At this point, it should be noted that with increasing knowledge of the relationship between cardiovascular diseases and the consumption of saturated fat and cholesterol, dietary trends have changed, and more people are opting for less fatty foods. As a result, beef and pork producers have changed feed formulas to lower the fat levels in the beef and pork that they produce, and they have asked the USDA to publish the new lower fat and cholesterol values for beef and pork.

Generally, animals are allowed to rest for a period just before slaughter. This is done in order that they not use up all the muscle sugar (glycogen). When the animal is slaughtered with an amount of glycogen in its body, the glycogen is converted under anaerobic conditions to lactic acid. The presence of this acid has a preservative effect on the meat. On the other hand, if the animal is not rested before slaughter and it uses up its glycogen, there will not be enough lactic acid formed to have the preservative effect, and thus the meat will spoil sooner.

After slaughter, the carcass portions are aged in cold rooms at low temperatures just above freezing, usually 35°F (1.7°C). This process may take up to 1 month, depending on temperature, humidity, and other conditions, but ordinarily, meat is held for about 12 days prior to consumption. The aging is accelerated if the beef is stored at higher temperatures, but the higher temperatures permit the growth of bacteria and surface spoilage. However, because the spoilage is on the surface, this problem can be resolved by storing the meat at the higher temperatures in the presence of ultraviolet light, which has a bactericidal effect.

Sometimes a processor is willing to accept the microbial growth in order to hasten the tenderization. Because the bacteria and mold growth create a certain amount of spoilage on the surface of the meat, the processor must then trim the surface to remove the effects of microbial growth and spoilage. The main purpose of aging is, of course, to tenderize the meat. Meat cuts can be tenderized just prior to consumption by the addition of commercially prepared enzymes, which have a proteolytic effect; that is, they tend to break down the proteins. However, this is a slow process if the enzymes are added only to the surface, and it is believed that injection into the meat or into the bloodstream of the living animal just before slaughter is a more effective way of producing tender beef.

When meat is heated in the presence of water, such as in boiling or in the making of stew, the connective tissue is converted to a more tender gelatin and becomes more palatable. On the other hand, when the meat is heated without water, such as in an oven in dry heat, the connective tissue tends to become tougher. The amount of connective tissue present in the meat depends on the location of the meat in the animal. The tenderloin, for example, has very little connective tissue, especially in the younger animals.

Although boiling tends to make meat less tough, one of the disadvantages is that taste components are extracted from the meat, and, therefore, boiled meat is less tasty. Of course, the water in which the meat has been boiled acquires the meat flavor components and can be concentrated and sold as a meat extract, which can be used as a base for soups and stews. Also, little browning (which enhances meat flavor) occurs during boiling.

Some points to remember in the cooking of meat are that the higher the temperature during cooking, the greater the amount of shrinkage; the lower the temperature during cooking, the higher the quality and the more uniformity of doneness throughout the meat. A certain amount of aging is beneficial from a quality standpoint. In fact, some chefs will hold the meat at room temperature for a considerable period before they finally put it into the oven. In the case of roasts, it has been found that an oven temperature of 375°F (190.6°C) should be maximum and about 325°F (162.8°C) minimum.

Of the different types of protein of which meat is composed, the contractile proteins are of some importance. These include myosin, actin, and tropomyosin, plus some other minor ones. When these proteins are extracted from the cells in the meat, either with salt solution or with mechanical agitation, such as chopping or grinding, they form a sticky substance that has binding qualities. When a mass of comminuted (ground) meat is heated, the contractile proteins coagulate, and they hold the small pieces of meat together. This is the way that some formed products, such as sausages and meat loaves, are produced.

Unfrozen meat should be stored in a temperature range of 28 to 38°F (–2.2 to 3.3°C). The relative humidity should be in the range of 85% to 90%. The high humidity will prevent excessive drying and shrinkage. In addition, high humidity tends to preserve the white color of the fat in the meat. The storage room should have a reasonable circulation of air to ensure heat removal and more uniform distribution of the humidity. Meats stored for long periods are generally frozen and stored at temperatures in the range of –10 to 0°F (–23.3 to –17.8°C).

Because many diseases can be transmitted from animals to humans, all meats shipped interstate in the United States are subject to inspection for wholesomeness under the authority of the Meat Inspection Act. "Wholesome" meat is safe to eat and is without adulteration. In addition to the wholesomeness inspection, meat may be graded for quality. Although the lean is more valuable than fat, the best cuts of beef have some fat, and the fat is well distributed, giving the marbling effect cited earlier. The prime grade is of the highest quality, but in general is limited in amount and sold only to hotels and the better eating places.

It should be noted that in federally regulated plants, the inspection for wholesomeness is mandatory. The grading of the meat is optional.

BEEF/VEAL

The most important beef animal breeds in the United State are the Angus, Hereford, Galloway, and Shorthorn. Less important are the Santa Gertrudis; the Chambray, a cross between a Charolaise bull (French) and a Brahma cow; and the Brangus, a cross between an Angus bull and a Brahma cow. The latter breeds were developed to produce cattle resistant to screw worm (an insect infesting cattle), Brahma cattle being resistant to this parasite.

Much of the cattle population of the United States is raised on the Western range, which includes the Great Plains, the Rocky Mountains, and the intermountain and Pacific Coast regions. The raising of calves produced in these areas may be finished (final stages of feeding before slaughter) elsewhere. These feeding areas include the best pulp feeding area of Colorado; the cottonseed cake section of Texas; the corn belt area (especially Iowa) and certain southern states, especially Florida; and California.

Discarded dairy animals supply a considerable portion of the cattle slaughtered in this country. Veal calves are produced mainly in dairying areas (especially Wisconsin), and slaughter calves (3 to 12 months old) are also produced in southwestern areas.

Farmers and ranchers may sell cattle to terminal markets (who sell to various buyers), local markets, auctioneers, or packers. Usually, cattle are shipped by truck to stockyards near or at the point of slaughter.

Steers (males castrated prior to sexual maturity) and heifers (females beyond the veal and calf age that have never calved) make up the major portion of beef animals. Cows (females that have calved), bulls (mature males), and stags (castrated after reaching sexual maturity) make up the remainder of the beef animals.

At the point of sale, cattle may be graded in the live state. Grading is based on conformation (shape, build, and breed of the animal), finish (quantity and distribution of the fat), and quality (quality of the hair and hide, moderate bone mass, etc.)

The term *conformation* is used to describe the structural characteristics of the beef animal. For the best conformation, the animal should be short and compact. A rather blocky, or large-bodied, short-legged appearance is good.

Basically, the best set of structural characteristics or conformation is valuable to the retail butcher because apparently he is able to make more profitable cuts from the animals having the top conformation, and because there is a relationship between good conformation and good quality.

In the United States and many other countries, animals must be rendered insensible to pain before slaughter (see Fig. 14.1 for general slaughtering scheme). With cattle, this may be done by hitting them on the head with a bolt from a specially devised stunning gun. To conform to religious laws, cattle for kosher meats are slaughtered by first cutting the throat with a special knife, in one motion, severing the jugular vein, and then stunning. In ordinary slaughter, after the animal is stunned, it is shackled at the hind legs and raised from the floor is an upside-down position, after which it is bled by severing its jugular vein.

After bleeding, the head is skinned and severed at the neck. The body is lowered to the floor or to a cradle, the skin is cut ventrally (along the abdomen) and pulled back from the belly portion, and the lower legs are removed. Hooks are inserted in the cords of the hind leg section, and the body is raised to the half-hoist position. Most of the skin is removed in this position, after which the carcass is raised to full hoist and skinning is completed. Also, in this position, the sternum (breastbone) is opened with a saw, the belly is split, the bung and esophagus are tied off, and the viscera are removed. The viscera are inspected by veterinarians, and the carcass is split into halves with a saw. The hide is inspected, salted, and sent to the tannery.

The split carcass is washed, and, if it is not to be boned out soon after cooling, it is covered with a cloth to shape the surface fat. Sides handled in this manner are sent to a cooler where in 24 to 48 hours it will be required to bring the temperature of all parts to about 35°F (1.7°C).

Dressing, including skinning, may be carried out completely while the animal is

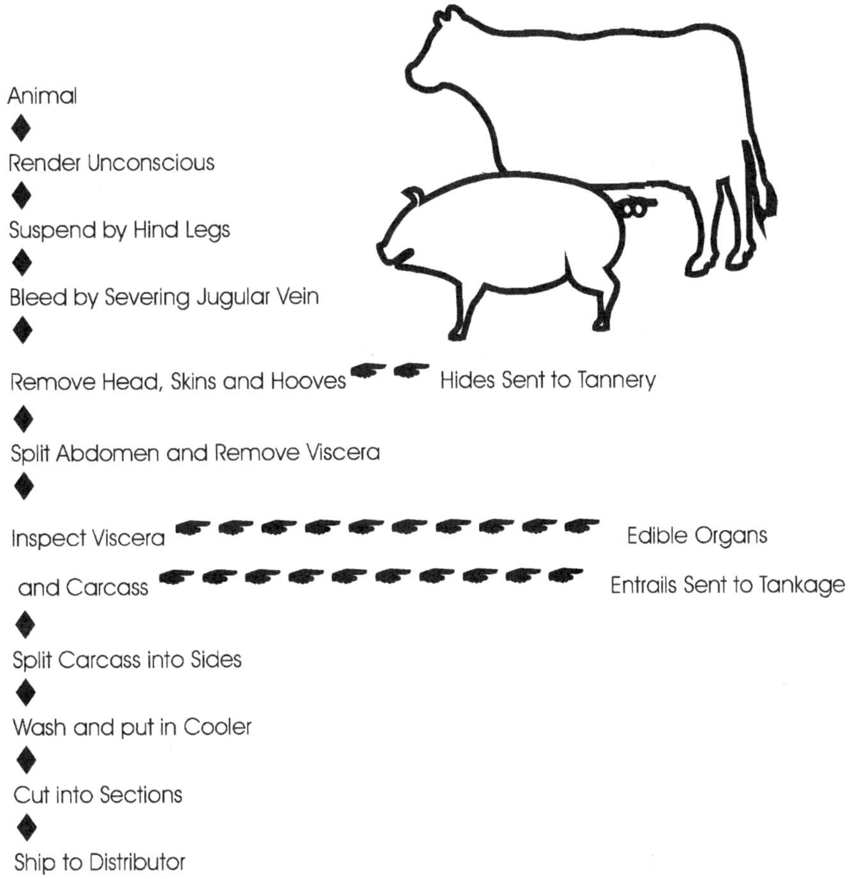

Figure 14.1. General scheme for slaughtering animals.

hanging from a rail. Mechanical instruments, such as hide pullers, may be used to increase the efficiency of rail dressing.

In the past, when slaughtered beef was not to be used in local areas, it was shipped out as sides or, more often, as quarters. Today, a considerable amount of beef is cut into various primal cuts, such as rounds, ribs, and loins (see Fig. 14.2). The various sections, usually without bones, are packaged, usually under vacuum, in a film that is impervious to moisture and oxygen. The packages are boxed and shipped to distributors and retailers. Some retail cuts are packaged in film under vacuum and shipped in cartons to retail stores. Because the meat has been boned and trimmed of excess fat, meat department personnel at the stores have little to do but cut the meat into individual portions, package the various cuts, and place them on display. Primal cuts packaged in moisture- and oxygen-impermeable film, under vacuum, and held at temperatures between 32 and 40°F (0 and 4.4°C), are said to have a storage life of at least 21 days.

Inspectors from the USDA evaluate beef carcasses and specify various grades. These are "prime" (the best beef from steers and heifers), "choice" (the next best from heifers, steers, and young cows), and "select" (the third best). There are grades of lesser quality, but they are generally used for purposes other than as steaks or roasts.

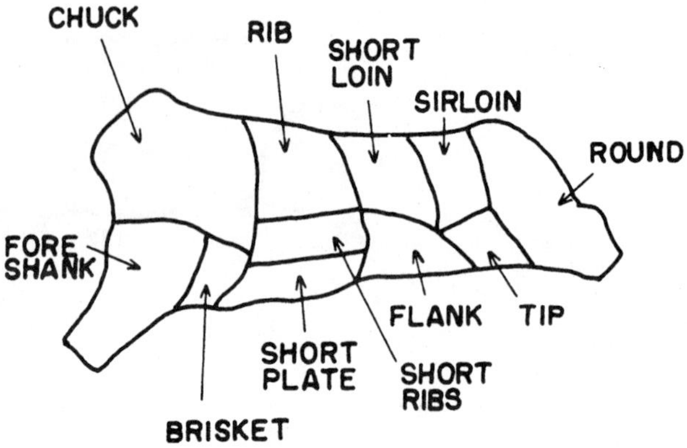

Figure 14.2. Cattle carcass.

A veal carcass is shown in Figure 14.3.

Grades cannot be assigned until the animal has been slaughtered. A new technique, the use of ultrasonic energy to show the structure of the meat prior to slaughter, has been developed and may influence grading techniques in the future. Meat, fat, and bone reflect ultrasonic energy differently so by exposing a live animal to this energy and recording the reflection pattern, an image similar to an X-ray cross-sectional view of sections of the animal carcass can be viewed.

PORK

There are many breeds of hogs that, in this country, have become pretty well mixed in recent years in order to produce meat-type animals (more lean and less fat) rather than lard-type animals. Part of the reason for getting away from the high-fat type of hog has been that vegetable oils have largely replaced lard, which previously was the chief fat used for cooking. Some breeds of hogs, such as the Poland-China, Berkshire, Chester White, and Hampshire, are not mixed.

The north-central states, especially Iowa, produce most of the hogs raised in the United States. Hog production is said to be increasing in the southern states.

Hogs are usually fed soybean meal, meat byproducts, fish meal, or milk, or combinations of these as a source of protein. Carbohydrate foods consist mainly of corn, but other

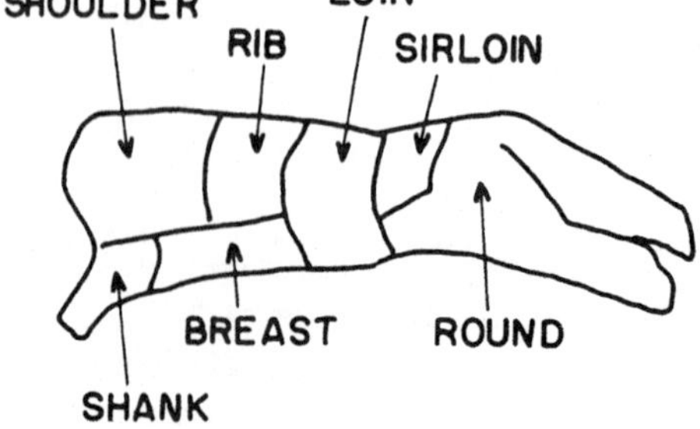

Figure 14.3. Veal carcass.

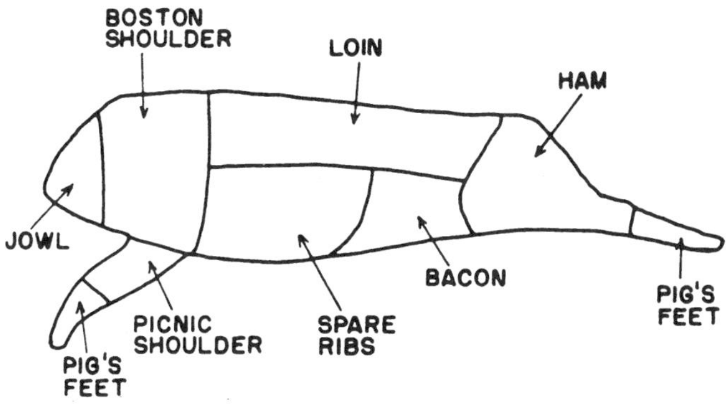

Figure 14.4. Hog carcass.

grains are used. Minerals may be added as supplements. Some vitamins, especially B_{12}, may be added to hog feed, as they may serve as growth factors.

Slaughter hogs are classified as barrows (male castrated prior to sexual maturity), gilts (young females that have not produced young pigs), sows (mature females that have produced young or that have reached an advanced stage of pregnancy), stags (males castrated after reaching maturity), and boars (uncastrated males).

Hogs are graded according to conformation, finish, and quality as U.S. No. 1, U.S. No. 2, U.S. No. 3, medium, and culls, this grading being listed in the order of desirability.

Prior to slaughter, hogs are rendered insensible by electric stunning or by exposure to an atmosphere of carbon dioxide. The hog is then shackled by the hind legs and raised to a conveyor after which a knife is inserted into the neck to sever the jugular vein. After bleeding for about 6 minutes the hog, still on the conveyor, is passed through a water bath (water temperature about 140°F [60°C]), where it is immersed for 5 to 6 minutes. This facilities removal of hair in a machine that has mechanically driven beaters. The loose hair is washed away with a water spray.

The carcass is then suspended by the hind legs by inserting the ends of a gambrel under the achilles tendon in each hind leg and attaching the gambrel to a moving conveyor. Remaining hair is scraped or singed off.

During dressing, the head is partially severed, the carcass is opened, and the viscera are removed. The edible organs are inspected by veterinarians and removed to refrigerated areas to be sold later as such or used for further processing. The carcass is then split and removed to the cooler, where the temperature of all parts should be brought to 35°F (1.7°C) within a period of 18 to 24 hours.

Hogs are generally not shipped as sides. Usually, they are cut up into hams, loins, shoulders, bellies, and back fat before shipment to distributors or to processing plants that manufacture various pork products. See Figure 14.4 for an example of a hog carcass.

LAMB/MUTTON

The bulk of the sheep produced in the United States is said to be made up of crossbreeds developed for the production of both meat and wool. Pure breeds include Cheviots, Lincolns, Oxfords, Shropshires, Rambouillets, and Merinos.

Sheep are raised on ranges in arid and semiarid regions and on farms. On farms, flocks of sheep are smaller than those raised on ranges. Where conditions allow, lambs may be fattened on the range, but during winter storms, the feed must be supplemented with protein concentrates, corn, hay, barley, and oats.

Fattening of lambs may take place in the corn belt states, especially during the fall and winter months. This may be done in corn fields, stubble fields, or pastures. Some fattening occurs while the animals are confined to barns. Most lamb fattening is carried on from September through May. Sheep are classified as ewes (females), wethers (males castrated prior to reaching maturity), rams (uncastrated mature males), and lambs about 3 to 12 months in age that do not show their first permanent teeth. Lambs are identified by the break joint—the temporary cartilage in the leg bone just above the hoof. As the animal grows older, the break joint loses its redness and sawtoothed appearance, and the cartilage is replaced by bone. Spring lambs are those from new crops marketed before July 1 and weighing 70 to 90 lb (31.8 to 40.9 kg). As with beef cattle and hogs, grading of slaughter lambs is based on conformation, finish, and quality.

Sheep are sold directly to the final buyer or indirectly through a commission agent. Because most sheep are produced west of the Mississippi and are consumed in the east, they must be transported for comparatively long distances for slaughter. Lambs and sheep are shipped both by truck and by rail, in double-decked cars.

Lambs are slaughtered after stunning by inserting a knife below the jaw to sever the jugular vein while the lamb is shackled by the hind legs to a conveyor. The hind legs are skinned and the joints of the forelegs are broken just above the feet. The whole pelt is then removed after which the carcass is opened and the viscera removed. After removal of the head, the carcass is washed and removed to the cooler. Inspection of the viscera, carcass, and edible organs is done by veterinarians. As in the case of beef and pork, diseased sheep are discarded and used for tankage. In coolers, the temperature of all parts of the carcass should be lowered to 35°F (1.7°C) within 24 hours.

The lamb carcass is not split prior to shipment to the distributor or retailer. Sheep carcasses are classified into two categories: mutton and lamb (the latter is under 12 to 14 months of age). Lamb and mutton carcasses are graded as prime, choice, good, utility, and cull. A sheep carcass is shown in Figure 14.5.

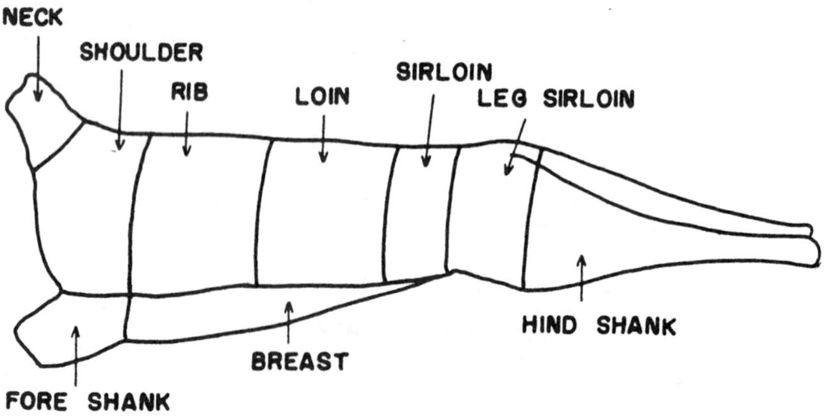

Figure 14.5. Sheep carcass.

CURED MEAT PRODUCTS

Much pork and some beef are cured or processed in some manner. (Veal, lamb, and mutton are ordinarily not cured.) During the cutting up of these animals, there are always some portions trimmed off from the carcass. Much of the trimmings from both beef and pork carcasses are used for the production of fresh or cooked sausage. Some of the fat that is trimmed off, especially from beef, is sold as such to be used for rendering. Fat rendered from trimmings of this kind is mostly used for the manufacture of soap. Bones may be used for the manufacture of bone meal and, in some instances, for the production of gelatin. Pork skin is used almost entirely for the manufacture of gelatin.

Curing agents for meat products may include only salt (sodium chloride), such as in the manufacture of New England-style corned beef. However, more often, curing agents include salt, sodium nitrite, and table sugar (sucrose) or glucose. Sometimes reducing agents, such as vitamin C or the related salt of iso-ascorbic acid, sodium erythrobate, are used to facilitate the coloring effect of nitrite, and in many instances, spices, and materials such as monosodium glutamate, are used as flavor enhancers. Nitrite also inhibits the growth of *Clostridium botulinum*, producer of a powerful toxin, and imparts a characteristic flavor.

When freshly cut, the color of meat is purplish red owing to its pigment (myoglobin). After cutting the meat, the myoglobin soon becomes loosely bound to oxygen to form oxymyoglobin (bright red), and, if exposed to oxygen for some time, the bond becomes more permanent and the meat takes on a gray or brown color and metmyoglobin is formed. If nitrite is allowed to react with myoglobin, oxymyoglobin, or even metmyoglobin, under good reducing conditions, it replaces the oxygen and forms a compound known as nitrosomyoglobin. When meat containing nitrosomyoglobin is cooked to coagulate the protein, this compound is called nitrosohemochrome. The latter two compounds are pink or red in color, and they are somewhat more permanent than oxymyoglobin and less subject to oxidation to form the brown or gray metmyoglobin.

In the curing of hams, a brine (55% to 80% saturated with salt, containing 20 to 50 lb [9.1 to 22.7 kg] of sugar, about 1.5 lb [0.68 kg] of sodium nitrite, per 100 gallons [378.5 liters] of pickle) may be used. Small amounts of ascorbic acid or iso-ascorbic acid, or sodium erythrobate may also be added. Ascorbic acid is added to provide reducing conditions in the meat, which facilitates the formation of nitrosohemochrome. Phosphate (for the retention of water) is allowed in cured meats in amounts not to exceed 0% to 0.5%. This pickle is pumped into the hams and they are placed in tierces (casks) and may or may not be covered with pickle. Curing in this manner may require 5 to 10 days, depending on the size of the product. The curing should be carried out at a temperature of 40°F (4.4°C) or below.

In smoking, moistened hardwood sawdust is usually burned in a mechanically controlled smoke generator (located outside the chamber where the product is hung) for purposes of treating with smoke and heat. Oak and hickory sawdust or logs (logs may be used in a special smoke generator) are mostly used for the production of smoke. The temperature of the smokehouse can be regulated by thermostatically controlled heating units within the smokehouse.

After curing, the hams are washed, hung to dry (with or without a stockingette cover), then smoked. The time and temperature of smoking must be sufficient to raise the temperature of all parts of the ham to at least 137°F (58.3°C). This is a USDA regulation that has been introduced to make certain that a roundworm, *Trichinella spiralis*, is destroyed. This parasite infests hogs, bears, and certain other animals and may infest the human. Cured hams to be boiled or canned may be smoked, but generally are not. Hams for canning are cured, skinned, boned, and placed in the can (some gelatin and various spices or flavoring agents may be added), and the cans are sealed under a slight vacuum and then heated in an agitated water bath at 165 to 180°F (73.9 to 82.2°C) until an internal temperature of 150 to 165°F (65.6 to 73.9°C) is reached. Boiled hams are placed in metal molds before cooking. Neither boiled nor canned hams are commercially sterile and must be refrigerated to prevent spoilage.

In preparing bacon, the belly portion of the hog is stitch-pumped with the curing solution. This is done by machines that through needlelike projections inject the curing solution simultaneously and automatically into the meat in a uniform pattern. The amount of pickle used for bacon is 5% to 10% of the green (uncured) weight. It is composed of a 65% to 75% saturated salt solution containing 20 to 100 lb (9.1 to 45.4 kg) of sugar, 1 to 1.5 lb (0.45 to 0.68 kg) of sodium nitrite, per 100 gallons (378.5 liters) of pickle. The pumped bellies may be placed in cover pickle (the pickle covers the product) or dry rubbed with curing salts during curing. Curing times are short, and in some cases, the product is smoked almost immediately after pumping with pickle. During smoking, the product is heated to an internal temperature of 120 to 125°F (48.9 to 51.7°C). This is done to develop the red color in the lean portion of the product. If the temperature were raised to 137°F (58.3°C) in all parts, as required for hams, much fat would be rendered out and lost. Apparently, regulatory agencies consider that bacon will be cooked prior to consumption so as to raise the temperature of all parts to 137°F (58.3°C) or higher.

After smoking and cooling, bacon slabs are squared up (uneven ends cut off), the product is sliced by machine, and it is packaged (usually under vacuum) so that a representative portion of the product can be viewed (through the plastic film) without opening the package.

Canadian bacon is prepared in a similar manner to that of regular bacon, expect that pork loins trimmed of fat are used instead of bellies. Also, curing times are somewhat longer than for regular bacon, and during smoking, the internal temperature must be raised to 137°F (58.3°C) or higher. Pork shoulders and shoulder picnics are cured and smoked much in the same manner as are hams.

Special pork products, such as capocollo, are prepared from boneless pork butts (part of the shoulder). They are dry cured for at least 25 days at temperatures not lower than 36°F (2.2°C). Special spices are used in the curing mixture with table salt and nitrite as the main ingredients (not more than 1 oz [28.3 g] of nitrite for each 100 lb [45.4 kg] of meat). After curing, the product is smoked for not less than 30 hours at a temperature not lower than 80°F (26.7°C), and then held in a drying room for at least 20 days at a temperature not lower than 45°F (7.2°C). These times and temperatures are specified to make certain the *Trichinella spiralis* has been destroyed, because this product is eaten without cooking.

Some beef products are cured, or cured and dried. Among these are corned beef, pastrami, and chipped or dried beef. Corned beef is usually produced from the brisket (lower forward portion of cattle near the foreleg). The fresh briskets are pumped with pickle and placed in cover pickle for 7 to 14 days at a temperature of 40°F (4.4°C) or below. In most cases, nitrite is used as an ingredient of the pickle, but in New England-style corned beef, only salt is used. In producing pastrami, table salt and nitrite (in limited amounts as stated previously), together with spices, are used in curing solutions.

Corned beef and pastrami are often sold as the cured product to be cooked in the home, restaurant, or delicatessen. However, they may be cooked and sold in the sliced (packaged) or unsliced form. Corned beef (containing nitrite) is also canned as corned beef hash, a mixture of cooked corned beef, cooked potatoes, and flavoring agents, such as onions. Some corned beef is cooked and canned as such.

Chipped beef is produced from beef hams or other portions that make up certain parts of the round section of the hind quarter. Usually cutter and canner grade cattle are used for this product. The beef hams are placed in tierces and covered with a pickle of high salt content containing not more (usually less) than 2 lb (907.2 g) of nitrite per 100 gallons (3.78.5 liters) of pickle. They are held in this manner for 50 to 65 days at a temperature below 40°F (4.4°C) (but above 34°F [1.1°C]). After curing, the product is hung in a room of low relative humidity and allowed to dry to the point where the salt content is 10% to 14%. It is then sliced into thin sheets and placed in glass jars that are capped under vacuum. This product is not heat-processed. The high salt content and the nitrite are sufficient to prevent the development of food spoilage or disease-causing bacteria. The vacuum packing is necessary to retain the pink color of chipped beef, and it prevents the development of mold growth.

Regarding the stability of cured meats, it should be pointed out that impurities in the salt used in the curing process may considerably affect the development of rancidity. For example, salt may contain traces of copper, iron, and other metals that are natural catalysts for initiating and accelerating rancidity.

SAUSAGE PRODUCTS

The meat ingredients used in a variety of sausage products may come, in part, from trimmings resulting from the cutting of beef quarters into rounds, loins, and other cuts and from the cutting of hogs into loins, hams, and shoulders. However, the chief source of meat for sausage comes from low-grade cattle and hogs. Bull meat is a prime ingredient of certain types of cooked sausage.

Fresh pork sausage is not cooked or smoked. The meat is ground and seasoned (usually only salt, sugar, sage, and pepper are added for seasoning, but ginger or fennel may also be used). Some ice, up to 3% by weight based on the meat ingredient, may be added. The product is mixed and stuffed into natural casings. In federally inspected plants, the meat ingredient is limited to contain not more than 50% of fat. The natural casings used for fresh pork sausage are made from the outer covering of the small intestines of sheep. Larger plastic or natural casings may be used for sausage.

Fresh pork sausage should be held at temperatures very near 32°F (0°C) after manufacture, because it is quite subject to bacterial spoilage, to oxidation of the fat (rancidification), and to loss of the typical pink color. Usually, when bacterial spoilage occurs, other types of spoilage do not.

Frankfurters or hot dogs (see Fig. 14.6), the most popular type of sausage, are comminuted, semisolid sausages prepared from one or more kinds of raw skeletal muscle meat (usually pork and beef) or raw skeletal meat and raw or cooked poultry meat (no more than 15% of the total ingredients, excluding water). The product is seasoned, cured, and usually smoked. The seasonings usually include sugar and spices

Hot Dogs

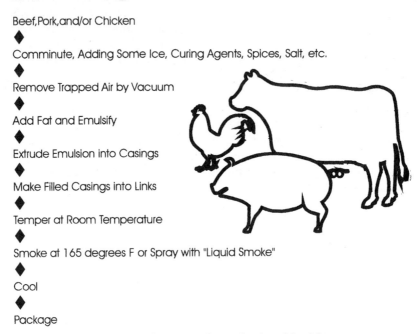

Beef,Pork,and/or Chicken

♦

Comminute, Adding Some Ice, Curing Agents, Spices, Salt, etc.

♦

Remove Trapped Air by Vacuum

♦

Add Fat and Emulsify

♦

Extrude Emulsion into Casings

♦

Make Filled Casings into Links

♦

Temper at Room Temperature

♦

Smoke at 165 degrees F or Spray with "Liquid Smoke"

♦

Cool

♦

Package

Figure 14.6. General sequence for production of frankfurters.

such as pepper, nutmeg or mace, mustard, garlic, and coriander to improve flavor. The curing salts usually contain sodium nitrite and sodium erythrobate (sodium iso-ascorbic acid) to prevent the growth of *Cl. botulinum* and to develop the color. There have been studies showing that nitrites, in combination with protein and heat, produce carcinogenic nitrosamines. The USDA sets limits on the amount of nitrites, nitrates, or a combination of the two at 156 ppm. This corresponds to about ¼ oz per 100 lb (about 7 g per 45.4 kg) of meat.

Work is being done to develop curing solutions that do not contain nitrites but still destroy *Cl. botulinum.* One such ingredient is sodium hypophosphite, which closely resembles nitrite in its preservative effect and is a GRAS substance. Approved phosphates may be added for water retention, and products such as dry milk, cereal, and soy flour may be added for a total of 3.5% of the finished product. It may contain meat byproducts such as hearts, snouts, and tripe but must be labeled as "frankfurter with byproducts" or "frankfurter with variety meats." Water or ice, or both, may be added to facilitate chopping or to dissolve the curing ingredients. If smoking is done, it may be with hard wood smoke or with liquid smoke spray. When the process is complete, the frankfurters may not contain more than 30% fat and not more than a 40% combination of fat and water.

Frankfurters may also be made from a single species and may be labeled "beef frankfurter" or "turkey frankfurter." The poultry franks have grown in popularity because they are often formulated so that there is less fat in the finished product. Some of the beef producers are also making lower-fat varieties. These products are quickly becoming some of the most popular items on the meat processors' production lines. Whatever the case, all ingredients must appear on the label in decreasing order, by weight.

By one method used in the production of frankfurters the meat is passed through a meat grinder and then finely comminuted in a silent cutter (a large metal bowl in which knives rotate at right angles to the plane of the bowl, which also rotates). While the meat is in the silent cutter, ice is added to prevent the temperature of the meat from rising above 60°F (15.6°C). It should be noted that the finished product may contain up to 10% added water (federal regulations), the purpose of which is to facilitate cutting and to provide more succulence to the finished product. Curing agents, spices, and fillers (if used) are added while the meat is in the silent cutter. The nitrite is added separately from other additives, because if it is mixed with the other additives before it is added to the frankfurter mix, the preservative effect of the nitrite will be diminished.

When ascorbic acid or iso-ascorbic acid is used to assist in color development, it is added about 1 minute before the end of chopping in the silent cutter. Colloid mills or similar equipment may be used instead of the silent cutter to comminute the meat. From the silent cutter, the meat emulsion is put into metal cars, which may be placed in a vacuum chamber to eliminate trapped air bubbles. The air may also be removed in a mixer under vacuum.

The meat emulsion consists of a continuous phase, which is the water, and a dispersed or discontinuous phase, which is the fat. The fat is dispersed more or less uniformly in the water. The agent that maintains this uniform dispersion of the fat (emulsifier) is actually the fraction of proteins that has been solubilized, especially in the presence of salt. In preparing the emulsion, one must heed the temperature rise during preparation, and the rate at which the fat is added as well as the speed with which the emulsion is mixed. The temperature should not be too high. If it exceeds about 68°F (20.0°C), the emulsion may break down. The optimum rate at which the fat should be added depends a great deal on the rate at which the fat can be emulsified.

If the fat is added more slowly than the emulsification rate, then the mixing time will be unnecessarily extended, creating an increase in temperature and a decrease in the amount of fat that can be emulsified. If the rate at which the fat is added greatly exceeds the rate at which it can be emulsified, then it is quite possible that little or no emulsion will be formed. The viscosity of the mix, the size of the fat droplets, the type of fat, and a considerable number of other factors affect the emulsion characteristics.

The meat emulsion is next placed in a stuffing machine where a powered piston extrudes the product into a stuffing horn and then into artificial (cellophane or other plastic) or natural (sheep or other animal intestine) casing. The stuffed casing comes out as a long cylinder, which is then made into link form, generally by machine, and the linked sausage is hung on racks and then smoked and cooked.

During smoking and heating, the frankfurters are hung in air-conditioned or natural draft smokehouses. During the first phase in the smokehouse, a temperature of 130 to 140°F (54.4 to 60°C) is used for 10 to 20 minutes. Smoke is then introduced, and the temperature is slowly raised to 165°F (73.9°C) and held until the internal temperature of the product reaches about 155 to 165°F (68.3 to 73.9°C). Immediately after cooking and smoking, the internal temperature of frankfurters should never be less than 150°F (65.6°C), and preferably should be at 155°F (68.2°C) or slightly higher. After cooking, the frankfurters are hung in a cooler.

Frankfurters stuffed into natural casings are not skinned after manufacturing, but those stuffed into cellulose casings are removed from the casing before they are sold.

The links of sausage are passed through a bath of warm water, then into the skinning machine, where the casing is slit with a blade, then rolled off or blown off.

After skinning, the frankfurters are packaged and immediately returned to the cooler, which should be held at a temperature near 32°F (0°C), until shipped out. Frankfurters, like other perishable foods, eventually spoil when held for long periods at temperatures near but above freezing. When spoiled, frankfurters either become slimy, as a result of the growth of bacteria or yeasts, or they turn green on parts of the surface. Greening is due to hydrogen peroxide produced by the growth of bacteria of the *Lactobacteriaceae* group. The hydrogen peroxide reacts with the nitrosohemochrome (color complex of nitrite and myoglobin) and oxidizes it to form a green color.

In continuous frankfurter manufacturing, automatic stuffers extrude the emulsion into molds, which pass through an automated tempering, smoking, cooking, and cooling unit. The product is packaged immediately after exit from the processing unit.

Bologna is another type of smoked and cooked sausage. The ingredients and manufacturing procedures for manufacturing bologna are similar to those used in making frankfurters, but the size of the casing, hence the size of the bologna, is much larger in diameter, and the bologna link is proportionately longer than the frankfurter link. When not sufficiently cooked, bologna may develop green cores after processing.

There are many other types of meat sausage produced. Among the smoked, uncooked sausages are country-style pork sausage and country-style sausage that contains beef as well as pork. These products contain moderate amounts of curing agents. Cooked sausages are made both with and without curing agents added, and are stuffed into natural cellulose or saran casings. In the manufacture of Braunschweiger or liver sausage, pork liver and jowls, with or without beef, are used.

Types of sausage vary, depending on the style of cutting or grinding of the meat, the size and shape of the product, the seasonings used, the degree of smoking, and the cooking procedure. A few of the different types include kielbasa, knockwurst, cooked salami, mettwurst, and Polish sausage.

There are various loaf-type specialty products, such as chicken loaf, luncheon meat, meat loaf, and head cheese, and these may be processed in casings or in metal containers. In the latter instance, they are usually placed in casings after cooking. This type of product may be cooked in hot water or baked in an oven. The products may be deep-fat fried, after baking, in order to brown the surface.

Some types of sausage are deliberately handled to permit a bacterial fermentation to occur in the product during processing. During the bacterial fermentation, lactic acid is produced, lowering the pH of the product to around 5.0. The lactic acid assists in the preservation of the product and contributes to the particular flavor of this type of sausage.

Fermented sausages are either of the semidried or dried type. Most fermented sausages are smoked, and the semidried type are heated to an internal temperature of at least 137°F (58.3°C). Many dried types of fermented sausage are not heated to temperatures above 90°F (32.2°C).

Curing salts are among the ingredients used in the production of fermented sausages, but nitrate sometimes replaces nitrite. During processing, bacteria remove oxygen from the nitrate, reducing it to nitrite. As stated earlier, nitrite provides the typical red or pink color in the meat.

After mixing with the curing agents, the meat for semidried sausage is held in pans for 48 to 72 hours in a refrigerated room at 38 to 40°F (3.3 to 4.4°C). The meat is then remixed, stuffed into casings, and held at 50 to 60°F (10 to 15.6°C) for 12 to 48 hours, then smoked. Smoking is accomplished in stages: first, at a temperature of 80 to 90°F (26.7 to 32.2°C) for 12 to 16 hours, then, at 100°F (37.8°C) for 24 hours, and finally, at 137°F (58.3°C) for 4 to 5 hours. During the processing of semidried sausage, conditions at first favor the growth of bacteria that reduce nitrate to nitrite. This is followed by the growth of lactic acid-producing bacteria. To ensure the type of fermentation desired, it is possible to obtain cultures of appropriate bacteria, which may then be added as an ingredient of the product. Typical semidried sausages are thuringer, cervelat, Lebanon bologna, and pork roll.

Dried sausages, such as Italian salami and pepperoni, are not heated to temperatures above 90°F (32.2°C), but they are stabilized against microbial spoilage because of their low pH (resulting from the action of lactic acid-producing bacteria), their low moisture content, and their high salt content. Spices and curing salts contribute to the preservation of the product. Some dried sausages are smoked.

During the manufacture of dried sausage, the meat, curing salts, and spices are mixed, and the mixture is passed through a grinder. The ground product is then placed in pans and held in a room at 38 to 40°F (3.3 to 4.4°C) for 2 to 3 days. After being stuffed into casings, the sausage is held on racks in a room at 70 to 75°F (21.1 to 23.9°C) and at a relative humidity of 75% to 80% for a period of 2 to 10 days. Smoked varieties are then smoked at low temperatures (not above 90°F [32.2°C]). Finally, the sausage is dried by hanging in a room at 45 to 55°F (7.2 to 12.8°C) (relative humidity of 70% to 72%) where there is an adequate air movement (at least 15 air changes per hour). Drying times vary between 10 and 90 days, and the moisture loss during drying is 20% to 40% of the weight of the freshly smoked product.

15

Dairy Products

Dairy products are produced from milk. In the United States, essentially all dairy products are produced from cow's milk, although minor quantities of goat's milk products may also be manufactured.

Because milk and milk products have traditionally been priced on the valued component, the fat, standards have been established mainly for this component but include provisions for other components, as well, at both the federal and state levels. The standards for milk and milk products may be found in the Code of Federal Regulations (CFR) Number 21, part 131. This is available from the Office of the Federal Register, National Archives and Records Administration, Washington, D.C. The Grade A Pasteurized Milk Ordinance Recommendations of the U.S. Public Health Service/Food and Drug Administration is another excellent source for setting microbial and sanitary standards.

FLUID MILK

The Holstein breed outnumbers all others used in the United States for the production of milk. Jersey and Guernsey breeds tolerate hot weather better than Holsteins, and hence may be the predominant types used for the production of milk in hot-weather areas. Some Ayrshire, and Brown Swiss or Shorthorn breeds are used in certain areas.

Cow's milk varies in the amounts of nutrients depending on diet and species but whole milk is adjusted to contain 3.3% fat (this may vary slightly in different states), 3.3% protein, 4.8% carbohydrate (mainly lactose, a disaccharide), 0.7% ash (minerals), and almost 88% water. Milk also contains vitamins and other nutrients in small amounts, making it the most complete of foods. The young of mammalians survive on it exclusively. However, components of milk from different species vary, and occasionally the young of one species may be unable to tolerate the milk from another species, mainly because of differences in the lactose contained therein.

The fat content of milk from Ayrshire and Brown Swiss, and especially from Guernsey and Jersey breeds, is slightly higher than that from Holstein cows, but the latter breed generally produces much more milk than the others. Most milk is produced on farms that primarily raise dairy cattle.

Microorganisms in Milk

As drawn from the cow's udder, milk seldom, if ever, is free from microorganisms; bacteria, molds, and yeasts are usually present in small numbers, among which the bacteria are most significant from the standpoint of quality and the transmission of foodborne disease. The control of microbial activity in milk and milk products, especially the control of bacteria and bacterial growth, is the most important function in the handling and manufacture of dairy products.

Raw milk, when improperly handled, may undergo any of several adverse changes. It may become sour as a result of the growth of bacteria that produce lactic acid, or it may become foamy as a result of the growth of gas-producing coliform bacteria or yeasts. Raw milk may also be subject to peptonization (digestion of the casein), or the formation of rope (a viscous polymer of sugars), when bacterial growth is not controlled.

Dairy herds are tested for tuberculosis and tested for and vaccinated against brucellosis, and most milk is pasteurized to prevent the transmission of a variety of foodborne diseases. However, under rare conditions, some pathogens may survive the pasteurization process, possibly because they may be protected by fat in which they may be encapsulated. *Listeria monocytogenes* is one such species (see Chapter 3).

Most, but not all, certified milk is pasteurized. However, certified milk must be produced from herds that have been inspected, tested, and found to be free from disease, and the milk must be drawn and handled under the best conditions of sanitation. There are bacterial standards for certified milk that limit the number and types of bacteria that are allowed. These requirements are set by the American Association of Medical Milk Commissions, Inc., 1824 North Hillhurst Ave., Los Angeles, CA 90027.

Sources of Bacteria in Milk and Methods of Limiting Bacterial Contamination

Some bacteria are normally present in the udder of the cow, and these may contribute to the bacterial flora of milk. However, unless the udder is infected, it is not considered to be an important source of such microorganisms. Other sources of microbial contamination of milk are the body of the cow; milking machines and other equipment and utensils; the air in the milking barn; and the hands, nose, and throat of those attending to the milking process. In the handling of milk on delivery to the processing plant or dairy, further sources of contamination may be encountered.

To limit the number of bacteria present in raw milk, certain precautionary procedures are ordinarily applied. The flanks, udder, and teats of the cow should be washed, treated with a sanitizing solution such as iodine, and dried before milk is drawn. Large dairy farms often have a special wash pen for cows to be milked. Utensils, including the milking machine, should be cleaned and disinfected either with live steam or with a solution of chlorine (about 200 ppm of available chlorine). Bulk milk tanks may be cleaned manually with detergent and water at about 130°F (543.4°C), then sanitized with chlorine solution, or cleaned mechanically with detergent and water at 150°F (65.6°C), and finally sanitized with chlorine solution. Outlet valves and the outside of the tank must be cleaned and sanitized manually. Cleaning in place (CIP) may be used to clean, sanitize, and rinse the milk pipeline, the teat cup assembly, and the bulk milk tank if a vacuum or pressure system is available.

The water supply used on dairy farms must be potable and located so that there is no possibility of contamination with animal or human discharges. Where municipal sewerage systems are not available, human wastes and floor washings from milk handling areas should be disposed of in septic tanks or cesspools. Manure should not be allowed to accumulate near milk handling areas and is best disposed of by spreading in thin layers on pastures.

Flies may be controlled in milk-handling areas, at least to some extent, by flytraps that utilize entrapment liquids; by poisons, such as formaldehyde; or by electric fly killers. The control of other insects, such as cockroaches, may require the use of an approved insecticide.

The walls and floor of the milking area should be kept clean, and the room should be reasonably well ventilated and free from dust. Personnel involved in the handling of milk should not have a history of intestinal disease. Flush toilets, in rooms that do not open directly into the area where cows are milked, should be provided for personnel, and hand washing and sanitizing facilities should be available at or near the area where milk is drawn. The hands of the milker or milking attendant should have been cleaned, sanitized, and dried before milking is started. Sanitary, surgical-type gloves may also be worn.

Handling of Milk on Farms

Milking is almost always carried out by machines. At first, a small amount of milk is drawn and examined for impurities, after which the cups of the milking machine are applied to the clean, sanitary teats of the cow. The cups are connected to a hose that leads to a holding container, and milk is collected by suction and a rhythmic squeezing action. After milking, the cups are immersed in a nonirritating bactericidal agent before they are applied to the teats of another cow. Milk from the cow passes through the milking machine from which it flows through a glass or stainless steel pipe to a bulk cooling tank cooled by refrigerated water or refrigerant sprayed directly onto the tank or expanded into a jacket that covers the outside of the tank. During cooling, the milk is slowly agitated mechanically to provide for faster heat transfer. Milk should be cooled to approximately 40°F (4.4°C) within 2 hours after it is drawn.

Milk may undergo a number of adverse changes during its handling. If held too long without adequate cooling, it is subject to various types of spoilage resulting from the growth of microorganisms. In addition, milk may develop various off-flavors owing to the feed consumed by the cow, especially if fed wild onions, french weed, or ragweed. Large proportions of beets, beet tops, potatoes, cabbage, or turnips in the fodder may also cause the development of off-flavors in the milk. Lipase, an enzyme present in cow's milk, may cause a hydrolysis of fat, splitting off butyric acid, which causes an off-flavor and off-odor. Milk that has been cooled, then warmed to about 85°F (29.4°C), then recooled or homogenized in the unpasteurized state, is subject to this kind of off-flavor development. Off-flavors resulting from the oxidation of some of its components may occur in milk, especially if traces of copper are present, because copper catalyzes this type of reaction. Milk, therefore, should be kept out of contact with equipment that contains copper.

Transportation of Fluid Milk

Milk is transported from the farm to the receiving station or to the fluid milk processing plant in pickup tankers. Pickup tankers are insulated, stainless steel tanks, usually having a holding capacity of more than 5000 gallons (18,925 liters) on a trailer handled by a motorized vehicle. No refrigeration is provided for many pickup tankers, because the insulated container prevents a significant rise in the temperature of the

Milk

Raw Milk
♦
Clarify
♦
Adjust Butterfat Content
♦
Fortify with Vitamin D
♦
Pasteurize
♦
Enhance Flavor (Flash Heat with Injected Steam and Cool with High Vacuum)
♦
Homogenize
♦
Cool
♦
Pour into Containers

Figure 15.1. Production of whole milk.

milk during the period required for transportation and delivery to the processing plant. The production of whole milk involves a series of steps, the important ones of which are shown in Figure 15.1.

When receiving the milk, the operator of the tanker first tests the product, which has been stored in a bulk tank, for odor and flavor and, if not suitable, the milk is rejected. If acceptable, the volume of the product in the bulk tank is measured with a rod. It is then agitated, after which a sample is taken in a glass or plastic bottle from which the butterfat content will be determined, because the farmer is paid on the basis of butterfat content. A commingled sample is also taken to be tested for the presence of antibiotics using the *Bacillus stearothermophilus* disc assay or equivalent as required by the Pasteurized Milk Ordinance of the U.S. Public Health Service. Quick tests approved by the FDA such as the Charm II Beta-lactam Tests developed by Charm Sciences Inc. of Malden, Massachusetts enable farmers, processors, and quality laboratories to quickly and accurately examine milk for antibiotics. If there is no problem, the samples from the individual farmers will not be tested. If there is a problem, however, each individual farm sample must be tested, and the farmer responsible for the contaminated milk must pay for the entire load. Because of such severe financial consequences, farmers are very careful not to attempt to sell milk from cows that have been recently treated with antibiotics.

The milk is then pumped from the bulk tank into the tanker truck through a sanitized plastic hose after which the hose is capped. The final step is to prepare a weight ticket for the farmer and tabulate the weight, temperature, and other data of the product on a record sheet.

Pickup tankers, including auxiliary equipment such as hoses, must be cleaned and sanitized, as are dairy farm milking and milk-holding equipment, after delivery of the product to the processing plant.

Processing of Fluid Milk

Milk, as delivered to the processing plant, is first clarified while cold. Clarification consists of passing the milk through a centrifuge similar to a cream separator but operated at low speed. This treatment is sufficient to separate out dirt and sediment that might be present, depositing them as a layer on the inner surface of the centrifuge bowl. The clarifier is not operated at sufficient speed to separate the cream from the milk. After clarification, the milk is usually pumped into a storage tank equipped with an agitator. While in the storage tank the milk is sampled, and the butterfat content is determined. It is then standardized by adding enough cream or skim milk (milk from which cream has been removed) to provide the fat content required by state regulations. The milk is then fortified with vitamin D at the rate of up to 10 mg as cholecalciferol (400 IU) per quart (0.95 liter). Each serving of milk (8 oz or 236 ml) will then contain 25% of the Daily Value (DV) for vitamin D, which is 10 mg. Milk, especially low-fat and skim, will usually be fortified with vitamin A also. The DV for vitamin A is 5000 IU or 1500 retinal equivalents (RE). Milk is often fortified to give 10% of the DV per serving, so this would correspond to 150 RE (500 IU) per cup or 600 RE (2000 IU) per quart.

The next step in fluid milk processing is to pasteurize it. In the old "vat method," milk must be heated in all parts to 145°F (62.8°C) and held at this temperature for 30 minutes. The method now used in most commercial operations utilizes the high-temperature–short-time (HTST) method in which the milk must be quickly heated to 161°F (71.7°C) and held at this temperature for 15 seconds, heated to 191°F (88.3°C) and held for 1 second, or heated to 194°F (90°C) and held for 0.5 seconds, then cooled. These temperatures ensure the destruction of the most heat-resistant pathogen in milk, *Coxiella burnetti*, the bacterium that causes Q fever. Milk may be pasteurized in insulated vats heated by coils carrying hot water, or in vats heated by hot water sprayed within a jacket surrounding the sides and bottom of the vat. With low-temperature pasteurization, the milk is ordinarily agitated during heating and cooling. Plate heaters and coolers (see Fig. 15.2) or tubular heaters may be used to pasteurize and cool milk. When tubular heaters are used, the product travels in one direction through an inner tube while hot water for heating or the refrigerated liquid for cooling passes in the opposite direction through an outer tube surrounding that which carries the product.

Milk may be given what is called a flavor treatment to provide a product that is uniform in odor and taste. During flavor treatment, milk is instantly heated to about 195°F (90.6°C) with live steam (injected directly into the product) after which it is subjected to a vacuum of about 10 in. (25.4 cm) in one chamber and to a vacuum of about 22 in. (55.9 cm) in another chamber. The high-vacuum treatment serves to regulate flavor, to cool the milk to about 150°F (65.6°C), and to evaporate water that may have been added through the injection of steam.

After the flavor treatment, while the milk is still hot, it is usually homogenized by passing it through a small orifice that breaks up the fat globules to a small size, preventing the separation of cream from the milk. The milk is then quickly cooled to about 35°F (1.7°C). This is done by the same general procedures used in heating, except that refrigerated water or brine, or directly expanded ammonia is used in the coils, vat jacket, outer tubes of the pasteurizer, or tubes of the plates on the cooling side of the plate pasteurizer. During HTST pasteurization and during flavor treatment and

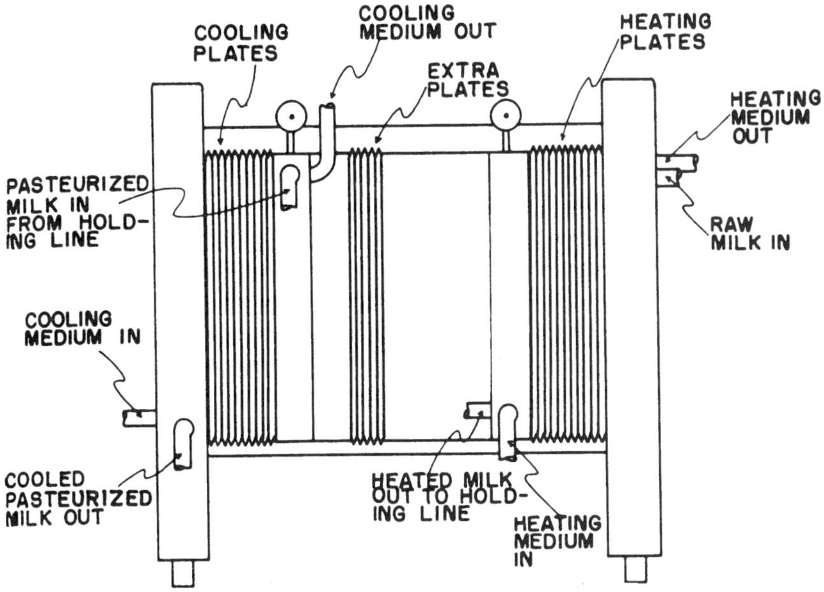

Figure 15.2. A plate pasteurizer.

homogenization, milk is passed through heating and cooling cycles at such a rapid rate that at no time is it held for long periods at high temperatures.

After processing and cooling, milk is filled mechanically into containers made of waxed or plastic-coated cardboard of different volumes up to 2 quarts (1.9 liters) and of semirigid plastic containers of 2 quarts (1.9 liters) or 1 gallon (3.8 liters), and the containers are sealed. In this state, milk should be held as close to 32°F (0°C) as possible until consumed.

Some large dairies are unable to process all milk immediately after delivery. Sometimes, milk may be stored for hours or even days before processing. In these cases, even deep chilling will not prevent some quality deterioration so the plants employ a thermization step. This step raises the temperature to prepasteurize the milk. The milk is brought to 145.4 to 149°F (63 to 65°C) for about 15 seconds and then chilled to 40°F (4.4°C) or below rapidly. The spore-forming spoilage organisms are not eliminated in this process and after it is complete, they may revert to their vegetative state. This may be an advantage, for when the milk is finally pasteurized, the organisms may be in the more vulnerable vegetative state and destroyed more effectively. The law in some countries prohibits double pasteurization so in these countries, an enzyme test (phosphatase test) that tests for the presence of an enzyme that is destroyed in normal pasteurization, must be done on the milk to ensure that it hasn't been totally pasteurized before it goes through the second heat treatment and final pasteurization step.

In plants producing fluid milk or milk products, all equipment, including tank or vat, pasteurizes and coolers, homogenizers, pipelines, and pumps, should be of sanitary design. There should be no threaded pipes. Joints should be of the clamp type that can be easily disassembled for cleaning and sanitizing. All surfaces contacting milk or milk products should be readily accessible for cleaning and sanitizing regardless of whether they are to be cleaned in place (CIP) or disassembled. The suitability of equipment for cleaning and sanitizing, and the frequency with which this is done, are equally important.

Some plants employ the newest technologies in milk production, Ultra-High-Temperature (UHT) treatment and aseptic filling. In these systems, milk is exposed to brief, intense heat treatment at temperatures of 275 to 284°F (135 to 140°C) for a few seconds. The milk is then homogenized, cooled, and filled aseptically into sterile containers. This method may be used for sterile milk products or, if not packaged in sterile containers, will increase the shelf-lives of pasteurized products substantially. Some dairies have extended these shelf-lives to as much as 40 to 45 days at 40°F (4.4°C) in situations where the shelf-life using conventional HTST methods and packaging in nonsterile containers is often less than 20 days. These UHT plants can use two major types of heating to reach these high temperatures, indirect heating or direct steam heating.

In indirect heating, the product is pumped from the holding tank to the balance tank to the regenerative portion of the plate heat exchanger (PHE). Here in the PHE it is brought to about 167°F (75°C) with sterilized milk and is then homogenized and sent to the PHE to be heated to 278.6°F (137°C) by a closed hot-water circuit with the temperature regulated by steam injection into the water. After heating, the milk goes through a holding tube for about 4 seconds and enters an aseptic section in the PHE where it is water cooled. The final cooling to about 68°F (20°C) is achieved by regenerative heat exchange with the incoming unpasteurized product as the cooling medium. The product then leaves the cooler and is aseptically filled into a sterile package or an aseptic tank for intermediate storage (see Fig. 15.3).

In direct steam heating, the product is pumped from the balance tank to the preheating section of the PHE, where it is heated to 176°F (80°C) and the pressure is increased by the positive displacement pump. Steam is then injected and the temperature instantly reaches 284°F (140°F) (the increased pressure keeps the product from boiling). The product is held in the holding tube for a few seconds and then passes to the condenser-equipped expansion chamber in which partial vacuum is maintained with

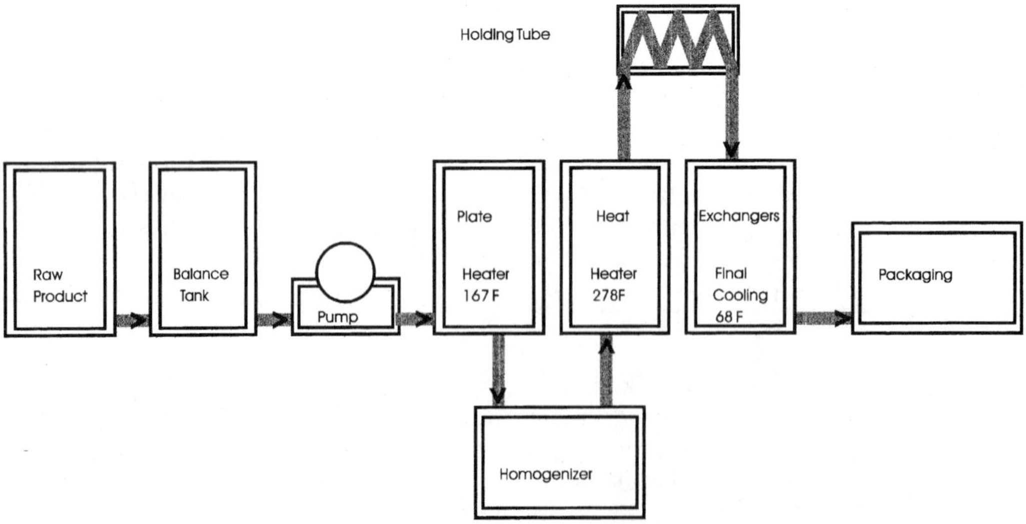

Figure 15.3. UHT processing with indirect steam heat.

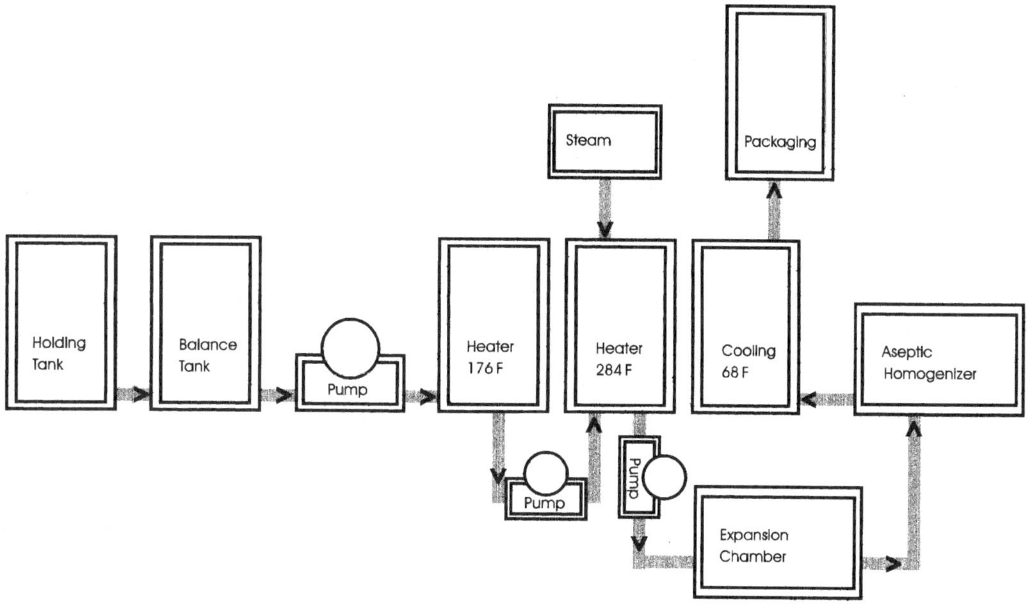

Figure 15.4. UHT processing with direct steam heat.

a vacuum pump. The vacuum is controlled so that the amount of vapor flashed off from the milk equals the amount of steam injected earlier in the process. A pump now moves the sterilized product to the aseptic homogenizer and then the product is cooled to 68°F (20°C) in the PHE and continues to the aseptic filling machine or to intermediate storage before filling (see Fig. 15.4).

Both of the methods described produce a sterile product that can be held at room temperature for long periods of time. The methods can also be modified to produce a nonsterile product if it is not packaged aseptically.

A number of new alternative methods to thermal processing for the food industry are being explored. These include utilization of pulsed electrical fields, oscillating magnetic fields, high hydrostatic pressure, intense light pulses, and irradiation. Raw skim milk has been pasteurized at Washington State University using short-duration, high-intensity pulsed electrical fields (PEF) and the temperature of the product does not exceed 131°F (55°C) in the process. Treatment of the liquid food takes place between two electrodes and lasts for less than 1 second. Continuous processes have been developed (see Fig. 15.5) and raw skim milk processed in this manner was tested and there were no apparent changes in the physical and chemical properties of the milk and a sensory panel found no difference between heat-pasteurized milk and the PEF milk. The shelf-life of the product held at 40°F was 2 weeks. Work in this area of low temperature pasteurization seems to have future applications in the food industry. Products other than milk that have been pasteurized using this method include apple juice, beaten eggs, and green pea soup.

Skim milk (0.5% fat) and low or reduced fat milk (0.5% to 2.0% fat) are produced from whole milk passed through a centrifuge at high speeds, after the milk has been heated to 90 to 110°F (32.2 to 43.3°C), to remove the butterfat as cream. These products are usually fortified with vitamins A and D prior to pasteurization and cooling. In some cases, sodium caseinate (a derivative of casein, the main protein in milk) is also

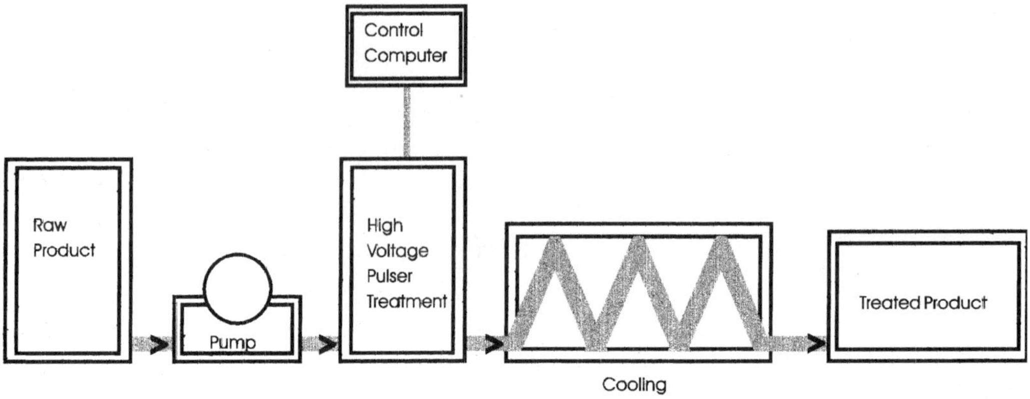

Figure 15.5. Continuous PEF process for liquid foods.

added. The cream from the centrifuge may be separated as approximately 40% butterfat (heavy cream), 30% butterfat (all-purpose cream), or 20% butterfat (light cream). The creams higher in butterfat may be diluted with skim milk to provide the various fat densities or to produce a product known as half-and-half (about 10.5% butterfat).

Because cream tends to spoil more quickly than milk, during pasteurization it is given a more drastic heat treatment than that given to milk. When batch pasteurization is used, cream is heated to 150 to 155°F (65.6 to 68.3°C) and held at this temperature for 30 minutes prior to cooling. When the HTST method is used, cream is heated to 166 to 175°F (74.4 to 79.4°C) and held at this temperature for 15 seconds prior to cooling. Table cream (light cream or half-and-half) is usually homogenized after pasteurization. All cream, after pasteurization, should be quickly cooled to 35°F (1.7°C) and containerized. It should be held at 35 to 40°F (1.7 to 4.4°C) until consumed or subjected to additional processing.

Chemical, bacteriological, and temperature standards for grade A milk and milk products, as issued by the U.S. Public Health Service, are listed in Table 15.1.

Large quantities of skim and low-fat milk are dried. This may be done by spraying atomized droplets of milk into a chamber through which heated air is circulated (spray drying; see Fig. 11.3) or milk may be dried by allowing it to flow over the surfaces of two heated metal drums that rotate toward each other (drum drying; see Fig. 11.2). With the latter method, the dried milk is scraped from the drums, as they rotate, by metal scraper blades. Dried milk (usually the spray-dried type that contains about 5% moisture) may be rehumidified to slightly higher moisture content after drying. This treatment agglomerates the fine milk particles to form clumps of milk powder, which results in a product that dissolves or disperses in water much more readily than the finely powdered dried milk. It is therefore considered to be an instantly soluble product.

Milk is used for the manufacture of a wide variety of popular dairy products, some of which are described in this chapter.

OTHER DAIRY PRODUCTS

Ice Cream

Ice cream is one of the most popular milk products. It is made simply by whipping air into an "ice cream mix" as the product is frozen. Ice cream contains about 10% to

Table 15.1. Chemical, Bacteriological, and Temperature Standards for Milk

Grade A raw milk and milk products for pasteurization, ultra pasteurization or aseptic processing	Temperature	Cooled to 45°F (7°C) or less within 2 hours after milking. The blend temperature after the first and subsequent milkings should not exceed 50°F (10°C)
	Bacterial limits	Individual producer milk not to exceed 100,000 per ml prior to commingling with other producer milk
	Drugs	No zone greater or equal to 16 mm with *Bacillus stearothermophilus* disk assay method in Appendix G, p. 195, 1993 PMO
	Somatic cell count[a]	Individual producer milk: not to exceed 750,000 per ml
Grade A pasteurized milk and milk products and bulk-shipped heat-treated milk products	Temperature	Cooled to 45°F (7°C) or less and maintained there
	Bacterial limits[b]	20,000 per ml
	Coliforms[b, c]	Not to exceed 10 per ml. In the case of bulk milk transport tank shipments, shall not exceed 100 per ml.
	Phosphatase[c]	Less than 1 microgram per ml by the Scharer Rapid Method (less than 500 milliunits/liter by Fluorometric Procedure) or equivalent
	Drugs	No zone greater or equal to 16 mm with *Bacillus stearothermophilus* disc assay method in Appendix G, p. 195, 1993 PMO
Grade A aseptically processed milk and milk products	Temperature	None
	Bacterial limits	No growth as specified in Section 6, 1993 PMO
	Drugs	No zone greater or equal to 16 mm with *Bacillus stearothermophilus* disc assay method in Appendix G, p. 195, 1993 PMO

[a]Goat milk 1,000,000.
[b]Not applicable to cultured products.
[c]Not applicable to bulk shipped heat-treated products.
Source: Public Health Service / Food and Drug Administration Publication No. 229.

14% fat, and consists of many ingredients including cream; milk or skimmed milk; sugars (sucrose, dextrose, corn syrup or high-fructose corn syrup); stabilizers which may be gelatin, vegetable gums, or modified food starches; pasteurized eggs or egg whites; flavorings such as vanilla; fruit juices or extracts; cocoa; chocolate; candy; cookies; nuts; or just about anything the imagination and food science will allow. Low-fat ice cream can be made by adding less cream and adding more milk solids, sodium caseinate, or both. Generally the butterfat content of ice cream is regulated by individual states and low-fat products have be to labeled as such.

There are various systems to manufacture ice cream but a general method is shown in Fig. 15.6.

"French" or custard ice creams have egg yolks added to improve texture. These must be pasteurized and use of homemade recipes that call for raw eggs is not recommended. The risk of the presence of salmonella exists and unless the eggs are cooked to at least 145°F to destroy this organism, the ice cream could be a carrier of this pathogen.

Quality ice cream is highly regarded not only for flavor but also for its velvety smooth texture. Milk fat (often called butterfat) from the milk products is responsible for this texture and the Food and Drug Administration (FDA) has set some minimum standards for amounts of milk fat and total nonfat milk solids in ice cream. As stated in Title 21 of the Code of Federal Regulations, ice cream must contain not less than 10% milk fat and not less than 10% nonfat milk solids. This may change if the milk content is raised providing the combination of the two ingredients equals 20%. This may also be changed if bulky flavors such as chocolate are used, but in no case can the level of milk fat go below 8% and total milk solids below 16%.

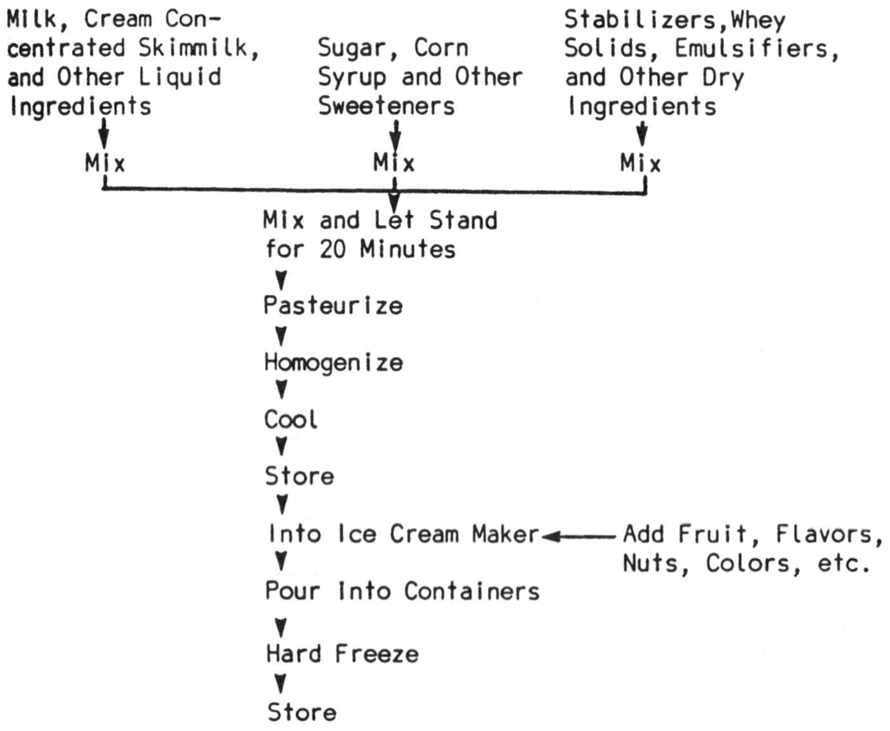

Figure 15.6. A method of ice cream manufacture.

There are various systems and procedures for the manufacture of ice cream but, in general, liquid dairy ingredients are mixed together and liquid sweeteners are mixed separately. These are then combined and blended together with the dry ingredients. They are allowed to "soak" for a period of time (usually about 20 minutes) and then are pasteurized, homogenized, and cooled with agitation to 36°F (2.2°C) and held in storage tanks. When the ice cream is made, the mix is combined with flavors and is added to the ice cream freezer, which is a tube with walls refrigerated to 5 to 15°F (–15 to –9.4°C) and blades rotate and scrape the mix from inner walls as it whips air into the mixture. The amount of air that is whipped into the ice cream mix is also under government regulation. Ice cream cannot contain less than 1.6 lb of total solids per gallon and the total weight must be at least 4.5 lb.

The amount of air incorporated into the mix is called overrun and can be calculated by the following formula:

$$\%\text{ overrun} = \frac{100 \times (\text{weight per gallon of mix} - \text{weight per gallon of ice cream})}{\text{weight per gallon of ice cream}}$$

Ice cream, unfortunately, does not rank with the top dairy products regarding its contribution to a wall-balanced diet. It has been recommended to reduce the amounts of fat to the diet, especially saturated fats, and cholesterol. Ice cream is both high in fat and contains primarily animal fats that contain saturated fatty acids and cholesterol. This has led to the development of low- and nonfat ice cream-like products. Nonfat dry milk solids and skim milk are combined with flavors, sweeteners, and gums, alginates, or modified starches to trap air and give a fatlike, smooth texture. See Table 15.2 for texture enhancing ingredients in nonfat dairy products.

In the dairy quality control laboratory, many of the tests done on other dairy products are also done on ice cream. These include butterfat, protein, total solids, antibiotics, and a variety of microbial tests including standard plate counts and coliforms.

As previously mentioned, with the large consumer demand for low-fat, low-cholesterol products, there has been an increase in the production of low-fat or nonfat ice cream-like products. Dairy soft-serves and ice milks use low amounts of butterfat and fat-free products are made with skim milk, water, and products such as modified food starches, cellulose gels, and gums, for example, cellulose, locust bean, and carrageenan

Table 15.2. Texture-Enhancing Ingredients in Some Popular Nonfat Dairy Products

Nonfat yogurt	Cultured pasteurized nonfat milk, whey protein concentrate, modified food starch, whey, gelatin
Nonfat ice cream	Skim milk, sugar, corn syrup, maltodextrin, egg whites, whey protein concentrate, cellulose gel, cellulose gum, mono- and diglycerides, carob bean gum, carrageenan
Nonfat frozen yogurt	Nonfat milk, sugar, corn syrup, nonfat yogurt, whey, mono- and diglycerides, guar gum, carob bean gum, polysorbate-80, calcium sulfate, carrageenan
Nonfat cheese	High-moisture skim milk cheese (skim milk, cheese culture, salt, enzymes), whey, skim milk, dried corn syrup, sodium phosphate, cellulose gel, sodium citrate, carrageenan

(see Table 15.2). These will retain large amounts of water and also will entrap air when it is whipped into the product during the freezing process, thus giving the product a fatlike mouth feel without the fat calories and cholesterol.

Cultured milk products comprise an ample portion of the dairy market with yogurt, buttermilk, and sour cream being the most utilized.

Yogurt

Yogurt is a fermented dairy product that has grown in popularity in the United States. It is used in developed countries as a dessert, between-meal snack, complete lunch, and diet food. People eat yogurt as part of a balanced diet because they like it and feel that it favorably affects their health. Flavored yogurts made from low-fat or skim milk are very popular and unflavored yogurts are eaten plain or are used as substitutes for some of the more fatty dairy products such as sour cream or some cheeses.

Most yogurt is made as a result of the growth of *Lactobacillus bulgaricus* and *Streptococcus thermophilus*. The bacteria may be naturally present in milk but to ensure consistent quality, pure cultures, preferably in a 1:1 ratio, are added to heat-treated and cooled milk (the process is further discussed later in the chapter). Large amounts in the viable state remain in the product until consumed. Other harmless lactic acid bacteria such as *Streptococcus lactis* and *Lactobacillus acidophilus* may also be present. The flavor of yogurt is a result of the symbiotic relationship between the bacteria. Besides lactic acid, components produced include acetic acid, diacetyl, and acetaldehyde (see Fig. 15.7). These add to the unique flavor achieved in yogurt.

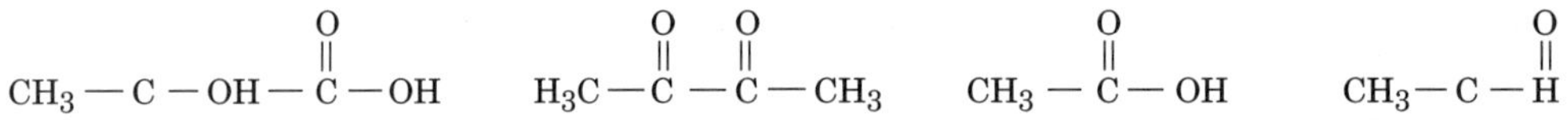

lactic acid diacetyl acetic acid acetaldehyde

At peak freshness just after making, plain yogurt may contain up to 1 billion live *L. bulgaricus* and *S. thermophilus* cells per milliliter. On storage at 39.2°F (4°C), these numbers decline, with *S. thermophilus* dying more rapidly. The two bacteria may be kept as a mixed culture or individually. Frequent transferring of a mixed culture may upset the desirable 1:1 ratio of the two bacteria.

The symbiosis exhibited by the two organisms is interesting in that the *L. bulgaricus* breaks down proteins, liberating amino acids that stimulate the growth of *S. thermophilus*. In return, *S. thermophilus* promotes acid production and reduces oxygen levels, both of which enhance the growth of *L. bulgaricus*. As the pH levels are lowered to about 5.0 primarily by *S. thermophilus*, the production of acetaldehyde by *L. bulgaricus*

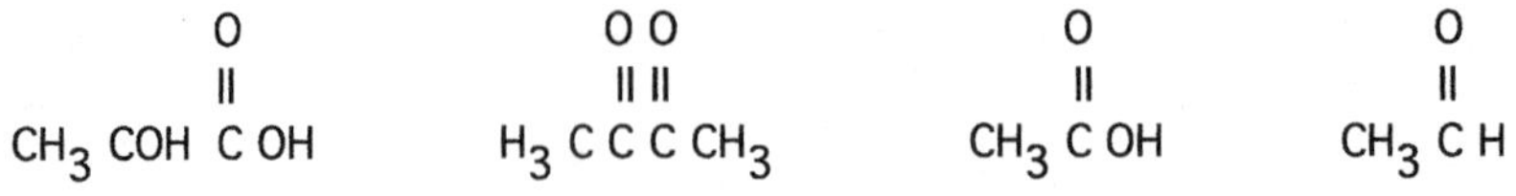

lactic acid di acetyl acetic acid acetaldehyde

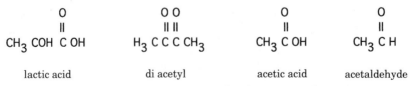

Figure 15.7. Some components produced in a yogurt fermentation.

is inhibited and the lactobacilli will reduce the pH to about 4.0 to 4.4. Milk used for yogurt must be penicillin-free, for *S. thermophilus* is highly sensitive to it. Excellent yogurt requires good bacterial cultures and these can be lyophilized to powdered form and transported dry. They may also be frozen as a concentrate and stored or shipped under liquid nitrogen at –310°F (–190°C).

Yogurt may be produced as a smooth viscous liquid, a soft curd, or a solid frozen dessert. The soft curd, one of the most popular products, is easily spooned out of its container without free whey. Acceptable colors and stabilizers may be added to improve appearance and texture but, unlike in cultured buttermilk, no salt is added. Sugar may also be added to neutralize the sour taste of acid and many yogurts have fruit added. The amount of fat in yogurt varies depending on the milk used in the process. Most yogurt in the United States is made with skim or low-fat milk and has a fat content of up to about 1.7%. Some yogurt is still made with whole milk and will contain about 3.25% fat.

Modern cultured yogurt manufacture is associated with a high degree of mechanization, but its center point is the unique bacterial fermentation. The only dairy product needed for yogurt production is milk but skim milk, condensed milk, nonfat dry milk solids, modified starches, alginates, and gums may be added to make low- or nonfat products. These mixes must be blended, homogenized, and pasteurized. The pasteurization temperatures are higher than those used for milk to ensure destruction of thermophiles. Vat pasteurization can be achieved at 190.4°F (88°C) for 30 minutes or HTST at 203°F (95°C) for 38 seconds. The mixture is then cooled to optimum incubation temperatures of 113°F (45°C) and the liquid culture is added at a rate of about 2% to 5% of the total mix. The inoculated warm mixture is then filled into containers that may contain fruit and is incubated until a pH of 4.4 (0.9% to 1.2% titratable acid as lactic) is attained. The finished product is then cooled and held at about 39.2°F (4°C).

The yogurt just described is called "sundae style" and requires mixing by the consumer. Yogurt may also be made in a "Swiss" style where the fruit puree, sugar, and stabilizer is blended with the fresh fermented yogurt at 60.8°F (16°C) and then packaged and cooled.

Fruit preserves are usually added at about a 15% level but have been added up to 27%. Stabilizers are not usually required for plain whole-milk yogurt but the flavored, low- and nonfat varieties may require the addition of gelatin, agar, modified food starches, or gums at levels of about 0.4% to 0.5%. The stabilizers can be added prior to pasteurization but care must be taken that excess amounts are not used or a sticky, gummy, hard mouth feel and free whey may result.

Typical flavored low-fat yogurts may have more calories than expected because of the amounts of sugar added (see Table 15.3).

There are some variations of normal yogurt that are made by the addition of *Lactobacillus acidophilus* to the fermentation to supplement or replace *L. bulgaricus*. A flavor change results from this as the acetaldehyde produced is reduced, as is the creamy character of the yogurt. *L. acidophilus* has an optimum set temperature of 96.8°F (36°C) which is much lower than that for *L. bulgaricus* which sets at 120.2°F (49°C). The advantage of acidophilus yogurt is the increase in numbers of viable acid-producing bacteria which implant themselves in the intestine of man, making them therapeutically effective. Other products include frozen flavored yogurts. These are made from a mix that is prepared and pasteurized much like ice cream mixes. Frozen yogurt is then frozen in much the same manner as ice cream with various levels of overrun.

Table 15.3. Some Nutritional Information for One Cup (227 gm) of Various Yogurt Selections

	Calories	Fat (g)	Sat. Fat (g)	Calories from Fat
Nonfat plain yogurt	126	<1	0.3	3
Low-fat plain yogurt	143	4	2.3	36
Whole milk plain yogurt	139	7	4.8	63
Nonfat flavored yogurt (fruit added)	204	<1	0.3	3
Low-fat flavored yogurt (fruit added)	232	2	1.6	18
Nonfat flavored yogurt sweetened with aspartame (fruit added)	100	<1	<1	0

NOTE: These values vary with different brands. Read the labels for accurate amounts.

Liquid yogurts made from whole and skim milk, and low-lactose yogurts that have been produced with milk treated with a lactase enzyme prior to fermentation, are examples of other yogurt-based products.

Nutritionally, yogurt is an excellent food, giving all of the protein benefits from milk but, especially in the case of low-and nonfat yogurts, with less fat and cholesterol.

Some quality problems that arise with yogurt include separation of whey because of uneven incubation temperatures, insufficient cooling, or improper use of stabilizers. Other problems include too much acid produced which is caused by too rapid a growth of *L. bulgaricus*, which can produce up to 2% to 3% acid. Normal yogurt has 0.9% to 1.5% acid. Careful maintenance of fermentation temperatures and use of pure cultures help to solve this problem. Premature curdling of milk has occurred also with yogurt that has had fruit added, especially Swiss style yogurt. When the fruit rests on the bottom, this rarely occurs. Another problem occurs owing to the presence of thermophilic bacteria or yeast in the milk. Off-flavors, such as bitterness caused by the bacteria, and alcohol and gas produced by the yeast will result from low-quality or improperly handled milk. Use of high-quality milk and proper quality control will solve these problems.

Cheese

There are many types of cheeses but the basic steps in cheesemaking include precipitation of the curd with bacterial fermentation or addition of rennet, draining of the curd, pressing, combination of cheeses (process cheese), aging or further fermentation involving bacteria or molds, and packaging and storage.

Cheese, along with buttermilk, sour cream, and yogurt, is one of the controlled fermentations of milk. Which product is yielded depends on the character and intensity of the reactions involved. Milk fermentations generally involve the breakdown of lactose into lactic acid through biochemical reactions initiated by streptococci and lactobacilli. Different results can also be achieved by incorporating post fermentation reactions into the process that produce peptones, peptides, amino acids, and fatty acids which add distinction to particular cheese flavors.

It must be noted that bacteria are not performing fermentations to provide food for humans. The fermentations are providing energy for the maintenance and production of more bacteria. The resulting unused compounds from the fermentations are what change the basic character of the resulting foods and make them interesting and attractive to the human palate. The lactic acid fermentation is the most important one in milk. The acid forms an isoelectric casein curd at pH 4.6. The sharp "prickly" taste of cheeses and yogurt is from the lactic acid. The reaction shows glucose or another monosaccharide being converted to lactic acid:

$$C_6H_{12}O_6 \rightarrow 2\ CH_3\text{—}CHOH\text{—}COOH$$

The reaction can be performed by many bacteria including *S. lactis*, *L. bulgaricus*, and *S. diacetilactis*.

In some cheeses such as Swiss and Gruy'ere, the propionic acid fermentation leads to the typical flavor and eyes. The bacterium *Propionibacterium shermanii* is usually the starter culture used and it gives best results when held at 75.2°F (24°C) for 3 weeks or longer. The reaction shows lactic acid going to propionic acid, acetic acid, carbon dioxide, and water.

$$3\ CH_3\text{—}CHOH\text{—}CHOH \rightarrow 2\ CH_3 - CH_2\text{—}COOH + CH_3\text{—}COOH + CO_2 + H_2O$$

Fresh milk contains about 0.2% citric acid and the citric acid fermentation is responsible for the delicate aromatic flavor of buttermilk, sour cream, and some cottage cheese. Organisms responsible for this activity include *Leuconostoc citrovorum*, *Leuconostoc dextranicum*, and *S. diacetilactis* and they grow at about 69.8°F (21°C). In the reaction, citric acid is converted to pyruvic acid and then to acetyl methylcarbinol:

$$CH_2\text{—}COOH\text{—}HOCCOOH\text{—}CH_2\text{—}COOH \xrightarrow{-\frac{1}{2}O_2} 2\ CH_3\text{—}CO\text{—}COOH \rightarrow CH_3\text{—}CO\text{—}CHOH\text{—}CH_3 + 2\ CO_2$$

The acetyl methylcarbinol can be oxidized to diacetyl:

$$CH_3\text{—}CO\text{—}CHOH\text{—}CH_3\ [\text{oxidation } (-2H)] \rightarrow CH_3\text{—}CO\text{—}CO\text{—}CH_3$$

or reduced to 2,3-butylene glycol. Carbon dioxide is also produced.

$$CH_3\text{—}CO\text{—}CHOH\text{—}CH_3\ [\text{reduction } (+2H)] \xrightarrow{+CO_2} CH_3\text{—}CHOH\text{—}CHOH\text{—}CH_3 + CO_2$$

A distinct flavor results when a fine balance between diacetyl, propionic acid, acetic acid, and other related compounds exists. If the Leuconostoc are overactive, excess CO_2 may be produced and the cottage cheese curds may have fine pinholes, causing them to float. This may be a problem when cooking the curds.

Some milk is fermented to produce products such as Kefir and Koumiss which contain about 1% to 3% ethyl alcohol. *Torula* or *Candida* yeast can grow with the lactic acid bacteria and can ferment lactose. In the fermentation, glucose is converted to ethyl alcohol and carbon dioxide:

$$C_6H_{12}O_6 \rightarrow 2\ CH_3{-}CH_2{-}OH + 2\ CO_2$$

This fermentation may also occur in other fermented products and be undesirable (such as in yogurt) and precautions and quality control measures must be incorporated to prevent it.

Another unwanted fermentation in milk is that of butyric acid. In this fermentation, large amounts of CO_2 and some hydrogen gas are produced and the cheese body is distorted. Cheeses affected include Swiss and Gruy'ere and large blowholes occur several months after the cheese is placed in curing rooms. The bacteria responsible are of the *Clostridia* group, particularly *Cl. butyricum*, which are resistant to pasteurization temperatures and grow well at temperatures of 100.4°F to 131°F (38 to 55°C). The representation shows that glucose can be converted to many byproducts (acetic acid, butyric acid, ethyl alcohol, butyl alcohol, and acetone) as well as carbon dioxide and hydrogen gas (this is not a balanced equation but merely shows the possibilities of various products from glucose):

$$\begin{aligned} C_6H_{12}O_6 \rightarrow CH_3{-}COOH &+ CH_3{-}CH_2{-}CH_2{-}COOH \\ &+ CH_3{-}CH_2{-}OH \\ &+ CH_3{-}CH_2{-}CH_2{-}CH_2{-}OH \\ &+ CH_3{-}CO{-}CH_3 \\ &+ CO_2 + H_2 \end{aligned}$$

Another fermentation in milk that causes spoilage and deterioration of the food is caused by the coliform group of bacteria. *Escherichia coli* and *Aerobacter aerogenes* are included in this coliform group and they are usually present in milk of poor quality. Their fermentation produces large amounts of carbon dioxide, resulting in slits in the cheese curd and obnoxious and unclean flavors in the cheese. Rapid production of lactic acid by starter culture bacteria will prevent their growth. In the reaction, lactic acid, acetic acid, and ethyl alcohol, as well as carbon dioxide and hydrogen gas are produced.

$$\begin{aligned} 2\ C_6H_{12}O_6 \overset{H_2O}{\rightarrow} 2\ CH_3{-}CHOH{-}COOH &+ CH_3{-}COOH \\ &+ CH_3{-}CH_2{-}OH + 2\ CO_2 + 2\ H_2 \end{aligned}$$

Starter cultures are used for most cheese production and they are harmless, active bacteria that are grown in milk or whey to produce certain characteristics in the various fermented milk products. The cultures may be one strain of bacteria or a mixed culture that may produce many characteristics and may also work in a symbiotic manner such as in yogurt. Table 15.4 shows some bacterial cultures, their functions, and the products in which they are used.

Starter cultures may be lyophilized (freeze-dried) with milk components, nutrients, and energizers (usually nitrogen sources) and packaged in the dry state for distribution. They also may be frozen with liquid nitrogen at –320.8°F (–196°C) and held in this state. Both methods yield viable cultures but the dry method requires no special containers or storage facilities other than a freezer or refrigerator. Many cheesemakers transfer starter cultures and hold them in this manner, with the viable, active cells ready to be inoculated. Some cheesemakers have successfully held lactic starters for several years but most usually start a new one from a dry, lyophilized culture every 3 or 4 weeks. A specific rotation of starters is necessary to ensure consistency of quality, however, and back-up cultures must be kept after one culture is used for a batch. This rotation must be adhered to or problems could develop. Starters that are used over

Table 15.4. Some Bacterial Starter Cultures Used for Various Dairy Products

Culture	Major Known Functions	Product Use
Propionibacterium shermanii	Flavor and eye formation	Swiss and Emmental cheese
Lactobacillus bulgaricus *Lactobacillus lactis* *Lactobacillus helviticus*	Acid and flavor	Buttermilk, yogurt, Kefir, Koumiss, and Swiss, Emmental, and Italian cheese
Lactobacillus acidophilus	Acid and claimed health benefits	Buttermilk and yogurt
Streptococcus thermophilus	Acid	Emmental, Cheddar, and Italian cheese, and yogurt
Streptococcus diacetilactis	Acid and flavor	Sour cream, ripened cheese, butter, buttermilk, starter cultures, cheese
Streptococcus lactis *Streptococcus cremoris*	Acid	Cultured buttermilk; sour cream, cottage cheese; all types of cheese, domestic and foreign; and starter cultures
Leuconostoc citrovorum *Leuconostoc dextranicum*	Flavor	Cultured buttermilk, sour cream, cottage cheese, ripened cream butter, and starter cultures
Streptococcus durans *Streptococcus faecalis*	Acid and flavor	Soft Italian, Cheddar, and some Swiss cheese

too many times tend to deteriorate and if a batch doesn't work properly, the cheesemaker must have tested cultures available for back-up. Preparation for these situations and plans for their correction must be incorporated into the quality control program.

Finished cheeses are very different in their final flavor, odor, color, and texture characteristics. The basic steps in cheesemaking, however, regardless of the product, are quite similar. These basic steps include: setting the milk, cutting or breaking the curd, cooking the curd, draining or dipping, curd knitting, salting, pressing, and special applications. Table 15.5 gives the basic steps and their primary purposes.

Setting the milk means preparing warm milk with a starter culture or rennet extract to form a smooth curd block. The curd may be either an acid isoelectric casein curd set with a starter or a sweeter, calcium para-casein curd set with rennet and starter. The acid curd takes from 5 to 16 hours whereas the sweeter curd can be completed in 15 to 30 minutes.

Cutting the curd must be done carefully and will increase the curd surface area many times. This leads to effective whey expulsion and permits the equal-sized smaller curds to be cooked uniformly throughout. Larger cubes give a higher-moisture cheese.

Cooking causes the curds to contract and also drives off the free whey. Cooking also influences curd texture, gains time for lactic acid development, and arrests it. It also suppresses the growth of spoilage organisms and influences the final moisture content.

Draining simply separates the whey from the curd by passing it through a straining device. In industry, this is accomplished by lining the exit gate of the vat with a sieve or having a sieve lay the entire length of the vat. Dipping is usually done in smaller operations by scooping curds into perforated molds or colanders or by inserting a coarse

Table 15.5. Some Basic Steps in Cheesemaking

Setting milk	Prepare milk for acid or rennet curd formation and the incorporation of suitable microbial cultures
Cutting or breaking curd	Speed whey expulsion and assist in uniform cook-through of the curd by increasing the surface area
Cooking curds	Contract for more effective removal of whey, develop texture, and establish moisture control
Draining or dipping	Permanently separate the whey from the curds
Curd knitting	Transform the curd into the characteristic texture of the cheese desired, give time for acid development, and aid in moisture control
Salting	Influence flavor, moisture, and texture
Pressing	Shape the cheese and close up the body
Special applications	Incorporate characteristic microorganisms for the specific cheese types and establish the proper environment for their growth

cloth into the kettle and bagging all the curds. Regardless of the method, this step accomplishes separation of whey and concentration and coalescing of curds, and provides additional time for lactic acid development.

The knitting and transformation of curds usually takes time and includes processes such as "cheddaring" in cheddar cheese where slabs of the curd are piled one on top of another and acid is allowed to develop to about 0.5% and more moisture is expelled. Other knitting processes include preliminary packing and pressing of brick and blue cheese curd, and the pulling and processing of acid-ripened provolone and mozzarella cheeses. These processes result in the characteristic texture of the various cheeses.

Salting can be done by spreading salt over the curds or by dipping the pressed cheese into brine solutions. Salting improves flavor, texture, and appearance while slowing or stopping the lactic acid fermentation when it has reached its optimum level. Salting also depresses the growth of spoilage organisms and reduces moisture, thus controlling it in the final product.

Pressing removes moisture and gives the cheese its shape. Care must be taken not to press with too much force in some cheeses such as Roquefort or blue cheese. If they are pressed too compactly, no air will be able to seep into the cheese and carbon dioxide will not be able to escape easily. Both of these are necessary for the successful growth of the molds needed to develop the characteristic flavors, aromas, and textures in the finished cheese.

The special applications cover a wide range of processes, from creaming cottage cheese to the addition of special microorganisms to ripen cheeses such as brick or Limburger.

Process cheeses are very popular in the United States and are manufactured by combining different types of cheeses as the main ingredient. Generally, not fully aged (green) cheese is ground together with some aged cheese and water, emulsifiers (i.e., disodium phosphate), salt, and powdered skim milk are added. The ingredients are heated, mixed, and extruded into molds.

Cheesemaking is an art as well as an application of advanced sciences. The success in this industry depends on a logical sequence of fixed steps where the emphasis is varied and the incorporation of special applications gives cheeses their various similarities and differences.

Butter

Butter is made from cream having a butterfat content of 25% to 40%, into which the bacterium *Streptococcus diacetyllactis* has been added to produce diacetyl, the major flavor component in butter.

The butter is produced by a continuous churning process and then concentrated to 80% fat by centrifugation.

Butter contains some water in the form of small droplets in a water-in-oil emulsion. Some bacteria may grow in this water and produce lipase, which will deesterify some fatty acids and cause rancidity, giving off-flavors and -odors. Salt of about 1% to 2.5% by weight is added to butter and adds flavor as well as inhibits the growth of the rancidity-causing bacteria. Sweet butter, which contains no salt, is more perishable than salted butter.

Whey

Whey, the fluid byproduct of cheesemaking, is vastly underutilized, with less than half of it being used. Whey has about 5% lactose, 2% other milk components, and 93% water. Whey is used for the production of ricotta cheese. This cheese, also called recooked cheese, is made by high-temperature heating of acidified cheese whey from production of other cheeses such as cheddar, swiss, and provolone. To the whey, 10% milk or skim milk may be added. The whey is heated and the coagulated albumin is dipped out and allowed to drain. With its growing popularity, ricotta is often made from whole milk, giving it the characteristic of creamed cottage cheese. The moisture in this cheese usually ranges between 70% and 80%. "Ziger" is the German name for ricotta-type cheese.

Attempts have been made to find other uses for whey such as in drinks, baked goods, ice cream, sherbert, candy, fudge, and other confections.

Buttermilk

A method to prepare buttermilk is to pasteurize skim milk, cool to about 71 to 72°F (21.7 to 22.2°C), and inoculate it with *Streptococcus lactis* or *Streptococcus cremoris* (to produce lactic acid) and *Leuconostoc citrovorum* or *Leuconostoc dextranicum* (to produce flavor). The incubation continues until an acid content of about 0.8% (pH 4.5) is attained. The coagulum is then broken by agitation and the product allowed to sit to remove air. It is then packaged and held at 35 to 40°F (1.7 to 4.4°C) until consumed.

Sour Cream

The production of sour cream is similar to that of buttermilk. It is prepared from light cream (16% to 20% butterfat) to which 8% to 9% nonfat milk solids and sometimes stabilizers (e.g., gelatin, gums, or modified starches) are added. After pasteurizing and homogenizing, the same bacteria used for buttermilk are added for the conversion of sugars to acid and the production of curd. The product is packaged and cooled and should be held at 35 to 40°F (1.7 to 4.4°C) until consumed.

DAIRY PRODUCT SUBSTITUTES

A number of dairy product substitutes have been developed for various reasons including economics, religion, and health benefits. Some of the most common examples are the fruit sorbets and ices, tofu-based frozen desserts, and fluid milklike products and cheese substitutes (milk fat is replaced with vegetable fat and vegetable protein).

16

Poultry and Eggs

In the United States, most poultry used for food consists of chicken and turkey. Some ducks and geese are consumed, but they are relatively insignificant as food sources. The rise in poultry consumption in the United States is mainly due to its low production costs and its dietary health benefits. Its low production costs result largely from the high feed–conversion ratio for growing poultry. For every 1.8 lb of food, the bird will develop 1 lb of meat. This is a higher ratio than can be obtained raising beef, pork, or lamb. The public awareness of the relation between saturated fat consumption and cardiovascular diseases also has had an effect on the increase in poultry consumption. A 3 oz chicken breast with the skin removed has only 3 g of fat (less than 1 g saturated) while the leanest cuts of beef such as round roast with all the visible fat removed still have over 4 g of fat (over 1.5 g saturated). Other popular beef cuts such as prime rib with lean and fat have over 26 g of fat in a 3 oz serving, with over 10 g of saturated fat. The consumption of eggs decreased in the 1970s and 1980s because of concern about cholesterol content. There is conflicting research on whether the amount of cholesterol ingested is as significant as the amount of saturated fat in the development of heart disease. Because of this, coupled with new data on cholesterol levels in the average egg (see end of this chapter), consumption of eggs may actually increase in future years.

The low cost of poultry and, especially, the public's perception of its health benefits, accounts for the increased use of poultry in the production of a hamburgerlike product and in the production of all-poultry frankfurters. Poultry is also allowed in the production of conventional frankfurters in amounts of up to 15% (see Chapter 14), and poultry is also being used in the production of a bologna-type and other cold-cut products. In all cases where poultry is used, either wholly or partially, the fat and cholesterol contents of these products are usually lower than in their beef counterparts. Check the nutrition label to be sure.

Today's modern technology, by manipulating genetics, allows poultry grown for meat purposes to grow rapidly, to be disease-resistant, and to have good meat qualities including a tender texture, good flavor, and a light color. Chickens having white feathers are preferred over other types, because there are no dark pin feathers, which, if not removed, detract from the appearance. Also, the skin of white-feathered birds is much lighter and, therefore, more desirable.

Present poultry breeds have been developed from wild birds, jungle fowl of Southeast Asia, and wild turkey of North America. Capons, roasters, and broilers are mostly mixed breeds or hybrids.

POULTRY

Chicken

For meat production, almost all flocks are started from 1-day-old chicks, and, ordinarily, "straight-run chicks" (about half male, half female) are used. Broiler flocks of this type are usually raised in a housing system that provides 0.5 ft^2 (464.5 cm^2) per bird until they are 2 weeks old and 1.0 ft^2 (929 cm^2) per bird between 2 and 10 weeks old. At the end of 10 weeks, they will be removed for slaughter. For capons and roasters that are 10 to 20 weeks old, 2 to 3 ft^2 (1858 to 2787 cm^2) of floor space per bird is used. Older birds (fowl) require 4 to 5 ft^2 (0.36 to 0.46 m^2) per bird. Commercial growers commonly raise at least four flocks of broilers per year.

In raising birds for meat, the floor and walls of the chicken houses (brooders) must be cleaned and disinfected. Fresh litter (usually shavings) is then put down on the floors. The day-old chicks, which may have been debeaked to prevent cannibalism, are introduced. Lights are kept on continually, and the temperature of the brooders is brought to 95°F (35°C) in cold weather or 90°F (32.2°C) in hot weather. The temperature of the brooders is lowered 5°F (2.8°C) weekly until 75°F (23.9°C) is reached, and it is held there until the birds are well feathered.

Troughs or hanging feeders may be used to hold food for meat-type birds, and hoppers for grit and calcium supplement may be used. Feeders, which must be adequate in size and number, may be filled automatically. Water troughs or hanging waterers may be used with 16 ft (4.88 m) of watering space per 200 birds when the temperature is at 75°F (23.9°C). When the temperature goes above 80°F (26.7°C), 20 ft (6.1 m) per 200 birds is needed. Feeders and waterers should be kept clean, and waterers should refill automatically.

Food for meat-type birds, a complex mixture, is obtained from feed companies that are expert in formulating such rations. Recommended starter feeds often contain corn; fish meal; poultry byproduct meal; corn gluten meal; soybean meal; alfalfa meal; dried distiller's solubles; small amounts of calcium and phosphorus salts; iodized salt; and A, D, E, K, and several B vitamins. Trace amounts of antibiotics may be added to feeds to prevent diseases. Among the above ingredients, corn and soybean meals constitute the major portion of the starter feeds. Finishing feeds, which may also include dried whey and steamed bone meal, are used and the combination of technologies from breeders, animal nutritionists, and feed manufacturers has made it possible to raise a 2 to 3 lb broiler in as little as 6 weeks. Other larger birds such as capons and roasters take longer.

Predators, such as rats and mice, must be kept out of poultry houses. Floors, litter, walls, roosts, and even the birds themselves may have to be treated with malathion insecticide to get rid of mites, lice, or ticks.

Birds grown for meat are the same age, and once marketing size is reached, they are placed in cages and removed from the growing house. They are slaughtered and processed elsewhere. Prior to installation of new flocks, the house is cleaned and disinfected. In order for a growing operation to be economical, the number of birds in roaster and capon flocks must be no less than 2000, and in broiler flocks, no less than 6000.

Some poultry processing plants are small, but the trend is to process poultry in plants capable of handling at least 10,000 birds per hour. These plants are usually divided into at least two rooms that separate bleeding, scalding, and defeathering from the eviscerating and chilling operations (see Fig. 16.1).

The birds are not fed for about 12 hours before they are to be slaughtered in order that their crops will be empty. This is important because it makes the operation much cleaner. The birds, shackled by their feet, are carried in the upside down position by conveyors from one operation to another. After shackling, they are slaughtered by slitting one or both of the jugular veins in the neck. An electrified knife or a stationary electric stunner may be used to render them unconscious prior to bleeding. Rendering birds unconscious prevents broken wings and bruising resulting from the flopping around during bleeding. After their jugular veins are severed, the birds are allowed to bleed from one to several minutes. Still attached to the conveyor, to facilitate removal of feathers, they next pass through the scalding tank containing water at 135 to 140°F (57.2 to 60°C) for larger birds or 122 to 128°F (50 to 53.3°C) for broilers. Immersion times vary with the size of birds, but several minutes in the scalding tank are required even for broilers. If scalding time or temperature is too high, the skin may be damaged.

Automatic picking machines are used to remove the feathers from poultry. In some systems, the birds are beaten by flexible rubber fingers as they pass through the machine. In other systems, they are dropped into baskets where feathers are removed by flexible rubber fingers rotating on a central shaft.

After the birds are defeathered, their feet are cut off, and they are rehung on the moving shackles by the lower legs. They then pass along a line where workers remove pin feathers by hand with the aid of a knife. Then they pass through a gas flame for singeing residual pin feathers. They are washed externally by water sprays as they pass along the conveyor.

Evisceration is usually carried out as the birds pass along the conveyor. The oil gland may be cut out before or after evisceration. A circular cut is made around the vent and the intestine is then pulled out a few inches. Another cut is then made through the abdominal wall from the vent toward the breastbone for broilers. For larger specimens a horizontal cut is made. The gizzard, liver, heart, and intestines are pulled out and allowed to hang so that they may be examined by a government inspector (a veterinarian) for signs of disease. Diseased birds are removed and destroyed. Gizzards that pass inspection are opened, emptied, peeled, and washed, and then packaged together with the livers and hearts, either for insertion within the bird or for holding separately. Intestines are not used.

The heads, gullets, and crops are removed and discarded, and the neck may be cut off and allowed to hang by the skin. Suction tubes are used to remove lungs and traces of reproductive organs or these are scraped out by hand. Both the interior and exterior of the bird are thoroughly washed. The neck and neck skin are placed in the body cavity.

The birds are next cooled, either in an air-agitated ice-water slush in tanks, in continuous ice-water chillers, or in moving refrigerated air. During chilling, the temperature of the birds, which may be 80 to 95°F (26.7 to 35°C), is lowered to 35°F (1.7°C), and they lose some residual blood and pick up a few percent of moisture from the chilled water. After chilling, they are drained, sized according to weight, and graded for quality. Grading is based on conformation, fleshing; covering with fat; and the presence of pin feathers, torn skin, bruises, and so on.

Graded birds are packed in wooded or waxed-fiber-board boxes. They should be surrounded with crushed ice and held at temperatures below 40°F (4.4°C). When the product is held at 28°F (–2.2°C), a significant extension in the shelf-life may be realized.

Some poultry is bagged or wrapped with plastic and frozen, by methods such as

Poultry

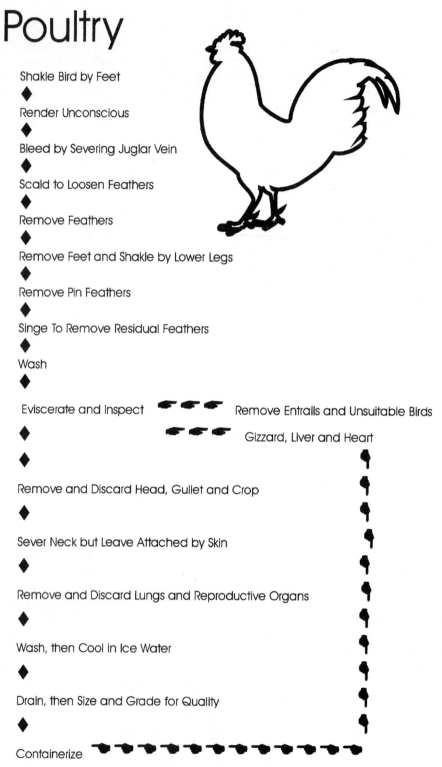

Shakle Bird by Feet

♦

Render Unconscious

♦

Bleed by Severing Juglar Vein

♦

Scald to Loosen Feathers

♦

Remove Feathers

♦

Remove Feet and Shakle by Lower Legs

♦

Remove Pin Feathers

♦

Singe To Remove Residual Feathers

♦

Wash

♦

Eviscerate and Inspect → Remove Entrails and Unsuitable Birds

♦ → Gizzard, Liver and Heart

♦

Remove and Discard Head, Gullet and Crop

♦

Sever Neck but Leave Attached by Skin

♦

Remove and Discard Lungs and Reproductive Organs

♦

Wash, then Cool in Ice Water

♦

Drain, then Size and Grade for Quality

♦

Containerize ← (Gizzard, Liver and Heart)

Figure 16.1. Chicken processing.

Figure 16.2. Inspection and grade shields for poultry. *(Source: U.S. Department of Agriculture.)*

cold-air blast, by contact with refrigerated plates or by immersion in liquid nitrogen at –320°F (–195.6°C).

Chickens are classified, according to age and condition, as broilers or fryers, roasters, capons (male castrated prior to maturity), stags (young uncastrated males), hens, stewing chickens or fowl (hens older than 20 weeks), and cocks or old roosters. Poultry may be also classified for quality as grade A, B, or C or No. 1, No. 2, or No. 3. A sample of grade shields is shown in Fig. 16.2.

Some chicken products are battered and breaded, deep-fat fried, and frozen. The high-quality storage life of uncooked chicken at 0°F (–17.8°C) can be more than a year if adequately packaged, while the high-quality storage life of fried chicken is less than 3 months at this temperature.

Some chicken is also precooked, deboned, and heat-processed in glass jars or cans. Some deboned poultry is used as an ingredient of canned soups. Chicken may also be precooked, deboned, cut into cubes or small portions, frozen, and then freeze-dried. This product is used as an ingredient of dried soups. As mentioned, chicken is quite popular in the production of frankfurters.

The edible portion of chicken is about 54%. This includes meat (about 39%), skin, and parts of the viscera that are also edible (about 15%). The nutritional content of all types of meat from a roasted chicken is about 28% protein, about 7% fat (2% saturated fat), and about 124 mg of cholesterol per one cup serving (140 g [4.9 oz]), about 1% ash (minerals), and the remaining 64% is mostly water.

Turkey

Turkey is the second most important poultry in the United States. In recent years, there has been a movement to develop strains that produce more meat, especially breast meat. This development has been limited mostly to white and bronze-feathered birds.

Turkey raising requires the same basic conditions as does chicken raising. For birds that are 1 to 3 weeks old, 1 ft^2 (0.09 m^2) of floor space is required. Birds that are 4 to 8 weeks of age require 1.5 ft^2 (0.14 m^2) per bird, and birds that are 8 to 15 weeks old require 2 ft^2 (0.18 m^2) per bird. Turkeys may be vaccinated to prevent erysipelas, salmonellosis, and other diseases.

In growing turkeys, a serious disease called blackhead has been known to break out. It was eventually learned that the disease was caused by a bacterial infection carried by chickens, and it was soon realized that if turkey flocks were going to be kept healthy, they would have to be kept away from chickens or from places where chickens had been, since the bacteria were found in heavy concentrations in chicken droppings and in areas that had been contaminated by chicken droppings.

Turkeys require a higher protein and vitamin content in their food than chickens, although, as with chickens, the protein content of the food is gradually reduced as the birds mature. Turkeys may be marketed as broilers 12 to 15 weeks old or as mature roasting birds 20 to 26 weeks of age.

Turkeys are slaughtered, defeathered, eviscerated, and further processed in much the same manner as chickens, except for some differences. Longer bleeding times are required for turkeys. They must also be scalded for longer times and at higher temperatures to facilitate the removal of feathers. With turkeys, the tendons in the legs are pulled out after defeathering and removing the feet. Further processing is essentially the same as for chickens. During cooling, turkeys pick up 4.5% to 8% water, the smaller birds absorbing the highest proportional amounts. Grading of turkeys is based on the same characteristics as those for chickens, and they may be graded for quality as A, B, or C.

A larger percentage of turkeys is frozen and sold to the consumer in the frozen state than is the case with chickens. However, a proportionally smaller amount of turkey meat than chicken meat is canned or dried. Some turkey is frozen, to be used later as an ingredient of pies or dinners. While the meat appears to be quite stable at 0°F (–17.8°C) as an ingredient of pies, it is somewhat less stable than chicken meat in frozen storage. As with chicken, turkey frankfurters are gaining in popularity and many varieties have total fat contents of less than 10% (the USDA standards allow up to 30%).

Ducks and Geese

Ducks and geese are both raised for human consumption, although the amount of the consumption of these species is insignificant in comparison to that of chickens and turkeys. Ducks and geese, for human consumption, have been developed from wild species. Their production and processing are similar to those of chickens.

EGGS

Except for some variations arising from differences in breeds, and sometimes even individual hens, the chemical composition of eggs is fairly constant. Of the total weight of the egg, the shell is about 10.25%, the white is about 59.50%, and the yolk is about 30.25%. The shell, largely calcium carbonate, has an outer coating (the cuticle) that protects the pores of the main part of the shell, as long as it remains intact. Inside the shell, there are two membranes, the one next to the shell being thicker and tougher than the one covering the contents of the egg.

The whole egg has about 12% protein, 10% fat, 1% carbohydrate, 1% ash, 75% to 76% water, and about 213 mg of cholesterol. The white of the egg is about 12% protein, less than 1% ash, about 88% water, and contains virtually no fat or cholesterol. Small amounts of sugar, carbon dioxide, and other constituents are present. The white is in two distinct parts: the thick, jellylike part that surrounds the yolk, and a less viscous "thin white" that spreads out when the egg is broken out of its shell. However, the thick white will also spread if it is cut, as its inner part is of thin consistency.

The yolk is about 17% protein, 49% water, 30% fat, and 1% ash. All of the 213 mg of cholesterol in the whole egg are contained in the yolk. Small quantities of numerous other constituents, including vitamins, are also present. The yolk is surrounded by the vitellin membrane, which when ruptured or cut, causes the yolk to spread when the egg is broken out of the shell. The yolk is much more complex chemically than the white, and accounts for the major nutritional composition of the egg, one of the most complete foods available to humans.

The egg contains a small air pocket that develops after the egg is laid when it cools from the body temperature of the hen, contracting the contents and pulling in the inner membrane. Because the eggshell is actually porous, the space formed by the contraction of the contents is soon filled by air that can be drawn in through the shell.

The raising of chickens for the production of eggs has similar requirements to those of raising chickens for meat, but there are some differences. The chief breeds used for egg production are white leghorns that lay white-shelled eggs and New Hampshires, Plymouth Rocks, and Rhose Islands Reds that lay brown eggs. Some cross-breeds are used.

In egg-laying flocks, all the birds are usually of the same age. Chicks are usually received at an age of 1 day, although they may be obtained as starter birds at an age of 6 to 8 weeks. For an egg-producing operation to be economically feasible, the minimum number of egg-laying birds required is at least 2500.

Space requirements for egg-laying birds are similar to those for meat-type birds. An inside temperature in the range 45 to 80°F (7.2 to 26.7°C) is considered satisfactory for egg-laying flocks. Floors and walls should be easy to clean. Feeders and waterers should be provided, as in the case of meat-type birds.

Laying nests may be constructed of metal or wood. Roll-away floors with egg trays are desirable, as this type of nest minimizes the number of dirty eggs. With this arrangement, once the egg has been laid and the hen leaves the nest, the egg rolls away from the area where it is laid to a collecting area, where it will not be dirtied by droppings. One nest for every four birds is considered adequate. An individual nest should be about 10 to 12 in. (25.4 to 30.5 cm) wide, 12 to 14 in. (30.5 to 35.6 cm) high, and 12 in. (30.5 cm) deep. To keep the nest clean, a perch is provided below the entrance of the nest.

Roosts (8 to 10 in. [20.3 to 25.4 cm] of space per bird) are 13 to 15 in. (33 to 38.1 cm) apart above dropping pits. Such pits should be cleaned periodically and should be constructed in such a way as to facilitate cleaning.

The recommended food requirements for laying birds seem to be somewhat less complex than for meat-type birds, although the vitamin requirements are somewhat more complex. Antibiotics may be included, and ground limestones or ground oyster shells are added as a source of necessary eggshell ingredients. Feed requirements amount to 85 to 115 lb (38.6 to 52.2 kg) per bird per year depending on bird size.

Poultrymen located near cities may sell eggs directly to consumers via home deliveries, or they may sell to produce dealers, cooperatives, shippers, or hucksters.

In smaller operations, eggs should be collected from laying houses at least three times daily in cool weather and four or five times daily in hot weather. They should be gathered in plastic or rubber-coated mesh baskets. Nests and baskets should be kept clean, as the interior of dirty eggs soon becomes contaminated and subject to spoilage. The larger egg producers cage the hens over conveyor belts, and eggs are collected continuously. After collection, eggs should be placed in storage at 40 to 45°F (4.4 to 7.2°C) and a relative humidity of 70%. No other materials, including other foods, should be stored in egg storage rooms, because pungent odors are readily absorbed by eggs.

Dirty eggs can be cleaned by buffing or washing, but unless this is done under rigidly controlled conditions, the interior may become contaminated with bacteria during cleaning. It is desirable, therefore, to emphasize production of clean eggs.

The egg shell presents a barrier to the entrance of microorganisms, but there are pores in the shell large enough to allow the entrance of bacteria and even molds. The number of pores that are found in the shell vary in the range 100 to 200 per cm^2. When the egg is just laid, the pores of the shell are sealed by a thin layer of protein, the cuticle. If buffed or washed, this protein coat is removed; during washing under improperly controlled conditions, contaminated water may enter the egg.

The logical reason for porosity of the shell is to allow for the flow of gases into and out from the developing embryo in case the egg has been fertilized. The inner membranes tend to prevent the entrance of microorganisms. In egg white, there is an enzyme (lysozyme) that tends to lyse or disintegrate some bacteria. There is also a substance in raw eggs, avidin, that ties up biotin, a required factor for the growth of some microorganisms. Finally, there is a material in fresh white that binds with iron, making it unavailable to several species of *Pseudomonas* bacteria that are responsible for more than 80% of the egg spoilage.

The cuticle may be lost not only by washing; it can also be dissolved by droppings. In any case, about 3 weeks after the egg is laid, the cuticle becomes brittle and particles chip off. Some bacteria are not affected by lysozyme and require little or no biotin. As they are held in storage, enzymes within the eggs cause chemical changes that deteriorate the iron-binding properties of the egg white. The defense mechanism against spoilage is, therefore, eventually lost, so that if microorganisms penetrate the shell, spoilage will occur. One of the most satisfactory treatments has been the application of mineral oil, which has increased the shelf-life of the egg by reducing the contamination by bacteria and molds through the shell pores.

After the eggs are cooled, they should be packed in clean, odorless containers (usually holding 30 dozen) held at 45 to 50°F (7.2 to 10°C) prior to shipment. They should be packed with their large ends up to prevent the air pocket from migrating upward into the yolk, which would increase the chances of spoilage due to airborne bacteria. All eggs should be candled (examined while in the shell under proper lighting) before selling to consumers. This should be done on the farm if they are sold directly to the consumer or by the distributor or retailer when marketed through retail channels. Candling is done to cull out specimens with such defects as blood spots, blood rings, meat spots, and germ spots (in fertile eggs).

Eggs are classified according to size as jumbo (30 oz [851 g] per dozen), extra large

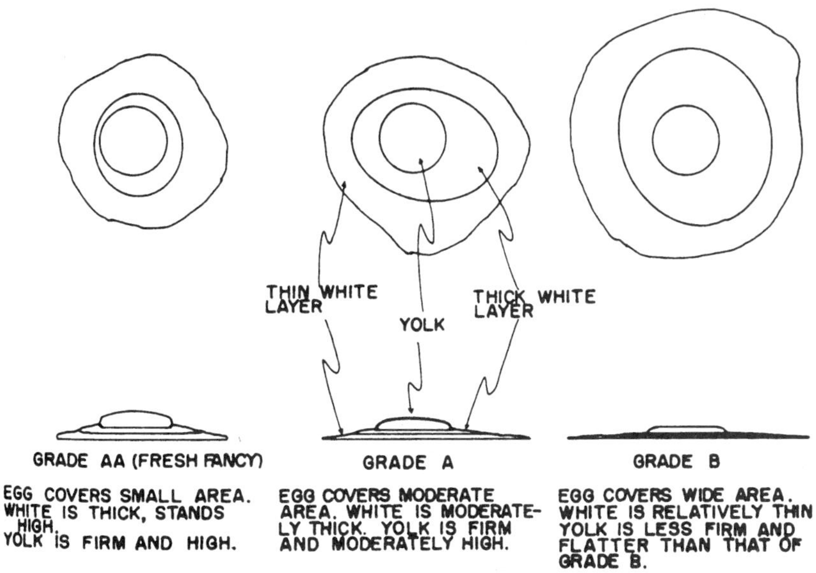

Figure 16.3. U.S. grades for eggs (broken out).

(27 oz [766 g] per dozen), large (24 oz [680 g] per dozen), medium (21 oz [595 g] per dozen, small (18 oz [510 g] per dozen), and peewee (15 oz [425 g] per dozen).

Eggs may be graded according to the interior quality and the condition and appearance of the shell (see Fig. 16.3). For grade AA, the shell must be clean and unbroken, and when broken out of the shell, the egg must cover a comparatively moderate area, the white must be reasonably thick and stand fairly high, and the yolk must be firm and high. For grade B, the shell must be clean and unbroken, the broken-out egg covering a wide area with only a small amount of white that can be considered thick, and the yolk somewhat flattened, covering a comparatively large area. Dirty or broken eggs may not be graded.

Processing of Eggs

Almost all eggs used in bakeries have been preserved either by freezing or by drying. The eggs should first be candled to eliminate defective eggs such as rots, blood rings, or other defects. They are then washed before they are broken out. At one time, eggs were broken out by hand. Today, this is done almost entirely by machine. Machines break the egg out into cups, which may or may not also separate whites from yolks, depending on whether the process is to produce whites, yolks, or whole egg magma (mixed white and yolks). While the separation of the egg is such that the white can be separated so that it is free of yolk, the yolk cannot be separated so that it is entirely free of the white. In fact, standards or definitions concerning yolk allow as much as 20% white. As the eggs in cups are carried along by an endless conveyor, they are examined by inspectors. If a particular cup contains a bad specimen, the cup with the egg is removed and replaced with a clean, chemically sanitized cup. The egg from the removed cup is then discarded and the cup is washed and sanitized by rinsing in a chemical solution (usually 50 ppm or more of chlorine). As the conveyor moves along, the contents of each cup are emptied into one container when eggs are to be kept whole, or the whites and yolks are emptied into separate containers when separation is desired. Eggs products, including whole egg magma, yolks, and whites produced in the United States must now be pasteurized prior to freezing or drying, in order to destroy *Salmonella* bacteria, because in the past the disease salmonellosis has been traced to contaminated eggs. After screening to remove chalaza (albuminous thread) and shell fragments, eggs are pasteurized by passing them through a plate-type heat exchanger. The product is heated to 140 to 145°F (60 to 62.8°C) and held at this temperature for 1 to 4 min prior to cooling. Cooling may be accomplished in tanks provided with cooling coils and paddles that agitate the product to facilitate cooling, or in thin-film heat exchangers.

After cooling to 40°F (4.4°C), the pasteurized eggs may be placed in metal cans holding about 30 lb (13.6 kg) of product. The filled cans are then placed in a cold room at 0 to –20°F (–17.8 to –28.9°C) until the product is frozen, after which it will be held at 0°F (–17.8°C) or lower until shipped out to the distributor or to the point of utilization.

Frozen whole egg magma and frozen yolks are subject to deterioration during frozen storage. Ingredients in the yolk tend to form a gummy mass during frozen storage. In order to prevent this, 5% to 7% salt or glycerin or 5% to 10% sugar may be added and mixed with the product. Citric acid may also be added to help extend shelf-life.

Whites, yolks, or whole egg magma may be spray-dried by being forced through a nozzle (to form droplets) into a chamber of heated air where most of the moisture is removed from the droplets to the heated air, which is vented to the outside. The dried product falls to the bottom of the drier and is collected. Spray-dried eggs have a moisture content of about 5%. This moisture content is not sufficiently low to prevent nonenzymatic browning during storage. Nonenzymatic browning involves sugars; thus, it can be prevented by removing sugars from eggs when they are allowed to undergo a natural fermentation at 70°F (21.1°C) over a period of several days. This is done when bacteria are allowed to grow and ferment out the sugar. Although this method is effective in removing sugars, it is considered to be unsanitary, as disease bacteria may also grow during fermentation.

Sugars may be removed from egg products by addition of yeasts, which utilize sugars, and holding the product at temperatures and for periods that will allow for an adequate growth of these microorganisms. The use of yeasts to ferment sugars, however, may produce undesirable flavors in egg products.

A mixture of two enzymes, glucose oxidase and catalase, is also used to remove sugar from eggs. By this method, the sugar (glucose) is oxidized to gluconic acid hydrogen peroxide by glucose oxidase, and the hydrogen peroxide, which is undesirable, is decomposed to water and oxygen by catalase (see Chapter 9). The enzyme method is probably the most satisfactory means of removing sugar from egg products. Treatment of these products to remove sugar must be carried out prior to drying.

The heating encountered during the pasteurization treatment and the physical forces encountered during spray drying have some effect on the functional characteristics (whipping quality, etc.) of egg products, especially those made from whites. Therefore, in some countries where pasteurization is not mandatory, egg white is allowed to undergo a natural fermentation and is then dried in cabinets on trays. In such instances, to eliminate disease-causing bacteria that may be present (because the product was not pasteurized prior to drying), the dried product is held at 130°F (54.4°C) or at higher temperatures for several days. This heat treatment after drying is said to destroy disease-causing bacteria.

Egg Consumption

Because eggs are among the most complete foods available to humans and have been relatively inexpensive, their consumption rate in the United States has been high. However, health warnings by the American Heart Association over the past 2 decades that eggs are high in cholesterol and should not be consumed at rates exceeding three per week, followed by the emergence on the market of no-cholesterol egg substitutes (discussed later in this chapter) that are just as nutritious as eggs, have caused a downward trend in egg consumption.

In the late 1980s, there were claims by some egg producers that their eggs had less cholesterol than "the average" egg and it was even advertised on the label. The truth is that older methods used to measure cholesterol were not as accurate as current ones. In 1976, it was determined that eggs contained an average of 274 mg of cholesterol per egg. The USDA tested eggs from some 200 suppliers in 1989 and found that the average egg contained 213 mg of cholesterol. The 20% drop is largely attributed to the scientific methodology rather than the change in the eggs themselves. It is known that the more chickens lay, the lower the cholesterol content of their eggs and today's chickens are laying more. It seems that *all* eggs have lower cholesterol than previously thought. The American Heart Association Nutrition Committee was prompted to raise its suggested egg consumption from three to four eggs per week because of the new findings. This revision, together with conflicting reports regarding the roles of cholesterol and saturated fatty acids in promoting cardiovascular disease, have apparently stopped the downward trend in the rate of egg consumption.

Egg Substitutes

If it is desired to eliminate cholesterol completely, egg substitutes may be purchased. These products are marketed under names such as "scramblers®" and "egg beaters®." They do not have to be labeled as "imitation" eggs because they have the same nutritional value as the product they copy. They contain mostly egg whites. The other ingredients vary but both contain either corn or soybean oil, emulsifiers (mono- and diglycerides), stabilizers (vegetable gums or modified starches), and a number of nutrients (vitamins and minerals) that are present in eggs. These products can be used as egg substitutes in most recipes and can be scrambled or used in omelets. There is, however, no substitute yet for the "sunny side up" or "over light" breakfast favorites for those who understand the threat of *Salmonella* in eggs not fully cooked and are willing to take the risk.

17

Fish and Shellfish

Although the word fish, much the same as the words meat, poultry, and cheese, is used to classify one type of food, varieties of fish are much greater in number than those of other foods. In the United States alone, at least 50 species of fish and shellfish are used as food for humans. Considering that the variations among aquatic species are relatively greater than those among species of meat animals, the magnitude of the time, space, and effort required to give even minimum coverage to fish as food must be appreciated.

Of the flesh foods raised to be eaten by humans, fish have the highest feed/conversion ratio. Every 1 lb (454 g) of feed yields about 0.67 lb (0.30 kg) of fish. Fish also require much less space than other animals (e.g., catfish space requirements are about 2500 lb [1135 kg.] per acre; silo systems can reportedly produce about 1 million lb [454,000 kg] of fish per acre). These facts suggest that fish for human use will eventually be produced largely by fish farming and in fact that situation is currently changing (see end of this chapter). Methods of culturing oysters, clams, mussels, abalone, shrimp, crawfish, crabs, northern lobsters, salmon, catfish, carp, buffalofish, milkfish, tilapia, shad, striped bass, trout, mullet, and plaice are being investigated in various countries. In some countries, several freshwater species (catfish, carp, trout, and tilapia) have been raised as a commercial enterprise for some years and milkfish have been raised (from the captured young fish) for many years in the Philippines. Oysters are now grown commercially in some areas, and the raising of shrimp in Japan is already commercialized.

As the availability of other animal protein sources decreases, a situation now existing even in affluent countries, greater efforts will be applied to the culturing of marine and freshwater species of fish and shellfish, and eventually these species may play a much more important part as a wholewide supply of animal protein.

Fish flesh is readily digested, and it is subjected to highly active bacterial enzymes. Therefore, fish tends to deteriorate rapidly and cannot be held at temperatures above freezing for long periods. A simple principle that applies to all fresh food, and especially to fish, is the 3/H rule: Handle the product under strict sanitary conditions (to keep the microbial contamination at a minimum). Handle the product at a cool temperature (microbes multiply rapidly and spoilage reactions proceed rapidly at warm temperatures but both proceed slowly at cool temperatures). Handle the product quickly (fish deteriorate as a function of time as well as temperature). To give some idea of the importance of temperature, fresh-caught fish will generally last about 12 days if held in ice (temperature at about 32°F or 0°C) whereas they will last only about 4 days at 46°F (7.8°C), a temperature commonly found in domestic refrigerators. There are at least three reasons why fish spoil so rapidly at refrigerator temperatures: first, because they are readily digestible; second, because the muscle glycogen is nearly depleted during harvesting, leaving little to be converted to acid, which would act as a preserva-

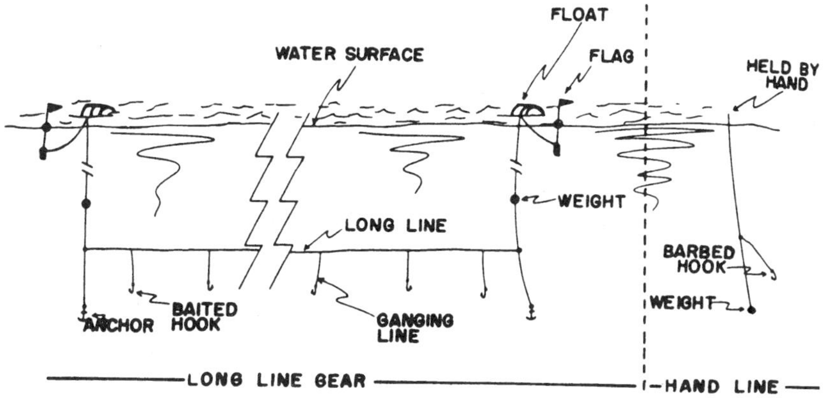

Figure 17.1. Line fishing.

tive; finally, because the bacteria found on fish are psychrophiles (they can grow well at low temperatures), and their enzymes are functional at low temperatures. Even among psychrophiles there is a range of optimum growth temperatures for individual species, and it is known that some of the psychrophilic bacteria found naturally on fish grow at such low temperatures that they are not reliably detected by standard bacteriological plating techniques that incubate at elevated (warm) temperatures.

Seafood is the only major food source that is still hunted rather than raised or grown. As mentioned, methods of fish farming and other aquaculture techniques are becoming more popular and some of these are discussed later in this chapter. Most marine species are still caught and a variety of methods are employed. Line fishing is one of the simpler methods used to catch fish. Different methods such as hand lines and long lines are shown in Figure 17.1. Fish such as halibut, cod, and haddock are caught using these methods.

Troll lines are used to catch certain species of salmon and other fish found near the ocean surface (see Figure 17.2).

Nets are used by many fisherman to catch popular species such as cod, haddock, flounder, and other bottom fish. Examples of different nets are shown in Figure 17.3. Gill nets are used to catch salmon and shad, and sometimes herring, mackerel, cod, and haddock. Otter trawls are used to catch the bottom fish, and purse seines are used

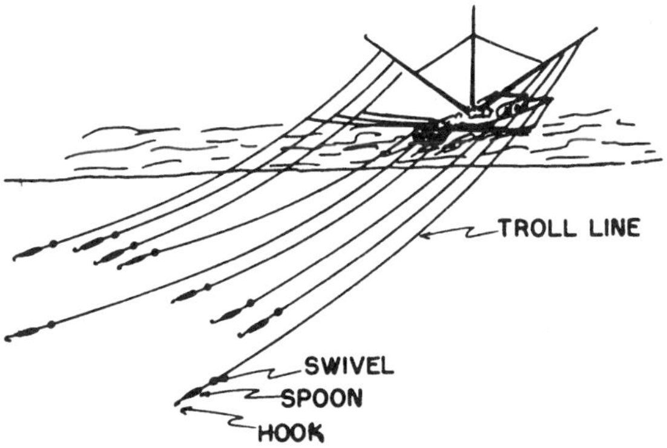

Figure 17.2. Troll line fishing.

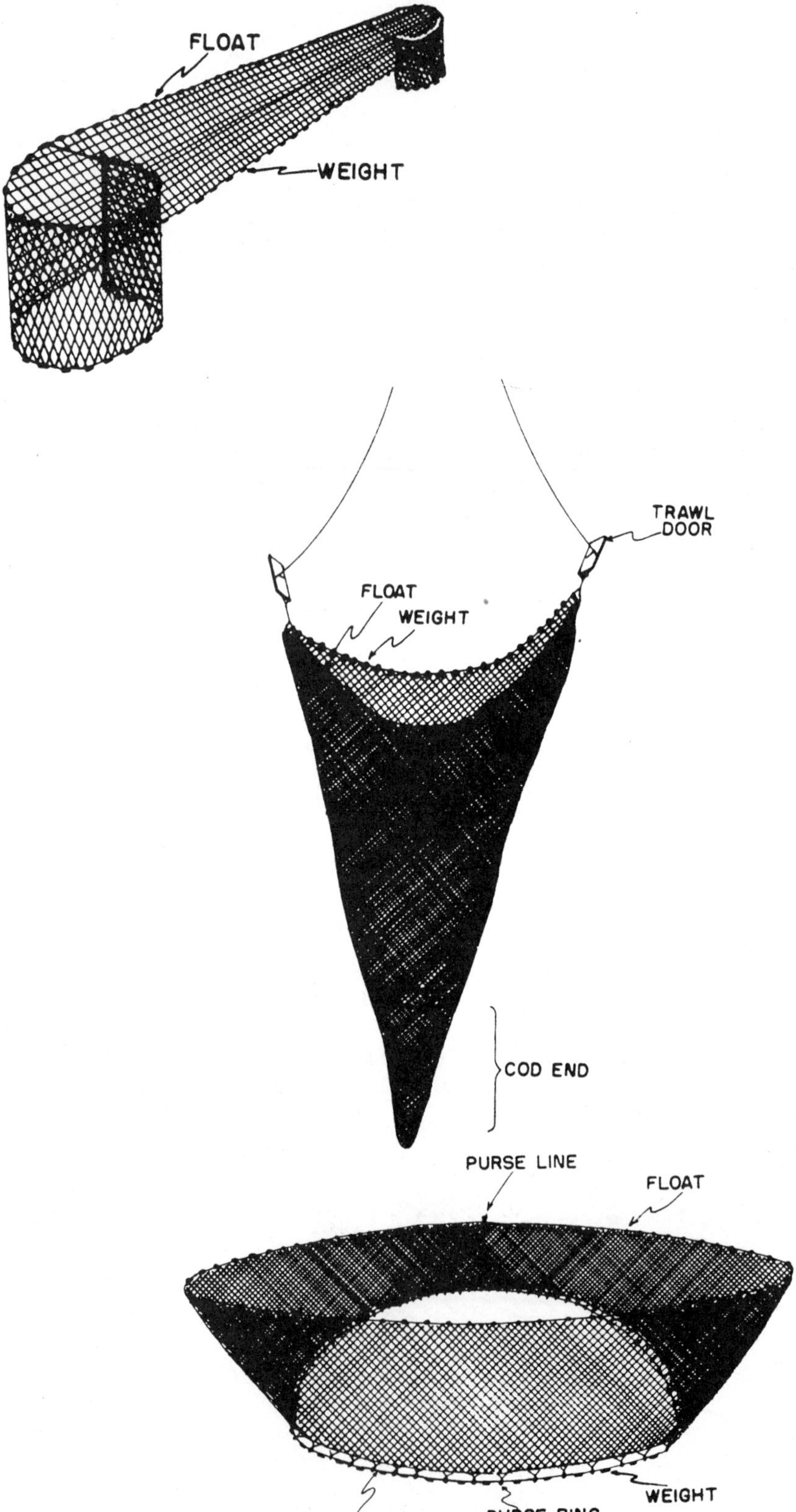

Figure 17.3. Drift gill net, otter trawl, and purse seine.

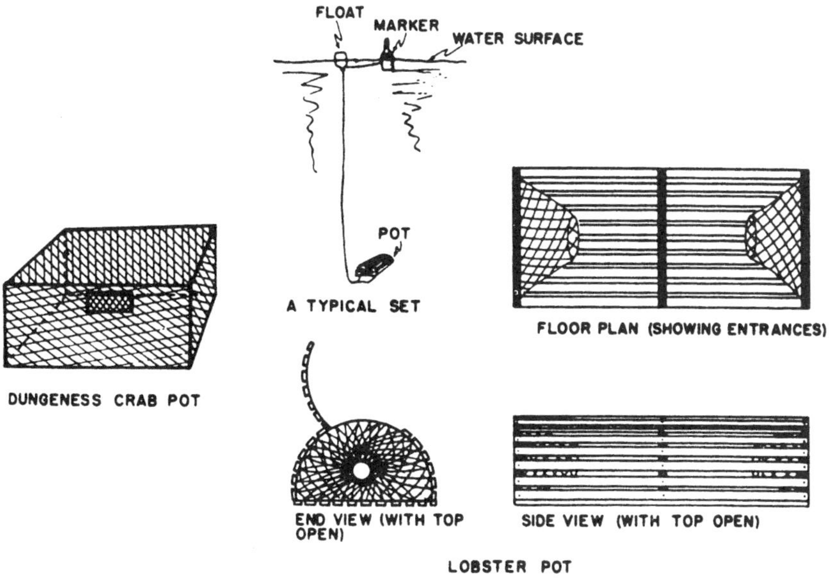

Figure 17.4. Entrapment devices.

to catch fish that swim together in large groups such as menhaden, tuna, salmon, herring, and mackerel. More fish are caught with purse seines than with any other method.

Crustaceans such as crabs and lobsters and some freshwater fish are caught in pots. Examples of these are shown in Figure 17.4.

Shellfish such as scallops, clams, and oysters are harvested by the use of dredges, tongs, rakes, or forks. Examples of these are shown in Figure 17.5.

IMPORTANT FAMILIES OF FISH

After fin fish are caught, deterioration starts very quickly if the fish are not handled properly; thus efforts must be made to get the fish out of the temperature danger zone (40 to 140°F [4.4 to 60°C]) as quickly as possible. This is usually done by adding ice to the fish and keeping them cold until further processing is done. Methods of processing the individual species are discussed in this chapter.

The Herring Family (*Clupeidae*)

Among the various categories of fish used by humans for food, one of the most important is the family *Clupeidae,* comprising pelagic species that travel in groups or schools near the ocean surface.

Sea Herring

In the United States sea herring (see Fig. 17.6) are found in ocean waters from Alaska to the state of Washington on the West Coast, and from Labrador to Cape

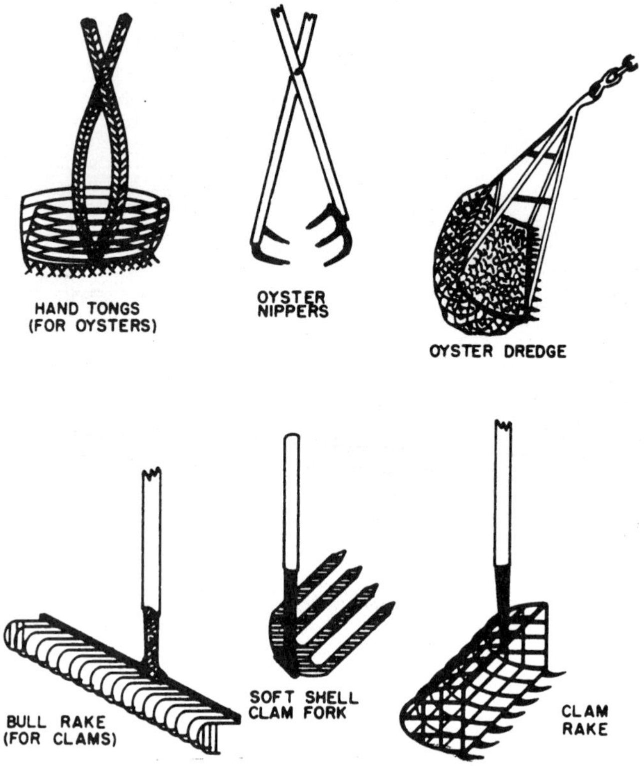

Figure 17.5. Shellfish harvesting devices.

Hatteras on the East Coast. Sea herring are plankton feeders, eating the various microscopic plants and animals (diatoms, larvae of various shellfish, etc.) when very young and, as adults, eating small shrimp, small fish, and so on. When unmolested, the sea herring may live to an age of 20 years or more.

Most sea herring are caught with purse seines, but some are caught with pound traps or weirs (similar to pound traps but constructed with poles and brush). Gill nets are sometimes used to catch these fish.

The larger fish may be used for export, as there is a foreign market for this species where it is used for human food. The smaller herring, which are canned as sardines, must be held in the purse seine alive (sometimes for longer than 24 hours) until the stomach is free from feed. This is done to prevent enzyme action, which would reduce the quality of the fish. Fish from the nets are loaded on the boat with a large vacuum pump and are salted. More salt is added at the processing plant after the fish are beheaded and eviscerated. The small herring are then preheated, placed in small rectangular cans, and subjected to steam for 18 to 20 minutes. The liquid in the cans is drained and replaced with vegetable oil, tomato sauce, or mustard sauce, and the cans are sealed, heat-processed and cooled. The resulting product is shelf-stable.

The larger herring are also canned whole but in larger oval cans, usually in tomato sauce. Some large herring are cut transversely into bite-sized pieces and packed in small cans in tomato sauce. This product is also shelf-stable.

Some herring are pickled (salt, vinegar, sugar, and spices) and packed in glass jars. Sour cream, onions, or other flavoring ingredients may be added before sealing the jars. Other pickled products include rollmops (pickled herring strips wrapped around a piece of pickle or onion) and kippered herring (salted and lightly smoked). These products must be refrigerated. Some herring are highly salted and heavily smoked, and this product is shelf-stable. Large quantities of scales from the skins of herrings are recovered and used for the manufacture of pearl essence.

In Alaska, and to some extent in eastern United States, the roe (eggs) is taken from herring approaching the spawning stage, salted, packaged in containers, and sold at high prices to certain Asian countries.

Large quantities of sea herring harvested in some countries are converted into fish meal, which is used as a protein supplement for cattle and poultry as a portion of the feed. When used for this purpose, the fish are sometimes brought to port in unrefrigerated vessels, although some vessels, especially those operating in southern waters, are refrigerated. At the processing plant, the fish is first cooked in live steam in a continuous cooker, then pressed in a screw-type continuous press. The press cake is dried with hot gases from oil-produced flames in a rotary drier, to a moisture content of 5% to 8%. The liquid from the pressed product is not discarded. It is first centrifuged to remove oil, which is collected and sold for industrial uses. The remaining liquid (stickwater), which contains proteins, peptides, and amino acids, is then vacuum concentrated to a solids content of 50% and acidified to prevent spoilage. This product may be sold as a protein supplement or it may be added back to the presscake before the latter is dried.

Shad

Shad are anadromous fish (ascend rivers to spawn) that spend the greater part of their lives in ocean waters as far as 50 miles (80.5 km) from shore. Shad are plankton eaters and are said not to eat fish. They range from the Gulf of St. Lawrence to Florida on the East Coast, but are caught in insignificant numbers only from New York southward. They were brought, some years ago, to the Pacific waters, and some are now caught in California. Shad are caught in rivers and their estuaries with drift gill nets. Aboard the boats, they are neither eviscerated nor iced, because they are brought to shore shortly after removal from the water.

Shad are used almost entirely as the fresh product, with only small amounts being frozen. They contain many small bones, but can be filleted to eliminate most of them from the flesh. The roe (unfertilized eggs), which prior to spawning is held together by a thin membrane, is highly prized. It is sold fresh, or packaged in moisture vapor-proof material and frozen to be sold to the restaurant trade and, in this state, may be stored at 0°F (−17.8°C) for 6 to 8 months. Longer storage under these conditions usually results in a rancid product resulting from oxidation of the fats contained therein.

Menhaden

Four species of menhaden, sometimes called "pogy," "bunker," or "mossbunker," are found in the Western Atlantic. They range from Nova Scotia to Brazil, feeding on microscopic plants and animals. They are caught with purse seines when schooling near the surface of the water and removed to the hold of the power boat or carrier vessel. Aboard the boat, they may be held without refrigeration, in which case they will be brought to port within a period of 24 hours after catching. Boats that keep these fish in refrigerated holds may remain at sea for several days prior to landing the catch.

Menhaden are not used for human food. They are processed to produce fish meal and oil in the manner described for herring. In the United States, larger quantities of menhaden are caught (several hundred thousand metric tons) than that of any other fish or shellfish.

The Anchovy

The anchovy (family *Engraulidae*) is a small herringlike fish found off the coasts of California and Mexico. It is caught in purse seines, and at one time was used as line bait for tuna fishing and later mostly in the production of fish meal and fish oil. However, a growing amount is used for human consumption in appetizers, garnishes, sauces, relishes, and especially in toppings for pizzas. It is packed in 2-oz (56-g) tins in oil for domestic use and in large tins for industrial use.

Other Clupeidae

Pilchards are members of the herring family, and at one time were plentiful off the coast of California, where they were caught with purse seines. They were used for canning as sardines and also to produce fish meal and oil. Pilchards are caught and canned by the South Africans.

The Cod Family (*Gadidae*)

The cod family (see Fig. 17.6) includes the cod, haddock, pollock, cusk, and several species of hake. The members of this family vary greatly in size. The cod may reach a length of 6 ft (1.8 m) and a weight of 200 lb (91 kg) and averages 10 to 12 lb (4.5 to 5.5 kg). The haddock may reach 3 ft (91.5 cm) in length and 24 lb (10.9 kg) in weight, but the average is much smaller than this. The pollock reaches a maximum of 3.5 ft (1.1 m) and a weight of 25 lb (11.4 kg). The cusk and the hakes are smaller than those listed previously.

The Cod

Cod (see Fig. 17.6) are found on both sides of the Atlantic and are most plentiful around Norway, Iceland, Newfoundland, Nova Scotia, and on Georges Bank off Cape Cod. In the Western Atlantic, they range from Greenland to North Carolina. Molluscs (clams, oysters, scallops, etc.) are said to make up an important part of the diet of the cod, but cod also eat small fish.

By far the largest quantities of cod are caught with otter trawls in waters ranging from 300 to 1500 ft (91 to 457 m) in depth. Small quantities of this species are caught with long lines, hand lines, or gill nets. When caught with otter trawls, the fish are gutted and washed on the deck of the boat. During summer months, the gills must be removed, but the head is left intact. The fish are stored in boxes or pens, in either case layered with ice.

At the processing plant, the fish are washed and filleted, skinless. The fillets are then candled (observed over a bright light) to locate and remove parasites such as

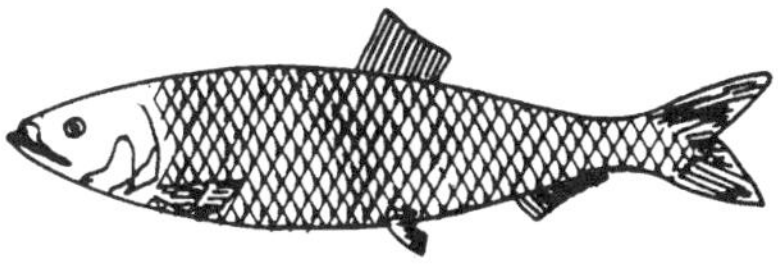
Herring (*Clupea harengus*).

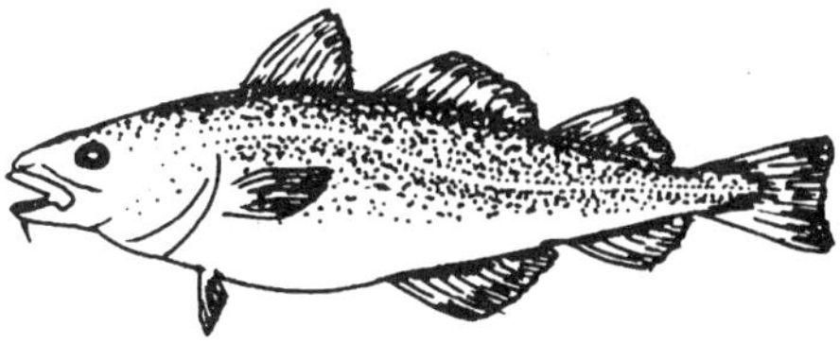
Cod (*Gadus morhua*).

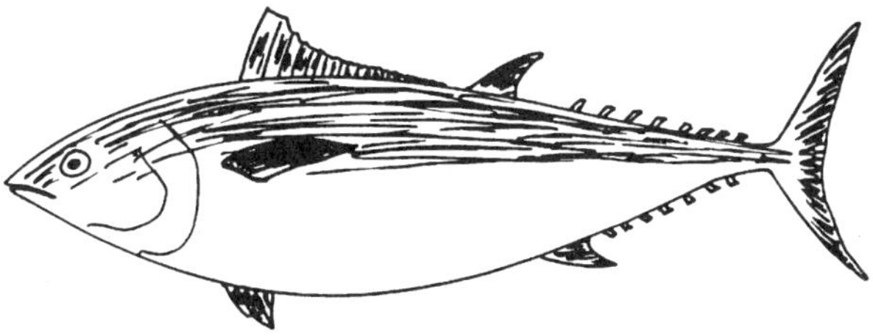
Bluefin tuna (*Thunnus thynnus*).

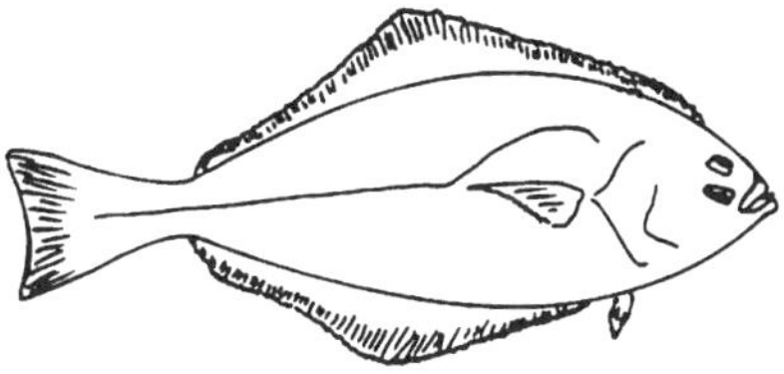
California halibut (*Paralichthys californicus*).

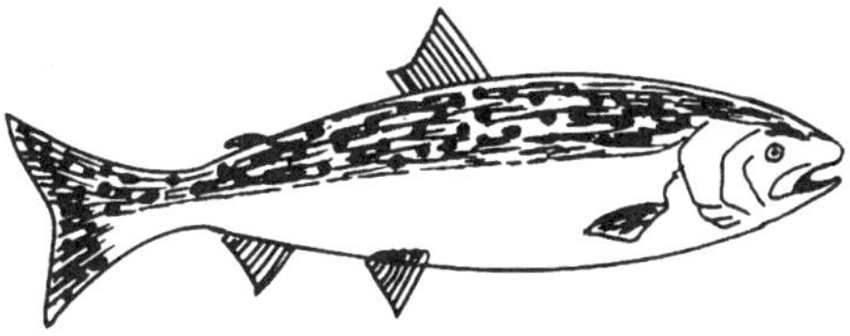
Pink salmon (*Oncorhynchus gorbuscha*).

Figure 17.6. Some important finfish species.

worms. Fillets to be sold as fresh are precooled, placed in metal tins of 10, 20, or 30 lb (4.5, 9.1, or 13.6 kg), and the tins are refrigerated (mechanically or in ice) until they reach their destination. Some fillets are precooled, packed in small trays, overwrapped with a transparent plastic, and shipped to their destination in insulated containers.

Fillets that are to be frozen for the retail trade are packed in 1-lb (454-g) waxed cartons with or without being first wrapped in moisture-vapor-proof plastic. The fillets may have been passed through a weak brine (10% to 45% saturated salt solution) prior to packaging. For purposes of freezing, retail-sized cartons of fillets may be placed on trays, the trays placed on racks, and the racks wheeled into a blast freezer where very cold air is blown over the product, or the cartons may be plate frozen by being placed on trays in contact with refrigerated plates at –28°F (–33.3°C). Some cod fillets are frozen, usually in a plate freezer into 16-lb (7.26-kg) blocks and then cut into fish sticks or fish portions that are then breaded in a batter of flour, dried milk solids, egg solids, spices, and flavoring. The product may be deep fried to the fully cooked or partially cooked state prior to being refrozen. The resulting product can be baked or microwaved from the frozen state by the consumer. Some battered and breaded product is not fried but frozen raw. This product is used by fast food restaurants and other food service facilities where it is deep fried to order.

Although salted cod was once the major export commodity of the United States, only very little cod is now salted in this country. To produce salted cod, the fish are beheaded and split longitudinally; the backbone and the abdominal cavity lining are removed, and the fish is washed and layer-salted in closed casks (brine salting). Most of the salted cod is produced by dry salting, which is done to produce either a lightly salted product or a heavily salted product. The lightly salted product requires only days to complete, has a relatively low salt content (less than 10%), has better organoleptic properties than the heavily salted one, but it is not shelf-stable. The heavily salted product requires weeks to complete, has a high salt content (about 30%), and is shelf-stable.

Fish cakes are prepared from salt cod by first cooking and freshening the fish to remove most of the salt, then mixing it with mashed cooked potatoes and small portions of oil, onions, and pepper. A proportion of about 40% shredded cooked fish and 60% cooked potatoes is used. This product may be canned without forming or it may be formed into small cakes, deep-fat fried to brown the surface, and frozen, or sold in the refrigerated state. Some manufacturers of fish cakes are using fish sawdust, from fish stick processing, and broken fish sticks as the fish component in fish cakes.

Haddock

The haddock is the second most important member of the cod family. Haddock are found on both sides of the Atlantic from Norway to New Jersey but are most plentiful in waters off Nova Scotia and Cape Cod (Georges Bank). The mature fish feed on crustaceans (crabs, shrimp, etc.), molluscs (clams, etc.), and on small fish. In recent years, the stocks of haddock have been greatly depleted as a result of overfishing.

Haddock are usually caught in areas where the depth of water is 150 to 360 ft (45.7 to 109.7 m). They are caught, handled aboard the boat, and processed to produce fillets, fish blocks, and fish sticks in the same manner as that described for cod. Haddock are not salted and dried although some are lightly salted and lightly smoked, without heat, to produce a product called "finnan haddie."

Pollock

Pollock are found on both sides of the Atlantic from Norway to the Chesapeake Bay but are most plentiful in waters off Nova Scotia, Cape Cod (Georges Bank), and in the Gulf of Maine. Pollock eat shrimp, crabs and other crustaceans, and small fish but do not eat bivalve molluscs.

Pollock are caught in waters at levels between the surface and a depth of 450 ft (137 m). They are caught, handled aboard the boat, and processed in much the same manner as that described for cod. Small quantities of pollock are salted and dried.

Hakes

There are several species of hakes, the most important of which is the silver hake or whiting. The whiting is most abundant in waters off Nova Scotia. It is caught and handled in much the same manner as are cod. Some whiting are headed, gutted, washed, and frozen in blocks without further cutting, for utilization as food.

With all members of the cod family, small fish, species not especially prized as food (red hake, etc.), and fish frames (the portion remaining after the fillet has been cut away) may be passed through mechanical meat/bone separators (machines that separate the flesh portion from bones and skin). This provides a significant yield of edible, ground fish flesh (resembling hamburger in texture) that may be used to produce frozen fish blocks to be further processed into products such as fish portions and fish sticks. Handled in conventional fashion, such products are not stable in frozen storage because the fat oxidizes and becomes rancid, and the tissues get tough at a faster rate than that of the corresponding fillets held under the same conditions. The faster oxidation of fats is probably due to a greater exposure to oxygen because of the great increase in surface area. The increased rate of toughening may be due to a wider distribution of the enzyme that decomposes trimethylamine oxide to form dimethyl amine and formaldehyde. The latter compound is known to denature proteins. The spoilage reactions may be slowed considerably by storing at lower temperatures, for example, –20°F (–28.9°C). Rancidity can be prevented altogether by protecting the product with a wrapper of gas-impermeable plastic film, for example, polyester, polyvinyl chloride, nylon-II, or aluminum laminate.

The Mackerel Family (*Scombridae*)

One of the most important segments of the seafood industry in the United States is that which processes fish of the mackerel family including the various tuna, the Atlantic mackerel, the jack mackerel, and the Spanish mackerel.

Tuna (see Fig. 17.6) are torpedo-shaped fish tapering to a pointed nose and a slender caudal peduncle (the portion near the tail). The more important species of tuna include the bluefin, the yellowfin, the skipjack, the yellowtail, and the albacore. Although some of these species can grow to a relatively large size, for example, the bluefin can attain a weight of 1000 lb (454 kg), the average size of the tuna caught is about 30 lb (13.6 kg); the smaller species average less. Mackerel are much smaller than tuna, but they are similarly shaped.

Tuna

The yellowfin tuna is found on the West Coast from southern California to southern Chile. Bluefin tuna range from Nova Scotia to Brazil on the East Coast and from southern California to northern Mexico on the West Coast. The skipjack is found in the Pacific Ocean from southern California to Central and South America. The albacore ranges from Puget Sound (state of Washington) to lower California. The yellowtail is found in Pacific waters from southern California to the coast of Mexico. Several other species of tuna are found elsewhere, especially in the Eastern Pacific Ocean.

At one time, fishing for tuna was carried out exclusively with pole lines. Today, tuna are caught mainly with purse seines which, owing to the size of the fish, are made with heavy twine. Some tuna may still be taken with pole lines and some with troll lines.

Most tuna fishing boats make trips lasting for several months, and for this reason, the fish are frozen aboard the boat. In freezing, the whole fish are cooled in a large well with refrigerated seawater (RSW) circulated through it at 28°F (–2.2°C). Once the fish are cooled, the RSW is pumped out of the well, and it is replaced with refrigerated brine at 10°F (–12.2°C) or lower. When the fish are frozen, the brine is pumped out of the well, and the fish are kept frozen by circulating mechanically refrigerated air.

At the processing plant, the fish are thawed in holding rooms at ambient temperature or in tanks with running water. When thawed, the fish are eviscerated and washed. They are then cooked in steam, under pressure, then cooled. When cooled, the heads and skins are removed. The fish are then cut longitudinally, after which all bones are removed. If all white meat varieties are being produced, the dark meat is also removed. The white meat is then shaped, mechanically, into a cylinder, fed to cans, and cut to length. To pack chunk-style tuna, pieces are filled into cans with an adjustable filler. Vegetable oil or a broth containing hydrolyzed vegetable protein in water is metered into the cans, which are then heated in steam, sealed, heat-processed, cooled, labeled, and stored.

Mackerel

Atlantic mackerel are found from the Gulf of St. Lawrence to Cape Hatteras in America, and from Norway to Spain in the Eastern Atlantic. Spanish mackerel range from Maine to Brazil in the Western Atlantic but are mostly caught in waters off the Carolinas and southward. The jack mackerel ranges from British Columbia to Mexico in the Pacific Ocean.

Mackerel may be taken in pound traps or with gill nets, but by far the greatest quantities are taken with purse seines. If the boat is to remain out of port after the fish are caught, they are held in ice in the round, uneviscerated state. Atlantic and Spanish mackerel are sold to retailers as the fresh product, either as fillets or as the round uncut fish. Some are frozen by placing the round fish in pans and holding at 0°F (–17.8°C) or below in rooms with or without circulating air. The fish, frozen as a block, are sprayed with water for purposes of glazing to prevent dehydration and held in the frozen state until defrosted for sale to restaurants or retail outlets.

Jack mackerel are canned in 1-lb (454-g) tall containers. The fish first pass on a conveyor belt under circular knives that cut off the heads and tails and also cut the fish to can-size lengths. The entrails are then removed, after which the fish are washed and flumed to a container feeding the packing table, where they are filled into cans by hand. The open cans are then heated in a steam box to raise the product temperature to 145°F (62.8°C), after which they are inverted to drain off liquid formed during heating. Oil, brine, tomato sauce, or mustard sauce is then added to cover the fish, and the cans are then sealed and heat-processed.

The Salmon Family (*Salmonidae*)

A number of commercially important species of the salmon family are found throughout the world. In the United States, the red, sockeye or blueback, the spring, king or chinook, the silversides or coho, the pink or humpback, and the chum or dog salmon are of chief importance. The steelhead trout, which behaves like a salmon, is caught in some volume. All the above-named fish are caught on the West Coast. Only small numbers of Atlantic salmon are caught on the American side of the Atlantic.

Salmon (see Fig. 17.6) have a deep body and a rounded belly. All have a wide, slightly lunate tail. Salmon eat small crustaceans and small fish when young and larger members of these groups as they become older. The Pacific salmon and steelhead trout are anadromous, ascending rivers sometimes more than 1000 miles (1609 km) to lay their eggs in the same rivers or streams in which they were hatched. These fish die soon after spawning.

West Coast salmon are caught, in greatest quantities, with purse seines in ocean waters near the coast. Purse seines cannot be operated in rivers or their estuaries, but significant quantities of salmon are caught in estuaries with drift nets. Coho and spring salmon may be caught with troll lines.

By far the greatest quantities canned are West Coast salmon. The fish are first conveyed to a machine, the "iron chink," which removes the head, tail, fins, and viscera. They are next trimmed by hand to remove extraneous material left by the iron chink, after which they are washed. The fish next pass under rotary blades on a slotted conveyor, where they are cut into can-sized lengths. The cut salmon pieces then pass on to a volumetric filling machine where salt is added to the can (about 1.25% by weight), and the cans are filled with fish. After filling, the covers are clinched on the cans, which are then sealed under vacuum, or they may be sealed without first clinching, a steam jet being used to remove air from the headspace in the can. After sealing, the cans are washed, heat-processed to provide commercial sterility, then cooled in the retort. The cans are then labeled, if not lithographed, packed in cases, and stored in a warehouse until shipped out. Heat-processing times and temperatures are applied according to the weight of the product in the can.

Some coho and spring salmon are sold fresh as steaks; some are frozen, but they must be protected from becoming rancid by sealing in gas-impermeable containers. Spring salmon are sometimes preserved by salting. The fish are split, trimmed, washed, covered with salt, and packed in casks, after which the casks are filled with saturated salt solution and held at 35 to 40°F (1.7 to 4.4°C) for 30 days. This product is usually shipped to processors who smoke the fish. In smoking, the salted fish is first soaked in water to remove salt, then smoked at temperatures below 90°F (32.2°C) or hot-smoked at a temperature of about 175°F (79.4°C).

The Flatfish Family (*Pleuronectidae*)

Many species of flatfish are utilized as food. On the East Coast of the United States, the halibut, the turbot, the sand dab, the fluke, the yellowtail flounder, the blackback flounder, the lemon sole, the plaice, and other species are edible. On the West Coast, the halibut, petrale sole, English sole, rex flounder, arrowtooth flounder or turbot, Dover sole, starry flounder, rock sole, and other types are caught as edible fish. All the above are flounder; none are true sole. In shape, flounder are flat, comparatively thin fish (see Fig. 17.6).

In the larval stage, the eyes of flounder are on either side of the head, as in other fish. Eventually, one eye migrates to the other side of the head, and the fish becomes reoriented so that the side having the eyes is uppermost. The fish swim or lie on the bottom with the eyed side uppermost. In some species, the left eye migrates to the right side, and in others, the right eye migrates to the left side.

Flatfish vary in size. Although halibut can grow to a very large size, the average weight of those caught today is about 40 lb (18.2 kg); the average size of turbot is about 7 lb (3.2 kg); plaice are about 10 lb (4.54 kg); other species are smaller.

On the East Coast, halibut are found from the Grand Banks (off Newfoundland) to the Gulf of St. Lawrence, and south to New York waters. The turbot is found on the Grand Banks and in Nova Scotian waters. Various other flounders are found from the Gulf of St. Lawrence to South Carolina. The yellowtail is taken only on Georges Bank. West Coast flounder are taken from waters that extend from California to North Alaska. The most important West Coast species, the halibut, is caught mostly in waters that extend from Northern British Columbia to northwest Alaska.

Flounder are found in waters that vary in depth from less than 50 ft (15 m) to more than 1200 ft (366 m). Halibut, lemon sole, and turbot are mostly found in deep water. Mature flounders feed on crabs, shrimp, worms, squid, and other molluscs but halibut, turbot, and dab are mostly fish eaters.

Small flounder are caught with otter trawls. Aboard fishing boats, they are held in pens or boxes in ice, as are cod, but these fish are not eviscerated prior to icing. On the West Coast, halibut are caught with long lines. The fish are eviscerated, the gills are removed, and they are placed in hold pens in ice much in the same manner as described for cod, but in this case, the "poke" (belly cavity) is also filled with ice.

Small flounders, to be sold either fresh or frozen, are handled, packed, and distributed similarly to cod fillets.

Halibut are handled in both the fresh and frozen state. As the fresh product, the fish are beheaded, washed, and packed in ice in boxes. Then they are shipped from the West Coast to the Midwest or the East Coast under refrigeration. If the fish are small, they may be sold by distributors to retailers as received. If the fish are large, they may be sold to retailers as portions. Frozen halibut are decapitated, washed, and placed on racks in freezer rooms at 0°F (–17.8°C) or below until shipped to distributors in the frozen state. Small halibut, after freezing, may be sawed into steaks, trimmed, and packaged in moisture-vapor-proof plastic film as 12, 14, or 16-oz (340-, 397-, or 454-g) portions.

OTHER FISH

Many species of fish are not discussed in this chapter. Some of these are bluefish, butterfish, croaker, red and black drum, eels, groupers, mullet, ocean perch (fairly important fish of small size, caught with otter trawls handled aboard boats in the iced uneviscerated state, and processed as fresh or frozen fillets), pompano, rockfish, sablefish, sea trout, red and other snappers, spot, striped bass, swordfish and other marine species, as well as freshwater fish, such as buffalofish, carp, catfish, chubs, cisco, trout, and whitefish.

Today, there is a considerable fish farming industry in the United States in which catfish and trout are grown in freshwater ponds. In many countries, carp and tilapia are grown in freshwater ponds and harvested as food for humans. See later in this chapter for more on aquaculture.

SHELLFISH

Bivalve Molluscs (Class *Pelyopoda*)

There are other molluscs, besides bivalves, that are used as food for humans; squid is among them. In the United States, the molluscs chiefly used for food are oysters, a number of clam varieties, and scallops. Bivalves have a calcareous shell that varies in thickness and outer smoothness and is lined with a smooth enamel inside.

Some bivalves are either male or female throughout their life spans; others change sex from male, in the early stages, to female in later years. When spawning occurs, millions of eggs and sperm are shed into surrounding waters where fertilization takes place.

Oysters

There are five species of oysters in the United States, three on the East Coast, and two on the West Coast, one of which was introduced from Japan. About 2 weeks after the eggs hatch, the free-swimming larvae attach and cement themselves to a hard surface (rock or shell) on the bottom. To provide for this attachment, oyster growers throw materials, such as the shells of quahogs (the cultch), into the water where spawning takes place. Some time after the set (attachment), the small bivalves may be removed to areas where tidal conditions provide a better supply of food. This also allows more room for growth.

Because oysters and other bivalves may be eaten raw or without sufficient cooking to destroy any disease-causing bacteria that might be present, and because they are grown near the shore, often near highly populated areas, great care must be taken to make sure that bivalve growing areas are not polluted with even traces of human excrement. Control of bivalve harvesting areas is supervised by a division of the Food and Drug Administration but must be effected by state authorities. This control consists of tests for disease-indicator bacteria on shellfish growing waters and on shellfish meats, sanitary surveys to determine that traces of sewage are not reaching the growing areas, and the licensing of shellfish dealers who must record the areas from which the bivalves were taken, from whom they were purchased, and to whom they were sold. Some bivalves may be taken from areas that do not meet the absolute specifications for approved areas but that are not grossly polluted, provided they are depurated in bacteriologically clean waters, either in the ocean or in tanks, under supervision by the state (held in clean waters for periods long enough to allow them to purify themselves or eliminate pathogenic bacteria by siphoning clean water).

Oysters are harvested with rakes, tongs, dredges, or with water-jet vacuum dredges. Oysters and other bivalves, except scallops, are able to live out of water at suitable temperatures for some time, as they can obtain oxygen from that which is dissolved in the water retained within the shell in contact with the gills. Aboard boats, oysters and clams must be held under sanitary conditions away from the bilges. The boat used to harvest bivalves should be outfitted with a chemical toilet so that oyster growing areas will not be polluted with human discharges.

At the processing plant, oysters to be marketed in the shell are washed in seawater, which may be chlorinated, packed in sacks or barrels, cooled, and shipped to restaurants. They should be held at temperatures between 32°F and 40°F (0°C and 4.4°C). Most oysters are shucked (meats removed from the shell) by hand with the aid of a knife. The meats are washed or agitated in fresh potable water by air blown into the wash tank, graded for size, and packed in glass or metal containers. The filled containers are cooled and shipped to market in crushed ice where the temperature of the product is about 33 to 34°F (0.6 to 1.1°C).

Some shucked oysters held in metal containers are frozen in moving air at –5°F (–20.6°C) and stored at 0°F (–17.8°C) or below until shipped to market. Oysters may also be breaded, packed in waxed paperboard cartons holding 10 to 14 oz (284 to 397 g) of product, frozen between refrigerated plates or in cold air, and stored at 0°F (−17.8°C) until shipped to market.

Oysters are eaten raw from the half shell or in stews (lightly heated in milk with some butter), or breaded and deep-fat fried.

The Hard-Shell Clam

The hard-shell clam is similar to the oyster in its internal structure. The shell is rounded, symmetrical, and relatively smooth on the outside, coming to a gradual peak near the hinge. The shell is quite hard and thick. Once the larvae have developed into clams that are 1/8 to 1/4 in. (0.3 to 0.6 cm) in diameter, they burrow into the mud and remain just below the surface of the ocean bottom. The hard-shell clam is found from the Maine coast to the Gulf of Mexico, but is most abundant off the Atlantic Coast from southern Massachusetts down to and including Virginia.

Hard-shell clams may be harvested by hand (feeling for them with the hands or feet and removing them by hand). They may also be removed from shallow water with clam rakes. The largest quantities of this clam are harvested with scratch rakes, with tongs similar to those used to remove oysters, or with dredges. Dredges are used in comparatively deep water and may be of the basket or water-jet type. Aboard boats, hard-shell clams should be handled in the manner described for oysters.

In preparing them for market, hard-shell clams are washed with seawater, graded for size, and cooled. Some clams may be taken from semipolluted waters, provided they are depurated.

In the shell, hard-shell clams are marketed according to size. The different sizes are "chowders" (large size), which are used to prepare chowders, fritters, or stuffed clams; "cherrystones" (medium size), which are used for baking; and "littlenecks" (small size), which are used as steamed clams or for eating raw on the half shell. Hard-shell clams are neither canned nor frozen in significant quantities.

Hard-shell clams may be cultured. In such instances, the bottom is first prepared by removing thick grass, stones, and other debris. Predators, such as starfish, cockles, conches, and welks, are removed by raking and by towing floor mops over the bottom (to entangle starfish). The young clams, raised in tanks, are then spread over the area by scattering them from the side of a boat. Once seeded, the area is left undisturbed to permit the clams to grow to harvesting size. Predators may be removed from time to time. After seeding, the area may seed itself naturally.

The Soft-Shell Clam

The soft-shell clam is found in the Western Atlantic as far north as the Arctic regions and as far south as Virginia, being most plentiful off the coasts of New England, New Jersey, and Virginia. In New England, soft-shell clams are harvested when the tide is low by digging into the mud with the short-handled clam hoe and removing them by hand.

In the Chesapeake Bay area, clams are harvested from boats using water-jet dredges and an escalator. They are placed in bags or baskets and brought to the processing plant, where they are washed with seawater and sorted according to size. Specimens 3 in. (7.6 cm) in length or smaller are usually cooked by steaming. The larger-sized clams are removed from the shell by hand and placed in metal containers, after which the containers are refrigerated by being surrounded with crushed ice. In this form, they are shipped to restaurants to be served as a breaded, deep-fat-fried product. Soft-shell clams may be removed from restricted areas and depurated.

Surf Clams

The surf clam or "skimmer" is large, reaching a length of 8 in. (20.3 cm). It is found just below the surface of sandy bottoms in waters 30 to 100 ft (9 to 30 m) deep off Atlantic Coast states from Massachusetts to Virginia. Most of the harvesting of this species is done off New Jersey with water-jet dredges having V-shaped scoops. Aboard the boats, the clams are placed in baskets or jute bags and brought to the processing plant without refrigeration.

Surf clams are used primarily for canning. The viscera are not utilized as food. At the canning plant, the clams are washed, then steamed lightly to cook the meat partially and so the shell will open. The meat is then removed by hand, the nectar (liquid left in the shell) being saved. The lower part of the neck (syphon), the mantle, the adductor muscle, and the foot (muscular portion that allows the clam to anchor itself in the mud) are then removed with scissors and diced into pieces about 3/8 in. (1 cm) wide. The diced portions are then filled into cans together with some hot nectar and salt, after which the cans are sealed and heat-processed.

Other Clams

Other species of clams used as human food include the butter clam and pismo clam harvested off the West Coast and the ocean quahog harvested off the East Coast. The latter is not as large, nor as much in demand, as the surf clam, but is used to help fill the demand for surf clam, which exceeds its supply.

Mussels that resemble soft-shell clams, except for the color of the shell and other minor differences, are used to some extent to help fill the demand for soft-shell clams, of which there is an insufficient supply.

It should be noted that all clams and mussels that feed mainly on algae may at times become toxic to humans (shellfish poisoning). This happens when bivalves feed on certain algae (dinoflagellates) containing substances that are toxic to humans but not to molluscs (see Chapter 3). Public health officials periodically test bivalves for toxin and close the shellfish beds when there is danger of shellfish poisoning outbreaks.

Scallops

There are several types of scallops, of which the sea scallop and the bay scallop are best known. The internal anatomy of the scallop is similar to that of the oyster, but the adductor muscle, the only part of the scallop that is eaten, is much larger than those of oysters and clams. Once beyond the larval stage, the scallop may attach itself temporarily to some object, but the adult scallop is quite mobile. By closing the opened shell with its adductor muscle, thus forcing water through two holes in the top shell, the scallop becomes jet propelled. These bivalves cannot be held out of water in the live state, as can clams and oysters, since the water drains from the shell, which cannot be tightly closed.

The bay scallop is generally circular in shape with a grooved upper and lower shell and a rectangular projection at the back near the ligament (the bay scallop is the logo that can be seen in any Shell Gasoline sign). The bay scallop reaches several inches (1 in. = 2.54 cm) in diameter, and the adductor muscle may be as large as 1 in. (2.54 cm) in diameter. Bay scallops are harvested with basket rakes in shallow water and with dredges in deeper water.

The sea scallop is much larger than the bay type. It may reach a size of 8 in. (20.3 cm) in diameter, and the adductor muscle may be as large as 3 in. (7.6 cm) or more in diameter. Unlike the bay scallop, the shell of the sea scallop is not grooved. Sea scallops are found in ocean waters 60 ft (18.3 m) or more in depth. While this bivalve ranges from Labrador to New Jersey, it is most plentiful on Georges Bank off Cape Cod. Sea scallops are harvested with dredges. Aboard the boat, the "eyes" or adductor muscles are removed from the bivalves with the aid of a knife, placed in muslin bags, and iced and brought this way to port. The remaining portions are discarded at sea.

Sea scallops are sold in the fresh or frozen form. If frozen for purposes of selling after defrosting, they are placed in freezer rooms in muslin bags and held until shipped to retailers. Some sea scallops are breaded and may be deep-fat-fried prior to packaging and freezing. At 0°F (–17.8°C), the storage life of scallops is longer than 1 year.

Other species of scallops include the calico scallop, found off the coast of Florida, and bay-type scallops, found off the coasts of Alaska and Australia and in the Irish sea.

Crustaceans (Class *Decapoda*)

Several types of crustaceans are used as food for humans, most of which are prized as delicacies. Included among these are shrimp (several species); lobsters (American, European, and Norwegian species); crabs (several species); and crayfish (several marine species and the freshwater species). Although the shells of crustacea vary in color, they all turn pink when cooked.

Crustaceans have a hardened external skeleton made up of a calcified polymer of glucosamine (a six-carbon sugar containing an amine [NH_2] group) called chitin.

The external anatomy of crustaceans consists of the mouth parts, the eyes, the antennae (varying greatly in size), the body or cephalothorax to which five pairs of legs are attached, and the abdomen or tail consisting of a number of jointed segments adjoined to the body. In some crustaceans, the first pair of legs is chelate or enlarged and developed into biting and crushing appendages called "claws." The end section of the tail has several parts, including the fan-shaped telson. In some species, the tail may be contracted or flexed to provide for movement in the water. On the underside of the tail there are a number of attachments called pleopods or "swimmerets," which for certain species are the main appendages providing for movement in the water. Some crustaceans will shed an injured claw, then generate and grow a new one.

Crustaceans grow by shedding the old shell (moulting) to become soft-shelled for a short period (the new larger shell soon becomes hard) and filling up the new larger shell, which allows more room for growth; moulting occurs most frequently in the early years of growth. Mating takes place when the female is in the soft shell stage. The fertilized eggs are attached to the swimmerets, are eventually hatched, and after several larvae stages the small crustaceans sink to the bottom and assume the general habits of the adult.

Lobsters

Lobsters have either a well-developed first pair of walking legs or biting claws. The European lobster is found around certain parts of the British Isles and mainland

Europe. The Norwegian lobster is found mainly around the coast of Norway and the west coast of Sweden. The American lobster ranges from Labrador to the coast of North Carolina in an area that extends seaward for a distance of 50 miles (80 km). However, there are deep-sea lobsters found more than 200 miles (322 km) from the coast. The depth of water where lobsters are taken is usually 30 to 150 ft (9 to 46 m), but deep-sea lobsters live at a depth up to 1200 ft (366 m). The American lobster is most abundant off the coasts of Maine and the maritime provinces of Canada. The average American lobster caught measures 9 to 10 in. (23 to 25 cm), weighs 1 to 2 lb (454 to 908 g), and is 4 to 7 years old. However, deep sea lobsters are larger, and specimens weighing more than 40 lb (18 kg) have been caught that are believed to be more than 50 years of age. The food of lobsters include fish, clams, and other molluscs.

Lobsters are caught in pots (see Fig. 17.5) and are held in the live state aboard boats without refrigeration, because they are brought to port shortly after harvesting. Lobsters may be held in the live state, out of water at low temperatures above freezing, for more than 1 week, if given sufficient air space, because they are able to obtain oxygen from what dissolves in the water on their gills (the gills must be kept moist). They may also be held in the live state for a month or more in ocean pounds, which allow the free flow of water, or in tanks in which seawater is filtered, aerated, and circulated. When lobsters are held in tanks, the biting claws may be immobilized by the insertion of a wooden plug into the flesh above the thumb or by an elastic band encircling the thumb and claw.

Lobsters are sold mostly to restaurants or to the consumer, in the live state. Lobsters should be cooked from the live state or killed and cooked immediately. The reason for this is that lobsters have a very active proteolytic enzyme system that soon digests part of the tissue of the dead lobster, partially liquefying the meat or causing it to become soft and crumbly (a condition known as "short meated"). Some cooked lobster meat is sold as a canned or frozen product but does not make up a significant part of the catch. The storage life of cooked frozen lobster meat at 0°F (–17.8°C) is at least 8 months. Whole lobsters in the raw, cooked, or partially cooked state cannot be frozen successfully, because when frozen in the cooked state, the tomalley (liver) becomes rancid and affects the flavor of the meat, and, when frozen raw or in a partially cooked state, the flesh undergoes proteolysis.

Shrimp

There are numerous species of shrimp used as food for humans. The edible types vary in size from very small, about 2 in. (approximately 5 cm), to more than 10 in. (25 cm). The larger shrimp are called prawns. The overall size of Gulf shrimp as caught is 7 to 8 in. (17 to 20 cm). Most shrimp caught by U.S. fishermen are taken from the Gulf of Mexico, and these consist of three main types: white, brown, and pink. Some shrimp are taken from Atlantic waters off the Carolinas, Georgia, and Florida, and some are taken off Alaska, Maine, and Massachusetts. Shrimp are imported from Mexico, India, Panama, Venezuela, Brazil, Guiana, Ecuador, Nicaragua, Colombia, El Salvador, Honduras, Thailand, Surinam, Malaysia, and other countries.

Shrimp are caught with otter trawls that are somewhat modified from those used to catch cod and haddock. In some instances, two boats may be used to tow the trawl attached to an outrigger. Aboard the boat, all but the tails of the shrimps are discarded. The tails are washed and stored in boxes or pens in ice.

At the processing plant, the shrimps may be peeled and deveined (the intestinal tract is removed), and they are washed. In some cases, shrimps are dipped in a solution of sodium tripolyphosphate to prevent softening of the texture and loss of water during storage. However, if excess amounts of this salt are absorbed, the cooked product will appear and have the texture of raw shrimp.

Shrimps may be frozen in a plate freezer where boxes of the product are frozen into shrimp blocks or the shrimps may be frozen individually using a liquid freezant (e.g., liquid nitrogen). Prior to freezing, some shrimps are butterflied (split longitudinally), and some may be cooked. Some frozen shrimp may be tempered (partially thawed) in order to coat them with a breading, and then refrozen.

Raw shrimp in the shell, when protected against dehydration, have a high-quality storage life of at least 2 years at 0°F (–17.8°C) or below. Cooked shrimp, especially those cooked in hot oil, have a storage life of 3 to 5 months at 0°F (–17.8°C). Uncooked shrimp, prepared and frozen in the butterfly form, are subject to storage changes. There is so much air space in the package that dehydration occurs through a continuing two-step process: (1) moisture from the product vaporizes and fills the voids and (2) moisture from the voids condenses on the inner surface of those parts of the package that are adjacent to the voids.

Frozen shrimp imported from other countries must be defrosted before they can be processed. This may be done by tempering the product at about 40°F (4.4°C) for 24 hours, then completing defrosting by holding the unpackaged shrimp in running water. A more sanitary defrosting method employing microwave heating is also used.

Considerable quantities of shrimp are canned. For this purpose, they may be delivered to the cannery with the heads on. The shrimp are first washed and separated from the ice. The tails are then removed from the heads, usually by machine. The shell is removed and the vein taken out by machine. Individual specimens are then inspected, and broken and decomposed shrimp are discarded and may be used for pet foods or fertilizers. The product is then blanched or heated in boiling saturated salt solution (25%) for a period of 45 seconds to 3 minutes. After blanching, the shrimp are graded for size and filled by machine into one of several can sizes. Hot, dilute salt solution is added to the product in the cans, and the cans are sealed immediately. Heat-processing to provide commercial sterility is carried out at 250°F (121.1°C) for various times depending on the size of the container.

In the Pacific Northwest and Alaska, very small shrimp may be canned without deveining. Shorter blanching times are used for this product and small amounts of citric acid are added to the brine used to cover the shrimp. The brine is added cold, and the cans are vacuum-sealed prior to the heat processing.

Crabs

Crabs have the general anatomy of other crustaceans, but the body is oval-shaped or disclike instead of cylindrical, as in lobsters, shrimp, and crawfish. Also, the abdomen (tail) is comparatively small, flattened, and permanently flexed under the body. Several species of crab are used as food by humans.

Blue Crab. The blue crab is found from Nova Scotia to Mexico, including the Gulf of Mexico, and is especially abundant in the Chesapeake Bay region. It is commercially important only south of New Jersey. The semioval body of the blue crab has spiked peaks near the back end. The first walking legs are well developed into biting claws and the last pair of legs (called back fins) is flattened and used to propel the crab in the water. When fully grown, these crustaceans measure 7 in. (18 cm) or more across the body. Crabs are caught with crab pots or traps or with trot lines.

Blue crabs are brought to the processing plant in the live state. They are then cooked in live steam or boiling seawater, or in steam at 240°F (115.6°C) for 10 minutes, or in steam at 250°F (121.1°C for 8 to 10 min. After cooking, the back shell, viscera, claws, and legs are removed; then the meat may be removed from the shell with the aid of a small, sharp knife. In hand picking, the body meat adjoining the back fin is separated from the finer body meat, as it is considered to be of better quality and of higher value. Certain machines are now available for removing crab meat from the shell. This may be done by impact, or a roller process may be applied to the cooked or partially cooked debacked body, legs, and claws. Somewhat better yields of meat are obtained by machine picking and much less labor is required to do the job, but machine picking does not provide for the separation of back fin lump meat from the other body meat, unless it is done by hand before the crabs are machine-processed.

After the meat has been picked, it is packed in metal cans that are then closed and heated in boiling water until an internal temperature (at the center) of 185°F (85°C) is reached. This temperature is maintained for 1 minute. The product is then cooled and held at 33 to 38°F (0.6 to 3.3°C) prior to distribution.

Blue crab meat is neither frozen nor heated-processed to the point of commercial sterility, because either treatment results in a product of poor quality. If blue crabs, when caught, are nearing the molting stage, they may be held in seawater pounds, until they shed their shells. They are then soft-shell crabs, and they are sold in the live state at a premium price, as soft-shell crabs are considered to be a delicacy. High mortality rates are usually encountered during the holding of crabs for molting.

Dungeness Crab. The Dungeness crab is found from the Alaskan peninsula to southern California but is most abundant in the area between San Francisco and southeast Alaska. It may attain a size of 9 in. (23 cm) across the back. It has well-developed biting claws.

Dungeness crabs are caught in water 12 to 120 ft (3.7 to 36.6 m) deep. The circular pot is used in deep water, the rectangular pot in shallow water. Ring nets may also be used. Dungeness crabs are brought to port in the live state aboard the boat, held in wells of seawater, and may thereafter be held in tanks of seawater until sold to restaurants in the live state.

In processing, to obtain meat from Dungeness crabs, the back shell of the live crab is first removed, the viscera and gills are then torn away, and the body is broken in half with the legs attached. The sections are then cooked in boiling seawater for 10 to 12 minutes; then the meat is removed by hand by shaking or by impacting against a metal container, or it may be removed by running the body and legs between mechanized rollers. Pieces of shell may then be separated by floating the meat in a salt solution of the appropriate specific gravity for the meat to float and the shell to sink. Fresh meat is packed in cans, the cans are sealed, and the product is held at 32 to 40°F (0 to 4.4°C) for purposes of distribution.

Some whole or eviscerated Dungeness crab is frozen in brine at 5 to 0°F (–15 to –17.8°C), packaged or glazed, and stored at 0°F (–17.8°C) for distribution in the frozen state. A larger amount of this type of crab meat is packed in hermetically sealed cans and frozen in moving air at 0 to –10°F (–17.8 to –23°C). This type of crab meat is not especially stable in frozen storage but may be held for as long as 6 months at –10°F (–23°C) with fairly good results.

Dungeness crab meat is also canned and heat-processed. The meat is packed in cans holding 6.5 oz (185 g) of product. A weak solution of salt and citric acid (pH 6.6 to 6.8) is then added, covering the meat to prevent discoloration. The cans are then sealed and heat-processed at 240°F (115.6°C) for 60 minutes, then cooled in the retort. The quality of this product is inferior to that of the fresh meat.

King Crab. This species is not a true crab but is similar to crabs in structure and habits. It is much larger than other crabs, attaining a spread of about 5 ft (1.5 m) and a weight of about 24 lb (10.9 kg).

King crabs are caught off central Alaska to the Aleutian Islands and off the islands of northern Japan. They are harvested with large rectangular pots. Aboard boats, the crabs are held in the live state in wells of circulating seawater.

King crab meat is either canned or frozen. In canning, the whole crab is cooked in boiling water, after which the meat is squeezed out between rubber rollers. The meat is then washed, packed in cans in a weak brine, and the cans are sealed, heat-processed, cooled, and stored.

The meat may be frozen in large blocks for the restaurant trade. The legs and claws are also frozen for retail outlets and restaurants. If properly packaged, and frozen to a temperature of 0°F (–17.8°C) or below and held at this temperature, King crab meat has a high-quality storage life of at least 12 months. Lower storage temperatures provide for an even longer storage life.

Snow or Tanner Crab. This species is relatively large, reaching a size of 5 to 6 in. (12.7 to 15.2 cm) across the back and 2.5 ft (76.2 cm) between the tips of the outstretched legs. The snow crab is taken in deep water off central and eastern Alaska and in the Bering Sea and some is taken off Nova Scotia and Newfoundland. It is caught in large, baited pots, as is the King crab.

Snow crabs are handled and processed in much the same manner as are King crabs, but most of the meat is canned and heat-processed. The meat of the snow crab is inferior to that of the King crab.

Red Crab. The red crab is found from Nova Scotia to South America but is taken almost entirely in deep waters off southern New England. Red crab meat is removed mechanically from the debacked, partially cooked specimens with machinery employing the roller process. Red crab meat is sold mostly as the fresh cooked refrigerated product but some is sold as frozen.

Jonah Crab. The Jonah crab is found in waters from Nova Scotia to North Carolina and is caught in lobster pots. The meat of this crab is difficult to remove from the shell, and the product is sold mostly as cooked refrigerated or frozen whole crabs or claws.

Marine Crayfish. The crayfish or spiny lobster has become a popular food in the United States. There are a number of different species of marine crayfish ranging from Florida and the Gulf of Mexico to Central and South America. They are also found off Australia, New Zealand, South Africa, and other countries. These species have the general anatomies of lobsters, but the first pair of walking legs is not developed into biting claws.

Because only the tail portion is eaten, this is removed from the live specimen, packaged, with shell on, in moisture-vapor-proof material, and frozen, for sale to restaurants or for the retail trade.

Freshwater Crayfish. Freshwater crayfish are grown in ponds. Although they have the general anatomy of the true lobster, with well-developed biting claws, they are much smaller, the maximum weight being about 8 oz (227 g). There is at present a small industry in which the small crayfish are placed in rice fields after the rice has been harvested. Here, they eat the rice roots and also serve to fertilize the fields. By planting time, the fields can be drained and the crayfish harvested. Generally, these specimens are handled in the fresh, refrigerated state and are processed only by cooking.

AQUACULTURE

Aquaculture or fish farming is not new. It is estimated that it was practiced in China as early as 2000 B.C. when common carp was raised. In the United States, channel catfish farming developed from near obscurity in 1970 to an annual yield of more than 165,000 tons in 1990. Farming of marine shrimp, primarily in South America, Central America, and Asia, supplied approximately 25% of the world's consumption in the early 1990s. Some major seafood processors have increased their use of farmed shrimp from approximately 15% in the late 1980s to nearly 75% in the mid-1990s. This is certainly the fastest growing aquaculture enterprise worldwide. In Norway and other areas of Western Europe, ocean pen salmon provide 90% of the salmon consumed and is also a thriving export commodity.

In the United States, many of the rich fishing areas have been overfished and legislation has been passed to limit fishing in these areas to allow natural replenishing of the depleted species. Aquaculture may be an alternative to supplying seafood to meet demands that can no longer be filled through fishing alone.

Some species that are currently being cultured include marine shrimp, channel catfish, rainbow trout, salmon, crawfish, tilapia, carp, turbot, sea basses, sea breams, and shellfish such as clams, oysters, scallops, and mussels.

The aquaculture of popular marine finfish such as cod and halibut has been limited because of problems that exist in culturing these species. Difficulties in larval production, poor acceptance of dry feeds by some species, and expensive pumping of seawater are all contributing problems that are being studied at universities such as Auburn University and the University of Rhode Island.

Aquaculture can be conducted quite differently where the fish are actually grown inside, in a "fish barn." A farm in Massachusetts raises a hybrid fish (saltwater striped bass and freshwater white bass) indoors 90 miles from the nearest ocean. Their annual revenues exceed $2 million, which is larger than those of most pond-based fish farms.

The system has a series of tanks that the fish travel through on their way to market. Fresh water (not salt) is used primarily and it is filtered and reused up to 250 times before it is discharged. When it is discarded, laboratory tests from state environmental officials show that it meets many drinking water standards. The fish feces is even recycled, as it is used as fertilizer by local plant farmers.

Pure oxygen is pumped into the water and the fish are fed a diet of pellets made of fish meal, canola, and soy. The pellets are also fortified with vitamins and minerals.

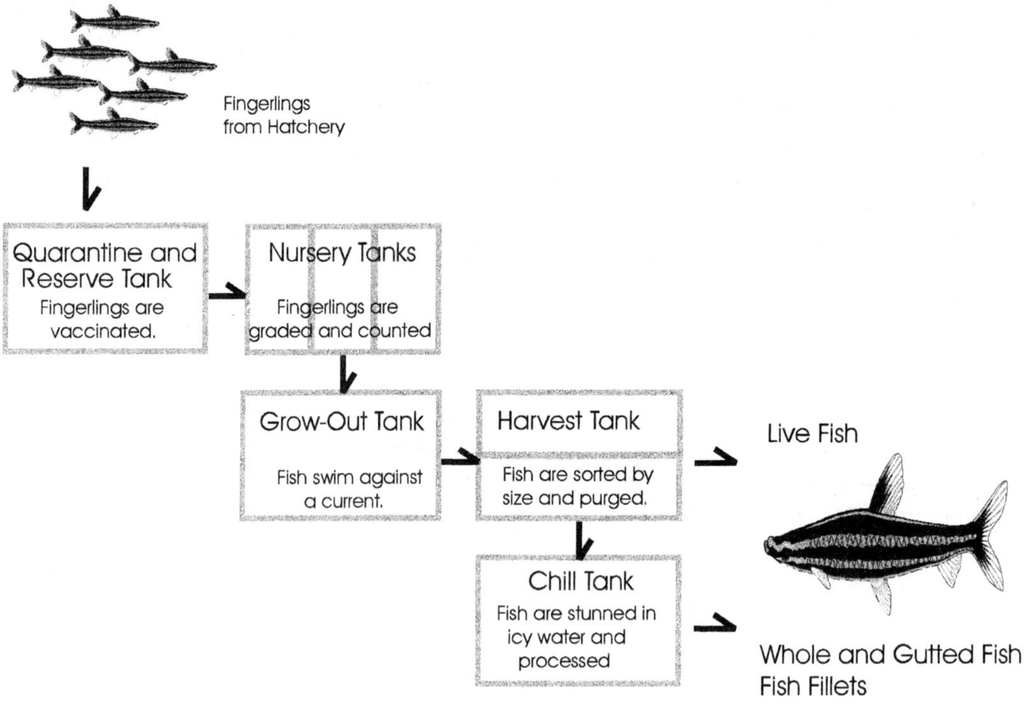

Figure 17.7. Aquaculture of striped bass.

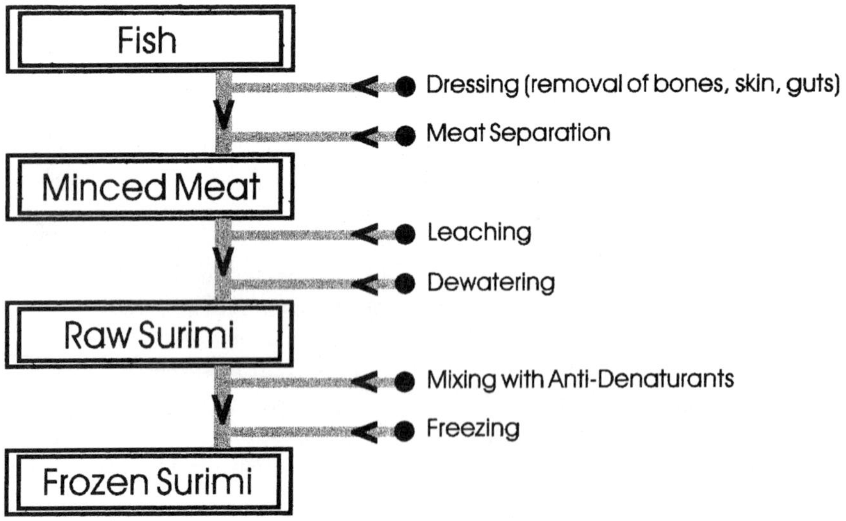

Figure 17.8. Production of surimi.

Crab Leg-Type Products

(Representative Formula)

Surimi (SA Grade)	1000 grams
Salt (NaCl)	25 grams
Ice Water	100–150 grams
Sugar (Sucrose)	50–70 grams
Potato Starch	50 grams
Monosodium Glutamate (MSG)	5 grams
Ribotide	0.1 grams
Glycine	3 grams
Sodium Succinate	0.3 grams
Milin	20–30 grams
Crab Extract Powder	10–20 grams
Crab Flavor	3 grams
Color	

Pictorial Evolution of Crab Leg-Type Products

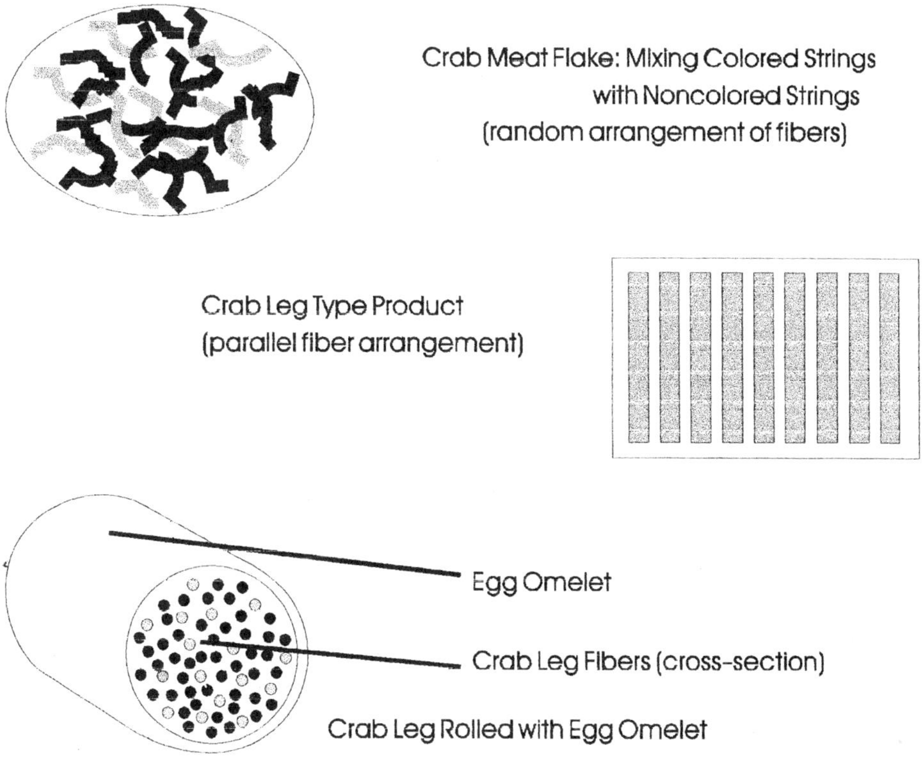

Figure 17.9. Representative formula for a crab leg-type product.

The tanks are equipped with pumps to create a man-made current for the fish to swim against to maintain proper muscle tone for proper texture.

The fish are spawned elsewhere and brought to the farm at 2 months of age. The fish are vaccinated against certain bacteria and stay in the reserve tanks (see Fig. 17.7) where their growth rate is controlled by temperature. Raising the temperature can increase their growth rate to three times what it is in the wild. The temperature may also be cooled if, for marketing reasons, later maturity is desired.

From the reserve tanks, they advance to the nursery tanks where they are graded and counted. When ready, the larger fish go to the grow-out tanks where the current is increased and they grow to market size. From here they go to the salt water holding tanks or "purge" tanks where they are sorted by size, not fed and are allowed to expel any excess food in their bodies which can alter the flavor of the fish.

At this time, the fish are about fourteen inches and about two pounds. Some fish

Manufacturing Process

Alaskan Pollock Surimi (frozen)
♦
Thawing in De-Freezing Machine with Hot Water
♦
Cutting in Silent Cutter
♦
Blend in Blender (Silent Cutter) with Salt and Starch
♦
Knead in Mixing and Kneading Machine (Silent Cutter) with Additives (flavor, sugar, etc.)
♦
Meat Conveying Pump
♦
Sheet Forming in Sheet Forming Machine
♦
Setting in Baking and Steaming Machine
♦
Cooling in Cooling Machine
♦
Cutting into Strings in Cutting Machine
♦
Color Paste Film in the Color Hopper
Wrapping and Forming Crab Leg-Type in Coloring Hopper
♦
Packing in Vacuum Packing Machine
♦
Fixation of Gel and Sterilization in Boil Style Sterilizer
♦
Cooling in Cooling Tank
♦
Drying in De-Watering Conveyor

Figure 17.10. Manufacturing of a crab-leg type produdct.

are sold live but those that aren't go to the chill tanks where they are stunned by the icy water and processed into fillets or sold as whole gutted fish.

The whole process takes about 9 months and a complete, sanitary quality control system must be strictly adhered to. The operation described is done in a 45,000 ft^2 building that houses 900,000 fish.

Food scientists are getting involved in aquaculture in terms of adding quality attributes to the finished products. Both color and flavor can be altered by additives to the fish feed. The Atlantic salmon aquaculture industry has found that the addition of carotenoids to the fish feed imparted a pinkish appearance to the fish flesh and this seems to be a positive quality factor to consumers. Taste can also be improved by the addition of bromophenols to the feed of aquacultured freshwater fish. This adds shrimp-like flavor that is preferred by the consumer.

SURIMI

It is estimated that over 900 years ago, the Japanese fisherman found that if they minced and washed the fish and mixed it with salt and spices, ground it to a paste, then cooked it by steaming or broiling, it would last much longer than fresh fish. The product is called surimi. Surimi is not itself a foodstuff but is an intermediate raw material from which the traditional Japanese kneaded foods called "kamaboko" are made.

The word "surimi" literally means "minced meat." Surimi is more than minced meat, as it has gel-forming capacity and has long-term stability in frozen storage. It gains these qualities from the addition of sugars as cryoprotectants. When fish muscle is separated from bones, skin, and entrails, and then comminuted, it is called minced

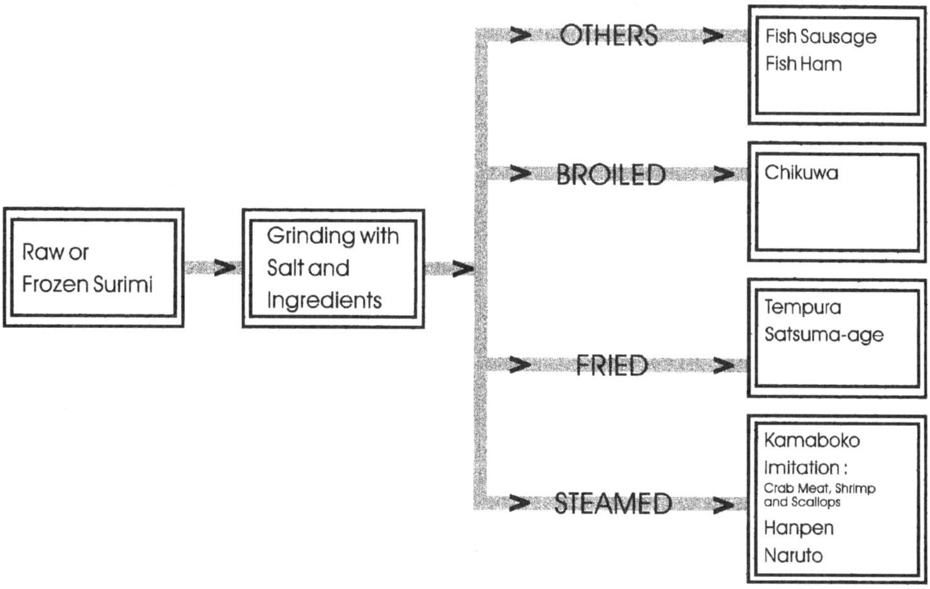

Figure 17.11. Different surimi-based products.

meat. It becomes raw or unfrozen surimi after it has been washed to remove fat and water-soluble constituents. The raw surimi is quite bland in flavor and the washing removes water-soluble components, thus isolating the meat's myofibrillar protein. This protein is water insoluble and gives the surimi the excellent gel-forming capacity.

When raw surimi is mixed with antidenaturants and frozen, the product is called frozen surimi. Sugar compounds such as sucrose and sorbitol are often added as antidenaturants or cryoprotectants. The production of frozen surimi is shown in Figure 17.8.

Surimi-based products have other ingredients added to give the product desired flavor, color, and texture. An example of the formulation of a crab leg-type product is shown in Figure 17.9 and the outline of a surimi-based product manufacturing process is shown in Figure 17.10.

Surimi has been made from Alaska Pollock but other fish such as sardine and mackerel have been used and some very high quality surimi and surimi based products have been produced. Some examples of different surimi-based products are shown in Figure 17.11.

18 Cereal Grains

Of all the plants humans have depended on for food, those that produce the cereal grains are by far the most important, as they have been since earliest recorded time. Cereal grains are the seeds of cultivated grasses that include wheat, corn, oats, barley, rye, rice, sorghum, and millet. There are a number of reasons why cereal grains have been so important in our diet. They can be grown in a variety of areas, some even in adverse soil and climatic conditions. They give high yields per acre as compared to most other crops, and, once harvested, their excellent storage stability combined with their high nutritional value make them the most desirable of foods for holding in reserve. They are easy to package and transport and they can be used to produce a large variety of highly desirable foods both for humans and animals, as well as beverages for human consumption.

Cereal grains are the most important source of the world's total food. Rice alone is reported to supply the major part of the diet for more than one-half of the world's population. Cereal grains are the staple food of the peoples of developing countries, providing them with as much as 75% of their total caloric intake and 67% of their total protein intake. The grains are eaten in many ways, sometimes as a paste or other preparation of the seeds, but more often milled and further processed into various products including flour, starch, oil, bran, syrup, sugar, and dried breakfast forms. They are also used to feed the animals that provide us with meat, eggs, milk, butter, cheese, and a host of other foods.

All cereal grains have three complete parts: the bran (a layered protective outer coat), the germ (the embryonic part of the plant), and the endosperm (the large starchy part, containing some protein) (see Fig. 18.1). Except for two amino acids, lysine and

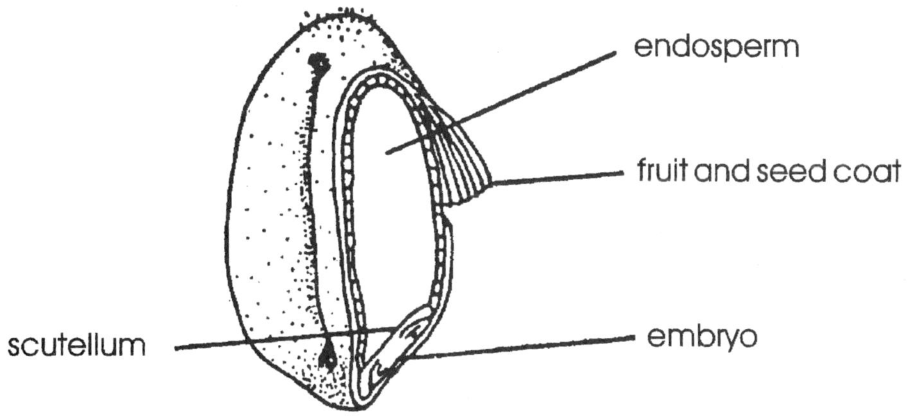

Figure 18.1. General structure of a grain kernel.

tryptophan, most cereals contain the essential amino acids required by humans, as well as vitamins and minerals. When they are consumed with other foods that can supplement the nutritional elements that are low in cereals, the minimum dietary requirements may be met. Research in cereal genetics may be expected to produce hybrid cereals that will be complete or nearly complete foods, containing more of the nutritional elements required by humans. A composite proximate analysis for cereal grains indicates that they have a protein content of about 11%, fat about 3%, moisture about 12%, carbohydrate about 68%, and fiber about 6%.

On a worldwide basis, rice is the most important cereal, being produced for human food in the largest amount, while in the United States, corn is produced in the largest amount, although it is used for animal food and other products as well as for human food. The grain grown in the largest quantity for human food use in the United States is wheat.

For most food uses in cereals, the bran and the germ are removed: the bran, because it is indigestible by humans and because of its adverse effect on the appearance and on some functional properties of flour, and the germ, because of its high oil content, which may subsequently become rancid. The germ is used to produce oil (e.g., corn oil). The bran goes mainly to feed animals. However, with dietary guidelines recommending more fiber, a growing amount of bran is being used in the production of breakfast cereals, bakery products, and other human foods.

The first ready-to-eat cereals were produced just before the turn of the twentieth century, with flaked and puffed cereals following within a decade. Ready-to-eat cereals, made from the endosperm of wheat, corn, rice, and oats, are convenient, nutritious, and they come in a very large variety of forms, textures, and tastes. The processing of cereals into breakfast commodities was started in the United States and is still largely carried out in the United States, with considerable quantities being exported throughout the world. The most popular of the breakfast cereals are those that are ready to eat. These are formed or puffed and oven-baked.

In the United States, grain is usually sold by the growers to operators of storage elevators, near the farms, where the grain is cleaned and stored. It is then sold directly to processors or to operators of storage elevators (near processors). They may sell directly to the processors or to their brokers.

A brief description of the handling, processing, and use of the more important cereal grains in the United States is found in this chapter.

WHEAT

Whole wheat, consisting of about 13% protein, can contribute considerably to the diet. The flour made from the whole wheat is higher in biological value than white flour (made from the endosperm only). Table 18.1 gives some examples of the higher nutritional value of whole wheat flour over white flour.

Wheat is the most popular cereal grain for the production of bread and cakes and other pastries. Wheat produces a white flour. In addition, the unique properties of wheat protein alone can produce bread doughs of the strength and elasticity required to produce low-density bread and pastries of desirable texture and flavor.

There are many varieties of wheat. They may be classified as hard red winter wheats, hard spring wheats, soft red winter wheats, white wheats, and durum wheats. Winter wheats are planted in the fall and harvested in the late spring or early summer. Spring

Table 18.1. Comparison of Some Nutrients in Whole Wheat and White Flours

Nutrient	Whole Wheat Flour[a]	White Flour[a]
Protein	13%	11%
Thiamin	2.3 mg/lb	0.3 mg/lb
Riboflavin	0.6 mg/lb	0.2 mg/lb
Niacin	26.0 mg/lb	3.5 mg/lb
Pyridoxine	2.0 mg/lb	1.0 mg/lb

[a]1 lb = 454g

wheats are planted in the spring and harvested in the late summer. Hard wheats are higher in protein content and produce more elastic doughs than soft wheats. Therefore, hard wheats are used for breads, and soft wheats are used for cakes. Durum wheats are used most for alimentary pastes (spaghetti, macaroni, etc.) and for the thickening of canned soups.

Wheat is harvested by combines that cut the stalk, remove and collect the seed, and either return the straw to the soil, to be plowed under with the stubble and thus provide humus, or compress and bale it for future uses such as for litter or ensilage.

Wheat may be bagged in jute sacks and stored in warehouses, or it may be stored in bulk in elevators. The latter method provides the best protection against rodent and insect infestation. The moisture content of bulk-stored wheat should not be higher than 14.5% and that of sack-stored wheat not higher than 16%. Otherwise, microorganisms may grow and cause heating and spoilage. When it is necessary to lower the moisture content of wheat, it may be dried in bins by blowing hot air (not higher in temperature than 175°F (79.4°C) across the bins.

In preparing wheat for milling, the wheat is blown into hopper scales that record the quantity of uncleaned wheat. Some of the coarser impurities are removed by this process. The grain then passes over a series of coarse and fine sieves that further remove contaminating materials, including chaff and straw. With the wheat still in the dry state, stones may be removed by passing wheat over short openings that allow the heavier stones to fall out of the mass and be trapped. The wheat is next passed over discs or cylinders containing indented surfaces that remove seeds shorter or longer than wheat, following a pass through a magnetic separator to remove any metals present. The next cleaning process is dry scouring to remove adhering dirt. The wheat is then washed in water, a process that both removes dirt and adds 2% to 3% water to the grain. The added water is necessary to provide desirable conditions for milling. A stone trap is included in the washer. Excess water is removed by centrifugation. A second wet cleaning with a light brushing action is ordinarily used, followed by aspiration (blowing air through the grain), which is the final cleaning operation. The grain is then carried into a bin from which it is fed to the milling operation. This bin is located on the top floor of the flour mill, the grain having been elevated to this position during the various cleaning operations.

In milling, grain is fed automatically through scale hoppers that regulate the flow of the seeds at rates corresponding to those of the following operations: milling may be carried out by passing the grain through a series of corrugated rolls rotating toward each other, which remove chunks of the endosperm from the bran. After each passage

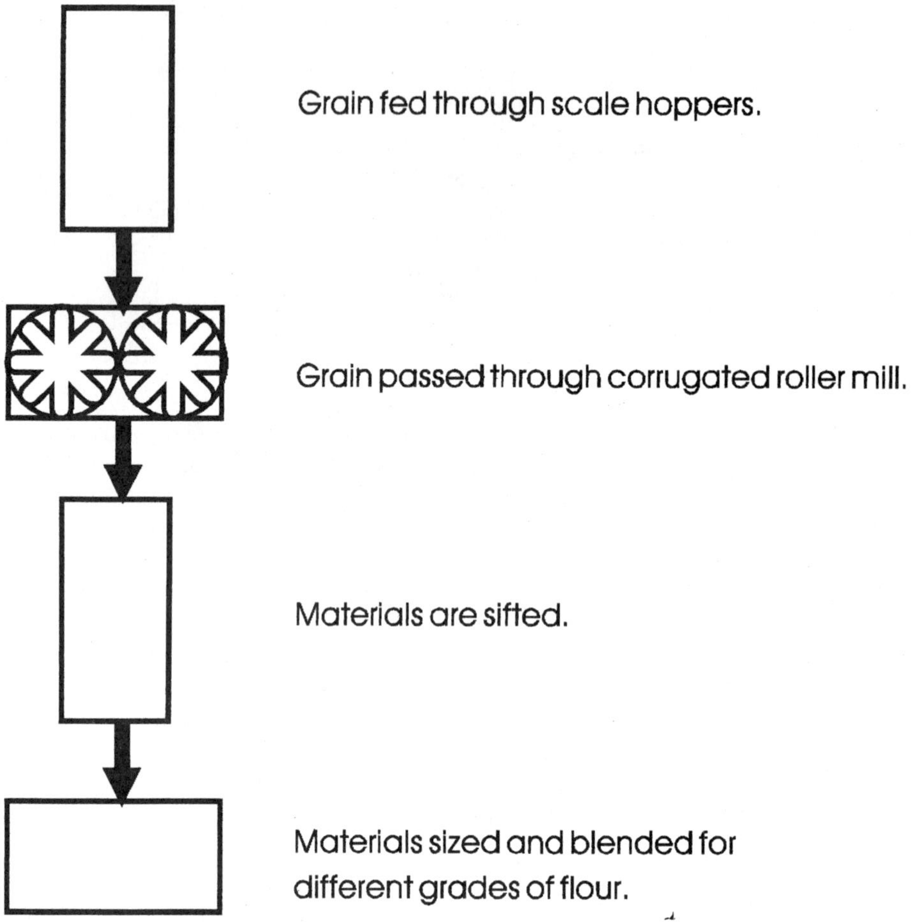

Figure 18.2. General procedure for milling of grain.

through the break rolls, the material is sifted through cloth, or wire sieves, and separated according to particle size. The various streams of different-sized flour particles are finally blended to provide the different grades of flour. The more finely ground flour is nearer to white but less nutritious than the coarser ground flour. This results from the more effective removal of bran and germ from finely ground flour (see Fig. 18.2). Impact milling is now used in some operations. With this method, the seed is broken open by impact in a machine called an Entoleter, first developed to control insect infestation. Flour particles of different sizes are separated by air classification or by centrifugation.

High-protein flour is desirable for some types of baked products; flour of moderate protein content for others; and high-starch, low-protein flour is desirable for still other baked goods. The smaller flour particles are higher in proteins; the larger flour particles are higher in starch. Through air classification in a turbomill it is possible to separate flour particles into various sizes, which can be blended to provide whatever protein or starch content is required by the baker or other users of flour. Turbomilling, developed in the late 1950s, was considered to be a significant milling innovation, because through this process, the variety of flour blends for different products is possible.

In the United States, wheat flour may be enriched with the minerals iron (as a salt such as ferrous sulfate) and calcium (also in the form of salts). Wheat flour is also enriched by the addition of small amounts of the vitamins thiamin, riboflavin, and niacin.

Wheat flour is used to make leavened products, such as bread, cakes, pastries, and doughnuts, and unleavened products, such as alimentary pastes (macaroni, spaghetti, noodles, etc.). Cake mixes are also prepared with flour, and flour is used for thickening canned and homemade stews, soups, gravies, and white sauces.

Various breakfast cereal products are made from wheat. Generally, in these products, the wheat is precooked and passed through heated rolls to form flakes or it may be shredded. An interesting technique also may be used to produce the "puffed" cereals which give a different appearance and texture. Either a grain or its composite dough can be used. The material is cooked to gelatinize the starch and then partially dried (see methods in Chapter 11). Then it is dropped into a pressure chamber that builds up to 100 to 200 lb/in.^{2} and a temperature of 500 to 800°F (260 to 427°C). The material is then ejected by the opening of a valve releasing the high pressure inside, thus "puffing" the dough or grain through water vapor expansion and escape. The result is a light, dry fragment. A similar texture can be achieved in "oven puffing" by cooking the grain and rolling it into a flat, thick flake. This is quickly toasted and the blistering lightens the texture. This method is usually used with rice.

Wheat bran may also be produced as flakes. High-protein cereals may be produced from wheat together with added wheat starch, sugar, malt, minerals (such as phosphates), vitamins, and other ingredients. Some wheat flakes are coated with very thin layers of sugar, honey, fruit, or other flavored syrups.

CORN

Many types of corn are grown in the United States. Sweet corn is produced as a vegetable and eaten fresh, canned, or frozen. Popcorn is also used as a food. However, the type of corn most utilized in the United States and considered as a grain rather than a vegetable is field corn. There are a number of varieties of corn usually classified as starchy or waxy, depending on the characteristics of the carbohydrate present. The development of hybrid strains has improved the yields of field corn. This corn is lower in protein than wheat, and, like all vegetable proteins, including wheat, is deficient in some amino acids and so does not provide a complete protein for humans. Corn is especially deficient in the amino acid lysine, but a variety of high-lysine corn has been developed that may eventually have a great impact on human nutrition in some parts of the world.

Ears of field corn are harvested by a machine that strips the matured ears from the stalks. If harvested in wet weather, corn may have to dried before it is stored. Usually, it is allowed to dry on the stalk in the field, is harvested, and is stored in small roofed bins or silos with metal or wire mesh walls. Much of the corn storage is done on the farm, as most of the corn crop is used as feed for animals. Stalks and leaves may be harvested, chopped, and placed in piles or in silos to form ensilage for animal feed. Stalks and leaves may also be chopped and returned to the soil for humus.

Corn milled for flour (corn meal) is cleaned, as is wheat, then moistened to a water content of 21%. The germ is removed mechanically. The endosperm is then dried to a moisture content of 15%, passed through crushing rolls, and sifted to remove the bran.

With use of sieves, milled corn is separated into grits (largest-sized particles) and meals and flours (smallest-sized particles).

Most of the corn crop is used for animal feed, but considerable amounts are used to produce cornstarch, corn syrup, high-fructose corn syrup, and other various sugar derivatives. The production of corn starch, modified starch, and uses of these is discussed further in Chapter 21.

Various types of sweeteners are made from cornstarch, as starch consists of a long straight or branched chain of glucose molecules that may be broken down to short chains of glucose molecules (dextrins), to maltose (two molecules of glucose), or to glucose (dextrose). The production of some of these products is also discussed in Chapter 21.

Corn is also used to produce popcorn. The variety used is a specific one. When the dried kernels are heated, internal moisture creates a vapor pressure owing to the rise in temperature, and when the pressure is sufficient, the hard outer shell is burst and the pressurized grain is expanded. Essentially, popcorn is puffed cereal.

OATS

Oats, one of the popular nutritious present-day cereals, was once regarded as useful for feeding only cattle. Oats can grow in colder and wetter climates than can wheat. Oats are harvested much in the same manner as is wheat. The moisture content at the time of harvest should not be higher than 13%.

Milling of oat kernels requires that they first be washed and cleaned and then dried in a rotary kiln or pan drier to a moisture content of about 12%. They are then hulled by impact, the seeds being thrown from a rotating disc against a rubber ring that splits off the hull and leaves the groat mostly intact. After the hulls are removed by passing the product through sieves, the oats are steam heated and passed between rollers to produce rolled oats, or they are cut into pieces about one-third of the original size and then steamed and rolled to produce quick-cooking rolled oats. Small amounts of oatmeal may be produced by grinding the steamed oats. Steaming facilitates cooking and inactivates enzymes that, if not inactivated, may cause bitter flavors to develop. Oat flour may be produced for use as an ingredient of bread or a thickener for soups. If made from unheated oats, lipolytic enzymes may remain and there may be a problem with the development of rancidity.

Oat bran has become a very popular item and much of this is due to the research that links its water-soluble fiber with reduction of blood cholesterol. Many articles have been published such as those by James Anderson of the University of Kentucky in the American Journal of Clinical Nutrition (1981) and by Jeremiah Stamler of Northwestern University School of Medicine in the Journal of the American Dietetic Association (1986) which show that consumption of oat bran can lower blood cholesterol (Anderson said that those who ate 100 g of oat bran a day lowered their cholesterol by 13% and Stamler said it dropped 3% if 35 to 40 g of oat bran was consumed daily). Frank Sacks of Harvard Medical School in the New England Journal of Medicine (1990) found that oat bran didn't lower the cholesterol level any more than refined wheat flour with a low bran content. He suggested here than just the replacement of foods with high levels of saturated fat and cholesterol with those low in these components will lower the blood cholesterol.

It is agreed by the vast majority that this replacement of fatty foods with lower fat varieties is beneficial. The American Dietetic Association and American Heart Association have recommended a decrease of fat to less than 30% of calories consumed and an increase of fiber in the diet. The Institute of Food Technologists (IFT) in 1989 cautioned against the overuse of such fibers. If fiber is greatly increased in a low-fat diet, consumers may not retain and adequately use numerous minerals required by the body. The IFT report said: "Recommendations for stepping up fiber in the diet—fiber supplements, oat or other types of fiber—should take into account a person's mineral intake and all other nutrient consequences for the best outcome."

BARLEY

Barley products do not bake as well as wheat products; thus, barley, containing little or no gluten, is not as popular as wheat when there is an option. However, barley has the advantage of growing in climates too cold and in soils too poor to grow wheat, and, in addition to being a hardy grain, its growth requires a shorter time than does that of wheat.

Some barley is produced in the United States. Spring and winter varieties are planted as in the case of wheat. In the United States, barley is used as feed for cattle and poultry, for the production of malt used in brewing, and as an ingredient of soups. Small amounts of barley flour are also produced.

For producing malt, the grain is soaked in water for several days or until the moisture content reaches approximately 50%. It is then removed from the steep tank and placed in containers where air at 65 to 70°F (18.3 to 21.1°C) can be drawn through it over a period of approximately 1 week. This allows the barley to germinate or sprout. The sprouted barley is then kiln dried over a period of 24 hours. Drying is begun at a low temperature that is gradually raised as drying proceeds. The purpose of malting barley is to produce enzymes that will hydrolyze starch to maltose, a sugar that can be utilized by yeasts to produce ethyl alcohol and carbon dioxide. Nondiastatic malt (will not hydrolyze starch) may be produced for its flavor components. Therefore, in drying the sprouted barley, the temperature must not be raised to the point where the starch-splitting enzymes, produced during sprouting, will be inactivated. It would appear, however, that temperatures are raised to the point where some of the sugars present in the sprouted barley are caramelized, hence the dark brown color of malt. Malt is used in the brewing industry for converting the starches present in barley, rye, rice, corn, or other grains to maltose, which can be utilized by yeasts.

Producing beer is an interesting process. The science of fermentation or zymurgy is utilized here. First, as mentioned previously, the barley is allowed to germinate to accumulate starch-digesting enzymes (amylases). Next, the partially germinated kernels of barley malt are dried to stop enzyme activity (but not to permanently inactivate them) and to develop color and some flavor. In this process, called mashing, the malt is mashed in warm water and another grain such as corn, rye, or rice may be added and the enzymes are reactivated by the gentle temperature and moisture and act to reduce the long-chained carbohydrates to sugars (glucose). The resulting sweet liquid produced by this enzymatic action is called wort. With beer and related beverages, hops are added here for flavor. This process, called brewing, takes place in the brew kettle at elevated temperatures and the enzymes are now permanently inactivated.

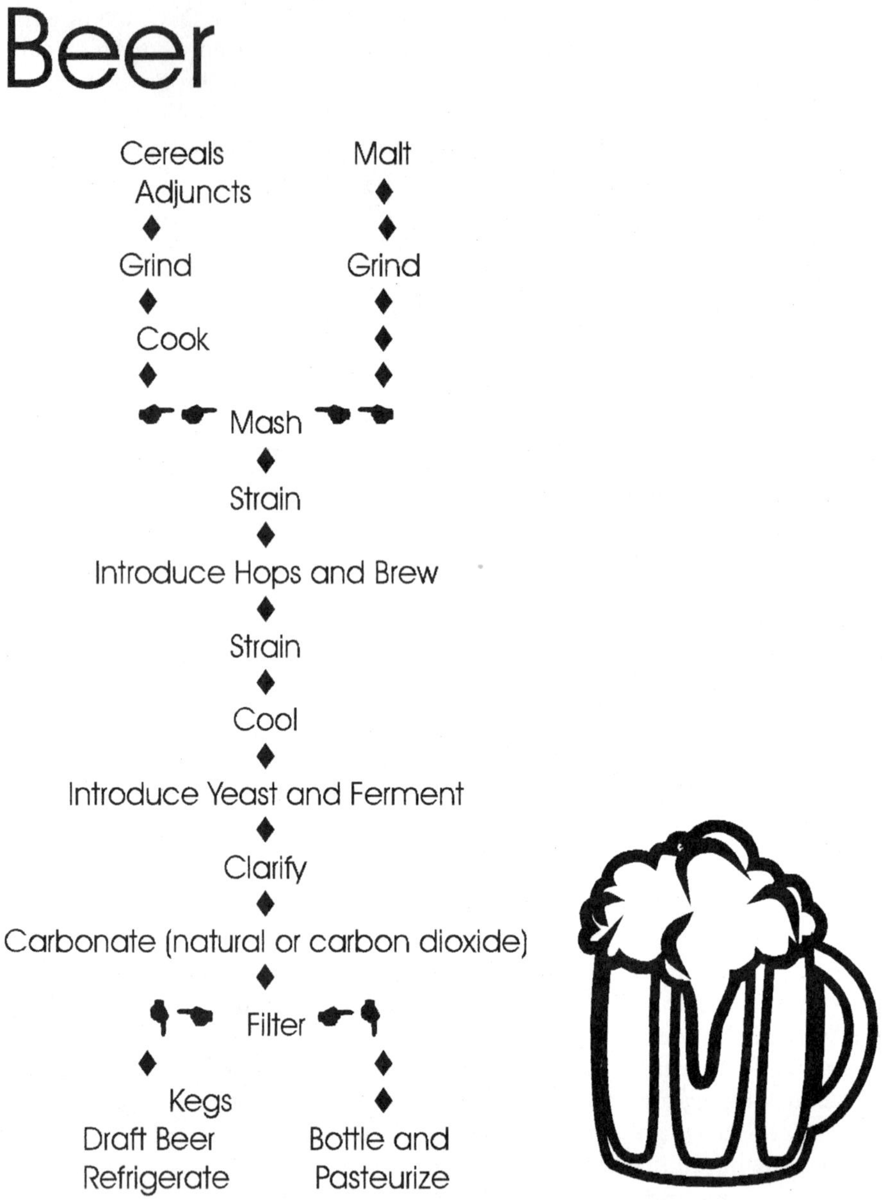

Figure 18.3. Method of domestic beer production in the United States.

The hops are now strained out and the bittersweet liquid remaining (hopped wort) can now be fermented by yeast to produce ethyl alcohol and carbon dioxide. Beer is then filtered through diatomaceous earth filters that remove almost all of the yeast. It is then clarified and bottled or packaged in kegs. Draft beer must be refrigerated whereas bottled beer is pasteurized in the bottle and then cooled. Some beer is "cold pasteurized," a process in which it is sent through fine filters to remove all of the yeast before it is bottled. Beer may be carbonated naturally in a secondary fermentation before packaging called kraeusening or carbon dioxide may be pumped in prior to packaging.

Malt may also be used in bread baking for much the same reason as in beer production, although in this case, the purpose is to have the yeast produce carbon dioxide for leavening (raising the dough), the alcohol produced being largely dissipated during the heating involved in baking.

Some quick-cooking rice is produced by precooking the kernels and redrying. This process provides for the preparation of rice for human consumption by merely bringing the water used for rehydration to the boiling point and allowing the mixture to stand for short periods.

Some puffed rice cereal is produced by heating the rice to a temperature above the boiling point of water in closed containers and suddenly releasing the pressure, which causes the kernel to increase in size, as the water vaporizes, allowing it to escape from the interior to the outside.

About one-third of the rice produced in the United States is used by the coffee brewing industry. This consists mostly of broken kernels, but some whole grain rice is also used for this purpose.

Rice flour is produced, and most of this is used by those who are allergic to wheat flour. Rice flour may be used, too, for the preparation of white sauces, especially for prepared frozen-food products, as certain types of rice flour produce sauces that do not curdle and weep (separation of liquid from the sauce) when frozen and defrosted.

Rice bran has been used for cereal and baked goods and its water-soluble fiber is said to have cholesterol-lowering effects similar to those suggested with oat bran.

Rice kernels may be enriched, as is wheat flour, by mixing with powder containing vitamins and minerals. This powder sticks to the surface of the kernels. The enrichment materials may then be coated with a waterproof, edible film to protect them from being washed off.

The protein of rice is comparable to that of wheat in composition, although rice is lower in total protein content than wheat. Neither of these grains contain a complete protein, that is, the proteins do not contain sufficient amounts of certain amino acids to provide for the requirements of the human, although the biological value of rice protein is reportedly equal to or superior to that of wheat protein.

OTHER CEREAL GRAINS

Sorghum

Sorghums, comprising four general classes (sweet sorghum, broom corn, grass sorghum, and grain sorghum), are grown in southern sections of the Great Plains and in parts of the Southwest. Some varieties of the grain sorghum class yield glutinous starch, similar to that of corn. During World War II, sorghum was used as a substitute for tapioca, because the importation of tapioca was impeded by the war situation. The deterrent to the use of grain sorghum for the production of starch is the pigmentation of the grain's pericarp, which complicates the production of a white starch. However, enough progress has been made in the development of desirable sorghums to warrant the consideration of sorghum for the production of starch in the future.

Buckwheat

Buckwheat is not a true cereal grain. All the cereal grains belong to the botanical family Gramineae, whereas buckwheat belongs to the family Polygonaceae. However, from a use standpoint, it is considered to be a cereal food. Although it is a minor crop in the United States, only the Soviet Union and France produce more buckwheat than the United States. It is grown mainly in New York, Pennsylvania, Michigan, Maine, and Ohio. Of the few varieties used, the Silverhull is used mainly for producing flour because of the higher yield of the endosperm. Buckwheat is dried to about 12% moisture, cleaned, graded by size, and milled similarly to wheat. Most of the flour is used for making pancakes.

Cottonseeds, Soybeans, and Peanuts

Although cottonseeds come from plants of the family Malvaceae, and soybeans and peanuts from plants belonging to the family Leguminosae, it should be mentioned that they have been used to produce edible flours and other food ingredients.

Cottonseeds are used the least in flour production because they contain the toxic pigment gossypol ($C_{30}H_{30}O_8$). The pigment glands that contain gossypol can be removed by a process that involves disintegration of the seeds in hexane and separation of the glands by centrifugation. Heating also destroys the toxin and when heat is applied, the flavor of the seed is improved.

Soybeans have been studied extensively and many soy byproducts have been produced. These include soy "meats" from the dehulled bran, oil, soy flour, and grits (about 50% protein). The grits can be washed at a pH of 4.5 and dried to produce soybean concentrate (70% protein). Grits can also be dissolved in alkali, filtered, and have acid added to a pH of 4.5. This can now be concentrated further to a simple protein isolate (90% to 95% protein). The flour can also be wet, extruded, and dried to textured vegetable protein (50% to 55% protein). The simple protein isolate can be modified further into a range of isolates, spun protein, or structured protein isolates.

Some of the extruded, dried soy protein doughs, on rehydration and cooking, have a very similar texture to that of meat and are being used in combination with meats in mixtures. They add protein, little fat, and are less costly than meat. Dried, flavored textured protein produced such as imitation bacon bits are quite popular also.

A problem exists with soybean in that the raw beans contain a trypsin inhibitor. This interferes with normal growth of animals and humans but it can be inactivated by the heat of cooking or by heat in processing.

Peanuts can also be processed into flour, protein concentrates, and protein isolates but their use in human foods is limited. Peanuts have lower amounts of lysine than do soybeans. Most of the world's peanuts are used for oil (about 67%). In the United States, however, about one-half of the crop is made into peanut butter. Peanuts have a high moisture content at harvest and are susceptible to mold spoilage. If the molds grow, they may produce aflatoxins as metabolites. Strict government controls govern peanuts and if levels of aflatoxins reach 15 parts per billion (ppb) in the raw peanut or 20 ppb in the finished product they must be rejected. This could be devastating economically to the peanut growers, so peanuts are stored under conditions to control mold growth and they are carefully inspected.

Millet

Millet is used for food in Asia and, to some extent, in Europe. In many parts of Europe, it is used for hay, as it is used in the United States. Some varieties are used as food seeds for caged birds and poultry.

Triticale

Triticale, a hybrid of wheat and rye, first produced in the late 1800s, combines the high total protein content of wheat with the high lysine content of rye. It is also more adaptable to unfavorable growth conditions and seems to resist wheat rust (a disease caused by molds). The improvement of this hybrid is continuing and can lead to more beneficial genetic changes. This cereal is now being grown in more than 52 different countries.

19

Bakery Products

Bakery products include those leavened (raised) by the carbon dioxide produced by the growth of yeasts (i.e., breads, rolls, etc.), items leavened by carbon dioxide produced chemically through the use of baking powder (including cakes, doughnuts, biscuits, etc.), items leavened by the incorporation of air (i.e., batter-whipped breads, angel food cake, etc.), and unleavened products (crackers, pie crusts, etc.).

Bread and other baked food products (cakes, cookies, rolls, pies, doughnuts, etc.) are important items belonging to the class of foods that is sold in ready-to-serve form. Some of these products are partially baked and require a final baking prior to serving. Growing amounts of baked goods are handled and sold in the frozen form, especially because the quality of baked goods is exceedingly well preserved by freezing. Some baked items, such as cookies and biscuits, are canned, and these products, which must be held under refrigeration until used, need to be baked before serving. The dry ingredients used for some baked goods, especially cakes, are premixed industrially and sold as prepared mixes, and although the user must add the fluid ingredients and bake the product, it is still a convenient system for the consumer. In self-rising flours, chemical leavening agents are added directly to the flour. Some of these mixes are formulated for baking in a microwave oven. These have the advantage of producing cakes in significantly shorter times than are required in conventional baking ovens.

Generally, the high quality of bread and other baked goods goes into a rapid decline soon after the products are removed from the oven. Freezing them is the only method now known to preserve them effectively for long periods.

BREAD

Bread is the oldest and most important baked product. It has been made from many grains, including wheat, corn, rye, rice, barley, oats, and even buckwheat. The growth of its popularity has been due to a number of factors, but an important one is that grains of one type or another have been grown in nearly all the inhabited parts of the world. An example of the composition of the ingredients of a loaf of white bread is approximately 57% flour, 36% water, 1.6% sugar, 1.6% fat or shortening, 1% milk powder, 1% salt, 0.8% yeast, 0.8% malt, and 0.2% mineral salts. The flour used for making bread is usually of the hard wheat type, which is higher in protein than that of soft wheat types. The reason for this is that in yeast-leavening products, the gluten (protein) in the unbaked loaf must be sufficient in quantity and of adequate elasticity to form a stretched mass that will entrap bubbles of carbon dioxide. This increases the volume and forms the structure of the loaf. It also allows retention of the structure until sufficient heating has occurred. When, as a result of heating that coagulates the gluten, a more rigid structure has been formed, the structure of the loaf of bread is fixed.

Maturing or oxidizing agents and bleaching agents are usually added to darker flour because the minerals in wheat are concentrated in the bran and adjacent layers (see Fig. 18.1). Even though the bran is removed, the adjacent layers are retained and because of their high mineral content they impart a darker color to the flour.

Flour millers are able to supply bakers with flour, which has essentially the same protein content from delivery to delivery, as specified by the baker, by blending a variety of flours that have been classified according to protein composition. Also, quality control laboratories test for protein content and quality. There are other tests that determine the characteristics of different flours by measuring the physical properties of the doughs made from them.

Water is a chief ingredient of the dough in baking. The amount of water added is such that the finished loaf does not contain more than 38% water. If the water available to a bakery is hard (contains minerals), the amount of yeast food (mineral salts) to be added may be modified. Also, during the mixing of the dough, a certain amount of heat is generated because of friction encountered during the forcing of the mixing bars through the dough, and from the motor that runs the mixing apparatus, as well. If the temperature rises above a certain point (82 to 85°F [27.8 to 29.4°C], the yeast may be destroyed and the gluten and starch of the dough adversely affected. Therefore, the water must be cooled prior to adding it to the ingredients in the mixer, or part of the water must be added as ice, which melts during the mixing and controls the temperature.

Sugar (cane or beet sugar) in small amounts is used in baking bread, to serve as a source of readily utilizable carbohydrate for the yeast, providing for a suitable fermentation that produces carbon dioxide required to raise the dough.

Some fat or shortening is added to bread mixtures. Ordinarily, this is a solid fat (an oil that has been made into a solid material through hydrogenation). This facilitates mixing, tenderizes the crumb of the loaf, and prevents staling of the bread. Today, as a rule, small amounts of monoglycerides are also added, as they are more active antistaling agents than fats. A monoglyceride consists of a glycerine (glycerol) molecule that has had a fatty acid combine with only one of its three alcohol groups; the other two of its alcohol groups remain unchanged. Milk powder may be added to bread dough, because it has a desirable effect on the texture of the crumb (inside the loaf) of the finished bread.

The yeast used in bread baking is *Saccharomyces cerevisiae*. This may be used as a dried material or as a moist compressed cake containing 70% water (the latter type must be held under refrigeration prior to use). Either type of yeast must be suspended in warm water prior to being added to the material in the mixer, in order to obtain an even distribution throughout the dough. A freeze-dried, vacuum-packed yeast is also available. This product has a greater number of viable yeast cells; therefore, theoretically, it is required in lesser amounts. It does not need to be suspended in water; therefore, it may be added with the dry ingredients. Because it is freeze-dried and vacuum-packed, it requires no refrigeration, and theoretically, has a longer shelf-life.

Small amounts of salt (sodium chloride) are used in making bread, because it is desirable for the flavor of the finished loaf and trace amounts may be utilized by the yeast during growth. During mixing, proofing, and the early part of baking, the yeasts grow and produce ethyl alcohol and carbon dioxide. The latter, a gas, causes the dough to rise, and provides for the volume of the loaf. Leavening action may result to a degree from the vaporization of water in the dough when the temperature of the mass is raised sufficiently in the oven. The increase in loaf volume is also believed to be affected in the presence of shortening, which can entrap air during mixing and release it when the air expands during baking. The ethyl alcohol is largely dissipated during baking, although residual amounts of alcohol, esters, and other components may remain and contribute to the flavor of the loaf.

When malt is used in breadmaking, it is usually of the diastatic type (contains active enzymes that will convert starch to maltose or glucose). As proofing continues, the small amount of sugar present in flour, and the sugar that has been added to the dough mixture, may be mostly used up by the yeast. Therefore, to continue the yeast growth, the action of the malt enzymes on starch can provide a source of sugars during the latter stages of proofing and the early stages of baking. Because a small amount of sugar is essential to the browning of the crust of the loaf of bread, the malt is also important in that it provides the sugar in the development of crust color.

The mineral salts added to dough mixtures are called yeast foods. Yeasts require small amounts of nitrogen-containing and phosphorus-containing salts for growth and production of carbon dioxide. For this reason, small quantities of ammonium salts and phosphates are added to the ingredients of the dough.

In addition to the above components, many special types of bread are produced that contain one or more additional ingredients (i.e., butter, extra milk powder, buttermilk or dried buttermilk solids, dried vegetable powders, honey, etc.).

There are two general methods of handling dough in baking bread: the straight dough method and the sponge method. With either process, flour (stored in bulk in bins), water, fat (usually melted), suspended yeast, and milk powder are weighed and added to the mixer automatically.

In the straight dough method (see Fig. 19.1), all the ingredients are added to the mixer and the material is mixed first at low speed (about 35 rpm), then at high speed (about 70 rpm). The mixed dough is then placed in large metal containers or troughs and held in an insulated room at about 80°F (26.7°C) and in an atmosphere of high humidity to allow fermentation. During fermentation, the mass of dough is kneaded several times to allow the escape of some carbon dioxide, which is produced continually during fermentation. In addition, working of the dough in this manner assists in stretching and conditioning the gluten, which is the important ingredient responsible for the formation and retention of the structure of the loaf.

In the sponge dough method, 50% to 75% of the flour, enough water for a moderately stiff dough, all the yeast, the malt, and the yeast foods are added to the mixer and combined. This sponge is fermented for 3 to 4 hours, then returned to the mixer and combined with the remainder of the flour and water, the shortening, the sugar, the milk powder, and the salt. The sponge method of baking bread produces a crumb of finer texture and with smaller gas holes than that obtained when the straight dough method of bread baking is used.

After fermentation, or after fermentation and mixing, the dough is divided into pieces that will eventually make up the finished loaves. This is done by a machine that measures the dough by volume and cuts off pieces of the desired size. When cut, the dough has an irregular shape with cut ends through which the leavening gas can escape. It is immediately dusted with flour and rounded. This is done by machine. Rounding dries the surface with flour and closes up the cut ends, thus preventing the escape of gas. The rounded dough is then carried on a belt to a proofer where it is transferred to another belt and held at 80°F (26.7°C) and 76% relative humidity for a period during which the dough relaxes and increases in volume as more carbon dioxide is produced. The pieces of dough are then molded and shaped by machine. In

Hard Wheat Flour, Water, Yeast
Sugar, Shortening, Monoglyceride,
Milk Powder, Salt, Malt, Mineral
Salts

↓

Mix

↓

Ferment at 80°F and High
Relative Humidity

↓

Divide into Loaf Size Portions

↓

Dust with Flour and Round

↓

Proof at 80°F and 76% Relative
Humidity

↓

Mold into Final Shape

↓

Put into Bake Pans

↓

Bake in Oven with Increasing
Temperature

↓

Cool

↓

Slice

↓

Package

Figure 19.1. Manufacture of bread (straight dough method).

this process, the floured dough is rolled out into a sheet, curled into a loose cylinder, again rolled, and the ends sealed. In some operations, two cylinders of dough are twisted together in the molding operation. The cylinders of dough then fall into pans conveyed to the oven for baking. During the first stages in the oven, the dough continues to ferment and increase in volume. As the dough passes through the oven, the temperature is increased, further expanding the dough (increase in volume of the gas due to temperature), and eventually the gluten is set by the heat, the starch first gelatinized and then set by the heat, and some water and ethanol are evaporated. Eventually, the outer layers of the dough become browned to form the crust. Browning is probably due to both the reaction between proteins and sugars, and caramelization of sugars (see Chapter 9). After baking, the loaves of bread are cooled as they are carried through air-conditioned tunnels. The loaves are cut into slices by machine, and the sliced loaves are packaged automatically by bread-wrapping machines.

Bread and other baked goods are subject to spoilage by molds. Therefore, small amounts of mold inhibitors, such as sodium or calcium propionate, are used and allowed at levels of 0.32 parts per 100 parts of flour (0.32%) in white bread and 0.38% in whole wheat products. Because water and other ingredients are used in bread, the actual concentration of these inhibitors in the finished loaf is much lower than in the flour. Mold inhibitors not only delay the growth of molds in bread, but they also inhibit the growth of certain bacteria that produce a slime in the crumb of the loaf, a condition known as "ropiness."

Another method of making bread is the batter-whipped process. In this method, a liquid mixture of water, sugar, yeast, milk powder, salt, and yeast food; small amounts of flour; and some vitamins is fermented from 2 to 3 hours. It is then cooled prior to mixing with liquid shortening and the bulk of the flour. The mixture is then agitated at high speed in a developer, which incorporates air. The dough is then extruded directly into baking pans which, after a short proofing period at 80°F (26.7°C), are conveyed directly to the baking oven. Automation in bread manufacture has led to higher production volumes, lowered production costs, shorter production time, and better control over the properties of the finished product, making it easy to produce large volumes of uniform bread that can meet whatever specifications are desired. Bread made by the continuous process is finer in texture, but less flavor is developed in processes of this type.

Standards for bread and flour, shipped in interstate commerce and labeled "enriched," were initiated by the federal government in 1952. These standards require the enrichment of bread with thiamin (vitamin B_{-1}), riboflavin (vitamin B_{-2}), niacin, and iron. Calcium and vitamin D may also be added as enrichment agents.

CAKES AND COOKIES

Leavening in cakes and cookies is produced chemically, instead of by yeast fermentation. The chemicals used to produce carbon dioxide are sodium bicarbonate (the source of carbon dioxide) and sources of acid to react with the bicarbonate, such as potassium hydrogen tartrate (cream of tartar), sodium hydrogen pyrophosphate, calcium hydrogen phosphate, and alum (sodium aluminum sulfate). The desirable chemical leavener produces small bubbles of gas at a constant rate consistent with the period involved in mixing and baking to temperatures that set the structure of the cake. Cookies are formed with dies or by extrusion, and, in many instances, the process is quite complex, as the dough may have to be extruded around a central component of fig paste or other type of filling.

Many prepared mixes are also produced, some of which may be used by bakers, although many are used by the home baker. Soft wheat flour or flour of low or moderate protein content is used for cakes. In premixes, the flour, egg powder, shortening, fruit or flavoring components, and leavening agents are combined in the dry state, although, when mixed, the shortening is in the melted or liquid state, and emulsifiers, such as monoglycerides, which improve air incorporation during mixing, may be used. Much of the successful preparation of cake mixes depends on the kind and quantity of chemical leaveners used. Angel cakes contain egg white and are leavened by the incorporation of air into the mixture. During baking, the air in the cake expands and acts as the leavening agent.

DOUGHNUTS

The ingredients for dough used in the manufacture of doughnuts are similar to those for cake, especially pound cake, except that some doughnuts are leavened with the use of yeast. After mixing, the dough is extruded, cut into doughnut form, and cooked in hot oil (370 to 380°F [187.8 to 193.3°C]). Fat absorption is reported to be about 15%. Fat absorption may be higher when processing parameters (e.g., temperature of the dough) are not controlled, resulting in greasy doughnuts. When fat absorption is insufficient, the keeping quality of the doughnuts is diminished. When doughnuts are to be sugared, the temperature and relative humidity must be controlled (70 to 75°F [21.1 to 23.9°C] and 85%, respectively) for optimum sugar pickup. Generally, doughnuts are prepared from materials already premixed elsewhere, and seldom do the manufacturers of doughnuts mix their own formula.

CRACKERS

Crackers are unleavened or only slightly leavened. The flour used for these products consists of wheat flour, although some rye flour is often used. For producing crackers, flour, liquid shortening, salt, and small amounts of a chemical leavening agent, with or without sugar, and sometimes a flavoring agent, such as onion powder, are mixed into a dough. The dough is then extruded into the desired shape and baked without proofing. Milk powder, whey powder, and emulsifiers, such as monoglycerides, may be used with some combination of the ingredients listed.

PIE CRUSTS

Pie crusts, which are also unleavened bakery products, can be made from all-purpose flour, but highest-quality pie crusts are obtained from unbleached soft whey pastry flour of low-protein content. When the protein content is too high, the desirable flakiness characteristic is minimized, and, to compensate, a larger amount of shortening must be used. The shortening (see Chapter 24) should be a solid or hydrogenated one (e.g., lard or hydrogenated vegetable fat), and it should be medium firm. Milk powder or fluid milk may be used in small quantities to enhance color. Eggs have the same effect. Salt and sugar may be added. The composition of a high-quality pie crust is approximately 47.5% flour, 1.0% salt, 3.3% nonfat milk powder, 2.0% glucose, 32.4% shortening, and 13.8% water. To ensure a high-quality crust, all ingredients should be mixed at a temperature of 60 to 65°F (15.6 to 18.3°C), and the ingredients combined with a minimum of mixing and handling.

LOW- AND NO-FAT BAKERY ITEMS

Food technologists and bakers have worked together to develop bakery products that have attributes such as mouthfeel, texture, and other physical and sensory qualities similar to many of the popular items that use fats and oils in their formulations. It is quite easy to find fat-free cookies, breads, cakes, and muffins in virtually any supermarket. The fat-free commercial muffin mixes are the most popular items on many bakery supply house product lines.

In formulating low-fat and fat-free items, the food technologist must first address all the functions of the fat system. For example, some functions of shortening (a fatty ingredient) in baked goods are to assist in aerating the batter and increasing its viscosity, both critical in development of a fine cell structure. Richness, tenderness, and prevention of starch retrogration are also important functions of shortenings. Modified starches have been used (see Table 21.1) to assist in controlling viscosity, building texture, increasing volume, aerating the batter, and controlling the moisture to extend shelf-life of the baked products. The proper modified starches are a good choice as they reduce the calories by 5 per gram and eliminate the fat. Formulations to eliminate all of the fat or part of it can be developed.

To achieve the desired end results, gums, fruit and cereal fibers, fruit purees, sugars, polyhydric alcohols, as well as modified food starches are among the ingredients that have been used in a wide variety of combinations as substitutes for the fats and oils. Emulsifiers such as lecithin are also added to some formulas to improve texture and volume of the finished product. Lecithin is well known as a constituent of eggs but it also is a byproduct of soybean oil refining. It is a lipid (a phospholipid) and must be labeled as a fat. In some formulas, it may add 0.5 g of fat per serving to the end product. This product cannot contain the fat-free claim but could still contain a low-fat claim (less than 3 g per serving).

20
Fruits and Vegetables

FRUITS

Fruits are botanically classified as those plant parts that house seeds; in other words, they are mature plant ovaries. Fruit includes tomatoes and a few others that are considered as vegetables in the supermarket. Because the popular definition of fruit applies only to what is naturally sweet and what is normally used in desserts, it is understandable that, for example, tomatoes and olives are treated as vegetables.

Berries belong to a class of fruit that are usually small and very delicate. On the other hand, melons as a class are usually large, often with a tough, and sometimes thick, outer skin. Fruits are vulnerable to a variety of diseases and infestations and, therefore, require spraying with protective chemicals during the growing season.

Fruit is often picked prior to maturity, and allowed to ripen in the distribution chain, reaching the consumer when about ready to eat. Fruit is considered to be ripe when it reaches the optimum succulence and texture and there is a desirable balance between sugars and acidity, as well as the subtle elements that contribute to aroma. Fruit that goes past its optimum ripeness enters senescence, a stage of overripeness and breakdown. In this stage, the texture loses its firmness; succulence is diminished; and sugars, acids, and aroma elements generally all decline in concentration. Some fruit, such as the banana, has an early senescence, and deterioration, once begun, is rapid. On the other hand, some fruit, such as apples, resists the onset of senescence as well as its progression. The onset of senescence can be controlled by keeping the fruit at the lowest temperature that it can tolerate and by increasing the amount of atmospheric carbon dioxide to a controlled level. Too much carbon dioxide, however, can be harmful.

All fruit preserved by canning should be heat-processed to attain commercial sterility. Whereas vegetables and certain other foods require the application of high temperatures (240°F, 250°F [115.6°C, 121.1°C] or higher) for significant lengths of time to attain commercial sterility, most fruit is sufficiently acid (pH usually below 4.6) that commercial sterility can be attained by heating the containers in boiling water to the point where all parts of the product reach a temperature of 180 to 200°F (82.2 to 93.3°C).

All fruit preserved by freezing should be brought to a temperature of 0°F (–17.8°C) or below during freezing and thereafter held at 0°F (–17.8°C) or below until sold to the consumer. Fruit may be packed in retail-sized containers and frozen in one of three ways: (1) the containers are placed on trays, the trays are placed on racks and frozen as the racks are moved through a tunnel in which blasts of cold air are circulated; (2) the cartons are placed on chain mesh belts that move slowly through a tunnel in which blasts of cold air are circulated; or (3) the cartons are placed on trays, the filled trays are placed between refrigerated metal plates, and the product is frozen with the containers in contact with the cold plates. There is a large variety of fruit used in the

United States with varying degrees of popularity. Following are descriptions of planting, harvesting, processing, and general handling of the more important fruits.

Grapes

In the United States, grapes are grown mainly in California, New York, Pennsylvania, Michigan, and Ohio. Washington, Missouri, and Arkansas also produce some grapes. Grapes are utilized to produce unfermented grape juice, vinegar, wine, raisins, jams, and jellies, and as the fresh product. Grapes are planted as vines or cuttings from older plants. The cuttings produce arms bearing fruit, the greatest yields coming after 3 years of growth. Properly pruned and cared for, vines produce fruit for many years.

Grapes to be shipped as fresh are packed in wooden crates, then precooled to about 40°F (4.4°C) in railroad cars or refrigerated rooms. Generally, the grapes will be fumigated with sulfur dioxide prior to or during cooling to prevent mold growth. Grapes that are to be stored for future shipment should be packed in crates precooled to 36 to 40°F (2.2 to 4.4°C), placed in refrigerated storage (29 to 32°F [–1.67 to 0°C]), fumigated with sulfur dioxide, and held in this manner until shipped. Periodic refumigation with sulfur dioxide may be required to prevent spoilage by molds. Under these conditions, grapes have a storage life of 1 to 7 months depending mainly on the variety. See Chapter 13 regarding the use of sulfite.

The Concord variety of grapes is chiefly used for the manufacture of grape juice. The grapes are washed in acid or alkaline solutions, then in water to remove spray residues, then destemmed and crushed by mechanical means. The crushed grapes are heated to about 180°F (82.2°C) to extract pigment from the skins, after which the heated material is subjected to mechanical pressure while enclosed in cotton press cloths. The juice is then filtered, pasteurized by heating to 170°F (76.7°C), and stored in bulk in covered tanks at about 40°F (4.4°C). This provides for the separation of tartaric acid salts (cream of tartar or potassium hydrogen tartrate). The juice is then siphoned off from the tartrate and treated with enzymes, which break down pectins, or with casein for purposes of clarification. It is then filtered and bottled. The bottles are capped and then pasteurized by heating in water at 170°F (76.7°C) for 30 minutes (see Fig. 20.1).

To produce grape juice concentrate, grape juice is subjected to heat and evaporation until concentrated. Volatile components can be recovered from the process by specific distillation procedures, and they can be added back to the concentrate. The concentrated juice may be frozen in retail-sized containers, or it may be shipped in large tanks to food processors for remanufacturing purposes.

Considerable quantities of wine are manufactured in the United States. The European varieties of grapes are mainly used for making wine. There are various procedures used in the production of wines but, generally, the juice is pressed from the grapes and usually treated with sulfur dioxide or compounds that liberate sulfur dioxide (e.g., sodium or potassium meta bisulfite) to destroy undesirable types of yeast and bacteria. Then a pure yeast culture is added after the sulfur dioxide treatment is complete (several hours to overnight). The juice is allowed to ferment and sugars are converted to ethyl alcohol and carbon dioxids until a level of 12% to 14% alcohol is reached. The wines are "racked" (transferred to different vessels to facilitate clarification) a number of times during the process and then may be filtered before bottling. The wines may then be aged in vessels (wooden, stainless steel, or glass) to develop characteristic

Fruit Juices

Fruit
↓
Wash
↓
Inspect (remove unsuitable fruit)
↓
Crush
↓
Press (remove residue)
↓
Juice (press again)
↓
Heat to 165°F
↓
Add Pectinase
↓
Clear Juice
↓
Pasteurize
↓
Fill Bottles Hot

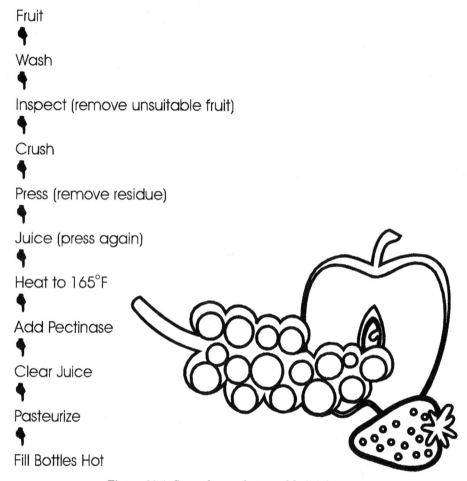

Figure 20.1. General manufacture of fruit juices.

flavors and aromas before bottling or they can be bottled and aged in the bottles (see Fig. 20.2).

Fruit brandies and fortified wines usually contain about 20% ethanol by volume (40 proof). Because natural yeast fermentation cannot yield beverages containing these high ethanol levels, more ethanol must be added. This ethanol can be obtained by distilling wine and using the nearly pure ethanol to fortify wines and brandies to obtain a higher alcohol content.

In the manufacture of wine vinegar, fermented grape juice (containing alcohol) is allowed to drip over wood shavings in an enclosed cylindrical container. The shavings had been previously soaked in a high-quality vinegar. Air may be introduced into the generator under pressure. Bacteria of the *Acetobacter* group present (from the vinegar)

Wine

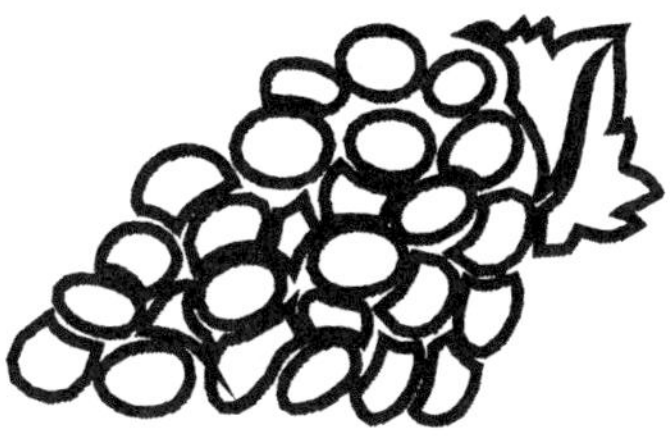

Grapes

↓

Crush and Remove Stems (Mixture called "Must")

↓

Acid and Sugar Measured and Corrected

↓

SO_2 Added (Sulfiting) and Allowed to Stand Overnight

↓

Culture of Selected Wine Yeast Added

↓

Primary Fermentation Starts and CO_2 Cap Formed

↓

Liquid Drawn into Settling Vats which Allow CO_2 to Escape but Does Not Allow Oxygen In

↓

Wine May Be "Racked" into another Vat Leaving "Lees" (Sediment) Behind

↓

Complete Fermentation

↓

Filter Wine

↓

Aging Can Be Done in Vats, Oak Barrels, or Bottles.

Figure 20.2. General method of wine production.

on the shavings convert the ethyl alcohol in the wine to acetic acid. The effluent from the vinegar generator may be collected and recycled to obtain a complete conversion of the ethyl alcohol. The finished vinegar may be stored for several months at 40 to 50°F (4.4 to 10°C), then filtered, bottled, and pasteurized.

Raisins are produced from grapes by sun drying and artificial drying. In sun drying, the grapes are picked as bunches and placed in a single layer on wooden trays between the rows of grape vines. The trays are tilted to face the sun. After they are partially dried, the grape bunches are turned and allowed to dry to the point where no juice can be pressed out. The trays are then stacked, and the air drying is continued in the shade until a moisture content of about 17% is reached. After drying, the raisins are

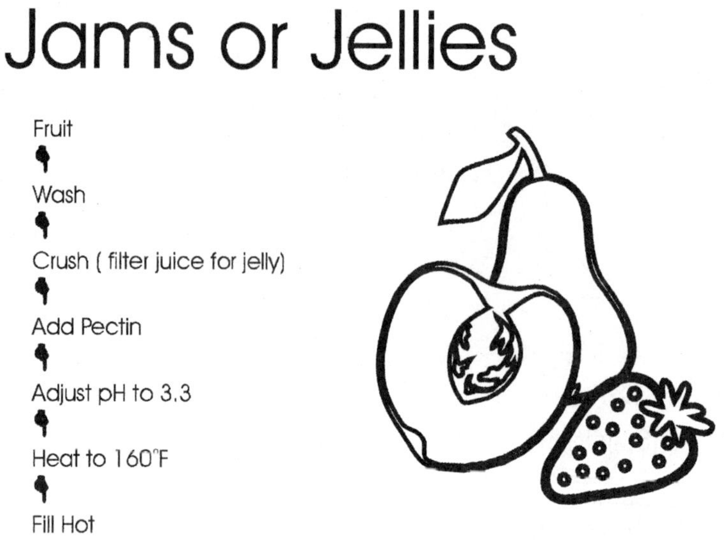

Figure 20.3. General method for the production of jams and jellies.

placed in sweat boxes to equilibrate, or to even out, the moisture that is present, and then they are packed in retail-size containers or in larger containers to be sold to the bakery trade.

In artificial drying, grapes are first dipped in 0.25% to 1% lye (sodium hydroxide) solution at 200 to 212°F (93.3 to 100°C) for 2 to 5 seconds to remove a natural wax that impedes drying and to check or crack the skin of the grape to facilitate drying. They are then washed, placed on trays, and treated with sulfur dioxide to prevent enzymatic and nonenzymatic browning during drying. Hence, instead of having the dark brown color of raisins that would normally be expected, the raisins will be of a light yellow color when dried. Raisins are dried at temperatures not exceeding 165°F (73.9°C) and at a low relative humidity (about 25%). After the moisture content has been lowered to about 16% to 18%, the raisins are packaged in containers of various sizes to be sold at retail or for use in the production of fruit cocktail. Muscat, Thompson seedless, and Ribier varieties are best for freezing. The grapes are precooled, washed, destemmed, sorted, and packed into 30-lb (13.6-kg) containers, using 18 lb (8.2 kg) of grapes and 12 lb (5.4 kg) of 55 Brix (% soluble solids) syrup, and then they are frozen.

Some grapes, for example, the Concord variety, are used for producing jelly and jam, with jelly making up the larger part. In jellymaking, sugar and about 0.25% to 0.3% of dispersed pectin are mixed with the clarified grape juice, and the mixture is concentrated in open kettles to a soluble solids content of about 65%. Citric acid solution is added to adjust the pH to 3.0 to 3.2. The product is then poured into glass jars, vacuum capped, and sprayed with hot water to bring the temperature of all parts to about 160°F (71.1°C), after which the product is cooled (see Fig. 20.3).

Apples

In this country, apples are grown in practically every state. They are grown commercially in 35 states, with the heaviest production taking place in Washington, New York, Virginia, Michigan, California, and Pennsylvania. There are hundreds of varieties of apples that may be grown on trees produced from seedlings that were grown in nurseries, or from grafts on existing apple trees. The fruit is developed on spurs formed by branchlets of 3 or more years of growth, the tree yielding fruit for many years thereafter. Fertilization of the soil and periodic pruning and thinning of apple trees are considered necessary for good apple crops.

Apples are used as the fresh fruit, and apples not suitable for fresh fruit are used for the production of juice, cider, sauce, vinegar, jam, jelly, pie filling, and as an ingredient in a variety of baked goods. Pectin is extracted from the peels and cores.

After harvesting, the fruit is washed in dilute HCl or NaOH solutions to remove spray residues and are rinsed. Apples are cooled to, and stored at, 32°F (0°C) until shipped. Increased storage life is attained by reducing the amount of oxygen and increasing the amount of carbon dioxide in the atmosphere around the apples.

For the production of apple juice see Figure 20.1. Some juice is not processed beyond the pressing stage and is sold as such or as unpreserved cider.

When vinegar is produced from apple cider, the juice in tanks is seeded with cultures of yeast and allowed to ferment for up to several weeks, depending on ambient temperatures. The fermented juice, containing ethyl alcohol, is then mixed with some vinegar and allowed to drip over wood shavings that have been soaked in high-quality vinegar and enclosed in a wooden cylinder (closed but not airtight). Bacteria on the wood shavings convert the ethyl alcohol to acetic acid. This is described in the section on the production of vinegar from wine. The effluent from the generator may be run through a second time. When the alcohol has been converted to acetic acid, the vinegar is filtered (if necessary), filled into bottles, and the bottles are capped and heated in water until a temperature (in all parts) of about 165°F (73.9°C) is reached.

For the production of apple slices, apples are size-graded, peeled, cored, sliced, and immersed in a 3% salt solution. Just before filling into cans, the slices are rinsed to remove salt. They are then filled into cans with 40% sugar solution. The cans are then heat-exhausted, sealed, and heat-processed.

For the production of frozen apple slices, there are a number of procedures, but, basically, the slices are immersed in brine, subjected to vacuum (to remove air), reimmersed in brine, washed, and packed with sugar in a ratio of 4 fruit to 1 sugar. The product is then frozen. One method differs in that a bisulfite dip is included to prevent nonenzymatic browning.

For drying, the peeled, cored, and sliced apples are first treated in a weak solution of citric acid and a bisulfite dip. The bisulfite provides sulfur dioxide which inhibits enzyme browning. The sulfured slices should be held in refrigerated storage for at least 24 hours to allow the sulfur dioxide to penetrate the apple slices. The apple slices are eventually spread on the slatted floors of natural-draft, loft-type kilns. In the kilns, heated air rises through the apple slices and removes moisture. After the moisture content has reached about 10%, the apples are packaged in moistureproof containers to be used in the bakery trade.

For the production of apple sauce, peeled and cored apples are sliced, cooked in steam, pulped, and heated to 190°F (87.8°C) with added sugar. The sauce is filled into glass jars that are then sealed and heated to 200°F (93.3°C), then cooled.

Pectin or pectin solutions may be manufactured or obtained from dried apple peels and cores. Fruit contains pectin substances in a form that is not water dispersible, and in this form, cannot be used for the manufacture of jams, jelly, and other foods. Some of the pectin substances in fruits are in the form of protopectin ([a more complex, longer chain of the carbohydrate galacturonic acid (a six-carbon compound]). To obtain pectin that is dispersible in water, the dried apple substances are heated in boiling water for about 40 minutes. The heat, together with the acid present in the fruit, hydrolyzes the protopectin to pectin, a substance that contains fewer galacturonic acid units than does protopectin. In the presence of large proportions of sugar (about 65% soluble solids) and acid, to provide a pH of around 3.0, pectin will form a gel such as that which provides the semisolid form for jams and jellies. After extracting the pectin from apple peels and cores, the solution is separated, and the residue is pressed between cloths to obtain more of the extract. The extract is then filtered, bottled, and pasteurized as a pectin solution to be used as such, or it may be spray dried to a moisture content of about 5% and handled in this form. When used, the dried material (pectin) must be dispersed by fast agitation in water, or mixed with granulated sugar to disperse in water, before addition as an ingredient of foods. If added to foods as such, pectin forms lumps that become caked with gelatinous material on the outside, preventing the pectin from dispersing.

Low-methoxy pectin may be produced from pectin by treating solutions of this material with an enzyme (pectin methyl esterase). This treatment removes methyl groups from the ester group of the galacturonic acid unit. Low-methoxy pectin has the property of forming gels in the presence of comparatively low concentrations of soluble solids (sugar) and at comparatively high pH (as high as pH 6.5), provided that a source of calcium is present. Low-methoxy pectin solutions are dried in the usual manner. Low-methoxy pectin (less than 50% methoxylated) permits the production of low-calorie jams and jellies, because of the lowered sugar requirement for obtaining the gel.

Some apple jelly is produced. Although the unclarified juice is high in pectin, after clarifying, most of the pectin has been removed; therefore, sugar, some water, and about 0.4% pectin are added and mixed. The mixture is then concentrated, by heating in open kettles, to a soluble solids content of about 65%. Citric acid solution is added to regulate the pH to 3.0 to 3.2, and the jelly is placed in glass jars, capped, pasteurized, and cooled in the usual manner.

Bananas

Bananas are not grown commercially in the continental United States, but some are grown in Hawaii and shipped to the mainland. In the western hemisphere, the chief production of bananas occurs in Mexico and Central American countries, in Cuba, Jamaica, Haiti, the Dominican Republica, Honduras, Colombia, and Brazil. Bananas are also grown in some Asian and Middle Eastern countries.

Banana trees are started from young plants that bud from the underground stem or bulb of older plants. The trees bear mature fruit 13 to 15 months after planting, depending on climate, and each plant requires an area of 100 to 400 ft^2 (9.3 to 37.2 m^2), depending on soil and water conditions. The trees develop flowering stalks with male and female flowers, and the female flower eventually becomes the fingers (single bananas) of the hand. Only one stem (bunch) of bananas is produced per tree.

The banana stem or bunch contains 6 to 14 hands (clusters of single bananas on the stem) and weighs 30 to 130 lb (13.6 to 59 kg). The stem is harvested when the single bananas are mature but green. The tree is then cut down. Bananas are handled mainly as the fresh fruit and shipped from the growing area while still green. At one time, bananas were shipped as bunches, on the stem, but today, they are handled mostly as hands or groups of single bananas, cut from the stem and packed in plastic-lined boxes. They may be treated with fumigants prior to boxing and should be precooled to 57 to 62°F (13.9 to 16.7°C). Bananas are subject to a chilling injury if held below 55°F (12.8°C); low temperatures kill certain surface cells and prevent normal ripening. Nor should bananas be held for extended periods at temperatures above 70°F (21.1°C); they must, therefore, be shipped under controlled temperatures. Bananas are ripened at 58 to 64°F (14.4 to 17.8°C), but the process can be accelerated by subjecting the fruit to ethylene gas.

Although bananas are handled mostly in the fresh state, some are peeled, pureed, and canned for the bakery and soda fountain trade. In such cases, the pureed banana pulp is quickly heated to about 280°F (137.8°C) in a heat exchanger, held at this temperature for a few seconds, cooled in a heat exchanger, filled into sterile, large-sized cans, and sealed under aseptic conditions. In such an operation, all parts of the process must be accomplished so that the cooling, filling, and sealing operations are done with the equipment presterilized and the containers sterilized just prior to filling. The cans must also be sealed under high-temperature steam or inert gas. Essentially no bananas are frozen but some are dried and sold as snack items.

Oranges

The orange is utilized as a food to a greater extent than any other citrus fruit. The trees are sent out from nursery stock and must be protected from freezing weather. As with other citrus trees, some pruning has to be done each year. The five states that produce oranges commercially are Florida, California, Arizona, Texas, and Louisiana, with Florida the greatest producer. In the United States, about three-quarters of all oranges are used for the production of frozen juice concentrate and for the so-called "fresh" orange juice. The oranges should be picked when the solids-to-acid ratio is between 12:1 and 18:1. Oranges may be dyed by immersing in a solution of certified food dye at 120°F (48.9°C) for about 3 minutes prior to waxing, polishing, and cooling, because the color of the skin is often green when the fruit is picked. Some oranges are cooled to 32 to 40°F (0 to 4.4°C) and others to 40 to 44°F (4.4 to 6.7°C), depending on variety. They should be held at these temperatures until sold to the consumer. Under these conditions, they have storage life of 1 to 3 months, depending on variety.

Much orange juice is frozen in a sequence as shown in Figure 20.4. The addition of orange juice is to add back flavor, which is lost during the vacuum step.

Orange juice may be frozen in slabs and eventually defrosted and shipped as a slush to be pasteurized and sold as the single-strength product. Some orange juice is canned, requiring no refrigeration from a public health standpoint. The process involves the extraction of orange juice from the fresh fruit and collecting it in tanks where it is adjusted for acidity and soluble solids content. The juice is then pasteurized to destroy microorganisms and to inactivate enzymes and it is filled into cans or glass containers while hot. Some plastic type bottles have been developed that will withstand the pressures of vacuum packaging and are being used by some processors. It is then rapidly cooled to 100°F (37.8°C). Although the product is microbiologically stable at room temperature, there are slight reductions in the nutritional and organoleptic qualities of the product during extended storage at room temperature. These losses can be prevented by storing at lower temperatures—the lower the better. Pectin and low-methoxy pectin may be manufactured from orange peel, as in the case of grapefruit.

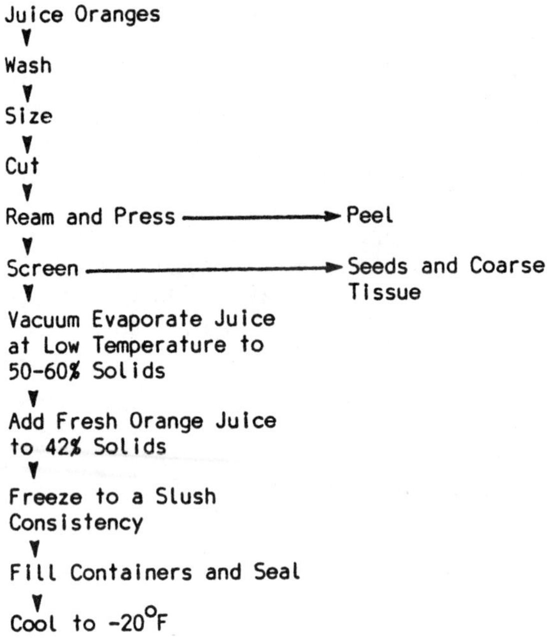

Figure 20.4. General production method for concentrated orange juice.

Aseptic Packaging

To preserve top quality and to supply a more space-efficient package, many juices and juice blends are packaged in composite paper cartons and processed and filled aseptically. In Chapter 15, aseptic packaging in the dairy industry is discussed. In the case of fruit juices, the juice is pasteurized and then cooled very rapidly before packaging. The cartons are made (from the outside in) of layers of polyethylene, paper, polyethylene, aluminum foil, polyethylene, and a coating of ionomer resin. The sheets to be made into the cartons are drawn through a hydrogen peroxide bath to start the sterilization process and then passed through squeeze-rollers to remove excess peroxide. The sheets are then formed into a tube and exposed to radiant heat to complete sterilization and remove any traces of peroxide. The tubes are then formed into a rectangular shape, end-sealed, and filled with the cool prepasteurized liquid. They are then top-sealed and separated into individual package units. The shelf-lives of these products can be as long as several months at room temperature with less loss of nutritional value and organoleptic properties.

Coffee

Coffee is included here because it is the fruit of the coffee tree that contains the beans that are processed into the beverage that is consumed by people in the United States at an estimated annual rate of more than 28 gallons per person. The fruit turns red as it ripens and is referred to as "cherries" (see Fig. 20.5). Each cherry contains two beans and it takes about 3000 beans to yield 1 lb (454 g) of finished ground

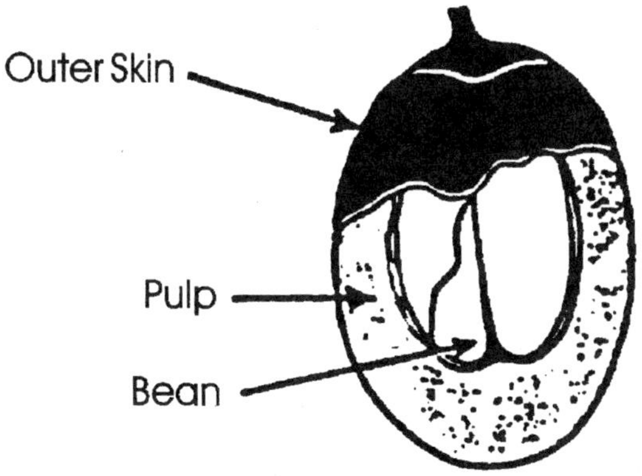

Figure 20.5. Structure of the coffee cherry.

coffee. In processing the cherries, the pulp is removed from the beans. This leaves a mucilaginous coating on the beans that must also be removed. This can be done in a number of ways including by microbial fermentation, use of pectinases, or various washing treatments.

The beans now contain only an outer hull and are dried to about 12% moisture. Drying is done outside in the sun or by machine. During drying, color and flavor attributes are developed and modified. After drying to 12% is complete, the hulls are removed and the beans are sorted for color and defects. The sorted beans are then graded for size and color and are "cup tested" for brewing quality. This is done by roasting, grinding, and brewing small samples of beans.

Beans are shipped in the green state (unroasted) and the processing by the coffee manufacturer includes blending, roasting, grinding, and brewing. After blending for desired types of coffee, the beans are roasted at temperatures of about 500°F (260°C) for about 5 minutes to develop their characteristic flavors. The bean temperature is raised to about 392°F (200°C) and all free moisture and some volatile chemical substances are removed.

The roasted beans are ground to the desired size depending on the intended use. Percolater, drip, large restaurant urns, vending machines, and that to be used for instant coffee all require different size grinds. After grinding, the aroma and flavor properties are highly unstable to oxygen and to loss of volatiles. To preserve the quality of the coffee, it may be packaged under vacuum in an inert gas in cans or jars. Ground coffee gives off much CO_2 so it must be allowed to outgas before packaging.

Brewing of the coffee involves a hot water extraction of the ground coffee. The strength and flavor of the coffee are controlled here.

Instant Coffee

The three major steps in the production of instant coffee are extraction, dehydration, and aromatization. Extraction is done with a series of percolators that are run at different temperatures. This ensures minimum heat damage, extraction of bitter components, and maximal extraction. The extract from the percolators is cooled rapidly and dehydrated immediately if possible because aroma and flavor can deteriorate quickly.

Dehydration is usually done by spray drying in driers specially designed for coffee. Freeze-drying is also used and a significant portion of all instant coffee produced in

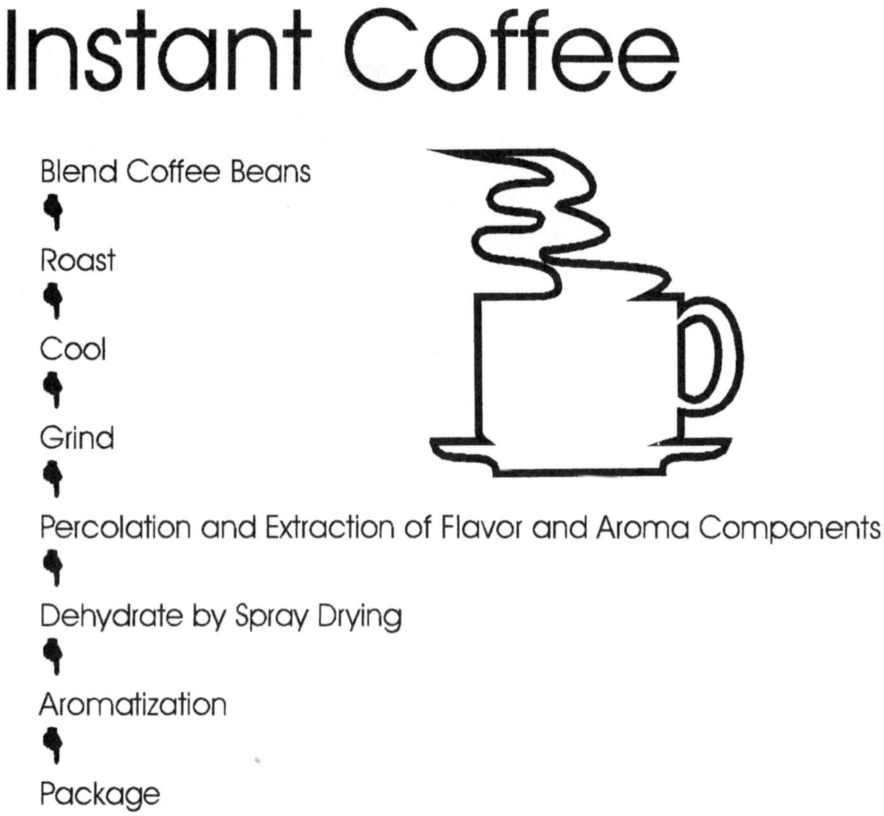

Instant Coffee
Blend Coffee Beans
Roast
Cool
Grind
Percolation and Extraction of Flavor and Aroma Components
Dehydrate by Spray Drying
Aromatization
Package

Figure 20.6. Simplified version of production of instant coffee.

the United States is freeze-dried. With this method, which is milder than the spray drying method, maximum retention of flavor and aroma is achieved. It is, however, a more expensive process.

During roasting, grinding, and extraction, escaping flavor and aroma constituents are trapped and recovered. These flavor and aroma components have also been obtained from oils pressed from the coffee beans. These components are added back to the dehydrated coffee (see Fig. 20.6).

Decaffeinated Coffee

Brewed coffee contains about 75 to 100 mg per 5-oz (150-ml) cup. Caffeine has a stimulating effect and also may produce insomnia, nervousness, and other physiological responses in some people. Decaffeinated coffee contains only about 3 mg per 5-oz (150-ml) cup, so the side effects attributed to caffeine should be reduced.

Coffee beans are decaffeinated by steaming green beans and extracting with water prior to roasting. Organic solvents such as petroleum ether may also be used to remove the caffeine but this leaves the problem of removing the residual solvent and recovery of the solvent vapors. A process involving the use of high-pressure CO_2 at a low temperature allows the removal of caffeine, as, under the proper conditions, CO_2 has the solvent power of a liquid but the penetrating power of a gas. Lower temperatures are used in this method and there is no need to remove any solvent residues.

Other Fruits

Other fruits include blackberries, blueberries, cranberries, raspberries, strawberries, apricots, cherries, grapefruit, lemons, melons, peaches, pears, pineapples, and plums.

Blackberries, raspberries, and strawberries are very popular fresh and in jellies and jams. They may also be found in fruit juice blends or packaged and frozen with or without sugar added. Apricots and plums are used fresh but many are also dried and some frozen. Blueberries and cherries are marketed fresh but are also quite popular as canned pie fillings. Peaches and pears are usually sold either fresh or are canned in light or heavy sugar syrups. Grapefruit, pineapples, and lemons are popular in juice production or used fresh. Pears can be processed much like apples and are used in ciders and juices, and may be canned.

All of the fruits can be canned, dried, and frozen using methods described in earlier chapters and adapted for the individual fruit.

VEGETABLES

Vegetables are plant foods that include various edible parts such as leaves, shoot, roots, tubers, flowers, and stems. They normally do not include fruit, which was covered earlier in this chapter. However, tomatoes and olives, which are technically fruit, are included in this section, because their culinary role is related more to vegetables than to fruit. Vegetables belong to an important class of foods that supply many nutritive requirements, including proteins, starches, fats, minerals, sugars, and vitamins. Vegetables also supply fiber to the diet, as well as a large variety of flavors, aromas, and colors that provide the knowledgeable chef with a repertoire of culinary tricks. On a worldwide basis, vegetables make up a considerable part of the human diet, with the largest proportion of the total vegetable crop being consumed in the fresh state (not preserved). In the United States, however, the proportion of processed vegetables consumed is quite high. In the fresh state, vegetables continue to carry on life processes after harvest, and they are susceptible to various forms of deterioration, hence requiring sanitary handling under controlled temperatures.

The holding of vegetables after harvesting is detrimental to their quality. They undergo microbial spoilage, they lose water, in many cases they lose sugar and other nutrients (e.g., vitamins), and they give up considerable energy in the form of heat (value reported is more than 100,000 Btus per ton per day [27,784 /Cal (kg)/MT/day]). Of course, by the heat they produce, vegetables hasten their own deterioration by the acceleration of microbial action as the temperature is increased (within limits). Enzymatic deterioration, especially at sites where bruises occur, is also accelerated by higher temperatures (again within limits).

Various procedures are involved in the production, preservation, and distribution of vegetables. The planted seeds may have been treated, the soil has been fertilized, the crops have usually been sprayed or dusted with insecticides or fungicides, and finally harvested and subjected to one of the various methods of preservation. The finished product is then distributed to the retailer or held in a temperature-controlled warehouse until it is distributed. Vegetables are sold at retail as fresh produce, as heat-processed and canned, as frozen, and occasionally as dried products.

Seeds are treated with fungicides or insecticides to prevent loss or decay to either insects or fungus prior to germination and growth of the plant. Bean and pea seeds may also be inoculated with bacterial cultures that take nitrogen from the air and make it available to the plant, which requires the nitrogen for growth.

Various growth fertilizers may be applied to the soil prior to planting. Generally, fertilizers consist of some combination of nitrogen compounds (ammonium salts, nitrates, or urea), phosphates, and potassium compounds. These chemicals, required for plant growth, are apt to be deficient in the soil. Soil is frequently treated with calcium-containing compounds to neutralize its acidity and provide a pH suitable for plant growth. Liquefied fish wastes are also sometimes used as fertilizers, and chemical fertilizers are sometimes applied in liquid form.

Vegetable crops may be produced by the organization that will eventually handle and process them or often the processor will contract farmers to grow the crops. Large processors have a field department that employs a number of horticulturists. The field department supplies the seed, specifies the kind and extent of fertilization and soil treatment required, and supplies the ingredients for these. Soil fertilization and treatment, and planting are done by the farmer. Weeding is done by actual removal of weeds or by chemical treatment, and the latter is done by the farmer or the processor. Spraying or dusting with insecticides or fungicides may be done by the farmer or the processor. Personnel from the processor's field department decide when the crop is at the right maturity for harvesting and arrange for it. If the crop cannot be handled quickly enough, the vegetables may pass their optimum maturity, resulting in a loss of product quality.

The heat unit system can be applied as the harvesting criterion for crops for which the required heat units for optimum maturity are known (e.g., beans, corn, peas). Heat units may be expressed as F-days or F-hours (C-days or C-hours). Heat units are those days or hours above the minimum growing temperature multiplied by the degrees that the ambient temperature exceeds the minimum growing temperature. For example, if the minimum growing temperature for a given vegetable is 40°F (4.4°C), and during a 2-day period the ambient temperature averages 60°F (15.6°C), then a number of heat units accumulated during the 2-day period can be calculated as:

$$(60°\text{F}-40°\text{F}) \times (2 \text{ days}) = 40°\text{F-days } (22.2°\text{C-days}) \text{ or}$$
$$(60°\text{F}-40°\text{F}) \times (48 \text{ hours}) = 960°\text{F-hours } (533.8°\text{C-hours})$$

Fresh vegetables are usually washed and held under temperature and relative humidity controls suitable for the various species. These conditions may be held during distribution, storage, and display on the retail level. If fresh vegetables are packaged for retail or wholesale distribution, special considerations must be made to allow them to "breathe" and avoid built-up condensation in the package.

Some fresh vegetables may be cut or shredded and packaged as preprepared salads or "slaws" (cabbage or broccoli). These are classified as potentially hazardous foods and must be prepared, packaged, and stored under strict sanitary conditions. Partial vacuums or other modified atmosphere packaging techniques are sometimes used here but extreme care must be taken to eliminate the possibility of *Clostridium botulinum* growth.

All vegetables sealed in cans or glass jars must be heat-processed (for typical production sequence see Fig. 20.7), usually at 240 or 250°F (115.6 or 121.1°C) and sometimes at higher temperatures to make them commercially sterile. Commercially sterile means that all disease-causing bacteria have been killed, and all bacteria and bacterial spores,

Vegetables
↓
Grade for Size
↓
Test for Suitability
(Tenderness, etc.)
↓
Wash
↓
Blanch
↓
Inspect
↓
Fill Cans
↓
Add Canner's Brine
↓
Seal Cans
↓
Heat Process

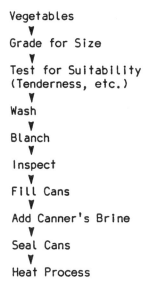

Figure 20.7. Production of canned vegetables.

which might grow out and cause spoilage under the conditions in which the product will be handled after processing, have been destroyed. The time for which a heat-preserved vegetable must be processed to attain commercial sterility depends on the specific vegetable, the size of the container, the temperature at which the product is processed, the type of container (glass, metal, or plastic), whether or not the product is agitated during heating, whether the vegetable product heats by convection or conduction, and other factors.

When a vegetable or vegetable product is preserved by freezing, it should be brought to a temperature of 0°F (–17.8°C) or below in all parts and held at this temperature or below until sold to the consumer. Vegetables are packaged and frozen on moving chain mesh belts in cold-air tunnels. Packages or vegetables are sometimes placed on trays, the trays placed on racks, and the product frozen as the racks are moved through cold-air tunnels. In other instances, vegetables are packaged, the packages placed on trays, and the trays placed between refrigerated metal plates for purposes of freezing the product. Some vegetables are loose-frozen on chain mesh belts in cold-air tunnels prior to packaging. Following are descriptions of planting, harvesting, processing, and general handling of the more important vegetables used in the United States.

Green and Wax Beans

Green and wax beans are planted from seed in the spring or when temperatures will allow. A period of 50 to 70 days is required from planting to harvesting. On harvesting, beans should be cooled to 40°F (4.4°C) and held at that temperature until sold as the fresh product, or until processed.

At the processing plant, beans are sorted for size, washed, and their ends are snipped off mechanically. They are then cut transversely or longitudinally (French-cut) and blanched. Some beans are frozen individually and packed in plastic pouches alone or as a component of mixed vegetables, or they may be first packaged and then frozen in a blast freezer. Some beans are canned in brine and heat-processed.

Broccoli

Broccoli requires 60 to 70 days from planting before it can be harvested, after which the broccoli must be cooled to, and held at, 35 to 40°F (1.7 to 4.4°C) until sold as fresh or processed. Cooling can be done by subjecting the product to cold-air blasts or by dehydrocooling (see Chapter 12). Most broccoli is sold as fresh, but some is frozen. Broccoli to be sold fresh is inspected to remove blossomed and insect-infested heads, washed, then bound at the stalk in small retail-sized bunches. The frozen product is prepared by splitting the heads to a smaller diameter and conveying to a station where they pass under a rotary cutter that trims the stalks to a length of 5 in. (12.7 cm). After washing, the trimmed stalks are blanched in free-flowing steam for 2.5 to 7 minutes (usually 3 to 5 minutes). The blanched product is cooled in water, then packaged and frozen.

Cucumbers

Cucumbers are used as the fresh product or to make pickles or relishes. A period of 55 to 60 days is required from planting to harvesting. Cucumbers, grown for the fresh market, are of a variety that produces relatively large specimens. Varieties grown for processing include relatively large cucumbers used for the production of relish and medium- and small-sized specimens for pickling.

At the processing plant, cucumbers to be used as fresh produce are washed, and may be covered with a thin layer of wax and polished by machine. They should be promptly cooled to 45 to 50°F (7.2 to 10°C) and held at this temperature until sold to the consumer. In this condition, they have a high-quality storage life of 10 to 14 days. At temperatures lower than 45°F (7.2°C), they may develop pitting or dark-colored, watery areas.

Many types of pickles are manufactured including sweet, sour, dill, and kosher-dill. In the manufacture of pickles, small- or medium-sized cucumbers are submerged in a solution of 10% salt (sodium chloride) and allowed to ferment. During fermentation, which takes several weeks, the sugar in the cucumbers is gradually utilized by bacteria, and salt penetrates the cucumber. The salt in the brine is gradually increased to 15%. (Dill-type pickles are fermented in a less concentrated salt solution.) After fermenting, the cucumbers are soaked in warm water, then packed in glass jars with some combination of vinegar, sugar, spices, and garlic. Pickles in jars should be pasteurized in hot water by bringing their temperature up to 160 to 180°F (71.1 to 82.2°C). Some cucumbers are sliced vertically or horizontally prior to packing them. Sliced pickles are covered with a weak solution of salt, vinegar, and spices, with or without sugar, and heat-processed as indicated.

Relishes may be made from fermented cucumbers that are freshened (some of the salt is removed), chopped, and mixed with vinegar and spices. Because they are used with other foods, some type of gum, such as locust bean gum, may be added to relishes for liquid retention.

Lettuce

Lettuce is used entirely as a fresh product, because if heated or frozen, the leaves wilt. This is unacceptable, because lettuce is expected to be crisp and firm when eaten.

There are a number of varieties of lettuce—the open head or loose leaf, the romaine, the iceberg, and others. Iceberg lettuce, used in the United States, requires a period of 75 to 80 days from planting to maturity. At the processing plant, lettuces are trimmed, washed, and placed in crates; the crates are placed in refrigerated freight cars or trucks. The cars or trucks are then moved into large metal chambers. The chambers are closed and subjected to vacuum, which accelerates the evaporation of moisture from the lettuce, lowering the temperature of the product to about 33°F (0.56°C) through evaporative cooling. During shipment and until sold to the consumer, lettuce should be held at 32 to 34°F (0 to 1.1°C). At this temperature, it has a storage life of 2 to 3 weeks.

Olives

Olive trees are started in nurseries as rooted cuttings and allowed to grow for several years before being set out in orchards. The desired varieties are grafted onto the seedling stock as 2- to 3-year stock. Trees are pruned and thinned yearly, and bear fruit after about 5 years. The trees may bear fruit for many years. Olives are used for the production of olive oil or they may be canned as black or as green or as green stuffed olives. The latter are usually packed in glass jars. The fruit is harvested when still green, just before turning pink. Large and unblemished olives are used for canning as black or green olives. Small and blemished fruit are used for oil or for chopped olives. Olives must be treated with lye solution (sodium hydroxide) to destroy or decrease the content of a bitter substance, oleuropein.

In the production of black olives, the sized, washed, and inspected fruit is placed in tanks and covered with a 1% to 5% lye (sodium hydroxide) solution at 60°F (15.6°C) for 4 to 8 hours, the mixture being stirred occasionally. The lye solution is then drained off, and the fruit is exposed to air for 3 to 6 hours with occasional stirring, to oxidize and set the color (to black). The fruit is then covered with water, agitated by compressed air, and held for 3 to 4 days. It is next drained and again treated with a 0.5% to 0.75% lye solution for 3 to 5 hours, then drained and washed once more. Black olives may or may not be pitted prior to canning. The prepared olives are sized, inspected, placed in cans volumetrically, and covered with a boiling 2.5% to 3% salt solution. The cans are then sealed and processed at high temperatures.

Green olives are placed in a brine (sodium chloride) solution (2.5% to 5%), the salt content being increased daily until it reaches 7.5% to 10%. The fruit is held in this manner for 30 to 45 days, during which time it undergoes a fermentation, owing to the growth of lactic acid bacteria. After washing without agitation in a tank of water for several days, the fruit is treated with lye solution, as in the case of black olives, then rewashed. The fruit is inspected, graded for size, pitted or not pitted, and canned in the same manner as black olives. Olives in large containers (No. 10 cans) must be exhausted (heated) to bring the temperature in all parts to 180°F (82.8°C) prior to sealing the cans, then heat-processed at high temperatures. Green olives may be pitted and stuffed with pimentos or nuts prior to canning. In some instances, olives (usually pitted) preserved in a 10% brine (or higher concentrations) are shipped in barrels to repackers who freshen the pickled product, stuff the individual fruit with pimentos or nuts, and repack it in glass jars. The fruit is then covered with a hot 2.5% to 3% brine, the jars sealed, and the product heat-processed at high temperatures.

Potatoes

White potatoes, of which there are many varieties, are used mainly as a fresh product. Some are dried, frozen, or canned. Other varieties are best suited to use as boiled or mashed potatoes. Some varieties make a superior baked product, while still others have good all-purpose qualities. Others are processed into snack items such as chips. Potato plants are grown from sprouted potatoes that have been cut so that each piece includes a sprout and some of the tissue. A period to 60 to 70 days is required from planting to harvesting.

Potatoes are dug by machine and trucked in bulk or in barrels to the processing plant. At the processing plant, potatoes to be sold as fresh are passed through a dry reel to remove soil. They are then washed, dried, and packed in paper or plastic bags of 5- or 10-lb (2.37- or 4.5-kg) capacity. White potatoes, held at temperatures above 50°F (10°C), lose sweetness as their sugar is converted to starch. At 40°F (4.4°C) or below they become sweeter, as their starch is converted to sugar. For ordinary purposes, when sold in the fresh state, high sugar content is not desirable. Also, if potatoes are to be canned as such or used as a component of corned beef hash or fish cakes, high sugar content is not desirable, because during heat processing, off-flavors and off-colors may develop as a result of nonenzymatic browning. On the other hand, if potatoes are to be processed as a frozen product, such as french fried or hash-browned, the sugar content should be high enough to provide some color without prolonged heating. The sugar content in potatoes that are used in large quantities for the manufacture of potato chips must be controlled. If the sugar content is too low, the desirable light brown color will not be attained during frying. If the sugar content is too high, the potato chips will burn or become black during drying.

Because of these changes in sugar content, white potatoes may have to be conditioned or held at a particular temperature for 1 to 3 weeks prior to processing. Potatoes handled as fresh may be held for 2 to 4 months at 40°F (4.4°C) without sprouting. Generally, they are held at 40 to 45°F (4.4 to 7.1°C) prior to shipment, because during transportation and subsequent handling, they will be held at higher temperatures that will cause the conversion of sugar to starch.

Frozen white potatoes are processed as french fried, baked stuffed, hash-browned, or as some other prepared product. Frozen french fried potatoes are prepared from the fresh product. The potatoes are either washed, peeled in lye, and washed in a neutralizing weak acid solution or they are heated in steam at 80 psi (5.6 kg/cm^2) for about 10 seconds and subjected to water sprays to remove the peel. They are then inspected to remove eyes, after which they are treated in a weak solution of citric acid and sodium bisulfite. This treatment prevents discoloration (browning) caused by enzyme action.

The potatoes are then cut into strips. If the potatoes have not been conditioned by holding at the temperature that regulates the sugar content to the desirable concentration, they may have to be blanched in water at about 180°F (82.2°C) for sufficient time to remove excess sugars or, if the sugar content is too low, they may be heated in weak sugar (glucose) solutions. Blanching or precooking is done in water or free-flowing steam for about 2 to 4 minutes, after which the potatoes are drained, then cooked for short periods in hot vegetable oil at 350 to 375°F (176.7 to 190.6°C). Deep-fat frying is done for only sufficient time to give the french fried potatoes a light brown color. After cooling, the potatoes are frozen.

Baked frozen potatoes are baked whole with the skin on. They are then cooled and cut in half lengthwise. The inner material is then mashed and mixed with cooked onions, margarine, and various flavoring materials. The half shells are then filled with the prepared mashed product, placed in cartons, and frozen.

Hash-browned potatoes are prepared in the manner of french fried potatoes except that they are frenched to a smaller size and cooked in oil for somewhat longer periods to provide a darker color. Hash-browned potatoes are packaged and frozen.

Some small-sized potatoes are canned. These potatoes are peeled by heating in a lye solution followed by washing in a weak acid solution. The peeled potatoes are then filled into cans, covered with a 1% to 2% salt solution, and the cans are then exhausted in free-flowing steam, sealed, and heat-processed.

Potatoes preserved by drying are mostly used as mashed potatoes by rehydrating them just prior to use. Some dried potatoes are produced as specialty products, such as scalloped potatoes.

Potatoes may contain toxic glycoalkaloid compounds. One of these is solanine, which appears as a green discoloration that may also involve the tissue below the skin. In such cases, the skin and tissue so affected should be removed. With the popularization of potato skins as entrees, the possible presence of these compounds and their side effects (headache, nausea, and diarrhea) should be considered.

Tomatoes

Tomatoes are handled as the fresh product, as well as processed products such as canned whole tomatoes, tomato juice, tomato puree, tomato paste, ketchup, soup, and sauces (spaghetti, pizza, and chili sauce). Tomatoes are frozen only as an ingredient of prepared products such as pizza or cooked lasagna. Tomatoes require 70 to 85 days from planting to harvesting. Usually, the seeds are planted in soil enclosed by glass or hot frames, and when the plants have reached a height of about 6 in. (15.3 cm) they are transplanted to the outdoor growing area. Tomatoes are handled in the fresh state mostly as the mature partially ripened product and sometimes as the firm ripe tomato.

At the processing plant the tomatoes are washed and sorted. Green specimens are separated from the firm ripe type, and diseased specimens are discarded. Mature partially ripened tomatoes are packaged in cardboard, cellophane-topped cartons. They should be cooled to 55 to 60°F (12.8 to 15.6°C) and held at this temperature until sold to the consumer. In this condition, they have a storage life of 2 to 3 weeks. Firm ripe tomatoes generally are held at 45 to 50°F (7.2 to 10°C), at which temperature they have a storage life of about 5 days. At 32°F (0°C) tomatoes may have a shelf-life of about 2 weeks. However, at this temperature, they are subject to chilling injury and may lose quality. Tomatoes to be canned should be in the firm ripe condition. They are washed and inspected, with green and rotten specimens culled out and mold and rot trimmed out from infected but otherwise good specimens. Trimming is very important because enforcement authorities have methods of detecting mold and rot in finished tomato preparations and can condemn tomato products on this basis.

After trimming, the tomatoes are cored, that is, the pistil and stamen section is cut out. Tomatoes to be canned as whole are scalded with sprays of hot water and cooled with sprays of cold water, which facilitates peeling off the skin. Some varieties have a skin that does not require peeling. The whole peeled tomatoes are then placed in cans after which a brine (2% sugar and 1% salt) is added to cover the tomatoes. In some instances 1% salt and juice from whole tomatoes or from peels and cores are added to fill the cans. The open cans are then heated in water at 175 to 180°F (79.4

to 82.2°C) until all parts of the product have reached 145 to 150°F (62.8 to 65.6°C). The cans are then sealed and heated until all parts of the product reach a temperature of 180 to 200°F (82.2 to 93.3°C) to provide commercial sterility. It should be noted that this process is based on the fact that tomatoes are classified as having a high acid content, which they normally do. However, as tomatoes are allowed to ripen their acidity tends to decrease. In some low-acid varieties, if ripening is allowed to proceed too far, the acidity may be lowered to levels that would make it dangerous to heat process them by the method described previously. This has importance where the canning of tomatoes does not include a check of pH, as in home canning. For this reason, the addition of lemon juice is often recommended in home canning procedures. For the production of tomato juice, cored and quartered tomatoes are heated in steam (hot break method) to inactivate pectinases, which would otherwise destroy pectin, causing a separation of solids from the juice, and to eliminate oxygen, which would otherwise destroy vitamin C. The juice is then filled into cans and heat-processed. There are alternative methods for producing tomato juice, but a superior process is to flash heat juice in a heat exchanger to 250°F (121.1°C), hold this temperature for 0.7 minutes, cool to 200 to 210°F (93.2 to 98.9°C), fill into presterilized cans, and seal and invert the cans to sterilize the covers, holding in this manner for several minutes before cooling. To increase the acidity slightly and add nutrients to the juice, small amounts of citric acid and vitamin C may be added.

Tomato puree and tomato paste are prepared by vacuum concentrating tomato juice. Tomato puree must contain at least 8.37% tomato solids, tomato paste must contain at least 22% tomato solids, and heavy tomato paste must contain at least 33% tomato solids. Salt may be added to these products and baking soda (sodium bicarbonate) may be added to heavy tomato paste to neutralize some of the acid. Generally, these products are added to the container at a temperature near 180 to 200°F (82.2 to 93.3°C). The containers are then sealed and inverted, allowed to stand for several minutes, and cooled. Owing to the low pH (high acidity) of concentrated tomato products, additional heating is not required to attain commercial sterility.

Tomato ketchup or catsup is made by concentrating tomato juice, and then adding sugar and salt and then a vinegar (10% acetic acid) extract of spices (headless cloves, black pepper, red pepper, cinnamon, mace, onions, garlic, etc.). The finished product should have a salt content of 3% and a total solids content of about 30%. This product is bottled hot and capped under vacuum to eliminate air. Removal of air prevents darkening from tannins extracted from the spice mixture. Tannins turn dark in the presence of oxygen. Further heating to attain commercial sterility is not required.

Chili sauce is made from vacuum-concentrated, finely chopped, peeled, cored tomatoes to which a vinegar extract of spices and red peppers has been added, together with onions and garlic. This product is bottled and handled much in the same manner as ketchup.

Tomato soup, Italian tomato sauce, with or without meat, and other tomato products are canned and bottled to some extent. These products may require various heat treatments to attain commercial sterility, depending on the pH (acidity) and added ingredients of the finished product.

Some tomato producers have incorporated genetic engineering into their products as they have developed a variety in which the genetic message to soften in the maturation process has been altered and the tomato turns red without softening or rotting as quickly as familiar varieties.

Cabbage

Most cabbage is handled in the fresh state or is made into sauerkraut. After harvest, cabbage should be held at 32 to 35°F (0 to 1.7°C) until sold to the consumer. Under these conditions, shelf-life may be 3 to 4 months.

Cabbage to be used for sauerkraut may be held out of refrigeration until the outer leaves wilt. The cabbages are cored and the outer coarse and green leaves are removed. The trimmed and cored heads are now washed with sprays of water and then sliced into strips that have a width of about 1⁄32 in. (0.08 cm). The cabbage is then placed in large vats and salted at the rate of about 2.25% of the weight of the cabbage in layers to ensure even distribution. In this condition, the cabbage is then allowed to ferment for 40 to 90 days to allow lactic acid to be produced at the rate of at least 1.5%. The bacteria responsible for the fermentation are listed below with their byproducts:

Leuconostoc mesenteroides: Forms acetic acid, lactic acid, alcohol, and carbon dioxide
Lactobacillus cucumeris: forms lactic acid
Lactobacillus pentoaceticus: forms lactic acid

The growth patterns of the bacteria are shown in Figure 20.8.

Most sauerkraut is canned in either metal or glass containers. It is heated before being added to the container and then heat-processed to an internal temperature of about 200 to 210°F (93.3 to 98.9°C). It is then cooled and can be stored at room temperature like other commercially sterile products.

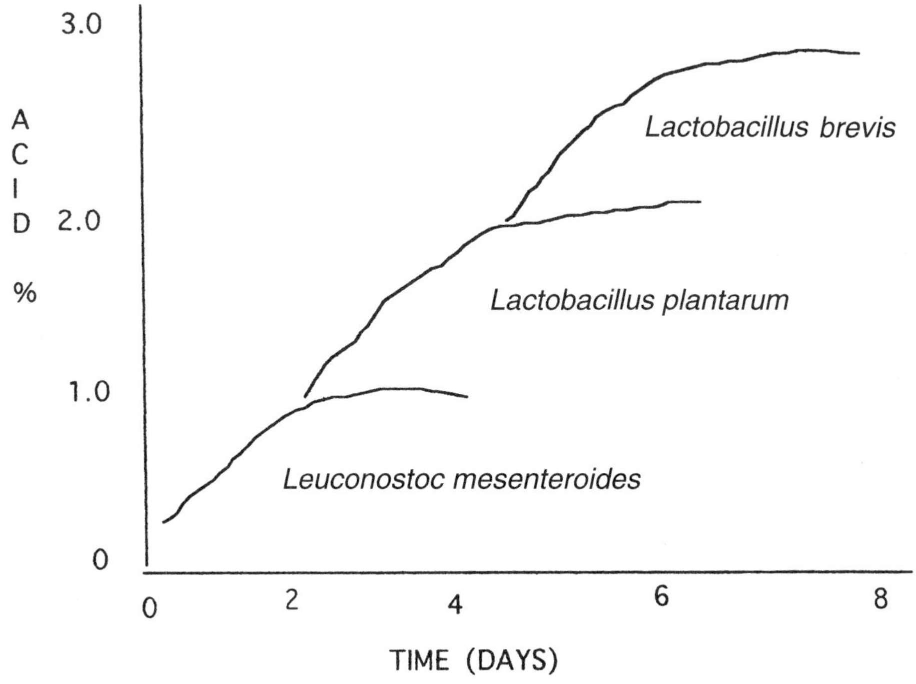

Figure 20.8. Bacterial growth in sauerkraut production.

Tea

Tea is included here because true tea does not come from fruit but from the young leaves of the tea plant which is a bush. These are different from the "herbal teas" that are extracted from other plant materials.

Tea leaves contain three major components that give the beverage its quality attributes. Caffeine adds a stimulating effect, tannins add body and astringency, and essential oils provide flavor and aroma.

Tea is classified into three major groups: green, black, and oolong. These types can be made from the same leaves depending on how long the leaves are "fermented." The fermentation of tea leaves is actually achieved through enzymatic oxidation of the tannin compounds of the leaf. As the enzymes act, they turn the leaf from green to black in a similar manner to the browning action in a cut apple. If the enzymes are inactivated by blanching, the leaf remains green. A partial fermentation results in the oolong type and fully fermented leaves yield the black leaves. Along with the color differences, there are also flavor changes.

Instant tea is produced in a similar manner as is instant coffee. First, a select tea leaf blend (usually fermented black types) is extracted by mixing about 10 parts water to one part leaves and then raising temperature to between 140 and 212°F (60 to

Instant Tea

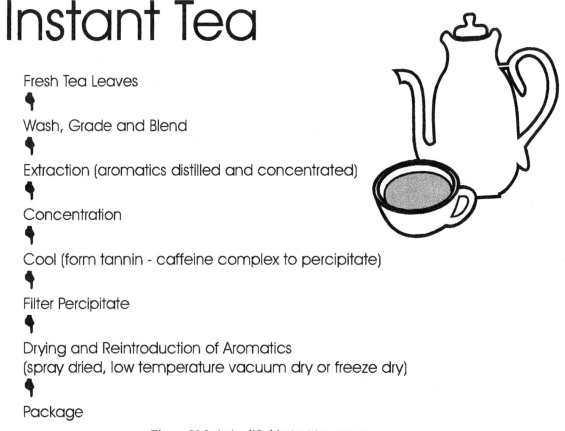

Fresh Tea Leaves

↓

Wash, Grade and Blend

↓

Extraction (aromatics distilled and concentrated)

↓

Concentration

↓

Cool (form tannin - caffeine complex to percipitate)

↓

Filter Percipitate

↓

Drying and Reintroduction of Aromatics
(spray dried, low temperature vacuum dry or freeze dry)

↓

Package

Figure 20.9. A simplified instant tea process.

100°C) and holding for 10 minutes. The extract (about 4% solids) has aromatics removed by distillation (these will be added back later) with specially designed flavor-recovery equipment and is concentrated in low-temperature evaporators to about 25% to 55% solids for more efficient dehydration.

Because instant tea is often consumed cold in a glass (iced tea), it must be "chill-proofed" to remove the haze (cloudiness) that results from a tannin–caffeine complex that is not very soluble in cold water. This is done by lowering the temperature of the tea concentrate to 50°F (10°C), which encourages the formation of the tannin–caffeine percipitate which can be removed by filtration. The extract now has the tea aromatics added back and is ready for dehydration.

Dextrins are often added to the tea before dehydration to yield a 50:50 tea/carbohydrate solids mixture. This protects the delicate tea aroma during drying and also yields a product that is more readily soluble in cold water.

Tea is dried with spray driers at lower temperatures than coffee and low-temperature belt driers. Tea may also be freeze-dried.

Tea has also gained in popularity as a bottled, single-service beverage. It is processed in a similar manner to other soft drinks but it may have to be acidified or retort processed if the pH is above 4.6. See Figure 20.9 for a simplified instant tea process.

Other Vegetables

Other vegetables that are popular include asparagus, lima beans, beets, brussel sprouts, carrots, cauliflower, celery, sweet corn, mushrooms, onions, green peas, sweet potatoes, spinach, squash, and turnips.

All are marketed fresh and some are frozen, canned, or dried by methods described earlier.

21

Sugars and Starches

Sugar, the common name for sucrose, is extracted and refined from sugar cane and sugar beets. There are many substances chemically classified as sugars, and when these are referred to, they are always used with a qualifier such as in milk sugar (lactose), corn sugar (dextrose), and malt sugar (maltose). When the word sugar is used without a qualifier, it generally refers to the common sweetener (sucrose). Other sugars have varying degrees of sweetness relative to sucrose, and some sugars differ from sucrose in that they lend varying degrees of bitterness whereas sucrose imparts only a sweet taste. Other important sources of sucrose include palm and maple trees and fruits. Chemically, and in every other way, cane sugar and beet sugar are the same. In addition to providing energy for the body and sweetness to foods, sugar performs numerous other roles in the food industry. It is used in baked products where it contributes to the desirable texture of baked goods, and it stabilizes the foam of beaten egg whites. When it caramelizes, it imparts a unique, desirable color and flavor to surfaces of pastries and cakes. It is used in ice cream and other dairy products, in beverages, and in other types of food; it is used in the home, in institutions, and in restaurants for foods and beverages. Sugar is also used in some nonfood products such as pet food and other animal feeds and baits. Sucrose ($C_{12}H_{22}O_{11}$) is the most important of the naturally occurring disaccharides. It may be hydrolyzed, yielding glucose (dextrose) and fructose (levulose), both six-carbon sugars.

SUGAR FROM CANE

Sugar cane is a giant grass belonging to the genus *Saccharum*. Although nearly all sugar canes are of the same species, differences in growing conditions affect the characteristics of the juices. For example, the sugar content of the juices from sugar canes grown in the tropics is higher than in sugar canes grown in cooler climates. In the United States, sugar cane is grown primarily in Louisiana, and some is grown in Florida and Hawaii. Cuba, Puerto Rico, the Virgin Islands, the Philippines, and other countries also produce sugar cane. Sugar cane is grown by planting cuttings from the stalk, each containing a bud. The length of time that the cane is allowed to grow before harvesting varies in different countries and may be from 7 months to 2 years. The yield of sugar from cane juice is about 14% to 17%. Sugar cane is harvested by cutting the stalks just above the ground. At this time, the tops of the stalks are cut off, because they contain high concentrations of an enzyme that hydrolyzes and greatly reduces the yield of cane sugar. Also at the time of harvesting, the leaves are stripped from the canes, although they may be burned off prior to harvesting. Parts of the sugar cane other than tops contain some of the enzyme that converts sucrose, so the cane must be processed shortly after harvesting in order to obtain maximum yields.

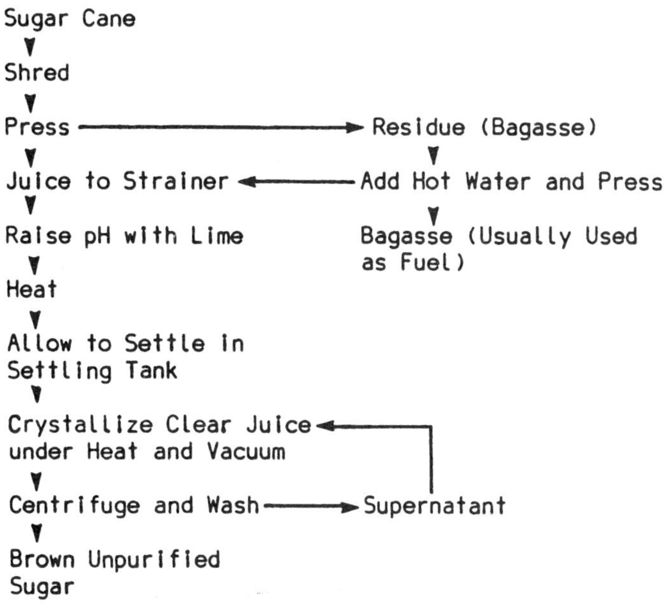

Figure 21.1. Production of raw sugar from sugar cane.

At processing plants producing raw sugar (see Fig. 21.1), the cane first passes through shredders and then through three to seven roller mills that press out the juice. After the first pressing, the bagasse (the pressed cane) may be mixed with hot water or dilute hot cane juice and again pressed to extract more sugar. Following the final pressing, the bagasse is usually brought directly to the boilers, where it is used as fuel. The wet bagasse is reported to have about one-fourth the fuel value of ordinary fuel oil. However, bagasse is a woodlike fiber, and it also is used to manufacture wallboard. It has been the object of study for use in other applications.

The juice, which is dark green in color, has a pH of about 5.2. After extraction, the cane juice is strained to remove pieces of stalk and other detritus, and a mixture of lime powder and water [source of calcium hydroxide, $Ca(OH)_2$] is added to raise the pH. The juice is heated to precipitate and remove impurities. The limed mixture is then held in tanks, where the lime–impurities mixture is allowed to settle out, and the clear juice is separated from the sediment. The clear juice is then heated (at temperatures below the boiling point of water) under vacuum to evaporate water and concentrate the sugar to the point where there is a mixture of sugar crystals and molasses. More syrup may be added to the evaporation pans as the syrup is concentrated.

The mixture (called massecuite) is centrifuged to obtain the brown purified sugar. The liquid centrifuged from the sugar crystals still contains dissolved sugar and is returned to the evaporation pans. When the molasses from the centrifuge treatment reaches such a low concentration of sugar that removal of sucrose is uneconomical, it is called blackstrap. This is not discarded, but is not returned to the evaporation pans. Generally, blackstrap is sold to the fermentation industries for the production of rum. Sugar may be purified or refined at the plant that manufactures the raw sugar, but usually the raw sugar is packed in jute bags and shipped to sugar refineries. Higher grades of molasses for domestic use may be made from juice obtained from cane that is not pressurized sufficiently to extract all the liquid. The molasses is heated to 160°F (71.1°C) and canned or bottled hot. The containers are sealed, heated in water at 185 to 200°F (85 to 93.3°C) for 10 minutes, and immediately cooled.

Raw sugar, as delivered to the refinery, contains 97% to 98% sucrose. The first step in refining (see Fig. 21.2) consists of mixing the raw sugar with a hot saturated sugar syrup. This softens the film of impurities that envelop the sugar crystals. This mixture is then centrifuged, and during centrifugation, the sugar crystals are sprayed with water to remove some of the impurities. The washed sugar crystals are then dissolved in hot water, treated with lime to bring the pH to 7.3 to 7.6, and the temperature is raised to 180°F (82.2°C). The coarser colloidal impurities in the hot mixture are filtered out through diatomaceous earth or paper pulp. The coloring material in the sugar solution is removed by filtering the hot liquid through bone charcoal, after which the sugar is crystallized out in vacuum pans and centrifuged to separate it from the liquid. During centrifugation, the crystals are washed with water. They are then dried, screened for crystal size, and packaged. The finished dried sugar, which has an indefinite shelf-life, is made available in different grades. Large-grained sugar is used for the manufacture of candy and other prepared sweet products. Ordinary table sugar is made up of fine-sized grains. Ultrafine sugar grains (for confectioner's sugar) are produced by grinding the crystals in pulverizing hammer mills. To prevent caking in confectioner's sugar, about 3% cornstarch is used. There are other intermediate grades. Sugar is also prepared as cubes or tablets by forming a mixture of sugar crystals and white sugar syrup under pressure, followed by drying. A variety of "soft sugars," ranging in colors from white through various shades of brown, are produced. These sugars are allowed to retain some of the molasses, which provides their unique flavor.

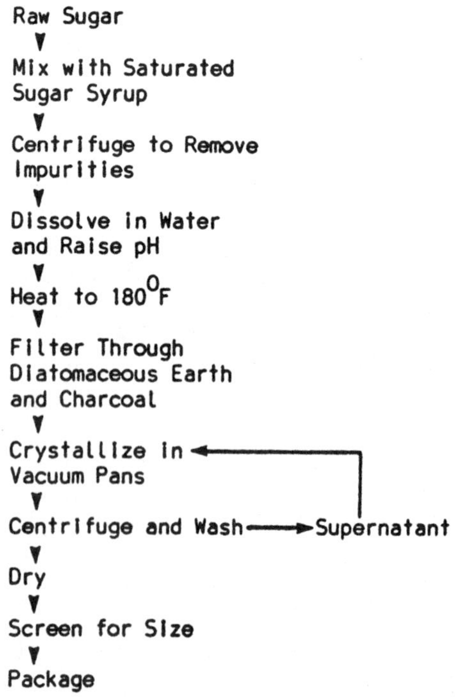

Figure 21.2. Refining of sugar.

Invert sugar is made by heating sucrose in the presence of an enzyme (invertase) and some acid, whereupon it combines chemically with approximately 5.25% of its weight of water. The temperature must not be so high as to inactivate the enzyme.

$$\underset{\text{sucrose + water}}{C_{12}H_{22}O_{11} + H_2O} \xrightarrow[\text{enzyme}]{\text{invertase}} \underset{\text{dextrose + levulose}}{C_6H_{12}O_6 + C_6H_{12}O_6}$$

The dextrose is not as sweet as sucrose but the levulose is sweeter than sucrose with the result that the final syrup, invert sugar, is slightly sweeter than the syrups having the same concentration of sucrose. Invert sugar has special uses in the food industry. Mixtures of invert sugar and sugar are more soluble than sugar alone. In a proportion of 1:1, a mixture of invert sugar and sugar is more soluble than any other combination of their mixtures.

It should be noted that raw sugar contains thermophilic bacteria (spore-forming bacteria that grow at high temperatures—as high as 170°F (76.7°C). If such bacteria are allowed to grow to high concentrations, they may serve as a source of contamination of such products as canned foods to which sugar is added. The spores of thermophilic bacteria are very difficult to destroy by heat, and canned foods can undergo some deterioration during heat processing. For this reason, during the various filtrations in which sugar solutions are held at high temperatures for comparatively long periods, they should be held at temperatures high enough to prevent the growth of thermophilic bacteria (about 185°F [85°C]).

BEET SUGAR

The sugar beet, *Beta vulgaris*, stores its sugar in the root, unlike the sugar cane, which stores its sugar in the stalk. Another difference between cane sugar production and beet sugar production is that the latter is a continuous operation and does not produce the intermediate raw sugar. Whereas sugar beets contain 16% to 20% of sugar as sucrose, the yield of sugar per acre is considerably less than that obtained from sugar cane, owing to the quantity of beets or cane harvested per acre. Sugar beets are planted as seed and require 70 days or more from planting to harvesting. Because the plants are subject to bacterial or mold infection and to infestation with aphids, maggots, and other insects, the plants may need spraying during the growing season. In the United States, beets for sugar are grown mainly in Colorado, California, Michigan, Utah, Idaho, Nebraska, and Montana. Sugar beets are harvested and topped mechanically and brought to the processing plant in bulk by freight cars or trucks. They may be stored outside the processing plant in large piles until treated to extract sugar.

In extracting sugar from beets, the beets are first thoroughly washed to remove mud and stones, and then passed through mechanical slicers that slice them into thin shreds called cossettes. They are then covered with hot water to extract the sugar. Extraction is done even more quickly and effectively by a continuous countercurrent extraction with water. The juice from the extraction process is treated with lime or calcium hydroxide and then with carbon dioxide. The lime removes impurities as a precipitate, and the carbon dioxide is used to precipitate the calcium hydroxide as calcium carbonate ($CaCO_3$). The juice is then filtered and again treated with carbon dioxide to precipitate residual calcium hydroxide. After a second filtration, the extract is treated with sulfur dioxide to bleach out colored components in the liquid. The liquid extract is then vacuum concentrated to 60% to 70% soluble solids and filtered through bone charcoal. Next, the filtered liquid is concentrated in vacuum pans to form crystalline sugar. The sugar is washed as it is centrifuged, dried, screened, and packaged, as in the case of cane sugar. In some manufacturing processes for beet sugar, the liquid from the third centrifugation is used for the manufacture of monosodium glutamate, a flavor enhancer used for many food preparations, including soups, gravies, Chinese foods, and meat dishes. Extracted beet pulp and the tops are dried and used as cattle feed.

OTHER SUCROSE SOURCES

Sucrose can be obtained from the sap of a variety of palm trees, one of the most important being the date palm (*Phoenix sylvestris*). Much of this sucrose is obtained in the Middle East by primitive methods that involve boiling in open kettles, after which there is separation of crystals from molasses, or the unseparated mass may be allowed to set into a whole sugar.

In the northern part of North America, sucrose is obtained from the sap of the hard maple tree (*Acer saccharinum*). Although the maple syrup sap is largely sucrose, it contains unique impurities that impart to it (when concentrated) a special delicate flavor that makes the natural maple syrup a valuable flavoring for certain preparations. Sucrose can also be produced from the cane of the sorghum plant, which is related to the sugar cane but resembles the corn plant.

CORN SYRUP AND SUGAR

Although corn is not a source of sucrose, the value of corn sugar as a sweetener in the food industry makes it worthy of mention. Corn syrup is produced by heating starch in water acidified with hydrochloric acid. The hydrolysis, in this case, is only partially completed so that the mixture contains some glucose, some maltose, and some longer chains of glucose (dextrins). A second method is also used in which the hydrolysis may be carried out by first heating the starch with acid followed by treating it with an enzyme that hydrolyzes starch. The latter method produces a syrup higher in maltose than does the straight-acid hydrolysis method. In this method, the use of amylolytic enzymes (alpha- and beta-amylases) is employed. Beta-amylase is specific in its action, attacking starch molecules at their nonreducing ends and causing progressive breaks in the molecule at 12-carbon intervals, with the resultant release of units of maltose (a 12-carbon sugar). Thus, by this conversion, the corn syrup will have a high maltose content. Other enzymes, glucosidases, may also be used to supplement either of the two conversion processes. In this case, the syrup will contain significant amounts of glucose.

After hydrolysis, the syrup is neutralized by addition of sodium carbonate, then filtered and concentrated to 60% solids, again filtered through bone charcoal, and finally passed through resins (ion exchange) that take out the salt (sodium chloride formed from the acid and sodium carbonate). The corn syrup may be spray or drum dried to about 3% moisture to obtain corn syrup solids. A more completely hydrolyzed syrup, after purification, may be concentrated, seeded with fine corn sugar crystals, and crystallized to produce crude corn sugar. Dextrose or corn sugar can be produced in a similar manner from a completely hydrolyzed starch.

In another process, corn starch is hydrolyzed into its glucose components, and then a percentage of the glucose molecules are converted into fructose. This high-fructose corn syrup (HFCS) has had a significant impact on the soft drink and confection industry. HFCS has been substituted either wholly or partially for sucrose in many formulas, allowing the manufacturers to meet their mammoth product demand with less dependence on the sugar cane industry. Corn syrup, corn sugar, and high-fructose corn syrup are used in bakery products, pharmaceuticals, carbonated beverages, confectioneries, ice cream, jams and jellies, meat products, and dessert powders. The cruder, less refined sugar products are used in tanning, for brewing, to produce vinegar, to produce caramel coloring, and in tobacco. Starch also has many other nonfood applications in industries such as cloth and textiles, adhesives, pharmaceuticals, and paper.

Corn syrups are divided into five commercial classes, depending on their DE value (dextrose equivalent value). By this classification scheme, pure dextrose is given a value of 100 DE.

Type	DE Value
Low conversion	28–38
Regular conversion	38–48
Intermediate conversion	48–58
High conversion	58–68
Extra high conversion	68–100

In addition to classification by their DE values, syrups are also classified by their solids contents.

Corn syrup and corn sugar are used to a large extent in the food industry. They are used to supplement sucrose because they are less expensive, while at the same time nearly as effective as sucrose in sweetening characteristics. In addition, they inhibit crystallization of sucrose, especially when their maltose content is high. They are especially useful in the baking and brewing industries because of the quick and complete fermentability of dextrose. Their use in preserves minimizes oxidative discoloration, and other unique properties make them useful in many other applications.

BEVERAGES

The carbonated, nonalcoholic beverage industry is huge in the United States and throughout the world. One major producer claimed in 1994 that if all of its cola ever produced were poured into one tremendous swimming pool with an average depth of 6 ft, this pool would be nearly 20 miles long and over 8 miles wide. Allowing 9 ft^2 of surface or 54 ft^3 each, this pool would hold over 512 million people at the same time.

The major ingredients in carbonated, nonalcoholic beverages include water, carbon dioxide, sugar, flavorings, colors, and acids. The amounts of sugar vary from about 8% to 14%. The most common sugar used in soft drinks is HFCS. Sucrose is still widely used but HFCS has replaced it in many formulas because it is sweeter, therefore less expensive on an equal-sweetness basis. Sugar not only contributes sweetness to the beverage, but also adds body and mouthfeel. When low- and no-calorie drinks are made,

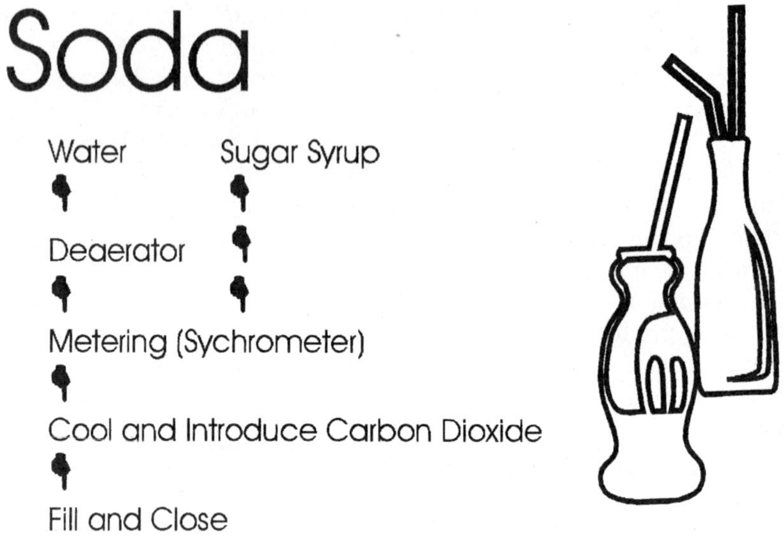

Figure 21.3. Procedure for soft drink production.

the sugar must be replaced with a texture-enhancing ingredient such as carboxymethyl cellulose or pectin as well as a sweetener. A simple layout of a soft drink plant is shown in Fig. 21.3.

CANDY

Candy or confections can be divided into two major categories, those in which sugar is the main ingredient and those that are based on chocolate. The sugar-based candies are discussed here and the chocolates are discussed in Chapter 22 with the fats and oils.

Candymaking can be described as "the science and art of manipulating sugar crystals to achieve desired textural effects." This is done by controlling the crystallization of sugar and the sugar/moisture ratio. Examples of sugar type candies include nougats, fondants, caramels, taffies, and jellies. The ingredient list used in candymaking is virtually endless and includes milk products, egg white, food acids, gums, starches, fats, emulsifiers, flavors, nuts, colors, and more.

The sugar in candies may be crystalline or noncrystalline. The crystals may be large or small and the noncrystalline structure may be glasslike or amorphous. Softness is another textural quality and soft and hard candies can be made from both crystalline and noncrystalline structures. To achieve softness, higher levels of moisture can be maintained, air can be whipped in, or other ingredients can achieve this textural attribute. Table 21.1 shows some major types of sugar-based candies.

The major sugars used in candymaking are sucrose, invert sugar, corn syrups, and HFCS. Sucrose syrup concentrations can be increased as temperature is increased. Textures can be controlled by the amount of moisture in the sugar solution. As the temperatures increase, more sucrose remains liquid. Boiling points of different sucrose concentrations are shown in Table 21.2. When the syrups are cooled, structures ranging from highly crystalline masses to amorphous glasses can be achieved.

Table 21.1. Some Major Types of Sugar-Based Candies

Texture	Example
Crystalline sugar	
Large crystals	Rock candy
Small crystals	Fondant fudge
Noncrystalline sugar	
Hard candies	Sour balls, butterscotch
Brittles	Peanut brittle
Chewy candies	Caramel, taffy
Gummy candies	Marshmallow, jellies, gumdrops

Table 21.2. Boiling Points of Different Sucrose Solutions

Percent Sucrose	Boiling Point (°F/°C)
30	212/100
40	213.8/101
50	215.6/102
60	217.4/103
70	222.8/106
80	233.6/112
90	253.4/123
95	284/140
97	303.8/151
98.2	320/160
99.5	330.8/166
99.6	339.8/171

In hard candy formulas, corn syrup, which does not crystallize as readily as sucrose, may be added to retard crystallization. Corn syrup, depending on how much the corn starch has been hydrolyzed by enzymatic or acid treatments, can contain varied amounts of dextrose, maltose, higher sugars, and dextrins. These can be added to different candies to increase viscosity, to slow dissolving rate of candies in the mouth, and to increase chewiness.

Invert sugar is a mixture of levulose (fructose) and dextrose (glucose) made from cleaving the linkage in sucrose holding these two sugars together in its disaccharide structure. Invert sugar is sweeter than sucrose, is less likely to crystallize, and is more soluble in water. These qualities encourage the formation of small crystals necessary for smoothness in fondant creams and fudges. Because of its hygroscopicity, it helps keep chewy candies from becoming dried out and brittle.

Other sugar products used in candymaking include maple sugar, brown sugar (a less refined sugar from the cane sugar refining process), molasses (also from the cane sugar refining process), and honey.

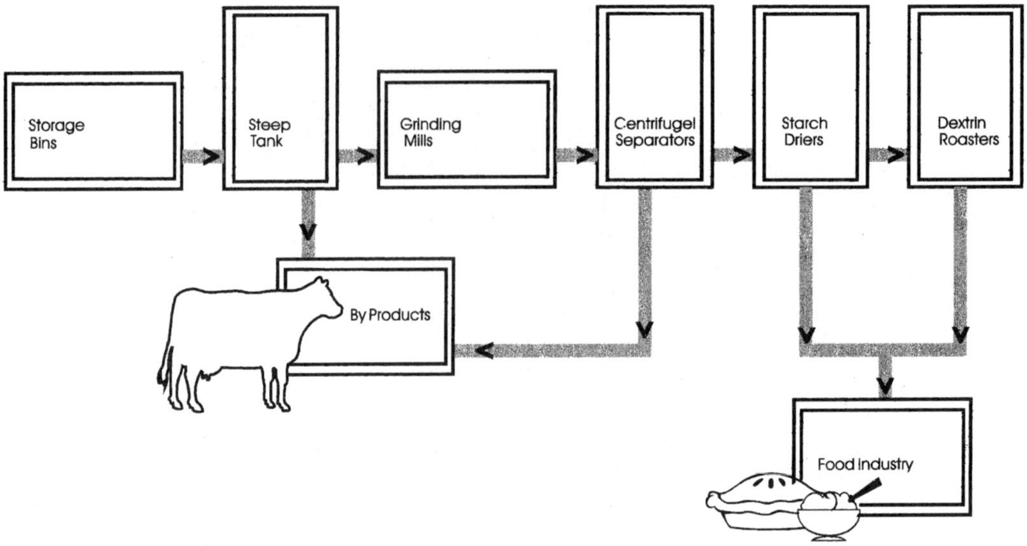

Figure 21.4. Wet milling of corn and production of starch.

STARCH

In producing cornstarch, the corn is cleaned, then placed in vats, where it is steeped in water (slightly acidified with sulfur dioxide to prevent fermentation) for about 40 hours. If not previously degerminated, the steeped kernels are passed through mills that separate the germ and loosen the hull. The mass is then passed through tanks of water where the germs (being lighter) float and are skimmed off. The remaining endosperm, containing starch, corn gluten, and hulls, is then finely ground in steel mills. The finely ground material is then passed through sifters to remove hulls, the starch and gluten passing through. The starch is separated from the gluten by centrifugation. The starch particles, being heavier, are separated at the outer region of the centrifuge, the lighter gluten migrating to the center. Starch may be produced from potatoes, rice, tapioca, or wheat by methods similar to that described for corn (see Fig. 21.4), except that in the case of potatoes and tapioca it is not necessary to degerm the product.

Starch may be modified chemically to provide properties suitable for various manufactured products. Table 21.3 gives some of the functions of modified starches in various foods.

In the unmodified form, starches have limited uses in the food industry. Waxy maize is a good example as it has much amylopectin with many branches and will hydrate easily and swell rapidly. On exposure to heat, however, it ruptures, losing viscosity and producing a weak-bodied, stringy, cohesive paste. Modifying starch enhances or represses inherent properties as needed for specific applications. As shown in Table 21.3, these are varied and diverse.

Starches may be modified in a number of ways. The most basic is to simply precook the starch which results in a product that will swell in cold water. Starch may also be modified chemically to achieve different functional characteristics. If the starch is treated with acid or enzymes, its viscosity may be reduced as a result of hydrolysis of a

Table 21.3. Functions and Benefits of Modified Starches in Food Systems

	Functions	Benefits
CANNED FOODS		
Pudding	Stabilize, thicken, control texture	Produces smooth and creamy body
Pie fillings	Stabilize, thicken, provide smooth, short texture	Resistant to breakdown under low pH conditions, body, mouthfeel, fill
Soups, chowders	Stabilize, thicken	Viscosity, suspended solids
Cream style corn	Thicken, stabilize (counters curdling in retorting)	Excellent stability, no curds, smooth and creamy texture, good shelf-life
Sauces, gravies	Thicken, stabilize, texturize, fat enhancer	Smooth texture, cling, opacify, body, mouthfeel
FROZEN FOODS		
Fruit pies	Stabilize, thicken, provide cutability	Good freeze–thaw stability, low pH resistance, clarity and texture
Fat-reduced systems	Provide oily texture for instant mixes	Excellent mouthfeel, full fat properties
Pot pies, sauces, and gravies	Stabilize, thicken	High filling viscosity
Battered products	Good cohesion and adhesion	Firm, crisp, uniform coating for fried, baked, and microwave products
Desserts	Stabilize, thicken	Smooth, creamy texture
GLASSED FOODS		
Salad dressings	Stabilize, thicken, provide required texture, emulsion stabilizer	Improves texture, body, and mouthfeel
Reduced-fat dressings	Fat mimetics, oily texture; heat, fat, shear-resistant; shelf-stable	Excellent mouthfeel, full fat properties
Baby foods	Stabilize, thicken, provide short texture	Freeze–thaw stability, resist high temperature and low pH
Sauces, relishes	Stabilize, thicken	Low pH stability, good particulate suspension
Gravies	Thicken and opacify	Provide viscosity and "home made" opacity with good stability
BEVERAGES/ENCAPSULATIONS		
Beverage emulsions	Emulsion stabilizer for flavors and clouds	Replace expensive gums
Encapsulation	Encapsulation agents for flavors, clouds, and vitamins	Replace expensive gums, low viscosity, oxidation resistance
Liquid and powdered nondairy creamers	Functional alternative to caseinates	Excellent emulsion performance, replaces milk proteins

continued on next page

Table 21.3. *Continued*

	Functions	Benefits
BAKED FOODS		
Pies, tarts, fillings	Stabilize, thicken, provide smooth or pulpy, moist texture	Resistant to weeping, retards "boil out," more attractive fillings
Glazes	Attractive coating, moisture barrier, tackiness	Clarity, sheen, excellent adhesion, low viscosity
Low-no-fat danishes	Lubricious roll-in cream	Laminated dough, binds moisture, extends shelf-life
Low-no-fat cakes/ icings	Serration, fatlike mouthfeel	Extends shelf-life, rich low-no-fat product
DRY FOODS		
Instant puddings and desserts	Instant thickening	Creamy texture, smoothness, fast meltaway, low bulk density, bland flavor
Soup mixes	Thicken without lumping	Body and mouthfeel
Cake mixes	Thickening, moisture retention, shortening replacement	Texture, shelf-life, moister cake, better volume, lighter cakes
Gravy and sauce bases	Thickening without lumping, free flowing powder	Texture, cling, natural, homemade appearance, increased shelf-life
Low-no-fat mixes	Oily texture	Excellent mouthfeel, full fat properties
CONFECTIONS		
Jelly gums	Setting and gelling agent, rapid gelling agent	Clarity, easier processing, controlled texture
Panned items	Coat with a clear, appetizing film	Clarity, sheen, reduced cracking
Hard gums	Structure and stability	Clarity, gum replacement, mouthfeel
SNACK FOODS/CEREALS		
Hot extruded products	Binder and stabilizer, less oil pickup, short, dry texture	Better shaped pieces, excellent storage stability, smoother, less sticky on extrusion
Baked or fried puffed snacks	Volume control, crispness	No need for cooking during forming process
French fries, extruded french fries	Form, crisp coating, internal binder	Crisper, more appetizing products after frying
RTE (ready to eat) breakfast cereals	Textural enhancement, particulates, seasoning adhesion	Excellent expansion, excellent bowl life, improved eating quality, adhesion, and film forming

continued on next page

Table 21.3. *Continued*

	Functions	Benefits
DAIRY PRODUCTS		
Yogurt	Thicken, stabilize	Creamy texture, improved mouthfeel
Cheese products	Binder, thickener	Product texturizer, oil enhancer
Imitation cheese	Casein replacer	Functional equivalent to casein
Chilled desserts	Thicken, stabilize	Improved mouthfeel, controlled texture
Low-no-fat products	Mouthfeel enhancer	Rich, creamy texture, bland flavor
MEAT PRODUCTS		
Formed meats	Binding, moisture retention	Firmer bite, reduced shrinkage
Smokehouse meats	Low-temperature moisture binding	Excellent shelf stability
Pet foods	Thickener, gel former, texturizer	Improved stability and texture, nongelling, firm, cuttable, gum replacer
Low-fat meats	Bind moisture, increase yield	Product with full fat qualities

number of the glycosidic linkages in the starch molecule, resulting in smaller molecules which produces a paste with low viscosity when gelatinized. Paste clarity may also be achieved in starches by exposing them to oxidizing agents such as sodium hypochlorite.

In some starches, it is desired to have viscosity remain high in the presence of acid and under conditions of shear or high temperatures. To achieve this, the starches are cross-bonded by reacting the starch suspensions with chemicals such as epichlorohydrin, phosphorus oxychloride, or sodium trimetaphosphate in the presence of an alkalai catalyst. Figure 21.5 shows an example of how crosslinking helps to reinforce hydrogen bonding, strengthening weaker pastes so they will avoid breakdown during extended cooking times, increased acid conditions, or severe agitation. To demonstrate the effectiveness of crosslinking, Figure 21.6 shows a Brabender curve of the viscosity of crosslinked waxy maize and unmodified waxy maize.

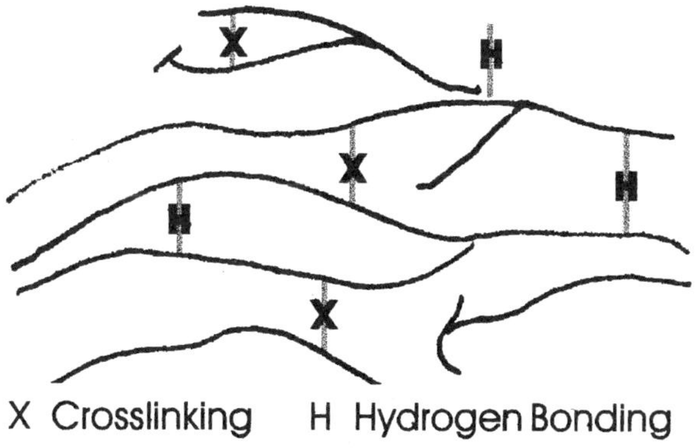

Figure 21.5. Crosslinking and hydrogen bonding in starches.

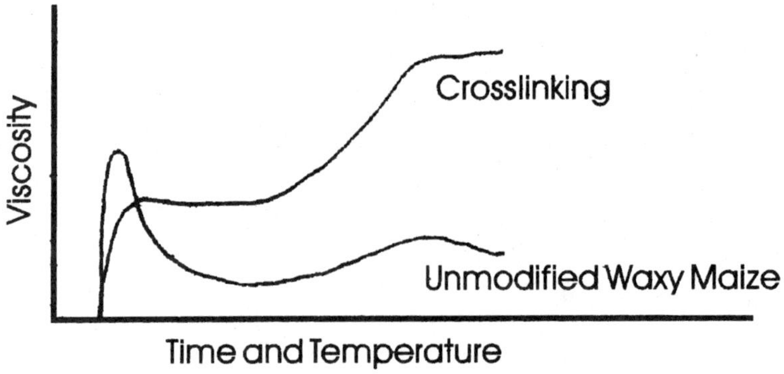

Figure 21.6. Effect of crosslinking on waxy maize starch.

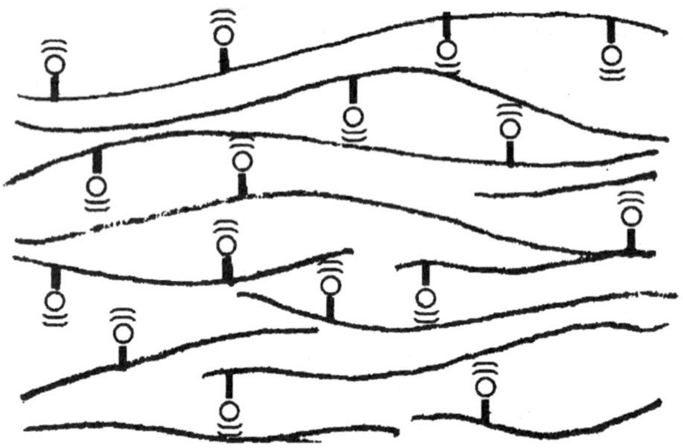

Figure 21.7. Stabilization of starches.

In some starches, stabilization that prevents gelling and weeping and maintains textural appearance is desired. When highly branched starches are frozen, their pastes may become cloudy, chunky, and will weep similar to those starches with more amylose. This is attributed to the lowering of the normal kinetic movement in amylopectin as the temperature drops, allowing outer branches of the amylopectin to associate through hydrogen bonding, bringing about less dramatic, but similar, conditions as those that occur with amylose. To avoid this, anionic groups are scattered throughout the starch granule to block molecular association (see Fig. 21.7). The result is a stabilized starch that has freeze–thaw stability and can be used in frozen foods without syneresis (weeping) occurring.

As mentioned in Table 21.3, modified starches are getting much attention in the development of low-fat and no-fat items. Acting as fat substitutes, the starches provide the oily texture and mouthfeel desired in such products.

Starch derivatives such as starch acetate, starch phosphate, starch succinate, and hydroxypropyl starch have also played consequential roles in improving functional and quality attributes such as freeze–thaw stability and paste clarity.

22

Fats and Oils

Fats and oils are classified as lipids, which comprise an important category of nutrients for humans; the other two important categories of nutrients, of course, are proteins and carbohydrates. As the reader may recall from Chapter 9, lipids are esters of glycerol and fatty acids, and fatty acids are mostly long, straight, hydrocarbon chains, with varying degrees of hydrogen saturation of the carbon atoms, having a carboxyl group linked to one of the end-carbon atoms. The esterfication reaction may be reversed with the addition of an alkali, which results in a combination of the fatty acids with the alkali to form soap, a reaction called saponification.

$$\begin{array}{lcccccc} CH_2OOCR^1 & & & & CH_2OH & + & NaOOCR^1 \\ | & & & & | & & \\ CHOOCR^2 & + & 3\ NaOH & \longrightarrow & CHOH & + & NaOOCR^2 \\ | & & & & | & & \\ CH_2OOCR^3 & & & & CH_2OH & + & NaOOCR^3 \\ \text{lipid} & & \text{alkali} & & \text{glycerol} & & \text{soaps} \end{array}$$

The specific gravity of lipids is less than that of water, and these compounds are generally insoluble in water; therefore, they float in water. They are, however, soluble in a variety of organic solvents (e.g., ether). They may be suspended in water as a stable emulsion in the presence of emulsifying agents (e.g., bile salts, alkali) that function by lowering the surface tension and by coating the lipid particles, preventing their coalescence.

The members of the fats and oils classification may differ from one another quite

Table 22.1. Common Names of Saturated Fatty Acids

Name	No. of C Atoms
Butyric acid	4
Caproic acid	6
Caprylic acid	8
Capric acid	10
Lauric acid	12
Myristic acid	14
Palmitic acid	16
Stearic acid	18
Arachidic acid	20
Behenic acid	22
Lignoceric acid	24
Cerotic acid	26

distinctly in physical, chemical, and dietary properties (melting points, caloric value, reactivity, mineral content, etc.). Some of the properties can be used to identify specific fats. For example, fats can absorb halogens at the point of unsaturation (a point of unsaturation where two neighboring carbon atoms in the chain each have only one hydrogen atom attached). Thus the iodine number of a fat refers to the amount of iodine (iodine being a halogen) in grams that 100 g of the fat can absorb. The saponification value of a fat refers to the amount of potassium hydroxide in milligrams that is neutralized by saponification of 1 g of fat.

The fatty acids occurring in natural fats may contain 4 to 26 carbon atoms (generally an even number; see Table 22.1), and they may or may not be completely saturated with hydrogen. The general formula for saturated fatty acids is $C_nH_{2n}O_2$. For unsaturated fatty acids, the number of H atoms is less than $2n$ (e.g., H_{2n-2}, H_{2n-4}, etc.) Most fatty acids are straight chains, although a few may contain a ring of carbon atoms in the straight chain. An example of this type is chaulmoogric acid, which has been used in the treatment of leprosy, and is obtained from the chaulmoogra tree of East India.

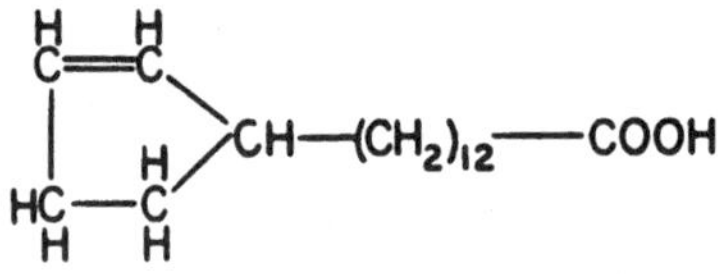

chaulmoogric acid

The unsaturated oils, such as some vegetable oils that are liquid at room temperature and are subject to oxidation (a deteriorative process), may be transformed to a solid at room temperature and at the same time stabilized against oxidative spoilage by hydrogenation. In hydrogenation, hydrogen is linked to unsaturated carbon atoms, though the process generally is not allowed to go to complete saturation. An example of a hydrogenation process follows.

$$\underset{\text{liquid lipid (oil)}}{C_3H_5(C_{17}H_{33}COO)_3} + 3\ H_2 \rightarrow \underset{\text{solid lipid (fat)}}{C_3H_5(C_{17}H_{35}COO)_3}$$

As the fatty acid part of fat molecules is the largest part, the fatty acids influence the properties of the fat, and the chemistry of fats and oils consequently is governed by the chemistry of the fatty acids they comprise.

Fats and oils have smoke-, fire-, and flash-points that define their thermal stability when heated in the presence of air. The smoke-point is reached when the temperature is sufficient to drive out decomposition products that can be seen as a smoke. As the temperature is increased to the flash-point, the decomposition products can be ignited but do not perpetuate fire. When the temperature is increased to the fire-point, the decomposition products will perpetuate a fire.

Lipids contribute to the diet in many ways. They are a primary source of energy, possessing more than twice the calories occurring in either proteins or carbohydrates. They act as vehicles for the fat-soluble vitamins A, D, E, and K and are important in the absorption of calcium, carotene, and thiamin. They also provide certain essential fatty acids that are required but cannot be produced in sufficient amounts by the body. Fatty acids, in the form of phospholipids, are essential to the body. These include phosphoglycerides (e.g., lecithin), phosphoinositides (e.g., diphosphoinositide), and phosphosphingosides (e.g., sphingomyelin).

Essentially, solid fats are of animal origin, and oils or fats, liquid at room temperature, are of vegetable origin. There are some exceptions, however. For instance, coconut oil is solid at ordinary room temperatures, having a melting point of 75 to 80°F (23.9 to 26.7°C). On the other hand, fish fats or oils are liquid at room temperature. Generally, the reason that some fats are solids and others are liquids at room temperature has to do with the percentage of saturated or unsaturated fatty acid in the fat molecules composing the fats. Stearic acid, CH_3—$(CH_2)_{16}$—COOH, is a saturated fatty acid. Linoleic acid, CH_3—$(CH_2)_4$—CH=CH—CH_2—CH=CH—$(CH_2)_7$—COOH, is an unsaturated fatty acid. It should be noted that these two fatty acids have an equal number of carbons but that all the carbons in stearic acid have been saturated with hydrogen or have taken on all the hydrogen that they can accept, while four of the carbons in linoleic acid are unsaturated or could each accept one more hydrogen. Oleic acid, CH_3—$(CH_2)_7$—CH=CH—$(CH_2)_7$—COOH, which has only two carbons that could accept another hydrogen, has a melting point of 61 to 62°F (16.1 to 16.7°C). When oleic acid is present in fats, even in comparatively large amounts, the fats are generally solid at room temperature if no more unsaturated fatty acids are present.

Fats containing only saturated fatty acids of short chain length (eight or fewer carbons) are liquid at room temperature, but generally such fats are not found in nature. However, such fats are found as components of some natural fats, such as butter made from cattle or goat's milk. It should be recognized that unsaturated fatty acids with four or more carbons that can accept another hydrogen are more reactive than saturated fatty acids and are especially apt to combine with the oxygen present in the atmosphere. This is especially the case when a carbon saturated with hydrogen is present between two groups of two carbons, each of which could accept another hydrogen: —CH=CH—CH_2—CH=CH—. When the unsaturated fatty acids in fats become oxidized, the fat generally becomes rancid and has an off-flavor.

Butterfat, which is discussed in Chapter 15, is not covered in this chapter. Nonedible lipid products, such as soaps, are also omitted.

LARD

Lard, a solid fat (at room temperature), is obtained from animals (mainly hogs) by rendering (see Fig. 22.1). Rendering involves the melting out of the fat from fatty tissues and removal of the nonfat material by mechanical means. In wet rendering, the fatty tissues are heated in water with steam under pressure. After heating, the fat layer (on top) is pumped into one tank and the watery portion into another, for purposes of settling and removal of water in one case and skimming of the fat in the other. Dry rendering is accomplished by heating fatty animal tissues under vacuum until all of the water has been evaporated from the mass of material. The fat, while still warm enough to be liquid, is then filtered off from the residual tissues. Low-temperature rendering may be used to obtain animal fats. In this case, the tissue is first ground, then heated to a temperature only slightly higher than that of the melting point of the fat. The nonfat tissue is then removed from the fat by centrifugation. Most animal fats to be used in food materials are deodorized by blowing high-temperature steam through the liquid material before packaging. Beef tallow is rendered from the

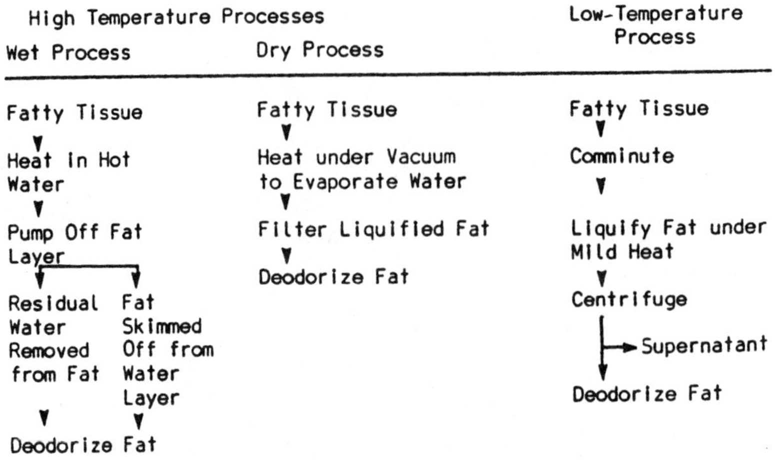

Figure 22.1. Recovery of animal fats by rendering.

various fatty trimmings from sides or quarters of beef. Comparatively few cattle fats are used in foods today, but beef tallow has various industrial uses.

Prime lard is rendered from fats deposited in body tissues, while ordinary lard is rendered mainly from the back fat portions, but also from the trimmings. It was once the custom to use lard in cooked foods or as cooking fats and oils.

Natural fats, including animal fats, tend to have some degree of organization with regard to the pattern in which the fatty acids are distributed on the glycerine molecules. This being the case, fat molecules containing three saturated fatty acids (fatty acids unable to accept more hydrogen) do not occur to any extent unless the content of saturated fatty acids exceeds two-thirds of the total fatty acids. Lard has about 40% saturated fatty acids. Interesterification is the changing of position of the fatty acids that are esterified to the glycerol molecule forming neutral fats and oils. It is also called randomization or ester interchange. Interesterification is performed on fats to improve their functional characteristics. Interesterification of fats tends to distribute the fatty acids in a random manner as they are attached to the glycerine molecules. Interesterification usually raises the melting point of fats and oils. Although the melting point of such fats as beef oleo and lard are not raised by interesterification, their characteristics and behavior in foods are changed. Interesterification can be brought about by heating the fat to the desired temperature, then adding a catalyst (sodium methoxide is often used as the catalyst). When the reaction is judged to be complete, phosphoric acid (about 85% strength) is added to destroy the catalyst. Natural lard occurs as clusters of crystals that grow larger as the lard is aged. Interesterification of lard changes the structure in such a manner that the crystals exist as finely dispersed particles. The grain of this lard is therefore much finer than that of the natural product. Fine-grain, interesterified lard has good creaming qualities and is desirable for the production of cakes and icings. On the other hand, the coarse-grained natural lard has properties that make it better suited for the manufacture of flaky pastries and pie crusts.

OILS

There are many oils, most of which are of vegetable origin. Included among these are coconut oil and palm kernel oil, in both of which the fatty acid composition consists

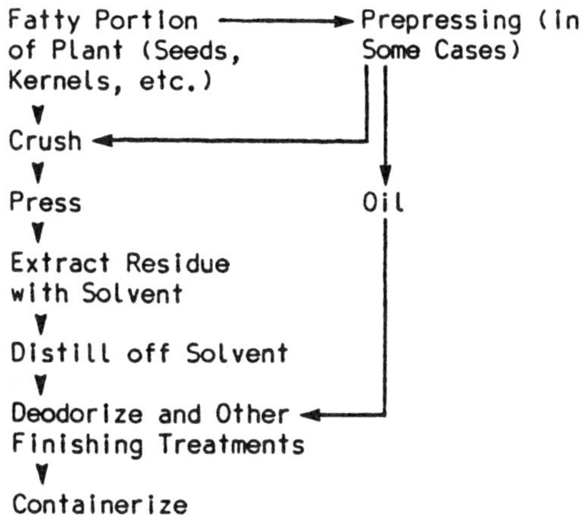

Figure 22.2. Recovery of plant oils.

of 45% to 80% lauric acid, a 12-carbon saturated acid. Other vegetable oils include cottonseed, peanut, olive, sesame, corn, soybean, and canola. These oils contain a preponderance of the unsaturated acids-linoleic (18 carbons, with 4 carbons containing 1 less hydrogen than they can accept, or 2 double bonds), or oleic (18 carbons, with 2 carbons containing 1 less hydrogen than they can accept, or 1 double bond). Remember that each carbon in the chain, except for the end-carbons, can accept two H atoms. Also among the oils are the marine oils (chiefly whale oil or menhaden oil), which contain some fatty acids with 20, 22, 24, or 26 carbons, and many of the fatty acids have 3, 4, or 5 double bonds (6, 8, or 10 carbons that could accept another hydrogen).

In the extraction of vegetable oils (see Fig. 22.2), some preparation of the material may be required. When corn oil is produced, the germ must first be separated from the corn kernels. The germs are then crushed prior to extraction. Some seeds, such as cottonseed, are pressed at low pressure to remove part of the oil prior to extraction. Other seeds are crushed or flaked before they are subjected to the extraction process. Soybeans are cut into thin flakes and extracted with a solvent. The extraction is started with solvent that has already taken up a high level of oil and is finished off with fresh solvent. To obtain the extracted oil, the solvent (which has a much lower boiling point than the oil) is distilled off and recovered for reuse in the extraction of more oil. Oil from olives and cacao beans is obtained by pressing the cooked pulp or beans. An expeller, which subjects the cooked material to high pressure, is used to squeeze out the oil. Oil from seeds may also be extracted in this manner. Owing to the pressure used in expelling oil this way, the temperature of the press cake may rise above 250°F (121.1°C). Expressed oils are usually darker than solvent-extracted oils.

Extracted or expelled oils contain phospholipids, gum, and other materials that are soluble in oil but settle out when wet with water. The natural oil may, therefore, be degummed by mixing with water and centrifugation to separate out the water and the insoluble material. If oil is to be deodorized, it must first be subjected to several degumming operations; otherwise the oil will become cloudy and difficult to filter.

Oil is usually refined prior to sale as a food product. This is done by mixing a concentrated aqueous solution of sodium hydroxide or sodium carbonate with the hot crude oil, and centrifuging out the separated gums and soaps. The oil is then washed with water to remove traces of soaps, and then dried under vacuum. Oils may be bleached prior to sale for use as food products. Neutral clays, sulfuric acid clays, or charcoal may be used in bleaching. Certain yellow colors or green chlorophyll colors may be removed by filtering oils through clays. The natural yellow color of oils (carotenoids) may be removed with activated charcoal or by heating to high temperature. In bleaching, the oil is mixed with the clay or charcoal and heated under vacuum. After sufficient time has elapsed to provide for contact of the oil and bleaching material, the oil is filtered off from the treatment material. Oils are often winterized to remove the saturated fats, in this case, by cooling and holding at 40°F (4.4°C), then filtering to remove the solid fats.

HYDROGENATION

Large quantities of oil are hydrogenated, usually for purposes of forming fats that are solid at room temperature. During hydrogenation, hydrogen is attached to those carbons that are still deficient in this element. Also, during hydrogenation, the more unsaturated fatty acids in the oils are the first to add hydrogen (first those with three double bonds or six unsaturated carbons). The result is that in a fully hydrogenated fat, the only fatty acids that may remain unsaturated have only one double bond, or two carbons, each of which can accept one hydrogen atom. Hydrogenation raises the melting point of the fat, but, in addition to this, the fat is changed. For example, oleic acid (one double bond) in the fat has been changed to elaidic acid, which has the same formula as oleic acid and still has one double bond, but the spatial distribution of the atoms differs as to the type of symmetry. Oleic acid is the "*cis*" form of the atom's molecule and elaidic acid is the "*trans*" form of the same molecule. Because the melting point of elaidic acid is much higher than that of oleic acid, the melting point of the fat is raised. Oils are usually not fully hydrogenated when solid fats are produced, and in the preparation of margarine, some polyunsaturated fatty acids are left. Soybean oil, which may be used for frying, is slightly hydrogenated to eliminate linolenic acid (three double bonds); otherwise, the oil would develop a fishy flavor when heated. This is probably a result of a primary or intermediate stage of oxidation.

In hydrogenation, the oil is placed in a container and the catalyst (usually finely divided metallic nickel) is added. The oil is then heated under vacuum to 200 to 400°F (93.3 to 204.4°C). When the proper temperature is attained, the vacuum is discontinued, and hydrogen, under a pressure of 5 to 50 psi (0.35 to 3.52 kg/cm^2), is forced through the oil. The mixture is whipped, to expose as much oil surface to the gas as possible. Hydrogenation is allowed to proceed to the desired point as determined by periodic tests made on the material. After hydrogenation, the material is cooled to a point where it is still liquid, then filtered through bleaching clay to remove the nickel catalyst and nickel soaps. Fats are ordinarily cooled in a heat exchanger after filtering. This is done to retain a homogeneous material, because, if allowed to cool slowly, the harder fats would crystallize out at the bottom of the mass, the more liquid fats solidifying at the top. After cooling, the fats are pumped through a unit that whips them. This is done to prevent large crystals from forming and to maintain a smooth creamy texture. The whipped fat (shortening) is then packaged in containers of various sizes (1 to 100 lb [0.5 to 45.4 kg]), depending on whether the product is to be sold at retail or to be used by a food product manufacturer.

After packaging, the fat must be tempered. This is done by holding it, in its containers, at 80 to 85°F (26.7 to 29.4°C) for a period of 24 to 72 hours. Tempering is done to provide suitable creaming properties in the shortening. Edible fats and oils are used as bakery shortening to make bread, cakes, icing, and pastries; as a frying vehicle used in shallow- or deep-fat frying; for ice cream-like foods; as oil for salads; for the addition to canned fish products (sardines, tuna, etc.); for the manufacture of margarine; as special coatings for meat; for greasing pans used in bakeries; and, in the home, for baking.

MARGARINE

Large quantities of margarine are manufactured. Actually, margarine has largely replaced butter, both in the home and in the manufacturing of foods. By law, margarine may contain edible fats or oils, whole milk, skim milk, cream, or reconstituted milk solids cultured with bacteria (for flavor) or combinations of these materials; also, emulsifiers (mono- or diglycerides or lecithin), citric acid or citrates to tie up metals that accelerate oxidation, salt or benzoates or both (to inhibit bacterial spoilage), vitamins A and D, artificial coloring (carotene or annatto), and artificial flavoring (distillates from cultured milk or cream). The amount of fat present in margarine must be at least 80% of the finished product. All the above ingredients are not necessarily included in margarine, but some combination is used. Blending of the materials is carried out in two steps. All the fat-soluble materials are mixed with the liquid fats in one container, and the water-soluble materials in another. The two batches are then mixed to form a loose emulsion with the aid of a high-speed agitator that whips and beats the mixture. The emulsified material is then solidified in a heat exchanger, after which it may be kneaded. In any case, the cooled material is extruded into sheets or bars; then the extruded material is cut to length, wrapped, and packaged for shipment.

LIPID EMULSIFIERS

The mono- and diglycerides used as emulsifiers are made by heating fats or oil and glycerine, with sodium hydroxide, under vacuum at approximately 400°F (204.4°C). Under these conditions, some of the fatty acids, attached to the glycerine in the fat, migrate and attach to the free glycerine present. Commercial monoglycerides consist of about 50% monoglycerides (one fatty acid attached to the glycerine molecule), 40% diglycerides (two fatty acids attached to the glycerine), and 10% triglycerides or fats (three fatty acids attached to the glycerine). Pure mono- or diglycerides can be obtained by distillation.

SALAD DRESSINGS

Salad dressings include mayonnaise, and other products that differ from mayonnaise mainly in that they do not contain sufficient oil to form a true emulsion. Mayonnaise is made from vegetable oil (cottonseed or corn), vinegar, sugar, salt, mustard, white pepper, and egg yolk. The proportion of components may vary among different formulations. In popular usage, mayonnaise is not called a salad dressing. Salad dressings other than mayonnaise are made from vegetable oils, vinegar, spices, and in many cases, starch. The oil content in salad dressings, about 40%, is about ⅔ the amount used in mayonnaise (65%).

CHOCOLATE

The history of chocolate goes back to the time of the famous Spanish explorer, Hernando Cortez. He first tasted it in 1519 at the court of the Aztec Emperor Montezuma. All history before this is based on conjecture, myth, and circumstantial evidence. The name itself comes from the term "chocolatl" which was a cold, bitter drink that the Mayan/Aztec Indians made by mixing ground cacao beans with liquid. In Mexico, chocolate beans were considered quite valuable; in fact, they were used as money. Ten could buy a rabbit; 100 a slave. Chocolate was also thought by many to be an aphrodisiac. Madame Du Barry is said to have given it to her men, and Cassanova said he used it instead of champagne as an inducement to romance. Chemically, chocolate does contain traces of caffeine and larger amounts of theobromine. Both are stimulants and can produce an irritating, diuretic effect. It was also common at the time to add pepper to chocolate. Pepper can also scratch tender membranes, which is what the supposed aphrodisiac cantharis does. Whatever the case, statistics have not shown an unusually high number of births 9 months after Valentine's Day nor did the population of Western Europe grow significantly because of chocolate.

Chocolate begins with cacao beans which grow in melon-shaped seed pods attached to the cacao tree. They may be white or pale purple and are slightly larger than coffee beans. After the beans are picked and removed from the pod, they are fermented. This process, which may take place under a heap of leaves, both microbiologically and enzymatically changes the beans and develops both flavor and color. The beans are then dried to about 7% moisture and are ready for exportation and further processing.

At the chocolate and cocoa manufacturing plant, the beans are roasted to further develop flavor and color. They are then passed through machines to remove the seed coats and separate the germ. At this stage the chocolate is called "nibs." The nibs are now passed through various mills and are torn apart and ground, releasing fat from the cells. The heat of grinding melts the fat and the nibs acquire a liquid consistency. This liquid is known as "chocolate liquor" and will solidify on cooling. The solid liquor is the familiar bitter chocolate that is used in baking and other applications. It can be processed further by adding sugar to yield sweet or bittersweet chocolate or sugar and milk may be added to yield milk chocolate.

Fat may be pressed from the chocolate liquor to yield cocoa butter and the raw materials for cocoa. Some of the cocoa is treated with alkali (the "Dutch Process") to darken its color and modify its flavor. This is the cocoa used in the popular devil's food cake.

Chocolate, as we know it, differs in the amounts of chocolate liquor, cocoa butter, sugar, milk, and other ingredients. In the United States, "sweet chocolate" must contain at least 15% chocolate liquor, "milk chocolate" at least 10%, and "bittersweet" at least 35%. A high-quality sweet chocolate might contain 32% chocolate liquor, 16% additional cocoa butter, 50% sugar, and minor quantities of vanilla and other ingredients such as lecithin to enhance texture qualities. See Figure 22.3 for the production of cocoa and chocolate.

LIPID SUBSTITUTES

The trend toward lowering dietary intakes of lipids has led to the formulation of a variety of products that can be used as substitutes for fats and oils in the production

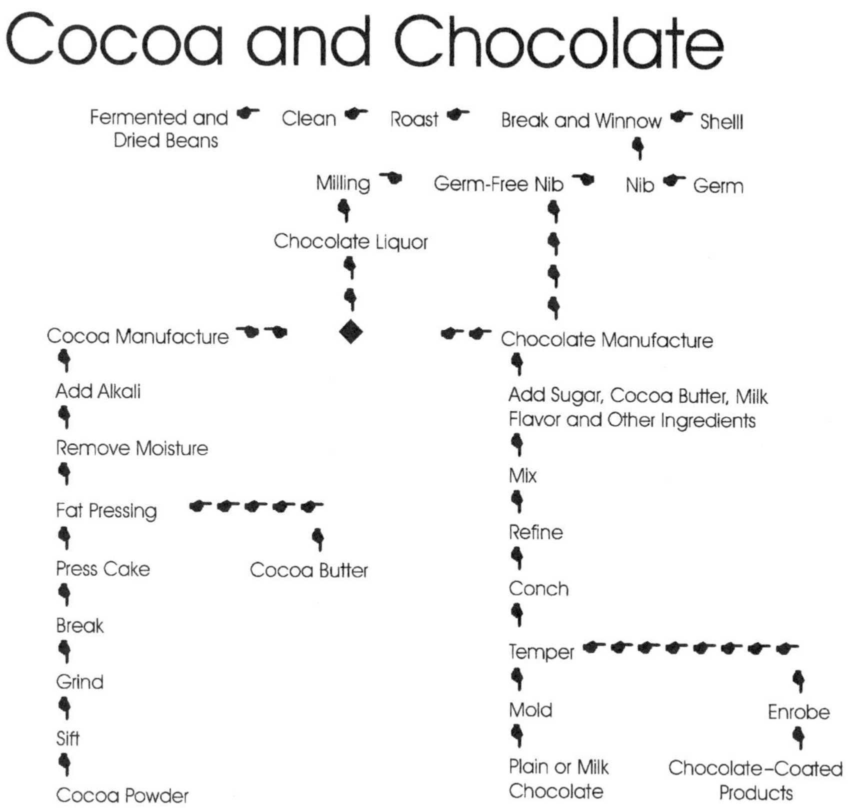

Figure 22.3. Production of cocoa and chocolate.

of a large number of food products that normally contain fats or oils. Some of these foods include baked goods, frozen desserts, and salad dressings, but the applications for lipid substitutes are expected to grow significantly as the relevant technology is expanded and improved.

One of the lipid substitutes is a product called Simplesse®. It is made from egg white and milk protein by a patented process called microparticulation. Another fat substitute is Oatrim®. It is made from oat bran or oat flour.

Other fat substitutes include nondigestible synthetic fats such as olestra, a class of sugar compounds from sucrose and naturally occurring fatty acids having 8 to 12 carbon atoms. Olestra was originally intended as a partial fat substitute in frozen desserts, shortenings, and cooking oils. It was approved for use in food by the Food and Drug Administration (FDA) in January of 1996 for "savory snacks" such as potato chips, corn chips, and crackers. Unlike other fat substitutes, olestra can withstand the high heat of frying so it can be used to fry some of these snack foods. One ounce of regular potato chips contains about 10 g of fat and 150 calories. Olestra-fried chips contain no fat and only about 60 calories. Some scientists claim that the use of olestra is potentially dangerous because it can cause diarrhea and also will absorb fat-soluble nutrients such as vitamins A, E, D, and K and carotenoids and remove them from the body before they can be absorbed.

Esterified propoxylated glycerols (EPGs) are a family of propylene oxide derivatives that are similar to natural fats but are not digestible. The use of EPG is as a total or partial substitute for fat in table spreads, frozen desserts, salad dressings, and bakery products.

Dialkyl dihexadecymalonate (DDM) is a minimally digestible fatty alcohol ester of malonic and alkylmalonic acids. It is used as a substitute in high-temperature cooking oils for producing potato and tortilla chips.

Carbohydrate-based fat substitutes include the gums (hydrophilic colloids), polydextrose (a polymer of dextrose with small amounts of sorbitol and citric acid), and a number of other modified starch derivatives having low caloric values and recommended as partial fat substitutes in a variety of fat-based food products.

Part IV

Food Science and the Culinary Arts

23

Equipment Used in Food Preparation

When considering "tools of the trade" in culinary arts, we must also include relevant scientific principles, a description of the components of this equipment, and how this equipment will function under certain conditions common to food processing and preparation. Here, we will discuss cutting tools, heating utensils, and sources of heat and equipment (stoves, etc.) to be used.

CUTTING EQUIPMENT

Under this heading will be included equipment that is used for making general cuts as well as those used for making specific cuts such as chopping and thin slicing.

There are a variety of knives available for general cutting purposes. When choosing knives, there are a number of things that must be considered. Among these are the type of knife, characteristics of the blade, durability, characteristics of the handle and, of course, cost.

Of the four most popular types of blades, the two most common ones to choose from are the carbon and the stainless steel varieties. Both are actually made from steel (an alloy of iron, carbon, and smaller portions of other elements), but the major difference is that the stainless steel type contains less carbon, more chromium, and often nickel. Other alloys have also been used to develop a high-carbon stainless steel blade and a third type, a superstainless steel blade. The latter contains more chromium and nickel in its plating alloy to give it a silvery look. Each blade has its strengths and weaknesses. The high-carbon blade will sharpen most easily and will hold a very sharp edge. Unless it is wiped immediately after use, however, it will rust or, if one is working with acid foods (e.g., citrus fruits or tomatoes), the acid may react with the metal, forming stains from which color and off-odor may be transferred to other foods. The stainless steel and superstainless blades have the advantage of not staining or rusting and are the least expensive. Both, however, are difficult to sharpen and will not hold a sharp edge. With the exception of expense, a fourth type of blade, the combination stainless steel–carbon, seems to contain the best combination of positive attributes. It doesn't stain or rust and it takes and holds a fairly good edge. The choice of blade is based on the needs of the chef or processor, but all the pros and cons must be weighed before investing. Once you choose the type of steel, the type of knife to be used must be carefully considered. In processing plants, a single type of knife may be sufficient (e.g., a 6- or 8-in utility knife for boning hams), but a chef will need different knives for different jobs. The three most often chosen are a 3- or 4-in. paring knife, an 8-in. chef's

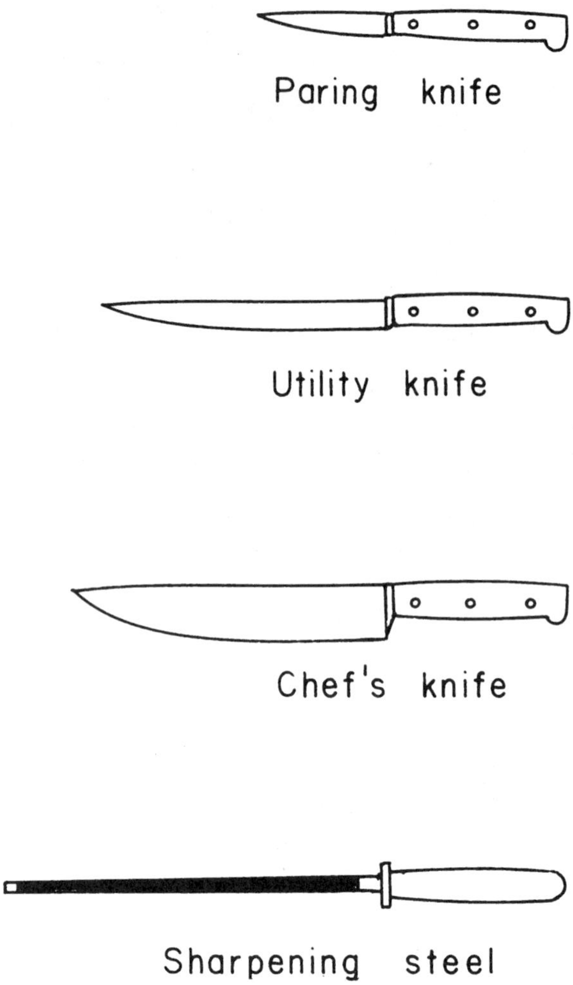

Figure 23.1. Basic tools for handcutting foods.

knife for chopping, and a 10-in. nonserrated slicing knife (see Fig. 23.1). A 10-in. steel for honing (to keep a sharp edge) would be needed in both the processing plant and kitchen. All knives, regardless of type, should be honed on a steel before each use session. Remember, more people are cut with dull knives than with sharp ones. If a knife is sharp, people tend to be more careful and it does not require as much energy to use. If a knife is dull, it is harder to work with, requires more force, and is more likely to slip. The latter two consequences of using a dull knife generally are the reasons that cutting injuries occur.

The next part of the knife that must be considered is the handle. It is claimed by some that wooden handles are less likely to slip. From a sanitation point, however, plastic handles are superior. Plastic can be molded so that the surface is not smooth but can still be cleaned and sanitized properly. If handled correctly, the slight difference in slipperiness is not noticeable and important differences in sanitary attributes will be gained. Wood will absorb water that has food residue, creating an environment suitable for the growth of bacteria (see Chapter 3). Wood has other disadvantages.

Wooden handles will eventually shrink after use and frequent washing. This will cause separation from the rivets, creating small crevices in which food and bacteria can accumulate.

The cutting surface on which the knife is used is important in maintaining its sharpness. Hard surfaces such as stainless steel, marble, glass, and enamel will dull a knife quickly. Other surfaces that would least dull a knife, such as a soft wood, do not meet the sanitary code for cutting boards. Hard wood cutting boards are still used in many establishments but they, like the aforementioned knife handles, will absorb water and when a crevice develops through use, a growth environment with moisture and food could be created. The best cutting surface to keep the knife sharp and to meet sanitary recommendations is unmolded white polyethylene or equivalent. Plastic boards do not absorb water, making them much easier to keep sanitary. Cleaning and sanitizing of cutting boards and knives should be performed after each use. The danger of cross-contamination exists if this is not done. Cross-contamination occurs when raw foods that contain microbes, parasites, etc. are prepared with equipment that will later be used in the preparation of cooked foods or foods that will be eaten raw. The equipment may be *cleaned* after the first use but not *sanitized!* Remember, something can be clean (free of visible dirt) without being sanitary (free of pathogens). To ensure sanitary conditions, carefully follow recommended methods described in Chapter 4 after each use and especially if using the same equipment for raw and prepared foods.

For fast chopping, thin slicing, and mixing, there are a number of commercial food processors, choppers, and slicers on the market. In these tools, the blades are all made from stainless steel for two reasons. First, they are the easiest to keep clean and stain-free. Because of the contoured shape of a high-carbon steel blade in food processors, it would be difficult to remove stains from it. This would significantly raise the risk of a cutting injury. Second, even though stainless steel does not hold the edge as well as high-carbon steel, there is no great loss here because these blades are not subjected to the dulling that is encountered by cutting knives (cutting surfaces, bone, and other hard tissues). Also, the high speeds at which food processors operate allows them to cut and chop even if they have lost some of their original sharpness. The slicers, however, do not operate at very high speeds. Their slicing ability is maintained as a result of the serration of their cutting edges. The general rules of sanitation apply to this equipment. Purchase equipment that is easily cleanable and has National Sanitation Foundation International (NSF) approval. The NSF is a nonprofit research and testing organization that evaluates foodservice equipment and materials. The food processors should be disassembled, cleaned, and sanitized as described in Chapter 4 after each use. Slicers may be in a location where they will be used periodically over a longer period of time (deli, sandwich shop, etc.) and it would not be practical to disassemble, clean, and sanitize after each use. In these cases, remember the growth patterns and requirements of bacteria (see Chapter 3) and prevent them from getting the opportunity to grow and reproduce in large numbers. Use small amounts of food to be sliced and keep the rest at temperatures out of the "danger zone" (less than 40°F [4.4°C] or more than 140°F [60°C]) until ready for use. The slicer should be disassembled, cleaned, and sanitized every 3 to 4 hours or whenever you switch from raw to prepared foods (remember cross-contamination). Be sure to purchase slicers that are easily disassembled and these precautionary sanitation measures will take only a few minutes. The same considerations mentioned with knives (type, frequency, and capacity of the task to be performed) apply when choosing the proper food processor or slicer. Be sure to select one with enough horsepower in the motor and, with food processors, capacity in the bowl. Also, belt-driven models can balk and slip when processing more difficult products (i.e., chopping meat). It is better to invest the extra money and buy the models that have the motor's drive shaft directly rotating the cutting blade (direct drive).

HEATING UTENSILS

The way heat reaches food varies. It involves phenomena described as radiation, conduction, and convection. In radiation, units of energy that are described as "quanta" travel as electromagnetic waves and are a source of heat. These infrared rays, known as radiant energy, go directly from the source to the object heated, unassisted by another medium. Sources of radiant energy in food preparation are the broiler, toaster, coals in a charcoal pit, a gas flame, and the coils of an electric stove. It should be noted, however, that all objects with any space around them, even cold ones, emit radiant heat.

In conduction heating, the transfer of heat occurs by contact, from molecules at any temperature to adjacent molecules at lower temperatures. Thus the heated molecules of a hot plate heat the molecules of the lower surface of a pan bottom which in turn heats adjacent molecules until heat from the upper surface of the pan transfers heat to the food with which it is in contact. Thereafter, the food is heated by conduction (contact) and convection (heated fluid in motion described below).

In convection heating, a fluid (gas or liquid) undergoes rise-and-fall patterns. In this manner the spread of heat throughout a container is accelerated as opposed to heat transfer by conduction only. Most cooking is done in pots and pans, and the transfer of heat always involves conduction and usually convection.

When food in a pan is heated on a stove, energy from the heating coil or gas flame is transferred directly to the pan by conduction because they are in direct contact and heat is transferred from one molecule to another throughout the pan. The hot pan will transfer heat by conduction to the food. In liquid foods, water molecules are heated and a portion of them get hot and become less dense than cooler water molecules that have not yet been warmed. The heated, less dense water rises while the cooler, more dense molecules will flow down toward the heat source and a circular flow of energy transfer (top to bottom and bottom to top) is developed. This movement, called convection, allows faster transfer of heat from molecule to molecule (through conduction) because the cooler ones come into contact with the hotter ones more often. Convection heating is also attained when heat arrives by air, such as in an oven, or in deep fat frying. More on conduction and convection was described in Chapter 10.

When choosing pans, especially those for stove-top cooking, one must consider heat distribution. If heat cannot spread quickly through the entire bottom of a pan, "hot" and "cold" spots will develop. These will cause uneven cooking and may even burn a portion of the food and not thoroughly cook others.

The fastest heat conductor among the most popular pan materials is copper, with aluminum not far behind. Cast iron and carbon (rolled) steel (used in traditional woks and crepe pans) are next, with stainless steel still slower. The slowest of all are glass, porcelain, earthenware, and pottery in general. The thickness of the material is also a factor in heat transfer. The thicker the gauge, the more evenly the heat distribution throughout the pot's or pan's interior surface. Each substance has its advantages and disadvantages and some of these are as follows:

Tinware (tin-plated iron or steel)
Advantages: Fair conductivity.

Disadvantages: Will mar and then rusts quickly. It turns dark after use and is affected by acid foods.

Aluminum

Advantages: It heats quickly and has good heat diffusion and heat conduction, and in heavy gauges, heats relatively evenly. Light gauge aluminum, however, it will develop hot spots.

Disadvantages: It pits and will discolor food. This is seen when stains on the pan develop when foods such as potatoes are cooked or the pan is washed with a high-alkali cleanser. When acid foods such as tomato sauce are cooked in the stained pan, the color will be absorbed. To avoid this, the stain may be removed by boiling in a weak food acid solution such as cream of tartar (2 teaspoons to 1 quart of water) for 5 to 10 minutes.

Copper

Advantages: It heats very quickly and is best in the heavier gauges because it gives a quick, even heat distribution when kept clean.

Disadvantages: It stains easily and must be well tinned or lined with stainless steel on surfaces contacting foods for copper will leach into acid foods and could prove poisonous. Copper is also a catalyst and may enhance spoilage.

Cast iron and carbon rolled steel

Advantages: They are sturdy and can withstand high temperatures without warping.

Disadvantages: They rust easily and can discolor food. The pans should be "seasoned" before using for the first time. This can be done by rubbing with vegetable oil and heating in an oven at 300°F (149°C) for 30 to 60 minutes. (Some recommend 450°F [232°C] for 30 minutes). What happens here is that microscopic jagged peaks in both nonstainless iron-based metals are coated with oil and appear to be smoothed, which helps to keep water from seeping in and creating rust.

Stainless steel

Advantages: This is the easiest to keep clean and it does not stain or impart impurities to food. In heavy-gauge material, pots and pans are sturdy.

Disadvantages: It has poor heat conductivity. This can be remedied by getting lighter gauges but then the development of "hot spots" could be a problem. Also, by constructing the pan so that the bottom thickness is decreased and clad with copper underneath, heat transfer is improved.

Glass, porcelain enamel, earthware

Advantages: If the gauge is heavy enough and they are covered, they hold heat well (especially earthenware) and will keep food warm during service.

Disadvantages: They break if dropped or hit by a hard object. They have poor heat conductivity and as a result they can break easily with quick temperature changes. When very hot water, for example, is poured quickly into a cold glass container, the bottom of the container will expand. Because conduction is poor, the top will not expand as rapidly, resulting in a structural stress that causes the glass to crack. The porcelain enamelware may chip and, unless treated to resist acid, the glaze can react with the acid. With the high degree of breakability, there is a danger of chips that may enter food. There are treated glasses such as Pyrex and Corning Ware that are less vulnerable to heat and breakage but they too have their limits.

There are also pans available with nonstick surfaces. These are coated with polytetrafluoroethylene (a solid, chemically inert [will not react with or absorb other substances] plastic that is baked on the surface). The plastic, better known as Teflon (the trademark of one manufacturer of this substance), covers the jagged peaks of the metal, making it impossible for the food to cling to them. The metal usually used in these pans is aluminum.

Cleaning and sanitizing of pots and pans can be achieved as described earlier. Some cookbooks recommend never washing with soap or detergent pans that have been "seasoned." This advice can lead to off-odors and -flavors from oxidizing fats (see Chapters 8 and 9) and bacterial growth. Methods of cleaning and sanitizing described earlier will not destroy the "seasoned" pan. The only difference is that the pans should be dried with a disposable towel to remove excess oil. If pans get scratched with a metal utensil, they will often be "reseasoned" the next time they are used for frying by the method described earlier.

The choice of cooking utensils depends on individual needs but the best general recommendation is to choose a pan of fairly heavy gauge, the bottom of which will diffuse heat evenly.

TYPES OF HEAT AND HEAT SOURCES

When choosing a source of heat and method to be used for food preparation, a number of things must be considered. First, the function of heat itself should be considered because heat changes the flavor, aroma, and color of foods. They may be either depleted or intensified which can enhance a food's palatability and appearance or, if not carefully controlled, lower the quality appreciably. Heated food also may become more chewable, digestible, and, as a result of its increased temperature, have a higher psychological appeal. Texture changes are quite evident as heat is increased and proteins coagulate. Foods may become more firm or the proteins may play a role in the processes of emulsification (Chapter 14), gelatinization (Chapter 14), or leavening (Chapters 19 and 24). Heat also plays a role not often mentioned with regard to cooking—its effect on the microorganisms that are present. Water activity levels may be lowered (see Chapter 11) and pathogenic and spoilage bacteria can be destroyed, thus preserving the food (see Chapter 10). Care must be taken, however, when heating and holding food in the warm or hot state. As mentioned in Chapter 4, improper heating can lead to rapid microbial growth, resulting in food spoilage or poisoning.

The functions you wish to achieve determine whether you use moist heat or dry heat. Heat transfer in food has been discussed earlier in this chapter and methods to achieve this are numerous.

With moist heat, food is heated by convection currents in hot water. There are advantages to this method. Foods that are naturally tough such as meats with a large amount of connective tissue or plants with generous amounts of fiber can become quite tender using moist heat. The water medium prevents drying of the product. The temperature of cooking is quite easy to maintain and hold. Remember, water boils at about 212°F (100°C) and will not get much hotter or cooler. Many natural juices and nutrients in foods prepared with moist heat migrate into the cooking medium and can be incorporated into sauces or gravies. Some moist-heat methods are:

- Boiling/Simmering/Poaching
- Steaming
- Pressure-cooking

Boiling, simmering, and poaching are very similar. Rapid boiling does not raise the temperature of the water. The extra energy incorporated to cause the rapid boil is lost as the water changes phases from liquid to gas (steam). Rapid boiling can adversely affect the texture of some foods (e.g., vegetables in a stew) but is desirable in others such as pasta to help keep it from sticking together. Simmering is probably the most important form of moist heat. The temperature is controlled so that some of the water molecules convert to steam but a rapid boil is not obtained. As the heat from the boiling water molecules is being transferred to the food, the temperature of the total product is lower than 212°F (100°C). Simmering temperatures should be at least 140°F (60°C) to ensure safety from pathogenic microorganisms (see Chapter 6) and commonly go as high as 160°F (71.1°C). This method protects fragile foods and tenderizes tough ones. Poaching is an interesting application of moist heat. The temperature of the water is just below boiling and the method incorporates basting or self-basting. Food is placed in simmering water and basted with the cooking liquid. Self-basting can be achieved by using a lid so that as steam condenses on the lid, it performs this action.

Steam is less dense than water and will make less frequent contact with the food. To compensate for this loss of efficiency, steam has a gain in energy. The vaporizing molecules are more energetic than those in the liquid phase and the pressure produced in a closed container will raise the temperature slightly.

As mentioned in the chapter on thermal processing (Chapter 10), higher temperatures can be achieved by boiling water under pressure. Pressure cooking incorporates this concept and can lower the cooking time significantly. Flavors and nutrients that may be destroyed by long periods of simmering or boiling are preserved to a higher degree in pressure cooking. This holds true especially for vegetables. Some meats are cooked with moist heat because they are not very tender.

Dry heat may be generated in a number of ways but all employ temperatures higher than those used in moist heat. Certain cooking goals, such as browning, searing, and crisping, can only be achieved at these elevated temperatures. Dry heat cooking methods include:

- Roasting/baking
- Broiling/toasting
- Pan-frying
- Deep-frying
- Microwave

Roasting and baking basically describe the same method. The difference exists in the actual definition. If the process involves a whole bird or piece of meat (other than a ham or minced-meat preparation) that will later be divided to serve, it is defined as roasting. Whether called baking or roasting, radiant heat of the oven, air, and heat from the pan is conducted and convected throughout the food. Moisture is lost from food as vapor and is circulated in the oven, thus somewhat moistening the "dry" environment. Temperatures vary, but rarely get much above 400 to 500°F (204.4 to 260°C). Some baking and roasting methods use high temperatures at the beginning to sear the outer surface, thus sealing in juices, then lower the temperature to more evenly distribute the heat throughout the food without overcooking outer layers before the internal surfaces cook.

Broiling and toasting are controlled versions of the oldest culinary technique, roasting over an open fire or glowing coals. Broiling and toasting depend largely on infrared radiation. All heat sources used in broiling and toasting emit visible light, and so are intense radiators of infrared energy. Nickel–chrome alloys used in electrical appliances reach temperatures of 2000°F (1093.3°C) and a gas flame is about 3000°F (1648.9°C). Temperatures of an oven wall rarely exceed 500°F (260°C). The production of these extremely high temperatures is the great advantage of broiling but, at the same time, it is the major problem. The development of intense, familiar flavors and colors in broiling result from the high-temperature cooking, but there is a huge disparity between the rate of radiation at the surface and the rate of conduction, via water, within the food. This explains the "burned" steak that is cold in the middle. Infrared radiation weakens quickly with distance because the rays spread out in all directions and become less concentrated. Thus, the skilled broiler cook can find the ideal distance from the heat source where the radiation and conduction can produce a rare, medium, or well-done steak whenever desired.

Pan-frying depends mostly on conduction and convection. Oil is often added to the pan and it serves several purposes. It brings the uneven surface of the pan into more uniform contact with the heat source, it lubricates and prevents sticking, and it supplies flavor. The problems encountered with broiling (burned outside and cold middle) can occur in pan-frying. Cooks will sear the surface at high temperatures and then lower the temperatures to allow the heat to penetrate by conduction and convection more evenly.

Deep-frying resembles pan-frying except that it uses enough oil to immerse the food completely. As a technique, it is similar to boiling but the temperature of the oil can get twice as hot as boiling water and therefore cooks food much more rapidly and can brown foods. The type of oil that is used determines many attributes of the final product such as flavor, color, and amounts of saturated fats and cholesterol absorbed by the fried food (see Chapter 2). Temperature of the oil is important and a frying thermometer should be used. If the oil is too cool when food is added, the food will cook slowly and will absorb more oil, becoming "greasy." If the oil is heated too high, smoking will be produced which indicates the oil is breaking down and will be no good for reuse. Also, if the oil is too hot, the same problem could develop that developed with other high-temperature methods—the outside may burn while the inside is still raw. Frying is certainly an art and results usually improve with practice.

Microwave heating and how it works is described in Chapter 10. It is included in the "dry" heat section because water is not used as the cooking medium. Actual heat, however, is not really generated by the oven or the microwaves themselves. The water molecules in the food vibrate when exposed to microwaves and this movement generates heat. The microwaves penetrate only a couple of inches into most foods, so the term "heating from the inside out" is not totally correct. This penetration is much greater, however, than in traditional heating, where the energy is absorbed almost entirely on the surface. Once the microwaves have activated the water molecules in the outer 2 inches, further heating is accomplished by conduction as mentioned previously. Non-water-containing materials such as paper, glass, and plastic will not get hot unless they contain foods that heat them through conduction. There are advantages and disadvantages to microwave cooking. The main advantage is speed. Cooking time can be cut by 50% to 75%. Microwave heating uses electrical energy more efficiently, and therefore costs less. With the generation of steam from within, baked goods may rise higher and with temperatures not getting much above that of boiling water (212°F [100°C], there is less splattering.

There are several disadvantages of microwave cooking that should be mentioned. Only "microwavable" containers may be used. These include glass and Corning Ware because microwaves can penetrate through them. No metal containers or metal foil can be used because microwaves cannot penetrate and will be reflected. Microwave-cooked meat can have a drier texture because the quicker heating can cause greater fluid loss. This also makes it harder to control the doneness of a roast. Even heat penetration in microwave cooking has been a problem. In larger casseroles, roasts, and whole poultry, cold spots may develop, leaving a possible microbial safety problem. To avoid this, use thermometers and check in several places making sure that you have reached at least 160°F (71.1°C) for beef; 170°F (76.7°C) for veal, pork, and lamb; and 185°F (85°C) for poultry. Stirring, rotating, deboning meat, and allowing food to "rest" after microwaving will also help to avoid cold spots. Covering your cookware will trap escaping steam and use this energy in the heating process. Foods do not brown in the conventional microwave oven. Some manufacturers have incorporated heat sources or a convection fan to raise the temperature for browning and development of flavors that accompany the process. Packaging technology has also helped to solve the browning problem. "Heat-susceptor" packages contain thin, gray strips or discs of metalized plastic. These absorb microwave energy and can reach temperatures sufficient to brown and crisp foods. Safety of microwave ovens is always a concern because microwaves can penetrate body tissue, resulting in injuries such as cataracts. Accordingly, it is important to be sure the door is checked for tight fit and the seal is not damaged. The seal should be conscientiously cleaned and free of spills and splatters. "Microwave leak testers" can be purchased to check the door and seal. This is extremely important for people who have coronary "pacemakers" because the microwaves can alter their proper function.

Equipment to cook food on or in basically includes three choices: microwave ovens, gas or electric appliances. There are advantages and disadvantages of each. The advantages and disadvantages of microwave ovens were mentioned but there are differences among microwave ovens. The power potential, expressed as watts, may vary. Some ovens may have a wattage as low as 400 while others operate at 700 watts. The required heating time will vary. Ovens having the lower wattage take longer time to cook food. Gas and electric cooking also has advantages and disadvantages. For stove-top units, gas has the advantage, primarily because of the quickness that temperature adjustments may be made and the range of actual cooking temperatures available. Electric burners take several minutes to reach the required temperature and to cool, and you are limited to the settings (e.g., low, medium, high) on the dial. Electric ovens and broilers, however, seem to have the advantage over gas. Ovens reach desired temperatures more quickly, maintain a relatively steady temperature, and, generally, are more accurate. Some electric ovens can also reach very high temperatures (1000°F [537.8°C]) which allows them to "self-clean." At these high temperatures, organic compounds (most food and grease that accumulates in ovens) will be burned (oxidized) to carbon dioxide and water (steam) which is exhausted. What remains is a white ash that consists of the inorganic portions of food (i.e., calcium, sodium, and other minerals). This ash can be easily wiped away with a damp cloth after the oven has cooled. A combination of gas and electric would probably be best but the choice of one or the other will probably have to be made. Weigh the pros and cons, determine what the needs are, and the best choice can be made.

24

Food Preparation—An Important Application of Basic Chemistry and Physics

When food is prepared for consumption, many of its characteristics such as texture, appearance, taste, and nutritional value are altered. These result from changes in the physical and chemical structure of foods. An understanding of the basic physical and chemical nature of food is an invaluable tool for chefs, bakers, and anyone involved in food preparation, and this knowledge enables one to control quality more effectively. A change in methodology or substitution of an ingredient may improve the final product or provide savings in time and money. This chapter gives short explanations of why certain changes occur in a variety of foods.

MEAT

The basic quality factors of any food are appearance, texture, and flavor. A high-quality meat looks good (rare is good to some and well-done to others), is tender, and tastes and smells good. The composition, structure, and method of storage and preparation will determine the quality. Chapter 14 discusses different types of meats and how they are processed.

The composition of meat varies among different species and within the same species. The protein contents vary but usually fall between 15% and 20%, but the fat contents are more inconsistent, having a range from 5% to 40%. Nutritionally, the protein from meat is of high quality and the percent calories from fat varies depending on the cut and species. Also, meat contains cholesterol and saturated fats, and it is recommended by nutritionists to limit the amount of fatty meats consumed. The lean tissue of meat contains mainly water and protein. The protein consists of bundles of muscle fibers which are the basic structural unit of lean tissue. Other protein materials include connective tissue that holds the muscles together and the oxygen-containing pigments and enzymes associated with muscle fibers.

The texture of meat or its degree of tenderness depends on a number of factors. The amount of fat in the connective tissue within the muscle is thought by some to be an accurate measure of tenderness. It certainly gives the impression of tenderness and juiciness because when melted, it lubricates the lean, but it is not considered a reliable predictor of tenderness. Tenderness appears to depend on two factors: the nature of the connective tissue and the condition of muscle filaments after slaughter. The nature of the connective tissue is closely related to the age of the animal. Young animals have collagen that is more water soluble and more heat labile than that of older animals. When cooked, the collagen degrades and does not contribute to toughness. As the animal gets older, the collagen becomes more heat stable and therefore is not broken down by cooking as readily. Muscles that have done much work have been shown to contain even more of the heat-stable collagen. Meat that is not affected by collagen may also become tough through the condition of muscle fibers. A few hours after slaughter, meat goes into rigor, a condition that leaves the muscles rigid and inelastic. Aging the meat after slaughter allows the rigor to pass and the muscles to become soft and pliable again and for innate enzymes to tenderize the meat. During aging the temperature is held at 34 to 38° F (1.1 to 3.3°C) to prevent microbial growth and the relative humidity at about 70% to prevent drying. The optimum period is about 11 days for beef and 1 day for pork. This can be quickened for beef by raising the temperature to 70°F (21.1°C) and 2 days at a relative humidity of 85% to 90% under ultraviolet light to control microbial growth. This fast-aged beef is usually what is sold at retail markets. Cooking methods and tenderness are discussed in Chapter 23 but, basically, moist heat and lower cooking temperatures lead to more tender meat. Other ways to make meat more tender are to marinate it or to apply commercial meat tenderizers. The acid in the marinade from vinegar (acetic), lemon juice (citric), or wine (tartaric) chemically softens the collagen and the meat tenderizers soften the connective tissue through enzymatic action (see chapter 8). Another method to help tenderize is to cut across the grain and in thin slices. This helps to shorten the fiber lengths, thus making chewing easier. Grinding, scoring, and pounding also help to tenderize in the same way although scoring only affects the surface.

Appearance in meat is mainly the result of the color developed during cooking or in a curing process. The pigment responsible is not hemoglobin in blood and the liquid from a rare steak is not blood, because most of this blood is removed through the arteries at slaughter. The red color is due to the pigment myoglobin which is in muscles and stores oxygen. With heat, the pigment color changes and can be used as an index of doneness. To cook a steak rare, it is heated to an internal temperature of 140°F (60°C) and the pigment is bright red with a thin layer of denatured myoglobin that is brown in color. Many cooks do not heat rare meat to this level but it has been shown that this temperature will ensure the destruction of *Salmonella* and is the basis for this recommendation (see Chapter 3). Medium rare, reached at 160°F (71.1°C), has a pink color because more of the myoglobin has been denatured. Well-done meat is cooked to 170°F (76.7°C) and it will be brown colored throughout. When sugars are present, such as in cured meats, a brown pigment is also produced from a reaction of the sugar and the protein (see Chapter 9). This also happens when cooking liver because of its glycogen content. The pink color in cured meats, such as ham, results from myoglobin combining with nitric oxide from the sodium nitrite that is in most cures (see Chapter 14).

The amount of myoglobin in the muscles also determines color of meat in poultry. The muscles that require more oxygen because they will be exercised more, such as the leg muscles, will become darker during cooking while the others, such as the wings and the breast, which do little work in poultry will be white. Game birds, on the other hand, will have dark meat in the wings and breast because these muscles are used more and need more oxygen.

Flavor in meat depends upon a number of factors, but, generally, well-exercised, tough meat is more flavorful than that which is more tender and less exercised. Cooking decomposes some factors in lean meat to give the basic cooked meat taste. It seems that a long cooking time is needed to develop a full, meaty flavor, as tests done on meats prepared by quicker methods, that is, in microwave ovens and pressure saucepans, did not have as desirable flavors as those prepared in the conventional manner. The meat that is closer to the bone also contains more flavor because it is in the most favorable position to absorb flavorful compounds from bone which are also used to give stocks their rich flavor. The aroma from heated adipose tissue (fat) with some water-soluble components distinguishes the aroma of different species such as beef, pork, and lamb. Another flavor that may be familiar is that of reheated meat that has been refrigerated. This "warmed-over" flavor is a result of oxidative rancidity (see Chapter 9). Meats with relatively higher amounts of polyunsaturated fatty acids (see Chapters 22 & 2) such as pork are more susceptible to this flavor development.

DAIRY PRODUCTS

All dairy products are made from fluid milk that is either chemically or physically altered. Chapter 15 discusses how milk is handled, processed, and used in different products. In food preparation, milk and milk products can be added to many foods to improve texture, appearance, and flavor. Whole milk contains a complete protein, some carbohydrates, and fat of about 3.8% (individual states control what the fat content of whole milk will be). It also contains about 88% water so the percent calories from fat in whole milk is quite high (over 50%). Milk contains saturated fats and cholesterol. Although drinking milk daily is recommended, a large proportion of it should be skim milk. When cooking with milk, a few problems exist, but these can be corrected when the causes of the problems are understood. One problem is the tendency for the proteins in milk, as with many other proteins, to coagulate when they are heated. When milk is heated, water is being evaporated at the surface, and the protein content is concentrated and it complexes with calcium salts, coagulates, and forms a film. When one removes this, much high-quality protein is lost. This can be avoided if evaporation is slowed down by covering the pan, whipping up a little foam, lowering the heat, or stirring. Another problem is scorching. Milk proteins are relatively dense, and casein and whey proteins will fall to the bottom, stick, and burn. The best ways to avoid this are to use gentle heat such as that from a moderate flame or double boiler or stirring. Milk can also become more susceptible to curdling in the presence of any acids from fruits and vegetables or phenolic compounds in foods such as potatoes or tea. Fresh milk and careful control of the cooking temperatures are the best defense against this. Heat also affects the flavor and color of milk. Some protein derivatives such as hydrogen sulfide and methyl sulfide contribute to the flavor of cooked milk. Heated milk fat also gives a compound (delta-decalactone) that gives the flavor found in foods cooked in butter. Other milk products are produced without heat. Butterfat is separated from whole milk and concentrated in varied amounts to make cream which exists in three major grades: light cream, light whipping cream, and heavy whipping cream (see Chapter 15). The names of the cream and their actual weights are inversely proportional. The heavy cream, because of its higher fat content, is actually lighter in weight than the others and the light cream is the heaviest. "Half and Half" is an intermediate between milk and cream and contains at least 10.5% butterfat. Whipped cream is a favorite among bakers and chefs alike. It is made when air is whipped into the liquid, some proteins are caught in the walls of the bubbles, and the imbalance of forces here causes the normal shape of the proteins to become distorted. They react with each other and form a thin film of coagulated molecules. This film gives the liquid foam a solid, delicate reinforcement. The foam in cream is much stronger than that in milk because of the higher fat content. The fat globules apparently cluster in the bubble walls where surface forces rupture some of their membranes. The soft fat now sticks together, forming a quasi-rigid but delicate network that reinforces the foam. This cannot be formed by protein alone. Cream whipped to optimum volume is soft and glossy but there is some leakage. A stiffer cream can be achieved with more whipping but care must be taken not to over do it because the stiffness indicates clumping of fat globules and if whipping is continued, butter will be formed. The temperature of the equipment and cream used is important. Cool equipment is much better because the fat globules get soft when their temperature rises 5 to 10°F (2.8 to 5.6°C) and they will be deformed by the weight of the foam, and the whole structure will be weakened. The bowls used should be cooled in a freezer for 20 to 30 minutes. Cooling the cream itself in the freezer is a good idea, especially in the summer, but it should not be allowed to freeze. The water leaves the solution to form ice and prevents the even dispersion of fat, and a good foam will be difficult to achieve. To attain a sweet whipping cream, add the sugar near the end of the process. Sugar decreases the potential volume of the cream because it probably interferes with the clumping proteins on the fat globule membranes. If a thicker cream is desired, gelatin (adding more protein) or lemon juice (adding acid to denature protein) may be added.

POULTRY AND EGGS

The handling and processing of poultry and eggs are discussed in Chapter 16. In the kitchen, the basic guidelines for cooking meat can be used for poultry, with some changes depending on the tenderness and fat content of the bird which are mainly determined by age. Older birds are cooked by methods suitable for tough cuts of meat (see Chapter 23). Poultry, like beef and pork, is a good source of protein but contains less fat, especially if the skin is removed. White meat is less fatty but the dark meat is more flavorful (see meat section). Poultry is a good addition to any diet attempting to reach the recommended 30% or less calories from fat. Young chickens of either sex are called broilers if they weigh about 2.5 lb (1.1 kg) and fryers if they weigh 2.5 to 3.5 lb (1.1 to 1.6 kg). Roasters of either sex are 8 months old and weigh 3.5 to 5 lb (1.6 to 2.3 kg). Capons are castrated males and weigh 6 to 8 lbs (2.7 to 3.6 kg). Fowl are hens aged 10 months or more and stags and cocks are males too old to roast but add flavor to any "crock pot" meal.

Much of the poultry sold contains viable *Salmonella* organisms (see Chapter 3), so it is essential to clean and sanitize any cutting boards or utensils and to wash one's hands after handling poultry to prevent cross-contamination. After choosing the method of cooking, doneness of poultry can be determined in a number of ways, but use of a thermometer to measure internal temperature is the most reliable. All poultry is cooked to the well-done stage. Temperatures in the thigh and breast muscles are the best indicators of doneness and tests have shown that, for poultry cooked in a 325°F (162.8°C) oven, optimum tenderness was reached at 170 to 175°F (76.7 to 79.4°C) for white meat and 175 to 180°F (79.4 to 82.2°C) for dark meat. If a stuffed bird is cooked, the stuffing must be cooked to at least 165°F (73.9°C) to ensure destruction of all pathogens. For appearance, the bird may be browned by raising the temperature at the end of the cooking period or by basting with butter. (See browning in Chapter 9). The flavor in poultry, as in meat, intensifies with the age of the animal. Chemicals called volatile carbonyls (see Chapter 9) are responsible for the "chickeny" aroma and without them, a beeflike aroma exists. Cooked chickens and turkeys contain fat that is susceptible to oxidative rancidity and may, in some cases, account for a "warmed-over" flavor as described in the meat section.

The egg (also discussed in Chapter 16) is an interesting mixture of chemicals that can be changed in appearance, texture, and flavor by addition of heat or chemicals or by physical manipulation. Nutritionally, eggs have the most complete protein. The egg white is almost all water and protein (see Chapter 16), and the yolk contains a number of nutrients including fat and cholesterol. It is recommended that whole egg consumption be limited to three to four per week, but egg white consumption is not limited. The behavior of the egg in the kitchen is mostly a matter of protein chemistry and, in particular, the chemistry of coagulation. Albumen proteins are long chains of amino acids (see Chapter 2) that are folded into a globular shape. Each protein holds its shape by a number of bonds between parts of its chain. Different molecules are prevented from bonding to each other because, owing to the environment in the albumen, a net negative charge is attained by each, thus resulting in their repelling one another. Acidity, salt, temperature, and even air bubbles can disturb this and cause the molecules to join together or "coagulate." The perfect example is when an egg is cooked. The addition of heat disturbs the bonds in the individual molecules and the proteins unfold. As the heat is increased, the unfolding molecules are now longer and have more areas exposed that may bond with other unfolded proteins. When this happens, water in the white of the egg may be trapped in the mass and the liquid has suddenly become a solid with the ability to deflect light rays, turning it from clear in appearance to white (see Fig. 24.1).

The chemistry of the egg can be examined further in the different cooking processes. Eggs are prepared in a number of ways including boiled in the shell, poached, fried, scrambled, and as a component in custards.

Eggs can be boiled either soft or hard, depending on the length of time they remain in the cooking water. Soft boiled eggs are not recommended because *Salmonella* have

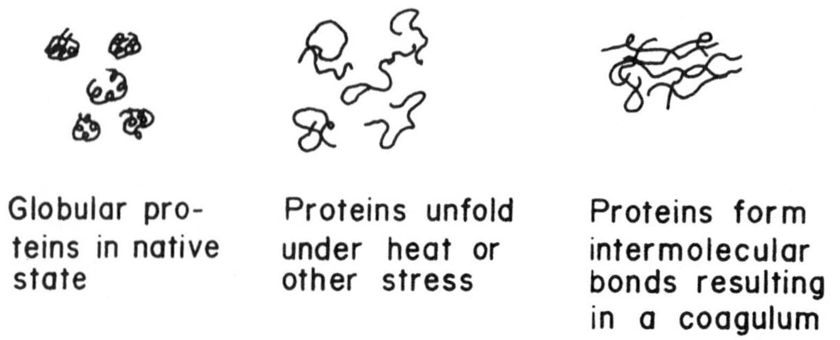
Globular pro-
teins in native
state
Proteins unfold
under heat or
other stress
Proteins form
intermolecular
bonds resulting
in a coagulum

Figure 24.1. Molecular depiction of coagulation in eggs.

been found in the yolk of supposedly "safe," uncracked eggs (see Chapter 3). The temperature of cooking will determine the quality of the fully cooked egg. If high temperatures at or near boiling are used, the white may become very rubbery (see protein coagulation in this chapter) before the yolk is completely cooked. To avoid this, cook at lower temperatures (185°F, 85°C) even though it will take longer (25 to 35 minutes as opposed to about 12 minutes in boiling water) but the white will not be rubbery and the yolk will be fully cooked. Owing to the expansion of air in the egg being boiled or just the temperature change, the shell often cracks and some of the white oozes into the water and cooks. To avoid this, use a little chemistry and coagulate the white before it leaks out. This can be done by adding salt or vinegar to the cooking water and the proteins are denatured and coagulate at the crack. The problem can also be solved by physics. Puncture the shell at the large end where the air pocket is so that the expanding air during heating can escape. Another problem in boiled eggs is the development of a greenish-gray color on the surface of the yolk. This is a result of more chemistry; a harmless chemical called ferrous sulfide is formed from iron in the yolk and sulfur from the protein in the white. When the protein is heated, some of its sulfur atoms are liberated and combine with hydrogen, forming hydrogen sulfide which, in small quantities, gives cooked eggs a pleasant odor but, in large quantities, smells like rotten eggs. As this forms, it spreads and reaches the yolk where it finds the iron and forms the darkly pigmented ferrous sulfide. The way to avoid this is to limit the amount of hydrogen sulfide formed. This can be done by cooking only long enough to just harden the yolk and then place in cold water. This lowers the pressure of the gas in the outer regions of the white (cool gasses generate less pressure) and cool protein loses less sulfur. Peel the eggs as soon as they are cooled as this will stop gas diffusion toward the yolk and reverse it.

To prepare poached eggs, remove from the shell and drop in hot water. The proteins begin to coagulate immediately and usually form their final shape. To control this shape, use egg poachers to eliminate too much spreading of the white. Hot water can be used to "baste" the egg to fully cook the yolk as recommended earlier.

To prepare fried eggs, place on a hot pan to prevent spreading of the white. The pan should have a nonstick surface (see Chapter 23) or be coated with a layer of fat or oil to prevent sticking. The cooking temperature should be about 280°F (137.8°C). If the temperature is too hot, the egg will splatter; if it is too cold, the white will spread undesirably. Sunny side up eggs are not recommended any more because of the possible presence of *Salmonella*, but if fresh eggs from a reputable dealer are used, the chances of infection are small. To avoid runny sunny side up eggs, add water to the pan and cover it to allow the heat from the steam to coagulate the protein at the egg's surface.

To make scrambled eggs, the yolk and white may be combined with a small amount (1 tablespoon) of milk, cream, or water. The mix should be heated slowly and turned when it starts to coagulate on the sides and bottom of the pan. This removes the cooked portion from the heat and exposes the uncooked portion, thus evenly distributing the heat and not overcooking any part of the egg. If scrambled eggs are overcooked, liquid is expelled and evaporated, leaving a shrunken, dry product. If an omelet is desired, the egg mixture is added to an oiled pan, and when the bottom coagulates, lift just enough to allow the uncooked portion to flow and contact the pan. The omelet should stay in one piece and take the shape of the pan. Low-cholesterol scrambled eggs and omelets can be made by removing some or all of the yolks.

Custards are mixtures of egg, sugar, and milk with salt and vanilla often added for flavor. One whole egg or two egg yolks supplies enough coagulable protein to gel one cup of milk. Mix the egg, sugar, and salt and then add scalded milk (the milk is scalded to shorten the cooking time and add flavor (see section on dairy products). When the mixture is heated, the egg protein is denatured and unites to form a network that traps the fluid milk, forming a delicate gel. Custards are baked in a moderate oven (350°F, 176.7°C) in a utensil that is placed in a pan of hot water (this gives a more even distribution of heat and the water molecules are in contact with the pan, thus conducting heat more evenly than hot air, and the temperature of the water will never get higher than 212°F (100°C) (see Chapter 23). The temperature needed here to form a gel is greater than that to merely coagulate protein because the egg protein is diluted with milk. When the custard gels, it should be removed from heat and placed in cold water. Overheating custards results in shrinkage of the gel and development of pores that fill with watery serum.

Eggs can also be whipped and will entrap air, forming a foam (see foam formation in dairy products section). The proteins in the albumen are responsible for this, but the yolk will ruin it. Lipoproteins (formed by combining proteins and fats) in the yolk interfere with the foaming potential of the proteins. The foam made from egg whites has a unique property in that it can be stabilized by cooking. Ordinarily, when a foam is heated, the trapped air and other gasses in the formed bubbles expand and break the bubble walls collapsing the foam. One protein in the egg albumen (ovalbumen) makes up more than half of the albumen proteins, and it does not unfold much during whipping but does coagulate readily when heated. What happens here is the coagulated ovalbumen creates a solid network in the bubble walls that resist collapse when air escapes or when water evaporates. This enables the baker to utilize chemistry and physics principles again and form a solid foam from a liquid one.

FISH AND SHELLFISH

Fish and shellfish handling and processing is covered in Chapter 17. Finfish is prepared in a number of ways including broiling, baking, frying, steaming, and poaching (see Chapter 23). The moist cooking methods are appropriate for lean fish while the fatty varieties are self-basting and may be broiled or baked. Assessing doneness in fish can be accomplished with a thermometer by inserting it into the thickest portion. Fish is edible when the internal temperature reaches 140°F (60°C) but at 150°F (65.6°C), tissues start to break down, allowing juices and flavor to escape. Fish is no longer translucent and flakes easily when done owing to the unique structure of fish muscles. The muscles in mammals and birds are composed of very long fibers arranged in longitudinal bundles. Fish muscle, on the other hand, consists of rather short fibers that are separated by large sheets of very thin connective tissue. The connective tissue is delicate and can be converted to gelatin with heat quite easily. The quantity of connective tissue is also less (about 3% of its weight) than in land animals (about 15% of its weight). In appearance, many fish have very little pigment in their muscle tissue. It was shown earlier that, in mammals, the white and dark meat was influenced by the amount of myoglobin in the muscles, the muscles that worked harder needing more. About 40% to 60% of the body weight of fish is contained in muscle tissue (more than mammals) yet much of their flesh is still white. This is explained when the chemistry of the muscle fibers is understood. One of the basic muscle proteins, myosin, exists in several different forms including red and white muscle myosin. Mammals have muscles that are needed for functions such as standing which require endurance.

These muscles need more oxygen and have more red myosin. Fish, with their specialized muscles needed for quick spurts, need less oxygen and therefore have more white muscle myosin. Those fish with more pigment either have a need for more muscle myosin with its myoglobin or have pigment from another source. Salmon, which contain a carotenoid (see Chapters 9 and 22) derived from its diet, is an example of such a fish. Many types of fish taste very bland but the fatty varieties and the darker flesh species are more flavorful.

Shellfish are divided into two major groups, molluscs and crustacea. The molluscs include clams, oysters, scallops, and mussels. These are prepared in a number of ways and some are eaten raw. If you choose to eat raw shellfish, be sure that they are alive (shell is closed tightly) and have been harvested from inspected waters. Consumption of raw shellfish from contaminated waters can lead to serious food poisoning (see Chapter 3). When steaming or boiling shellfish, be sure the shells open completely when cooked. They open because, when the fish is cooked, the muscle that holds the shells together can no longer function. If the shell doesn't open, the fish was probably dead before it was cooked. The second group of shellfish, the crustacea, includes lobsters, shrimp, and crabs. These are processed in a number of ways described in Chapter 17 and are often received at the restaurant in this form. Lobsters are the exception and must be cooked while still alive, because of the presence of a very strong enzyme in their digestive tract which is also described in Chapter 17.

CEREAL GRAINS

Cereal grains have gained more recognition in their importance in a well-balanced diet. Their handling and processing is discussed in Chapter 18. Cereal grains have proteins, carbohydrates, and fats and are a good source of many vitamins and minerals (see Chapter 2). Grains are used for cereals that must be cooked or for those that are dry and ready to eat. In preparation of cooked cereals, skim milk may be used to give a balanced protein source and lower fat and calorie consumption. Dry cereals are completely precooked and are eaten with the addition of cold milk. Here, again, skim milk is recommended. Grains are also the main source of carbohydrates in the brewing and liquor industry. Enzyme activity in the grains breaks the complex carbohydrates into simple sugars that are fermented into ethyl alcohol and carbon dioxide by yeast. In beer and ale, hops are used to develop the bitter flavors common to beer and ale. With other alcoholic beverages made from grains, a relatively weak alcoholic beverage is boiled and the alcohol condenses and is collected in the distillation process. Different strengths are made by mixing the different liquids. Aging is done with some liquors to produce desired flavors.

In addition to these uses, grains, especially wheat, are also used for flours used in baking (see bakery products section of this chapter) and to make pasta products. Pasta is made from the hardest of wheats, durum, and has a higher protein and gluten content than other wheats, giving it the capability of producing a very stiff dough. Pasta has two main ingredients, water and either flour or semolina (a coarser flourlike product). Semolina is the choice for commercial pasta makers for it has large chunks of protein and little starch. A dough made from it requires less water than flours and pasta dough contains only about 25% water as compared to about 40% in bread dough. In pasta making, after the dough is made, it is extruded into various shapes. The gluten matrix (see bakery products section of this chapter) of semolina is stronger than that of flours and can withstand the pressure of extraction into spaghetti rods or other shapes of pasta. The dough is then dried to about 10% moisture. This is a sensitive process and requires much care to ensure the timing and temperature are perfect to dry thoroughly but not too quickly. If the temperatures are too high, the pasta will become rigid on the surface and the escaping inner moisture will cause cracking when it is removed later in the drying process. Eggs, up to 5.5% egg solids by weight, may be added to pasta to produce egg noodles. The main purpose for the addition of eggs is for color and flavor.

BAKERY PRODUCTS

A number of bakery products are explained in Chapter 19. In this section, the chemistry of baking is discussed and the actions of some of the important ingredients are examined. The subjects covered are flour, leavening agents, shortening, emulsifiers, and other ingredients. Nutritionally, some bakery products such as bread are considered staples in many diets and supply a source of protein, complex carbohydrates (see Chapter 2), vitamins, minerals, and fiber while adding little fat. Others, however, are used mainly as desserts and add more enjoyment to a meal or snack than nutrition. The appearance, texture, and taste of baked items is of great importance and quality and type of ingredients and methods of preparation controlling the chemical and physical reactions determine the final results. The baker truly exploits the sciences of chemistry and physics.

Flour is the main ingredient in most bakery products and the understanding of some protein chemistry will explain why some flours are good for certain products and not for others. The key word here is gluten. Gluten is a combination of two proteins, gliadin and glutenin. Wheat is unique among the grains, because it is the only one whose endosperm proteins will interact to form a gluten strong enough to produce raised breads. Other grains such as rye can produce a weak gluten. Gluten is both plastic and elastic, that is, it can change its shape under pressure yet will return to its original shape when pressure is removed. Gliadin and glutenin molecules are large proteins and their interaction with each other and water is extremely complicated but can be explained fairly simply. Gliadin molecules tend to form compact oval-shaped balls while glutenin molecules are somewhat longer and more extended. Mix these together and the side chains or R-groups (see Chapter 2) will react, crosslinking the proteins into a tangled mass. The gluten formed does not dissolve in water (the reason why, in raw dough, they survive chewing) but does absorb about twice its own weight in water as a result of hydrogen bonding (a weak attraction type bond between hydrogen and other chemical components such as those in sugars, salts, and proteins). This formation of gluten explains why dough is so elastic, why kneading toughens it, and why over-kneading breaks it down. When flour is first mixed with water, the proteins begin to unfold somewhat and water tends to separate and lubricate them by forming hydrogen bonds. At first, the mixture is a thick liquid but as it is mixed, the proteins are drawn together into visible filaments. Kneading both compresses and stretches the protein–water complex. The constant movement and stress forces the long molecules into a more orderly pattern that can form more regular bonds between different molecules and the crosslinking causes the mixture to be less easily deformed. The dough now is stiff with a smooth, shiny surface. At this point, the dough is much more elastic because the proteins have been elongated and unfolded to a large extent but many kinks still exist due to attractions and bonds between side groups (R-groups) of the same molecule. This explains why dough can be stretched; the kinking bonds will resist but will eventually be broken. When the stress is removed, however, they will reassert themselves and return the dough to its original shape. If dough is overdeveloped, the bonds called disulfide, or sulfur-to-sulfur-bonds, which are important in crosslinking and "kinking," are broken and a sulfur molecule can combine with a hydrogen molecule, forming a chemical called a thiol (see Fig. 24.2). The presence of thiol interferes with the sulfur bonds "rekinking" after pressure is removed. The result is a broken dough that is a thick fluid with no elasticity. It is hard to break a dough by hand but it can be done with mechanical mixers or food processors.

Other components of flour that contribute to successful dough and batter formation are starch, lipids, other carbohydrates, and enzymes. Starch has two major functions in batters and doughs. Starch granules help form the mechanical structure of baked products by contributing a semisolid phase and by regulating the location of water in the cooking dough. In yeast-raised products, the damaged starch granules from the milling process are attacked by enzymes, forming sugars that yeast use for nutrients. Lipids are not present in large amounts (only about 1% of the weight) but play an important role. It is thought that lipids form bonds with both gliadin and glutenin and help bind these in the formation of gluten as well as bind the gluten to starch

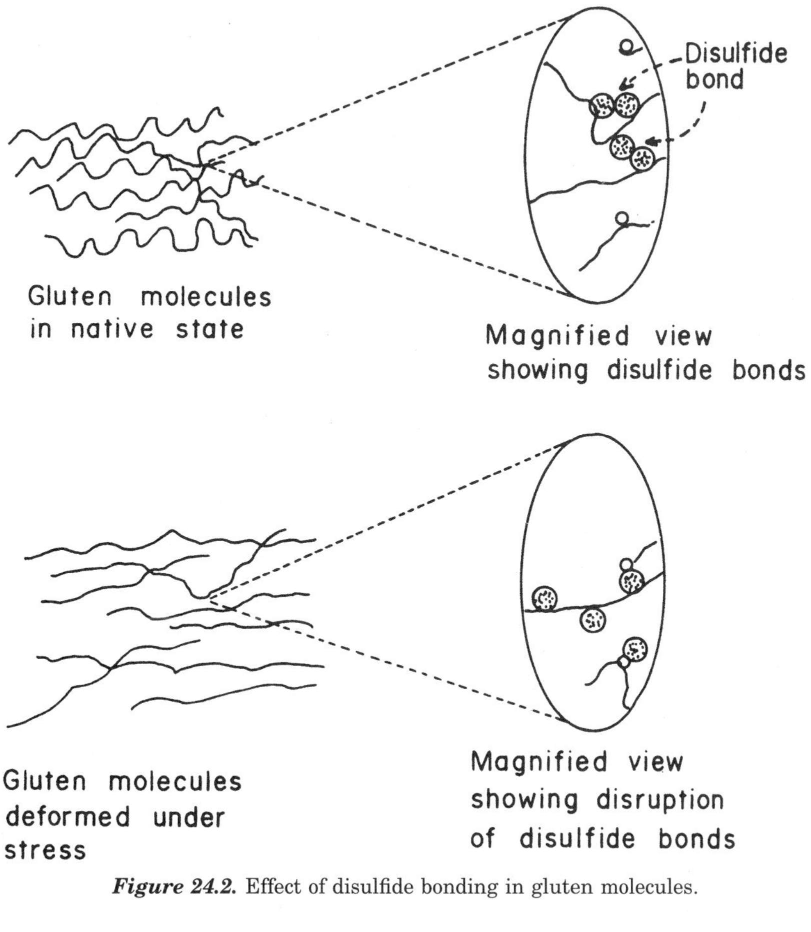

Figure 24.2. Effect of disulfide bonding in gluten molecules.

molecules. Lipids are thought to play one more important role; they seem to exist in very thin sheets that separate gluten layers. This helps slippage and plasticity in the dough.

Most baked products are leavened to make them light, porous, and more palatable and they may be made from a batter or dough. The major difference in these is the amount of water they contain; batters, which contain more water, have much more fluidity and are usually poured whereas the doughs, having less water, are stiff but can be worked by hand. The major leavens are steam, air, and carbon dioxide. Examples of steam-leavened products are cream puffs, popovers, and the flakes in pie crust which result from the vaporization of water from the mix. In air leavening, air is incorporated into batters in the preparation stage almost incidentally. When the product is cooked, the air expands and acts as a leaven. Carbon dioxide can be incorporated by yeast fermentation or chemically. With yeast, a source of sugar must be present and the temperature must be correct for the yeast to grow and reproduce (see Chapter 3). Yeast produces enzymes that convert glucose to carbon dioxide (CO_2) and ethyl alcohol (see section on cereal grains). The alcohol is of little use here and will be essentially eliminated in the cooking process, but the CO_2 acts as a leaven and the gluten–starch–lipid complex described earlier will hold the CO_2 and allow the product to rise. Chemically, carbon dioxide is usually produced by baking soda (sodium bicarbonate) or baking powder which is sodium bicarbonate in combination with a food acid such as tartaric acid (usually in its salt form, potassium acid tartrate or cream of tarter). The baking soda requires heat to react but the baking powder will react when water is added. The reactions of these are fairly simple and are shown below:

Baking soda:

$$2\,\underset{\substack{\text{sodium}\\\text{bicarbonate}}}{NaHCO_3} \xrightarrow{\text{heat}} \underset{\substack{\text{sodium}\\\text{carbonate}}}{Na_2CO_3} + \underset{\substack{\text{carbon}\\\text{dioxide}}}{CO_2} + \underset{\text{water}}{H_2O}$$

Baking powder:

$$\underset{\substack{\text{sodium}\\\text{bicarbonate}}}{NaHCO_3} + \underset{\substack{\text{potassium}\\\text{acid tartrate}\\\text{(bitrate)}}}{KHC_4H_4O_6} \xrightarrow{H_2O} \underset{\substack{\text{potassium}\\\text{sodium tartrate}}}{KNaC_4H_4O_6} + \underset{\substack{\text{carbon}\\\text{dioxide}}}{CO_2} + \underset{\text{water}}{H_2O}$$

Shortening is a fat or oil that "shortens" or breaks up the gluten. This is shown in pastries where the dough is folded and rolled giving alternating fat and gluten layers. This weakens the gluten and yields a flaky product. Solid fats work much better here for they don't seep into the dough. Heat generated in working the dough could melt the fats; therefore it is recommended in many pastry recipes to keep the shortening in the freezer prior to using. In a batter the fat has another role. It is usually "creamed" with sugar first and the sharp sugar crystals cut into the solid fat, creating air cells that help in the leavening process. Also, fat in batters helps in the cooking process. It separates starch granules from coagulated protein and also makes the cake seem moister and smoother in the mouth. The role of fat in yeast-leavened products was discussed earlier.

Emulsifiers are added in commercial shortenings and cake mixes for they, because of their polar and nonpolar ends, have the ability to combine with lipid and water at the same time. This property enables them to prevent small fat droplets from joining and forming larger droplets that would separate from the water phase and break the emulsion. In cake mixes and commercial shortenings, this property of insulating lipid droplets is used to prevent the fat from interfering with the protein–air foam and to maintain maximum air capacity. In sweet cakes, sugar tends to lower air capacity, but with the incorporation of emulsifiers, the sugar does not coalesce with the small, insulated fat droplets and the air capacity is kept and lightness and volume can be achieved.

A few other ingredients often added to doughs and batters are salt, sugar, milk, and eggs. Salt is added for taste but has other functions. It inhibits yeast activity somewhat and too much will be detrimental to the optimum leaven. Salt also can toughen gluten by forming bonds with side chains on protein molecules. (Remember salt ionizes completely in solution as described by the simple formula: $NaCl \rightarrow Na^+ + Cl^-$). This can help the amateur breadmaker who is forced to use the softer wheat flours. Salt also inhibits protein-digesting enzymes that can soften the gluten and inhibit its carbon dioxide holding capacity.

Sugar, as mentioned earlier, acts as food for yeast but if too much is added it can also lower water activity (see Chapters 3 and 11) and inhibits fermentation. Sugar can also affect the development of gluten by competing with the protein for water. This helps explain why it takes longer for these doughs to form and develop but the final product, such as in sweet bread, is moister and more tender. Added sugar will also enhance browning reactions (see Chapter 9) and will result in a darker crust.

Milk and eggs add three major ingredients: water, protein, and lipid. The moisture content of the added milk or eggs affects the amount of plain water to be added to the flour to develop the dough. The proteins will coagulate and add to the structure, and beaten egg whites, as mentioned earlier in the poultry and eggs section, can incorporate much air into a product. The lipids from milk and eggs have the same effect as added shortening. It is recommended to heat or scald milk (198°F, 92.2°C) for 1 minute to denature the serum proteins that can cause stickiness in dough. (This is a result of their interaction with flour proteins). It must be noted, however, that scalded milk must be cooled prior to addition to the mixture to avoid damage to the heat-labile yeast. Finally, eggs will add color to products to which they are added.

VEGETABLES

A botanist defines a fruit as the part of the plant that contains seeds (the ovary). By this definition, tomatoes, eggplants, cucumbers, and squash are fruits. According to the United States Department of Agriculture and most chefs and cooks, however, the definition of a vegetable is a plant food that is usually eaten as part of a meal's main course. This is the definition accepted by the majority of the population.

Vegetables are an important addition to any diet. They contain complex carbohydrates and fiber, some proteins, and, generally, very little fat. They also are a rich source of vitamins and minerals. Many varieties of vegetables are discussed in Chapter

20. When vegetables are prepared for service, some nutrients may be lost by leaching out into the cooking water, or by changes to structure resulting from heat (as with thiamin and vitamin C), or by oxidation during cooking (vitamin C). Some basic rules for cooking vegetables to minimize loss of nutrients are:

1. Bring water to a full boil before adding vegetables—this limits oxidation by inactivating an enzyme (ascorbic acid oxidase) that catalyzes it. Boiling also eliminates oxygen from cooking water in which it has dissolved and from the vegetables.
2. Return water to a boil as fast as possible after vegetables are added for the same reasons listed in (1).
3. Cook in just enough water to cover and avoid scorching. Putting a lid on the pan limits the amount of water needed and speeds cooking time because the lid tends to increase the vapor pressure above the water, hastening the rise in temperature and shortening the time it takes to produce steam which the lid helps to keep from escaping.
4. Cook until just barely done—don't overcook.
5. Serve immediately.

Microwave heating may limit losses of nutrients mainly because of the shorter cooking time and lesser amounts of cooking water used.

Methods for cooking vegetables include baking, boiling, steaming, and pan- or stir-frying, and cooking in a microwave oven. These are discussed in Chapter 23.

Texture in vegetables depends on two factors, the character of the cell wall and the amount of water in the tissues. The major component of cell walls is cellulose (see Chapter 2) which is the backbone of the structure and is made up of a long polymer of glucose. The special way in which the glucose molecules are attached to one another differentiates cellulose from starch and makes it indigestible to humans. Other components including hemicellulose, which is also a long-chain polymer that is slightly different than cellulose, and pectin, which is also similar to cellulose but consists of a chain of galacturonic acids molecules, also contribute to texture. Pectin is more prominent in some fruits (see Chapter 20). The second important factor in texture is inner water pressure or turgor of the individual cells. Water is contained in three major areas of the plant, the cytoplasm within the cell, the vacuoles (saclike spaces within cells), and the cell wall itself. These areas are separated by membranes that are permeable and water can cross them to attain equilibrium in the relative amounts of water and dissolved molecules and ions in each area. A drop in water content outside the cell will draw water out and vice versa. When a cell reaches its limit in water content, the vacuoles press against the cytoplasm which presses against the cell membrane and wall, resulting in a firm texture that we describe as crispness. When one bites into a crisp, raw vegetable, there is initial resistance, then a release of juices and flavor. Cooking the vegetables denatures the cytoplasm and cell membrane which contain much of the plant protein. The cells no longer retain water and become limp. The tenderness of a cooked vegetable depends on this loss of water and also on the effect of heat which changes pectin from an insoluble to a soluble form. The hemicellulose mentioned earlier is also partially dissolved and the resulting texture is softer than that of the fresh vegetable, and overcooking can produce a mushy product.

Appearance in vegetables has much to do with pigments responsible for the bright colors whose presence increases the pleasures of eating. An understanding of the chemistry of these pigments will help to explain the color changes that occur during cooking. Pigments may be either fat- or water-soluble and include chlorophyll, carotenoids, anthocyanins, and anthoxanthins. Chlorophyll, the major green pigment in plants, is fat-soluble but can leach into water during cooking. When vegetables are first plunged into boiling water, the green color seems to intensify. This is possibly due to the explusion of intercellular air and the resulting more transparent plant tissue. As cooking continues, however, some organic acids (see Chapter 13) will leach from the cells into the cooking water and contact the chlorophyll. When this happens, the magnesium, which is in the chlorophyll molecule, is displaced by hydrogen from the organic acids. The resulting compound is called pheophytin and has a drab green appearance. This color change can be lessened by cooking with the cover off for the first few minutes which will allow for escape of some volatile acids and limiting cooking time will also help. Cooking in an alkali such as sodium bicarbonate (baking soda) will help to retain green color because another bright green compound, chlorophyllin, is formed. The texture of these vegetables tends to be more mushy, however, because the alkali breaks down hemicellulose in the cells. Many water-soluble vitamins are sensitive to the change to an alkaline pH (see Chapter 13) and will be destroyed when cooked in this medium. Carotenoids are the yellow, orange, and red-orange fat-soluble pigments that are present in winter squash, carrots, and sweet potatoes. They are not affected as much as chlorophyll, but the color shifts slightly from orange toward yellow when exposed to boiling water for 2 to 3 minutes. Being fat-soluble, they can be absorbed by fats as in a beef stew, resulting in a fat that seems to turn orange. Anthocyanins include the blue, red, and purple colors in plants such as berries, grapes, and red cabbage. They are all water-soluble and are changed in cooking. An example of this is red cabbage which will turn blue when cooked. This can be avoided if acid (e.g., vinegar, wine, or lemon juice) is added to the cooking water. These pigments are also very sensitive to pH changes. To test this, add some baking soda to cranberry juice. You'll get production of carbon dioxide (see bakery products section) but the resulting color may surprise you. To return the color to red, the pH must be changed back by the addition of an acid. The anthoxanthins are colorless and often not noticed. If they are cooked in an alkali medium, they will turn to a yellowish color. Vegetables containing flavonols, a group of anthoxanthins, such as turnips and cauliflower can also change color if they are cooked in a covered pan for long periods of time. The sulfur-containing compounds in the vegetables will be released in the form of hydrogen sulfide which will convert the colorless anthoxanthins to anthocyanins as will be noted in the pink pigment in overcooked turnips and cauliflower. This happens because hydrogen sulfide is a reducing agent and will reduce the colorless anthoxanthins to anthocyanins which have the pink pigment.

Vegetables contain little acid or sugar; therefore their flavor is due to aromatic components they contain. Vegetables have relatively mild odors when raw, but if they are heated or cut up, the odors intensify. As the temperature increases, the substances become more volatile. An example of this is cooked cabbage. When the tissues of vegetables are ruptured, enzymes are released and can catalyze the production of new compounds. Good examples of this are when garlic is crushed or onions are cut, with the resulting compounds causing irritation to eyes.

FRUITS

Fruits, as well as having attractive colors, are an important part of a well-balanced diet as they add, flavor, vitamins, minerals, carbohydrates for energy, fiber (both water-soluble and insoluble), and almost no fat and can be used to replace or supplement fatty choices for desserts or snacks. A variety of fruits are described in Chapter 20. Fruits are eaten raw, canned, frozen, or dried (see Chapters 10, 11, and 12) and may be cooked in a variety of ways. Fruit products are also popular choices in the form of jams, jellies, preserves, juices, and fillings for bakery items (see Chapter 20). As with all foods, appearance, flavor, and texture are important quality attributes.

Texture in fruits is dependent on cell integrity and amounts of water in their tissue. This is described in the vegetable section of this chapter. Fruits, however, contain more pectin than do vegetables and this adds to their crispness. Mature but ripe fruits contain relatively high amounts of protopectin, a precursor of pectin that is water-insoluble. As the fruit matures further, more water-soluble pectins appear and the texture softens. This is thought to be a result of the action of groups of enzymes called pectinases and polygalacturonases. Fruits, like vegetables, are made up of three major types of tissues—dermal (protective), vascular (food- and water-conducting; supporting), and parenchyma (soft tissue in the pulp). The latter make up most of the edible parts. Some fruits, such as apples, also contain a fair amount of intercellular air spaces. These are formed when three or more cells adjoin and don't fit together perfectly. This texture makes "bobbing for apples" possible as the amount of air in these spaces allows apples to float. The texture of fruit is depreciated greatly when cooked mainly because of the breakdown of cellulose, hemicellulose, and conversion of pectin from its water-insoluble to water-soluble state. To minimize the loss of texture, fruits are often cooked in sugar syrups which, by their water attraction, help to strengthen the deteriorating cell walls and restore some turgor.

Appearance in fruit depends much on its color. A discussion of pigments is included in the vegetable section of this chapter. As mentioned in Chapter 11, much care must be taken to avoid browning in fruits. A big problem exists with the light-colored fruits (e.g., apples, bananas, pears) that contain phenolic compounds and the enzyme polyphenoloxidase. When fruit is undisturbed, no discoloration occurs but when it is cut or crushed, the enzyme can interact with the compounds and a brown pigment is produced. Some ways to stop this in most fruits (the skin of bananas is an exception) are chilling below 40°F (4.4°C) which will slow the enzyme action (see Chapter 8); boiling or blanching will destroy the enzyme but affects texture adversely. Chloride ions in salt will inhibit the action, but this has a detrimental effect on taste. Acids such as those from lemons will slow the enzyme action. The use of sulfur products, which is usually done commercially, will stop the reaction. A percentage of the population (asthmatics, especially) is allergic to sulfur so much care must be taken in this process, and the use of the sulfur must be declared on the label.

As with other foods, flavor in fruits depends on taste and aroma. Taste is determined mainly by the amounts of acid and sugar in the fruits. When preparing fruit products such as jams, jellies, and fruit fillings, the amounts are controlled by the food processor, and the U.S. Department of Agriculture has established standards of identity for many fruit products. The stages of maturity, storage practices, and species of fruit will determine the taste in fresh fruit. Aroma, the other characteristic of flavor, is due to a complex mixture of volatile chemicals that are often very unstable.

SUGAR

Sugars make up the simplest group of carbohydrates (see Chapter 2) and are used by the body for energy. Sucrose, or table sugar, obtained from sugar cane or beets, is the most common sugar used and the size and number of crystals formed determine the many types of candies that are made. This is controlled by the addition of agents to the sugar syrup that interfere with the formation of crystals. Other sugars such as glucose or fructose have different structures and therefore cause the formation of many small sucrose crystals rather than a few large ones. If enough interfering substances are added, a noncrystalline candy can be made such as lollipops which have about 25% corn syrup in the formula. Fat and proteins from milk also act as interfering substances in products such as fudge where a less crystalline, smoother texture is desired. Caramels, toffee, and taffy are other examples of candies with little or no crystallization of sucrose. It is prevented by a high portion of corn syrup in taffy, high fat content in toffee, and both in caramel. The chemistry of browning, as discussed in Chapter 9, comes into play in caramel as the sugar and protein from milk play a role in the color development. Fondant is another type of candy that is an important ingredient in making the huge variety of fillings for chocolates and chocolate candy bars. Fondant is made from sucrose syrup which is heated to dissolve the sugar and cooled to saturation at 104°F (40°C). Air is then beaten into the syrup which increases the number and decreases the size of sugar crystals, resulting in a mixture that is smooth and pliable. Fondants can have sugar contents of about 88%.

Brown sugar is a mixture of molasses (see Chapter 21) and refined, white sugar. All of these sugars have unique tastes and are used in a number of culinary preparations including sauces, syrups, and candies. Invert sugar is made by enzymatically converting sucrose to glucose and fructose (see Chapter 21) or by heating sucrose syrup in the presence of a weak acid such as citric. Invert sugar is sweeter and more water-soluble than sucrose, so it is used in baked items such as cookies and cakes. The final products tend to be softer and moister.

FATS AND OILS

Fats and oils are lipids (see Chapters 2 and 22) and are important in designing a well-balanced meal plan. The effect of heat on fats and oils is of great importance to culinary science and again can be explained if the properties and chemistry of fats are understood. Included are melting points, smoke points, deterioration, and use as a cooking medium. Most fats do not have sharply defined melting points but they soften over a range of 10 to 20°F (5.6 to 12.5°C). The reason for this is the presence of a number of different triglycerides in a given fat. As the temperature rises, different ones melt. The length and degree of unsaturation in the fatty acid components greatly determine the melting point. It is this response of fats to temperatures that facilitates the spreading of melting butter on toast and basting with melting fats. Fats and oils have much higher boiling points than does water because of their large molecular size. Fats are nonpolar so do not form hydrogen bonds. They do, however, form weaker bonds (van der Waals bonds) all along their large molecules. It takes relatively high temperatures (500 to 750°F [260 to 398.9°C]) to break all of them and convert the liquid to a gas. The smoke point is defined as the point when visible gaseous products are produced from fat breakdown. This point seems to depend on the initial amount of free fatty acids in the oil or fat. The amount of free fatty acids is generally much lower in vegetable oils (smoke point of about 450°F or 232.2°C) than in animal fats (smoke point of about 375°F or 190.6°C). Other materials in the fats such as carbohydrates, proteins, and emulsifiers will lower the smoke point. As the fat or oil is used, the smoke point lowers as a result of breakdown of the fats and the buildup of other materials. Caution must be taken when smoke appears because it is a warning that the ignition point is nearing. Many a fire in a restaurant resulted from overused cooking oil. Other factors that can deteriorate fats and oils are absorption of odors and rancidity (see Chapter 9). When using fats and oils as a cooking medium smoke-point, absorbed odors, and rancidity all must be observed closely or one risks a fire or the transmission of off-flavors and -odors to the foods. Foods are cooked in fats by sauteeing and deep fat frying (see Chapter 23). Liquid fats and oils are better heat conductors than air and foods heated in fat are not only cooked but also browned due to caramelization and the sugar–protein reaction (see Chapter 9). Texture is important, as these fried foods must be crisp. A number of changes take place in fats as they are used. The color gets darker, smoke-point is lowered, as is the ignition-point, viscosity increases, foams start to form, and the amount of fat absorbed by the food increases. All of these changes are warnings that the fat is getting old and continued use will lead to an increase in probability of fire and to a reduction in food quality. It is important for the chef or cook to watch for these warnings. Cooking oils and fats must be changed regularly.

Part V

Food Science Laboratory Exercises

Laboratory Safety Recommendations

The following suggestions and recommendations on safe practices in a science laboratory are brought to your attention. The order of listing bears no significance to the order of importance of the topic.

All injuries, skin rashes, illness from breathing chemical vapors, etc., should be reported promptly to your instructor.

Students of Food Science, because of their technical training, are aware of the hazards involved in manipulating the many highly toxic, explosive, inflammable, and infectious materials with which they work. The frequent use of these materials, however, can lead to a disregard of safety precautions. *ALL* Food Science students are cautioned not to let familiarity breed contempt for chemical or other potentially dangerous material in the laboratory.

RUBBER, POLYVINYL CHLORIDE (PVC), AND HEAT-RESISTANT GLOVES

Rubber globes or surgical PVC gloves should be used to avoid contact with irritating chemicals. Heat-resistant gloves should be used when handling any hot materials.

SOAP AND WATER

Hands should be washed with soap and water each time a student has finished a laboratory procedure and leaves the area. This is especially important when using the restrooms. Wash hands *before using the toilet* for safety as well as after using it for hygienic purposes.

TRANSPORTATION

Care must be taken when transporting any chemicals in the laboratory area. Always use two hands, wear protective gloves, and, if transporting glass containers to the other areas, use plastic buckets. Remember, *always* use *all* safety equipment provided and if you do not feel confident moving anything, consult the instructor.

FIRE EXTINGUISHERS

There are two types of fire extinguishers in the laboratory. There should be a type A extinguisher (water) for wood and paper fires and a type B/C (carbon dioxide) extinguisher for chemical fires. Know their location and the rules for evacuating the room in case of fire.

EMERGENCY SHOWERS

There should be an emergency shower in the laboratory. If clothing catches fire or if large amounts of corrosive chemicals are spilled on the clothing or skin, this shower should be used.

WATER FOUNTAINS

Each laboratory must be equipped with special water fountains for eye injuries. Know the location of the nearest water fountain. If any chemicals are splashed into the eye, flush with water from this fountain by directing the stream into the open eye.

FIRST AID KIT

A first aid kit must be maintained in the laboratory area. Know its location and report any injury to your instructor.

SUPPORTS AND CLAMPS

Some experiments will require use of clamps or supports. Be sure to use these carefully and make sure everything is secure before starting the experiment.

VACUUM PUMPS AND OVENS

Vacuum pumps must be used with caution. Flywheels must be facing away from the student and care must be taken that no clothing, hair, or any body parts are near the wheel when started. Always unplug the motor when working near the pump. Care must be taken when working with vacuum ovens and flasks. Remember that systems under vacuum are potential sources of explosion if any parts of the apparatus are accidentally hit with sufficient force to crack the glass. When disconnecting the system, be sure all vacuum has been released. Also, be sure to disconnect the hose connecting the pump and moisture collector if the oven is held under vacuum with the pump off.

PIPETTING

Always use pipetters (pipette fillers); *do not* pipette any chemicals with your mouth. Be sure the tip of the pipette is kept well under the surface of the liquid.

SPILLAGE AND RESIDUES

If chemicals are spilled, notify your instructor and proper measures will be taken to dilute or neutralize the chemicals before removing them. Do not panic! Methods are available to clean all chemicals used in the laboratory. Always clean the laboratory benches before and after using. Spray with clean water and clean with a paper towel. Allow to dry air. Clean all equipment after each experiment unless otherwise instructed. Clean water and detergent is usually sufficient. Do not allow residues to remain on equipment, tubing, glassware, etc. They can cause burns and unwanted reactions, and can cause glass fittings to adhere to each other.

LABORATORY BENCHES

Always wash benches before and after use. Never leave any equipment, glassware, or chemicals on the benches when experiments are completed. Always replace equipment and chemicals. Leave the lab *cleaner* than when you entered it.

HIGH SHELVES

Do not stand on stools, boxes, etc. to reach high shelves. Contact your instructor.

WASH BOTTLES

Wash bottles *must be labeled.* If they are not, *do not use them!* Contact the instructor immediately and the problem will be corrected.

INFLAMMABLE MATERIALS

Materials such as benzene, toluene, and ether should be handled well removed from electric ovens, burner flames, or hot plates. They should never be stored in refrigerators. They must be stored in the safety cabinet at the front of the laboratory. You will be instructed to use some of these chemicals in the fume hood only. Read the laboratory instructions carefully and ask the instructor if you have any questions about any chemicals.

SOLVENT VAPORS

Large quantities of the vapors of chemicals such as toluene, benzene, methanol, carbon tetrachloride, ethyl acetate, chlorinated hydrocarbons, and others should not be breathed. These should be handled in the hood or well-ventilated area. Consult the laboratory procedures and the instructor when using any of these chemicals.

CORROSIVE MATERIALS

Extreme care must be exercised when handling corrosive materials. Acids and similar chemicals should not be stored on high shelves, especially those in large containers. Bottles of ammonia, nitric acid, and others that readily give off fumes should be handled carefully in the fume hood. Corrosive materials should not be heated in fragile containers such as 2-liter beakers. On storage shelves, acids and alkalis must not be stored together.

HANDLING GLASS TUBING AND RODS

The accepted method of preventing cuts from broken glass rods and tubing is to wrap the glass in cloth while inserting into a stopper. Wearing protective gloves will also help. When inserting glass into stoppers or rubber tubing, lubricate with water. Always apply pressure away from your body and hold the glass tubing within a few inches of the point of insertion.

Cleaning and Sanitizing Procedures

Whenever food is prepared, all food contact equipment and utensils must be cleaned and sanitized (see Chapter 4 on food safety and sanitation). The methods described here utilize EPA-approved sanitizing chemicals and adapt them to practical situations that can be adjusted for all applications, from industrial to home use.

Chlorine Sanitizer Solutions

To figure the amount of EPA-certified bleach to add to get proper sanitizing action as recommended by the Food Code of the U.S. Public Health Service.

For 1 gallon of 50 ppm chlorine (hypochlorite or OCl^-) sanitizing solution

1. One gallon = 128 oz
   one gallon = 3.784 liters (3784 ml)
2. A 50 ppm solution = 0.005%
3. One gallon needs (0.00005) (128 oz) = 0.0064 oz OCl^- OR
   (0.00005) (3784 ml) = 0.1892 ml OCl^-
4. Find the strength of chlorine solution and divide the amount the OCl^- needed by the strength of the commercial solution. For example:

Bleach solution with 5.25% sodium hypochlorite (NaOCl)
Step 1: One gallon of solution needs 0.1892 g of effective OCl^-
Step 2: Sodium hypochlorite (mol wt of 74.5) has 69% of its weight in effective OCl^- (hypochlorite with a mol wt of 51.5), so 51.5/74.5 = 0.69 or 69%.
Step 3: 0.1892/0.69 = 0.2742 g of NaOCl to get 0.1892 g of effective OCl^-.
Step 4: 0.2742/0.525 = 5.22 ml of bleach with 5.25 NaOCl to get 0.1892 g of OCl^- (5.22 ml = approx. 1 tsp [1 tsp = 4.93 ml]).

5. For any amounts of more than 1 gallon, use appropriate ratios. For example:
   for 5 gallons, use 5×1 = 5 tsp or 1 2/3 tbsp
   for 1 quart, use 1/4 tsp
6. For greater strengths use appropriate ratios. For example:
   100 ppm use 2×1 tsp = 2 tsp (9.86 ml) or 2/3 tbsp per gallon

Manual Cleaning and Sanitizing Policy

1. For dishes, pots, pans, and utensils:
   a. Wash in hot water (75 to 110°F [23.9 to 43.3°C]) and detergent.
   b. Rinse with clean, hot water. If a three-compartment sink used, this may be done in the second compartment. If a two-compartment sink is used, this must be done in a separate container or under running water. Also see (d) below for alternate method.
   c. Submerge in warm water (75°F [23.9°C] or above) for at least 10 seconds with a chlorine sanitizer (50 ppm) or for 30 seconds with other chemical sanitizers. The water must contain about 1 tsp (4.93 ml) bleach (EPA-registered for this use) per gallon of water. Test with paper test strip[a] to 50 to 100 ppm OCl^- (chlorine). Solutions of iodophore sanitizers (I) of 12 1/2 to 25 ppm or quaternary ammonium of 200 ppm may also be used but items must be submerged for at least 30 seconds as mentioned above.
   d. If using a two-compartment sink, you may rinse in the second compartment and then sanitize in a third container containing 50 ppm chlorine or equivalent for at least 10 seconds with a chlorine sanitizer or for 30 seconds with other chemical sanitizers.
   e. Air dry.
2. For counters, cutting boards, tables, equipment, and any food contact surfaces.
   a. Using hot water and detergent, wash with a clean brush or cloth.
   b. Rinse with clean, hot water.
   c. Wipe with clean cloth soaked with solution containing the amount of sanitizer mentioned above in 1c (for EPA-registered bleach, use 1 to 2 tsp per gallon) or spray with the same strength solution.

      Test with strip should show 50 to 100 ppm OCl^- or 12.5 to 25 ppm I (quaternary ammonia is not recommended with this method).
   d. Air dry.

Equivalents: 2 tbsp = 1 oz
3 tsp = 1 tbsp

[a]Paper test strips to measure concentrations of the different sanitizers can be purchased from most sanitizer suppliers.

Acidity in Foods

INTRODUCTION

All foods have some acids in them. Some foods, such as fermented products such as pickles and sauerkraut or citrus fruits, have high amounts of acid. As the amount of acid increases and the pH is lowered, foods can be preserved and the shelf-life of the new food products is increased. Measuring the amount of acid in some foods is also used as a quality control tool to determine how old a product is or how it had been handled prior to delivery. The dairy industry uses acid measures often to test both raw and finished products for quality. In this laboratory, the students will determine, by titration, the amount of acid in some common food products.

MATERIALS

A 25-ml burette
0.1 *N* NaOH
1% Phenolphthalein solution
Distilled water
A 250-ml beaker
A 5-ml pipette
Samples of vinegar, sauerkraut, and wine
Safety glasses

PROCEDURE

Safety glasses are worn for this laboratory session.

1. Pour 100 ml of distilled water into each of three beakers.
2. Add 5 ml of wine to one beaker, 5 ml of sauerkraut juice into the second, and 5 ml of vinegar into the third.
3. Add 5 drops of phenolphthalein to each sample before titration.
4. Add NaOH, drop by drop, to swirling liquid until a pale pink color appears. When working with the wine sample, another color change may occur before the pink color change is noted. This is due to pH-sensitive pigments naturally present in the wine.
5. Calculate the % acid in each.

Find normality of the acids (acetic in the vinegar, tartaric in the wine, and lactic in the sauerkraut) by the following formula:

$$\text{Normality of acid} = \frac{\text{normality of NaOH} \times \text{volume of NaOH from titration}}{\text{volume of acid titrated}}$$

The formulas for the three acids are:

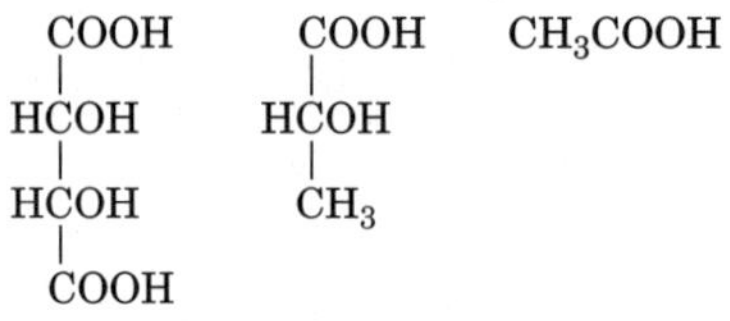

tartaric lactic acetic

For wine, multiply normality by GEW of tartaric acid (75) and multiply this by 0.1 to get % acid.

For sauerkraut, multiply normality by GEW of lactic acid (90) and multiply this by 0.1 to get % acid.

For vinegar, multiply normality by GEW of acetic acid (60) and multiply this by 0.1 to get % acid.

$$\text{G.E.W.} = \text{gram equivalent weight}$$

$$\text{G.E.W.} = \frac{\text{gram molecular weight}}{\text{valence (or acid hydrogens)}}$$

Salt (Sodium Chloride) in Foods

INTRODUCTION

Salts are made when an acid and a base react. They are formed by total or partial replacement of hydrogen with a metal. The most common of the salts is formed when sodium hydroxide and hydrochloric acid react, giving sodium chloride and water:

$$NaOH + HCl \rightarrow NaCl + HOH$$

In measuring the presence of salt in foods, a number of methods may be used. Salt dissolves in water; therefore it may be measured in aqueous solution by use of a hydrometer. Another method employs a titration using silver nitrate. In this method, a known normal strength of silver nitrate ($AgNO_3$) is allowed to react with the salt in a food product that has been liberated through a nitric acid (HNO_3) digestion. After the salt and silver nitrate have reacted, the nitrate that remains is titrated against a known normal solution of ammonium thiocyanate (NH_4SCN) in the presence of a ferric nitrate $Fe(NO_3)_3$ indicator. In an excess of ammonium thiocyanate, ferric thiocyanate, $FeSCN^{2+}$, which has a salmon-colored complex, will form, indicating the end point. The following reactions take place:

In food:

$AgNO_3$	+	NaCl	→	AgCl	+	$NaNO_3$
silver	+	sodium	→	silver	+	sodium
nitrate		chloride		chloride		nitrate

During titration:

$AgNO_3$	+	NH_4SCN	→	AgSCN	+	NH_4NO_3
silver	+	ammonium	→	silver	+	ammonium
nitrate		thiocyanate		thiocyanate		nitrate

NH_4SCN	+	$Fe(NO_3)_3$	→	$NH_4\ (NO_3)$	+	$FeSCN^{2+}$
ammonium	+	ferric		ammonium	+	ferric thiocyanate
thiocyanate		nitrate		nitrate		(indicates end point)

MATERIALS

Part I

Brine hydrometer
Cylinder
Salt
Graph paper
Laboratory balance

Part II

0.1 *N* Ammonium thiocyanate (NH_4SCN)
0.5 *N* Silver nitrate ($AgNO_3$)
5% Ferric nitrate ($Fe(NO_3)_3$
Nitric acid (HNO_3)
Burette (at least 25-ml capacity)
Erlenmeyer flasks
Laboratory beakers
Heating unit
Laboratory fume hood
Safety glasses
Latex gloves
Pipettes and pipette fillers

PROCEDURE

Safety glasses and latex gloves must be worn in this laboratory.

Part I

1. The brine hydrometer is calculated from 0 to 100 degrees saturation. 100 degrees = 100% saturation. At 60°F (15.6°C)
   100% sat. = 36 g per 100 ml H_2O
2. Make up solutions containing 0, 10, 20, 30 g of NaCl per 100 ml of water.
3. Take hydrometer reading of 0 g of salt, add 10 g of salt, take reading, add 10 g of salt, take the third reading, etc. This method saves salt and minimizes temperature adjustments. Remember, all readings are taken at 60°F (15.6°C).
4. Plot graph as shown below:

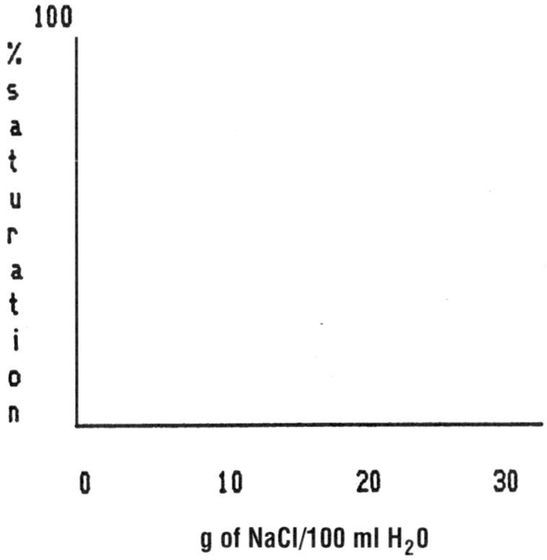

5. Take hydrometer reading of unknown salt solution. Plot on graph and determine the grams of salt per 100 ml of H_2O.

Part II

Duplicate samples and two blanks must be run.

1. Weigh 3.5 g of prepared sample on paper and add to Erlenmeyer flask (250 ml is usually sufficient but a larger flask may be needed if frothing is expected. This may occur in some dry products).
2. Pipette 5 ml of $AgNO_3$ into flask.
3. Add 15 ml of HNO_3 (more may be needed if using dry products).
4. Heat in fume hood until orange color dissipates and all food is digested. (A white precipitate will be formed and will not liquefy. Do not confuse this with undigested food).
5. Add 100 ml of demineralized water.
6. Add 2 ml of $Fe(NO_3)_3$.
7. Titrate with 0.1 *N* NH_4SCN until you a distinct change in color (an orange color will develop and hold for 15 seconds when equilibrium has been reached).

Calculation:

$$\frac{\dfrac{\text{ml } HN_4SCN \text{ in blank} - \text{ml } NH_4SCN \text{ in sample}}{\text{ml } NH_4SCN \text{ in blank}} \times \text{ml } AgNO_3 \times N\ AgNO_3 \times \text{MEW NaCl}}{\text{weight of sample}}$$

This calculation will divide the weight of NaCl by the weight of the food and this value × 100 will give you the % salt in the product.

M.E.W. = mili equivalent weight
MEW = GEW/1000

Microorganisms in Food

INTRODUCTION

Foods, unless heat-treated to sterility, contain many live microorganisms. If the numbers of these organisms reach certain levels, the shelf-life and safety of the food becomes questionable. The food industry regularly checks foods to be assured the numbers of microorganisms are within recommended limits. The standard plate count (SPC) is a method that is often used to test food for microbial quality. This test can also be used as part of a HACCP program to verify that the sanitary methods being used in food preparation are effective.

Examples of standard plate count guidelines for some foods are given below. Counts should be below the numbers shown.

Milk: 10,000 per ml
Cream: 40,000 per ml
Drinking water: 500 per ml
Prepared foods: 100,000 per g

MATERIALS

Sterile petri dishes
Sterile standard plate count or nutrient agar
Sterile 99-ml dilution blanks of buffered distilled water (SBDW)
180 ml of sterile SBDW
Sterile 1.1-ml pipettes
Sterile food blender
Food sample
Incubator held at about 98.6°F (37°C)
Bunsen burner
Laboratory balance
Metal spoons or laboratory spatulas
Pipette fillers
Safety glasses
Latex gloves

PROCEDURE

1. Safety glasses and gloves are worn for this laboratory.
2. The laboratory benches should be wiped with 70% ethyl alcohol solutions to sanitize area.

3. Flame spatula, cool in a portion of the food not to be tested, and then transfer 20 g of food to the blender.
4. Add 180 ml of SBDW and blend for 5 minutes.
5. Depending on food being tested, make dilutions that will determine if guidelines are being met (1:1000 and 1:10,000 for prepared foods; 1:100 and 1:1000 for milk). See method for making dilutions.
6. Add agar that is held at about 122°F (50°C) to the inoculated petri dish and swirl gently 25 times to the left and 25 times to the right to ensure even distribution of organisms.
7. Allow to harden, invert, and incubate for 48 hours.
8. Count the colonies, multiply by the dilution factor, and report as SPC per ml or g. For example, if your dilution was 1:1000 and you have a count of 30 colonies, report as 30 × 1000 or 30,000 colonies per ml or g.

Below is an example of how to make a serial dilution of a food sample:

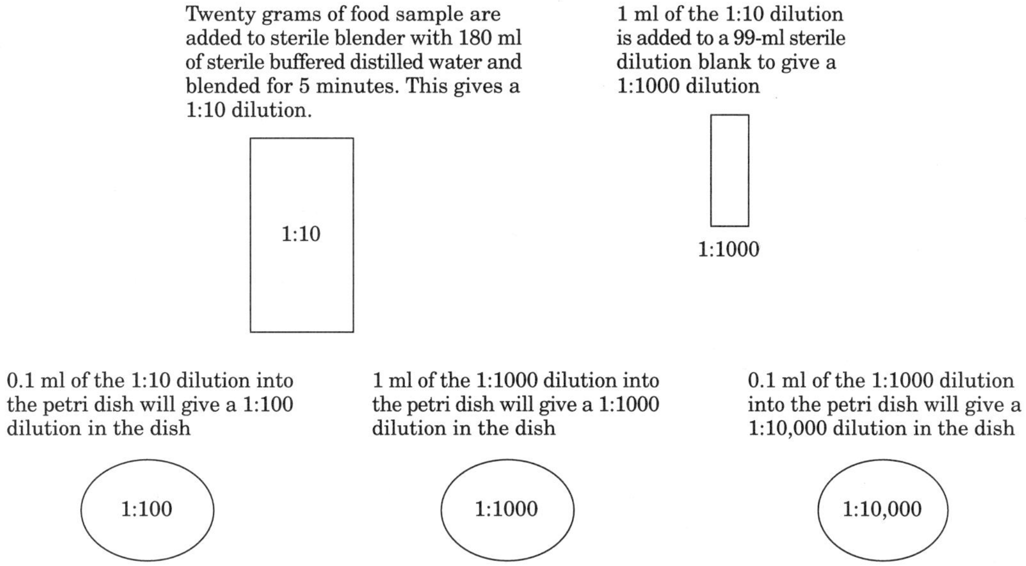

Sensory Evaluation of Food

INTRODUCTION

One of the most important tests a food must pass before a business will risk mass production is that given by a sensory panel. The food may be nutritious and meet economic criteria in production but if it does not smell, look, feel, and taste good, it will not be a successful product. Test panels may be trained in sensory evaluation or they may be untrained consumer panels. The results from both are extremely important to the food technologist in determining many factors such as formulation, processing techniques, and packaging.

In this laboratory exercise, students will rate their taste sensitivity, test their ability to rate different sugars based on sweetness, and try to identify some familiar flavored foods that have had their colors altered.

MATERIALS

Four taste solutions
- acid (citric acid solutions) in concentrations of 1.0, 0.1, 0.01, 0.001, and 0.0001%
- salt (sodium chloride solutions) in concentrations of 1.0, 0.1, 0.01, 0.001, and 0.0001%
- sweet (sucrose solutions) in concentrations of 1.0, 0.1, 0.01, 0.001, and 0.0001%
- bitter (quinine sulfate solutions) in concentrations of 0.1, 0.01, 0.001, 0.0001, and 0.00001%

Sugar solutions
- 1% solutions of fructose, glucose, maltose, lactose, mannose, and sucrose

Fruit sherbert samples
Food coloring
Small plastic cups, spoons
Beakers for waste solutions
Unsalted plain crackers

PROCEDURE

Part I: Taste Thresholds

1. Students will work in pairs; one student will be the taster and the other the server.
2. Select one taste at a time, with bitter being tested last.

3. Do not inform the taster which taste is being used. Give a small amount of water (at the same temperature as the test solutions) in a cup and inform the tester that it is water. This will establish a basis for evaluation and should be done at the beginning and whenever tastes are changed. Start testing with the weakest solution and advance to the strongest until the taster definitely identifies the taste. You may want to give secret water samples occasionally to establish that the taster is not guessing.
4. The taster should hold each sample in the mouth for a few seconds, agitate with the tongue, then expectorate the sample into the waste beaker. The taste should note if there is any sensation even if he or she cannot identify the taste.
5. Rinse the mouth out with warm water between tests and take a cracker if you want to clear your palate. Make comparative tests at each concentration level with water.
6. The server will note each response, whether it is a sensation or a taste identification. You may want to repeat weaker solutions until you are sure the correct threshold has been established.
7. Switch server and taster after all four thresholds have been identified.
8. Discard plastic cups after the taster is done with them.

Part II: Sweetness Rankings

1. Obtain samples of the five unknown sugar solutions and the sucrose solution.
2. Compare each unknown solution to the sucrose solution and rate them based on sweetness. Prepare a chart ranking each unknown solution and sucrose from least to most sweet.

Part III: Color and Flavor Identification

Students will work in pairs as previously with a server and a tester.

1. Obtain five common flavors of fruit sherberts and make ten samples, two of each flavor.
2. Stir each sample so that all will have the same consistency.
3. Alter the color of one of each of the flavor samples, for example, color one sample of lemon sherbert red.
4. Record each color alteration with actual flavor.
5. The tester should taste each sample and respond as to what the flavor is. (The testers are not told which flavors have been used).
6. Tester will rinse with warm water between tests and have a cracker to clear the palate between tests if desired.
7. Record all responses and calculate the percentage correct responses and which wrong responses were given.

Thermal Processing of a Low-Acid Product

INTRODUCTION

Thermal processing of a low-acid product to achieve commercial sterilization requires a temperature higher than that of boiling water (212°F or [100°C]). To achieve this high temperature, a pressure retort (cooker) must be used. With this, we can raise the temperature to 240°F (115.6°C) at 10 psi or 250°F (121.1°C) at 15 psi and process the food for shorter periods of time, while still reaching the lethality levels required for a safe, wholesome product.

MATERIALS

Low-acid food (green beans)[a]
Cutting implements
Measuring implements
Chlorine sanitizer
Thermometers
Glass containers and lids[b]
Pressure retort[b]
"Cook Chex"[c]
Incubator

PROCEDURE

1. Develop HACCP flow chart for canning green beans and identify CCPs.
   Some suggested safety—HACCP checks
   Before processing:
   a. Check raw materials for defects, contamination, freshness, weight.

[a]Methods for other products can be found in the "Ball Blue Book," Alltrista Corporation, Muncie, IN 47305-2326

[b]Glass canning jars can be purchased at many hardware and other retail stores and pressure retorts (cookers) can be purchased from various food processing equipment companies.

[c]"Cook Chex" is a paper strip impregnated with heat (steam)-sensitive ink that will change color when proper processing time and temperature have been attained. It may be purchased from PYMAH Corp., North Hollywood, CA.

b. Check blanching time and temperature.
c. Check containers for imperfections or faults, check cans to ensure proper seal, rinse with clean, hot water and invert to remove soil, etc.
d. Check for proper fill of containers and proper weight.
e. Be sure to add "Cook Chex."

During processing:
a. Check temperatures using recording thermometer or process data sheet (see example in this section).

After processing:
a. Check residual chlorine in cooling water with test strips to 50 ppm.
b. Check "Cook Chex" or recording thermometer to ensure proper processing times.
c. Check storage conditions—temperature, humidity, etc.
d. Check codes if recall needed.
e. Keep sample for incubation.
f. Laboratory quality tests such as sensory evaluation and microbial tests as described in this text that can be done on finished product.

2. Clean and sanitize all equipment and utensils as described in the section on sanitizers in this chapter.
3. Cleaning, washing, and preparation:
   a. Wash beans in clean, cold water, removing soil and any defective pieces.
   b. Remove ends of green beans.
   c. Slice into uniform size pieces.
4. Blanching:
   a. Blanching removes occluded or dissolved air or other gasses and shrinks the beans somewhat to ensure a good pack.
   b. Bring water to 180°F (82°C).
   c. Blanch by immersing beans in water for 2 to 3 minutes.
5. Filling:
   a. Place blanched beans into containers, shaking vigorously to ensure a solid pack. (The fill should be at least 90% volume capacity and weigh at least the can's capacity.)
   b. Cover with 200°F (93.3°C) to 212°F (100°C) water or brine (0.12 lb of salt per gallon of water [54.5 g per 3.784 liters]). The stream generated here will exhaust the jar, removing air. Beans should not protrude from liquid into head space or discoloration could develop.
   c. With glass containers, lids should be "softened" in hot water and should be "hand tightened."
6. Processing:
   a. Place containers into retort with water as recommended by the manufacturer, add "Cook Chex," and fit lid.
   b. Allow steam to exhaust from vent for at least 10 minutes. Close vent.
   c. Allow pressure to reach 10 psi (240°F [115.6°C]).
   d. Process for 20 minutes, adjusting temperature control to maintain 10 psi throughout process. *It is very important that the pressure is not allowed to go below 10 psi.*
7. Cooling:
   a. Allow to cool slowly to 0 pressure (do not open vent).
   b. Open vent. Open lid CAREFULLY SO THAT STEAM ESCAPES AWAY FROM YOU.

c. Remove jars from retort and allow to air cool for 5 minutes. Check "Cook Chex," following manufacturer's recommendations on color changes.
d. Place jars in cooling tank and turn on hot water (make sure it is hot). Gradually add cold water until finally all cold water in spraying onto jars. You will hear the jars "pop" as a vacuum is achieved. If you do not have a cooling tank, the jars can be cooled at room temperature.
e. Code all lids with date, batch number, and retort used.
f. One sample from each batch must be kept and placed into the incubator at 97°F (+ or – 2°F) (36.1°C [+ or – 1.1°C]) for 10 days.

Below is an example of a partial processing data sheet. This can be used to monitor the thermal process to ensure safety.

FOOD PROCESSING DATA SHEET

DATE:
PRODUCT:

CONTAINER	PROCESS TEMP	TIME STEAM FLOW STARTS	TIME PROCESS TEMP. IS REACHED	TIME PROCESS ENDS	"COOK CHEX"

Production of Wine and Beer

INTRODUCTION

Fermentations have been defined as the interaction between a microorganism and susceptible organic substrate. In wine and beer fermentations, the substrate is sugar, the microorganism is yeast, and the major byproducts are carbon dioxide and ethyl alcohol. In this laboratory, we will attempt to convert grape juice to wine, and malt, hops, and corn sugar to beer.

MATERIALS

1. Wine kit[a] containing grape juice concentrate, white granular sugar, yeast nutrient, tartaric acid blend, wine yeast
2. Beer kit[a] containing malt extract, malt, corn sugar, yeast, and any other special ingredients for flavor, color, and clarifying aids
3. For both fermentations: 6.5 gallon (24.6 liters) "primary fermentor," 5 gallon (18.92 liter) carboy with air lock and stopper, siphon hose, racking tube, hydrometer, and jar
4. For wine:
   wine bottles, corks, corker
   potassium or sodium metabisulfite tablets or powder to make sulfite sanitizer (a 50 ppm solution can be made by adding about 1.6 g of potassium metabisulfite [$K_2S_2O_3$] or about 1.4 g of sodium metabisulfite [$Na_2S_2O_3$] to 5 gallons of liquid). test kit for SO_2[b]
5. For beer:
   beer bottles (not screw tops), caps, capper
6. Bleach approved for sanitizing food contact surfaces

[a]wine and beer kits can be obtained from beer and wine hobby shops. Many have mail order services such as: Beer & Wine Hobby, Woburn, MA 01801.

[b]SO_2 test kits can be obtained from CHEMetrics, Inc., Calverton, VA 22016 or from many wine and beer hobby shops.

PROCEDURE

Wine

NOTE: ANYONE WHO IS ALLERGIC TO SULFITES SHOULD NOT ATTEMPT THIS LABORATORY.

1. Develop a HACCP flow sheet for the wine production and determine any CCPs.
2. Wash, rinse, and sanitize all glassware and equipment used in the process. The final rinse can be done with 1% sodium or potassium metabisulfite. This is a strong solution and should not be used in confined areas. Breathing the fumes should be avoided.
3. Follow directions within the wine kit and use hydrometer for total soluble solids and specific gravity testing and use method described in this laboratory section for determining acidity. To adjust total soluble solids (brix), add 1.5 oz (42.5 g) of sugar per gallon to raise the brix 1%. To lower total acidity, add 0.9 g of potassium bicarbonate per liter to lower acidity 0.1% as tartaric. To raise the acidity, add 0.85 g citric or 1 g of tartaric acid per liter to raise the acidity 0.1% as tartaric.
3. The general procedure for wine is shown below:

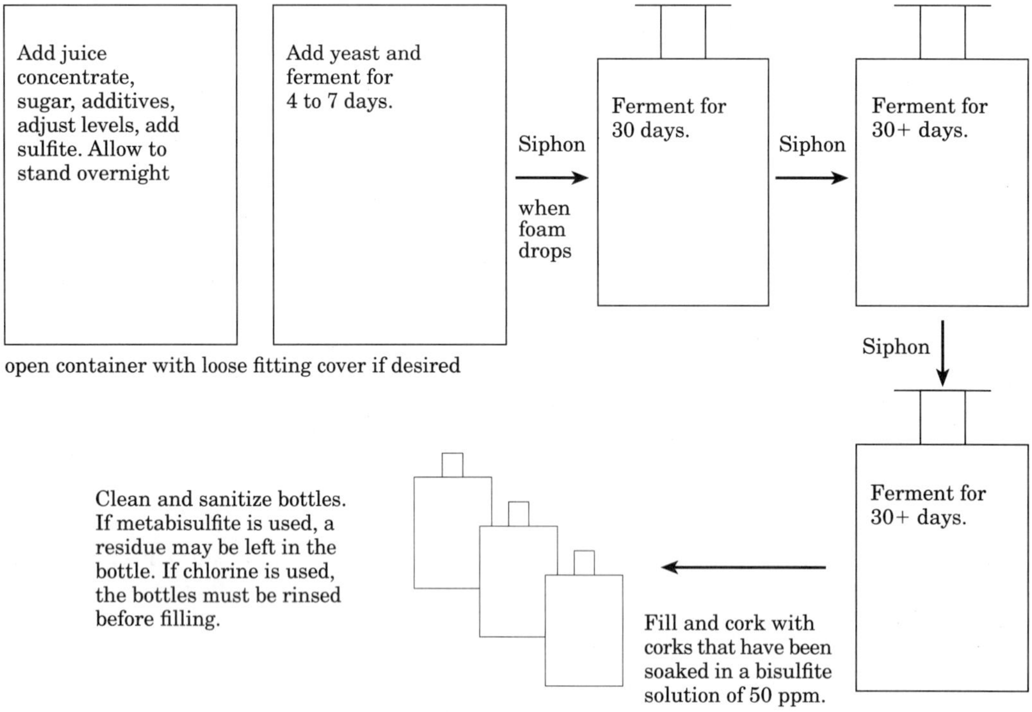

Beer

1. Develop a HACCP flow sheet for the beer production and determine any CCPs.
2. Wash, rinse, and sanitize with a chlorine solution of 50 ppm all glassware and equipment used in the process. After the sanitizing rinse, the chlorine residues

must be removed with another rinse with potable water if the equipment and glassware is to be used immediately.

3. Follow directions within the beer kit and use hydrometer for total soluble solids and specific gravity testing.
4. The general method for beer production is shown below.

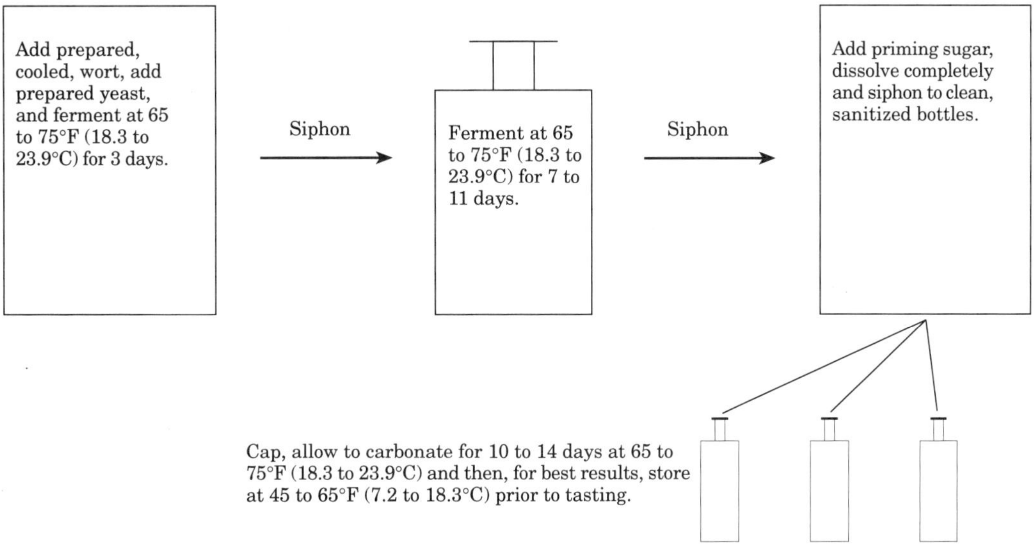

Production of Sauerkraut

INTRODUCTION

Sauerkraut is made by extracting sugar and liquid from shredded cabbage with salt and fermenting the sugar in an anaerobic atmosphere with the natural flora of bacteria present on the cabbage.

MATERIALS

Chlorine sanitizer
Plastic container for fermenting vessel
Cabbage (3.5% sugar ideal)
Noniodized salt (2.25% of weight of cabbage)
Food processor/shredder
Hand-cutting tools
Canning jars with lids
Boiling water bath for heat processing
Heavy-duty plastic bags

PROCEDURE

1. Make a HACCP flow chart and identify CCPs.
2. Clean and sanitize all tools and equipment.
3. Remove outer leaves and any undesirable portions of cabbage.
4. Wash cabbage and drain.
5. Cut into halves or quarters, remove core. Cut to sizes suitable for use in food processor.
6. Shred into thin threads about 3 mm (0.12 in.) thick or about thickness of a dime.
7. In 5-lb batches, combine salt and cabbage in plastic container.
8. Let wilt for 5 minutes, compress evenly until juices come to the surface. Repeat until all cabbage is used. Remove some juice for acid test.
9. Check heavy-duty plastic bag for leaks. Place over cabbage and add water into bag to create an air-free environment over cabbage.
10. Ferment at 68 to 72°F (20 to 22.2°C) and test every week until acid level reaches 1.5% as lactic.
11. Remove any mold and undesirable sauerkraut. Heat sauerkraut to 185 to 210°F (85 to 98.9°C). Pack into clean hot jars leaving 1/2 in. (1.27 cm) headspace.

12. Remove any air bubbles, adjust caps, and process in boiling water bath that covers containers.
    Times: pints 15 minutes; quarts 20 minutes.

The general method of sauerkraut production is shown below:

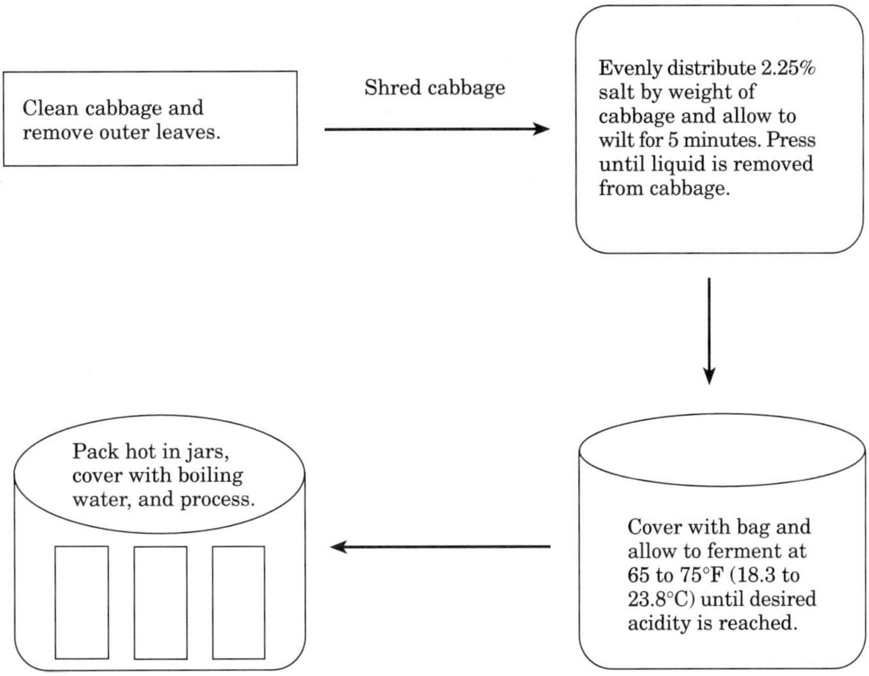

Production of Hot Dogs

INTRODUCTION

Cooked sausage products include hot dogs, bologna, liverwurst, and any other meat product extruded into a casing and cooked. Sodium or potassium nitrite is included in the formulation to stop the growth of *Clostridium botulinum.* In the process, a virtually oxygen-free atmosphere will be achieved in the finished product and the heat process will not reach temperatures sufficient to kill *Cl. botulinum.* The nitrite is also reduced in the process to nitric oxide which will combine with myoglobin, the pigment in meat muscle tissue, to form a pinkish color (nitrosomyoglobin). This is discussed in Chapter 14 and a general procedure for production is also shown. In this laboratory, students will produce hot dogs (or bologna loaves) with and without the addition of sodium nitrite.

MATERIALS

Food processor
Laboratory balance or scale
Oven or smokehouse with heat-resistant container for water
Sausage stuffer and sausage casings
Loaf pans if do not have sausage stuffer and casings
Lean beef (at least 85% lean), 45% of formula
Pork (50% lean), 55% of formula
Ice, 25% of beef and pork weight
Salt, 11.4 g per lb of meat
Sugar, 9 g per lb of meat
Sodium nitrite, 0.25 oz (7 g) per 100 lb (45.4 kg) of meat
Spice preparation (can be obtained from spice houses or hot dog producers), 12 oz (341 g) per 100 lb (45.4 kg) of meat

PROCEDURE

1. Develop HACCP flow chart for hot dogs identifying CCPs.
2. Clean and sanitize all food contact surfaces and experiment.
3. All meat and ice to food processor and mix to distribute evenly.
4. Add sugar, salt, and spice preparation to mixture while processor is running and evenly distribute.

5. Add sodium nitrite to half of formula and mix to evenly distribute.
6. Pack in casings or loaf pans.
7. Start oven at 140°F (60°C) for 20 minutes.
8. Add water to oven in approved container.
9. Increase to 150°F (65.6°C) for 20 minutes (if in smoker, smoke incorporated at this time).
10. Increase to 160°F (71.1°C) for 20 minutes or to internal temperature of 156°F (68.9°C).
11. If 156°F (68.9°C) not reached in step 8, oven temperature may be raised to 170°F (76.7°C).
12. Cool in water to 70°F (21.1°C).
13. Package and store at 40°F (4.4°C) or lower.

Production of Surimi

INTRODUCTION

Surimi is a semiprocessed wet fish protein product that can be made into "Kamaboko"-typc products and fish sausages. These products are familiar to us as imitation crab legs and other seafood products. In this laboratory session, students will produce a surimi-based product. In Chapter 17 of the text, there is a description of the process and an outline of the manufacture of a crab leg-type product. Using the methods in this laboratory, other flavored surimi such as beef and chicken have also been successfully produced. The amount of flavoring added will vary depending on preference of the processor. Experimentation must be done.

MATERIALS

White fish (e.g., pollock, cod, haddock, flounder)
Flavoring (e.g., crab, shrimp, chicken, beef)
Modified food starch
Food grade phosphate
Sugar
MSG (monosodium glutamate)
Salt
Food processor
Laboratory balance or scale
Steam bath and hot water bath for setting and pasteurizing products
Parchment paper and plastic wrap for packaging
Food color
Chlorine sanitizers and detergent
Ice

PROCEDURE

1. Develop HACCP flow chart for surimi and surimi-based product, identifying CCPs.
2. Clean and sanitize all equipment and food contact surfaces.
3. Surimi production
   a. Weigh fish and mince in the food processor (batch size depends on the size of food processor used).
   b. Wash fish three times in 5 to 10 times its weight of water.

c. Add 5% by weight sugar and 0.2% phosphate to fish and mince to evenly distribute.

4. Surimi-based product production

   General formula for surimi based product:
   surimi 1000 g
   salt 25 g
   ice water 100 g
   sugar 60 g
   starch 50 g
   MSG 5 g
   flavor 30 g

   a. Add salt and starch to surimi. Keep temperature below 39°F (4°C). Evenly distribute in food processor.
   b. Add other ingredients and knead in food processor. (If colored surimi is desired, the color can be added here.)
   c. Spread paste into uniform sheet on parchment paper (temperature here not to exceed 50°F [10°C]).
   d. Set in oven over steaming water for 20 minutes at 104°F (40°C).
   e. When set, can either form into different shapes (step 6) or package for pasteurization. (step 8).
   f. When set, cut into noodlelike strips, roll into cylinder, and wrap with colored set surimi (red is used in crab-leg type products).
   g. Package in plastic film and seal.
   h. Process (pasteurize) in water at 194°F (90°C) for 30 minutes.
   i. Quick cool at 32°F (0°C) for 60 minutes.
   j. Store at 40°F (4.4°C) until ready for use or freeze and store at 0°F (–17.7°C).

Production of Yogurt

INTRODUCTION

Yogurt is a fermented milk product that is growing in popularity. It is made from whole, reduced fat, or skim milk that is pasteurized at high temperatures, cooled, and inoculated with a fresh culture of yogurt that has a 1:1 ratio of *Lactobacillus bulgaricus* and *Streptococcus thermophilus*. These organisms, which grow in a symbiotic relationship, are incubated at 110°F (43.3°C) to 114°F (45.6°C) for 5 to 6 hours. A mixture of *Lactobacillus bulgaricus*, *Streptococcus thermophilus*, and *Lactobacillus acidophilus* may also be used but the incubation temperatures must be lowered to 96.8°F (36°C) and held for 12 hours. A number of combinations will be used in this laboratory exercise, making whole milk yogurt and acidophilus yogurt and also skim milk yogurt and skim milk acidophilus yogurt using added nonfat dry milk solids and gelatin as a stabilizer.

MATERIALS

Chloride sanitizing solution and detergent
Active *Lactobacillus bulgaricus* and *Streptococcus thermophilus* cultures
Active *Lactobacillus bulgaricus, Streptococcus thermophilus,* and *Lactobacillus acidophilus* cultures (Live cultures are present in yogurt purchased at the grocery store. Check label for presence of acidophilus)
Whole milk
Skim milk
Nonfat dry milk (NFDM) powder
Gelatin
Fruit preserves
6-oz containers
Incubators held at 110 to 114°F (43.3 to 45.6°C) and 96.8°F (36°C)
All containers, kitchen whips, steam kettles or double boilers, etc.
STEM thermometers.

PROCEDURE

1. Develop a HACCP flow chart for yogurt and identify all CCPs.
2. Clean and sanitize all equipment and food contact surfaces.
   Four types of yogurt will be produced:
   a. whole milk yogurt
   b. whole milk acidophilus yogurt

c. skim milk yogurt
d. skim milk acidophilus yogurt

3. Production steps:
For whole milk:
1. Measure 2 qts (1.9 liters) of whole milk into steam kettle and heat to 195°F (90.6°C) for 40 to 60 seconds.
2. Move to warm sanitized pan and cool to 116°F (46.7°C) and hold for 15 minutes.
3. Inoculate with starter culture using 2.5% of the volume of the milk.
4. Add fruit preserves that have been heated to 70°F (21.1°C) to the bottom of the containers at 15% total weight of the milk.
5. Add warm inoculated milk, cover container, and place in appropriate incubator depending on starter culture used (see introduction) and incubate for proper time.
6. When set, store at 40°F (4.4°C) until ready for serving.

For skim milk:
1. Add 75 g of NFDM solids (about 4%) and 0.5% by weight of gelatin to 2 qts (1.91) skim milk and stir to dissolve with sanitized whip. Heat to 195°F (90.6°C) for 60 seconds.
2. Repeat steps 2 to 6 above.

Appendix

SOME USEFUL CONVERSIONS

Weight

1 oz = 28.35 g and 1 g = 1000 mg = 1,000,000 mcg = 0.035 oz
1 lb = 0.454 kg and 1 kg = 1000 g = 2.205 lb
1 ft^3 water weighs 62.4 lb = 28.3 kg
1 ton = 2000 lb = 0.893 long tons = 0.907 metric tons

Volume

1 oz = 29.573 ml and 1 ml = 0.034 oz
1 qt = 0.946l and 1l = 1,000 ml = 1.057 qt
1 gal = 3.785l = 231 in^2 and 1l = 0.264 gal
1 cup = 8 oz = 236.8 ml = 16 tbsp = 48 tsp
1 tbsp = 3 tsp = 0.5 oz = 14.8 ml
1 tsp = 4.93 ml
1 ft^3 = 7.48 gal = 0.03 m^3
1 stick butter or margarine = 1/4 lb = 1/2 cup = 8 tbsp

Length

1 in = 2.54 cm and 1 cm = 0.3937 in
1 ft = 30.48 cm = 0.305 m and 1 m = 100 cm = 3.28 ft
1 yd = 0.914 m and 1 m = 1.09 yd

Area

1 in^2 = 6.5 cm^2
1 ft^2 = 0.092 m^2 and 1 m^2 = 10.76 ft^2
1 yd^2 = 0.835 m^2 and 1 m^2 = 1.19 yd^2

Other (see chapt 2)

1 I.U. = 0.3 R.E. units of vitamin A
1 I.U. = 10 micrograms of Vit. D
30 I.U. = 10 α-T.E. of Vit. E.

ABBREVIATIONS

kg = kilogram(s)
g = gram(s)
mg = milligram(s)
μg = mcg = microgram(s)
l = liter(s)
ml = milliliter(s)
m = meter(s)
cm = centimeter(s)
lb = pound(s)
oz = ounce(s)
ha = hectare

gal = gallon(s)
qt = quart(s)
tbsp = tablespoon(s)
tsp = teaspoon(s)
yd = yard(s)
ft = foot (feet)
in = inch(es)
IU = international unit(s)
RE = retinol equivalent unit(s)
ppp = parts per million
ppm = parts per million
α-TE = alpha-tocopherol equivalents

Suggested Readings

Alais, C. and Linden, G. 1991. Food Biochemistry. Chapman & Hall, New York.

Alltrista Corporation. 1993. Ball Blue Book. Alltrista Corporation, Muncie, IN.

American Council on Health and Science. 1995. Introducing ACHS. ACHS, New York.

Berk, Z. 1976. Braverman's Introduction to the Biochemistry of Foods. Elsevier, Amsterdam.

Brown, J.E. 1995. Nutrition Now. West, St. Paul, MN.

Cassens, R.G. 1995. Use of Nitrite in Cured Meats Today. Food Technol. *49*(7), 72–80.

Charley, H. 1982. Food Science, 2nd ed. John Wiley & Sons, New York.

Charm Sciences, Inc. 1995. Approved Charm Tests. Charm Sciences, Inc., Malden, MA.

Clydesdale, F. and Frederick, F. 1985. Food Nutrition and Health. Chapman & Hall, New York.

Clydesdale, F.M. 1989. Present and Future of Food Science and Technology in Industrialized Countries. Food Technol. *43*(9), 134–146.

Coca-Cola. 1994. Fabulous Facts about Coca-Cola. Atlanta, GA.

Considine, D.M. and Considine, G.D., (eds.). 1982. Foods and Food Production Encyclopedia. Chapman & Hall, New York.

Coultate, T.P. 1990. Food: The Chemistry of Its Components. The Royal Society of Chemists, Cambridge, U.K.

Daintith, J. 1981. The Facts on File Dictionary of Chemistry. Intercontinental Book Productions, New York.

Dairy Handbook. Alfa-Laval, Food Engineering AB, Lund, Sweden.

deMan, J.M. 1990. Principles of Food Biochemistry. Van Nostrand Reinhold, New York.

Derosier, N.W. 1966. The Technology of Food Preservation. AVI, Westport, CT.

Division of Microbiology, Center for Food Safety and Applied Nutrition, U.S. Food and Drug Administration. 1984. Bacteriological Analytical Manual. Association of Official Analytical Chemists, Arlington, VA.

Duffy, T. 1993. Bass in the Barn. The Boston Globe, Dec. 1, 1993.

Dziezak, J.D. 1987. Microwave Foods—Industry's Response to Consumer Demands for Convenience. Food Technol. *41*(6), 51–62.

Educational Foundation. 1993. HACCP Reference Book. The Educational Foundation of the National Restaurant Association, Chicago.

Educational Foundation. 1992. Applied Foodservice Sanitation. The Educational Foundation of the National Restaurant Association, Chicago.

Federal Register. 1990. 21 CFR Parts 101, 104, and 105—Food Labeling; Reference Daily Values, Mandatory Status of Nutrition Labeling and Nutrient Content Revision; Serving Sizes; Proposed Rules. *55*(139).

Fruits and Vegetable Processing–Past, Present, and Future. 1990. Food Technol. *44*(2), 91–104.

FSIS Directive. 1989. Grademark Labeling on Meat and Poultry Products. Food Safety and Inspection Service, Washington, D.C.

Ghenra, R., et al. (eds.). 1992. American Type Culture Collection Catalogue of Bacteria and Phages. American Type Culture Collection, Rockville, MD.

Gorga, C. and Ronsivalli, L.J. 1988. Quality Assurance of Seafoods. Van Nostrand Reinhold, New York.

Grosser, A.E. 1981. The Cookbook Decoder. Beaufort Books, New York.

Guthrie, R.K. 1988. Food Sanitation, 3rd ed. Chapman & Hall, New York.
Hall, C.W., Farrell, A.W., and Rippen, A.L. 1986. Encyclopedia of Food Engineering, 2nd ed. Van Nostrand Reinhold, New York.
Hall, R.L. 1989. Pioneers in Food Science and Technology: Giants in the Earth. Food Technol. *43*(9), 186–195.
Hamill, P.V.V., Drizd, T.A., Johnson, C.L., Reed, R.B., Roche, A.F., and W.M. Moore. 1979. Physical growth: National Center for Health Statistics percentiles. Am. J. Clin. Nutr. *32*, 607–629.
Herring, H.B. 1994. 900,000 Striped Bass, and Not a Fishing Pole in Sight. New York Times, Nov. 6, 1994.
Hill, J.W. and Feigl, D.M. 1978. Chemistry and Life. Burgess, Minneapolis, MN.
Hillman, H. 1983. Kitchen Science. Houghton Mifflin, Boston.
Hippleheuser, A.L., Landberg, L.A., and Turnak, F.L. 1995. A System Approach to Formulating a Low-Fat Muffin. Food Technol. *49*(3), 92–96.
Institute of Food Technologists, The Scientific Status Summaries of the IFT Expert Panel on Food Safety and Nutrition. 1983. Radiation Preservation of Foods. The Institute of Food Technologists, Chicago.
Integrated Microwave Packaging. 1990. Food Eng. *62*(4), 50.
Jackson, E.B. 1989. Sugar Confectionary Manufacture. Chapman & Hall, New York.
Jay, J. 1996. Modern Food Microbiology, 5th ed., Chapman & Hall, New York.
Josephson, E.S. and Peterson, M.S. (eds). 1983. Preservation of Foods by Ionization Radiation, Vols. I, II, and III. CRC Press, Boca Raton, FL.
Keane, P.A. 1992. The Flavor Profile. Manual on Descriptive Analysis Testing for Sensory Evaluation. ASTM Manual Series: MNL 13, Baltimore, pp. 5–14.
Lee, F.A. 1983. Basic Food Chemistry. Chapman & Hall, New York.
Lehninger, A.L. 1972. Biochemistry, Worth, New York.
Lemonick, M.D. 1996. Are We Ready for Fat-Free Fat? Time *147*(2), 52–61.
Lewis, R. 1989. Food Additives Handbook. Van Nostrand Reinhold, New York.
Liebman, B. 1990. Trans in Trouble. Nutr. Action *17*(8), 7.
Little, Arthur D. Profile Attribute Analysis. Cambridge, MA.
Lopez, A. 1981. A Complete Course in Canning, Vols. I and II. The Canning Trade, Baltimore, MD.
Lovell, R.T. 1988. Nutrition and Feeding of Fish. Chapman & Hall, New York.
Lovell, R.T. 1991. Foods from Aquaculture. Food Technol. *45*(9), 87–91.
Luh, B. and Woodroof, J. 1988. Commercial Vegetable Processing. Van Nostrand Reinhold, New York.
Lund, D. 1989. Food Processing: From Art to Engineering. 1989. Food Technol. *43*(9), 242–247.
Manci, W. 1995. Food Chemists Bound to Play Major Role in Aquaculture. The Aquaculture News, December 1995.
McGee, H. 1984. On Food and Cooking. Macmillan, New York.
Marriot, N.G. 1994. Principles of Food Sanitation, 3rd ed. Van Nostrand Reinhold, New York.
Martin, R.E. and Flick, G.J., Jr. (eds). 1989. The Seafood Industry. Chapman & Hall, New York.
Merlenstein, N.H. 1993. A New Era in Food Labeling. Food Technol. *47*(2), 81–96.
National Research Council. 1989. Recommended Dietary Allowances. National Academy of Sciences, Washington, D.C.
National Starch and Chemical Company. 1995. How to Choose, A professional Guide to Food Starches.
National Starch and Chemical Corporation. 1989. Food Starch Technology.
Newsome, R.L. 1987. Perspective on Food Irradiation. IFT Expert Panel on Food Safety and Nutrition. Institute of Food Technologists, Chicago.
Newsome, R.L. 1986. Food Colors. IFT Expert Panel on Food Safety and Nutrition. Institute of Food Technologists, Chicago.
North, M.O. and Bell, D.D. 1989. Commercial Chicken Production Manual, 4th ed. Chapman & Hall, New York.

Parkhurst, C.R. and Mourntney, G.J. 1987. Poultry, Meat and Egg Production. Chapman & Hall, New York.
Perlman, A.L., Jantz, C.J., and Mantick, N.S. 1994. A Taste Right Out of History. Res. Dev. March/April, 42–44.
Pierson, M.D. and Corlett, D.A., Jr. 1992. HACCP: Principles and Applications. Chapman & Hall, New York.
Plastic Packaging of Foods—Problems and Solutions. 1990. Food Technol. *43*(12), 83–94.
Pope, K. Technology Improves on the Nose as Scientists Try to Mimic Smell. The Wall Street Journal, March 1, 1995. Sec. B, p. 1.
Potter, N.N. and Hotchkiss, J.H., 1995. Food Science, 5th ed. Chapman & Hall, New York.
Qin, B.L., Pothakamury, U.R., Vega, H., Martin, O., Barbosa, Canovas, G.V., and Swanson, B. 1995. Food Pasteurization Using High-Intensity Pulsed Electrical Fields. Food Technol. *49*(12), 55–59.
Richardson, T.R. and Finley, J.W. (eds.) 1985. Chemical Changes in Food During Processing. Chapman & Hall, New York.
Rombauer, I.S. and Rombauer Becker, M. 1974. Joy of Cooking, Vols. 1 and 2. The New American Library, New York.
Ronsivalli, L.J. and Baker, D.W. 1981. Low Temperature Preservation of Seafoods: A Review. Marine Fisheries Rev. *43*(4), 1–15.
Roth, L.O., Grow, F.R., and Mahoney, G.W.A. 1975. An Introduction to Agricultural Engineering. Chapman & Hall, New York.
Ryser, E.T. and Marth, E.M. 1989. New: Food-Borne Pathogens of Public Health Significance. J. Am. Dietet. Assoc. *89*(7), 948–954.
Santroch, L. 1980. Canning and Cooking the All-American Way. Wisconsin Aluminum Foundry Co., Inc., Manitowoc, WI.
Schultz, H.W. 1981. The Food Law Handbook. Van Nostrand Reinhold, New York.
Simplesse. 1989. Food Eng. *61*(3), 71–72.
Stadelman, W.J. and Cotterill, D.J. (eds). 1986. Egg Science and Technology, 3rd ed. Van Nostrand Reinhold, New York.
Staff Report. 1989. Top 10 Food Science Innovations 1939–1989. Food Technol. *43*(9), 308.
Stefferud, A. (ed.). 1962. After a Hundred Years, The Yearbook of Agriculture. Superintendent of Documents, Washington, D.C.
Sultan, W. 1989. Practical Baking. Van Nostrand Reinhold, New York.
The Educational Foundation of the National Restaurant Association. 1985. Applied Foodservice Sanitation. John Wiley & Sons, New York.
Tetra Therm Aseptic VTIS Training Manual.
Tiexeira, A.A. and Shoemaker, C.F. 1988. Computerized Food Processing Operations. Chapman & Hall, New York.
Unique Aseptic Design. 1990. Food Eng. *62*(1), 95–96.
USDA. 1989. HACCP Principles for Food Production. Food Safety and Inspection Service, Washington, D.C.
USDA. 1993. Poultry Irradiation and Preventing Foodborne Illness. USDA, FSIS, Washington, D.C.
U.S. Department of Health, Education, and Welfare. 1978. Grade "A" Pasteurized Milk Ordinance. U.S. Government Printing Office, Washington, D.C.
U.S. Department of Health and Human Services, Food and Drug Administration. 1976. Food Service Sanitation Manual, Washington, D.C.
U.S. Department of Health and Human Services. 1993. Grade "A" Pasteurized Milk Ordinance. Washington, D.C. Public Health Service/Food and Drug Administration Publication No. 229, Washington, D.C.
U.S. Department of Health and Human Services. 1995. Food Code, 1995. Washington D.C. Public Health Service/Food and Drug Administration Publication PB96104401, Washington, D.C.

Vieira, E. 1984. Pollock or Cod; Can the Difference Be Told? Marine Fish. Rev. NOAA. *46*(1), 37–38.

Weiss, T.J. 1982. Food Oils and Their Uses. Van Nostrand Reinhold, New York.

Whitney, E.N. and Hamilton, E.M. 1987. Understanding Nutrition. West, St. Paul, MN.

Whitney, E.N. and Hamilton, E.M. 1993. Understanding Nutrition, 6th ed., West St. Paul, MN.

Williams, S.R. 1989. Nutrition and Diet Therapy. Times Mirror/Mosby College Publishing, St. Louis.

Wong, N.P. (ed.). 1986. Fundamentals of Dairy Chemistry, 3rd ed. Chapman & Hall, New York.

Woodroof, J.G. 1990. 50 Years of Fruit and Vegetable Processing. Food Technol. *44*(2), 92–95.

Woodroof, J. and Luh, B. 1986. Commercial Fruit Processing. AVI, Westport, CT.

Woodruff, S. 1994. Secrets of Fat-Free Baking. Garden City Park, NY.

Wrick, K.L., Friedman, L.J., Brewda, J.K. and Caroll, J.J. 1993. Consumer Viewpoints on Designer Foods. Food Technol. *47*(3), 94–104.

Index

Accum, Frederick, 3
Acer saccharinum, 328
Acetic acid (vinegar), 66
Acetobacter, 304
Acidulants, 189
  acetic acid, 189
  malic acid, 189
Actin, 211
Adulteration, 101
Aerobacter aerogenes, 240
Aggregation of proteins, 135
AIDS, 62
Alaska pollock, 282
Aldehydes, 24
Alkaline compounds, 190
  sodium bicarbonate, 190
  sodium carbonate, 190
  sodium hydroxide, 190
Alkaloids, 45
Amebiasis, 60
American Association of Medical Milk Commissions, 225
American Council on Science and Health, 107
American Dietetic Association (ADA), 9, 35, 106, 289
American Heart Association (AHA), 35, 255, 289
American Journal of Clinical Nutrition, 288
Amino acid, 13
  essential, 18
  methionine, 45
  nonessential, 18
  r-groups, 13
  structures, 13, 16, 17
Anchovy, 262
Anderson, James, 288
Anisakis, 61
Antioxidants, 168, 186
  amines, 187
  antioxidants, 187
  butylated hydroxyanisole (BHA), 187
  butylated hydroxytoluene (BHT), 187
  citric acid, 187, 189
  phenolic compounds, 187
  vitamin C, 187
  vitamin E, 187
Appert, Nicholas, 3, 139
Apples, 306
  cans, 307
  cider, 307
  drying, 307
  frozen apple slices, 307
  jelly, 308
  juice, 307
  low-methoxy pectin, 308
  nonenzymatic browning, 307
  pectin, 307
  sauce, 307
  slices, 307
  vinegar, 307
Apricots, 313
Aquaculture, 277, 281
Arsenic, 46
Arthur D. Little, Inc., 95
Ascorbic acid, 180
Ascorbic acid oxidase, 126
Aseptic packaging, 310
Avidin, 252, 45

Bacillus stearothermophilus (FS 1518), 147, 227
Bacillus thermoacidurans, 158
Bacteria, 47, 48, 50, 53
  Aeromonas hydrophila, 62
  Arizona hinshawaii, 62
  Bacillus cereus, 54, 62, 66
  Brucella melitensis, 61
  Campylobacter jejuni, 62
  cholera, 59
  Clostridium botulinum, 64, 65, 202
  Clostridium perfringens, 54, 65
  Corynebacterium tuberculosis, 61
  Escherichia coli (serotype 0157:H7), 62
  Listeria monocytogenes, 61
  Plesimonas shigelloides, 62
  Salmonella, 54, 55, 56
  Salmonellae, 56
  Shigella, 57, 58
  shigellosis, 57
  Staphylococcus aureus, 63, 64
  Vibrio parahaemolyticus, 58
  vibriosis, 58
  Yersinia enterocolitica, 61
Bagasse, 325
Bakery products, 295

baking powder, 368
baking soda, 368
emulsifiers, 369
gluten, 366
leavens, 368
lipids, 367
shortening, 368
starch, 367
Bananas, 308
canned, 309
dried, 309
Barley, 289
hops, 290
malt, 289
Basic information, 109
Beef/veal, 211
choice, 213
conformation, 212
grades, 213, 214
prime, 213
select, 213
Beer, 289
mashing, 289
wort, 289
Beet sugar, 327
Beta-carotene, 37
Beta-lactam tests, 227
Beta vulgaris, 327
Biochemistry, 11
Biotin, 252
Birdseye, Clarence, 4, 173
Bivalve molluscs, 268
Blackberries, 313
Blackhead, 250
Blackstrap, 325
Blueberries, 313
Blue nos. 1 and 2, 110
Botulism, 64
Brabender curve, 335
Bread, 295
batter-whipped process, 299
bleaching agents, 296
browning of the crust, 297
composition, 295
enriched, 299
monoglycerides, 296
proofing, 296
ropiness, 298
some fat, 296
spoilage, 298
sponge dough method, 297
sponge method, 297
straight dough method, 297
sugar, 296
white bread, 295
yeast, 296
yeast foods, 297
Broccoli, 316
Bromates, 22
Buckling, 142
Buckwheat, 292
Butter, 243
Buttermilk, 66, 243

Cabbage, 321
sauerkraut, 321
Cairncross, 95
Cakes and cookies, 299
angel cakes, 299
leavening, 299
prepared mixes, 299
Calories, 13
Candida, 239
Candy, 330, 373
fondant, 373
Canned foods, 84
Cans, 140, 156
Caramelization, 133
Carbohydrates, 11, 23
Caseinate, 110
Cassens, Robert, 202
Catalase, 160
Cathepsins, 125
Cellulose, 31
Center for Science in the Public Interest, (CSPI), 107
Cereal grains, 283, 365
beer, 365
flours, 365
milling, 291
pasta products, 365
proximate analysis, 284
semolina, 365
structure, 283
Cereals
breakfast cereal, 287
puffing, 287
Charm Sciences Inc., 227
Cheeses, 67, 238
basic steps and their primary purposes, 241
blue cheese, 242
cheddaring, 242
citric acid, 239
cooking, 241
cottage cheese, 239
gruy'ere, 239
knitting, 242
lactic acid fermentation, 239
L. bulgaricus, 239
Leuconostoc citrovorum, 239
Leuconostoc dextranicum, 239
mozzarella, 242
process cheeses, 242
Propionibacterium shermanii, 239
provolone, 242
ricotta cheese, 243
roquefort, 242
S. diacetilactis, 239

S. lactis, 239
starter cultures, 240
swiss, 239
Chemical additives
benzoates, 203
benzoic acid, 203
benzoyl peroxide, 205
chlorine, 203, 205
chlorine dioxide, 205
hydrogen peroxide, 203
iodine, 203
nitrosyl chloride, 205
oxidizing agents, 203, 205
parahydroxybenzoic acid, 204
sodium nitrite, 202
sorbic acid, 202
Chemical preservatives, 199
benzoic acid, 200
calcium propionate, 200, 201
fatty acids, 201
nitrites, 200
sodium, 200, 201
sodium benzoate, 200
sodium chloride, 200
sodium diacetate, 200, 201
sorbic acid, 200
sulfur dioxide, 200, 201
Cherries, 313
Chicken, 245, 246
evisceration, 247
grade, 249
grade shields, 249
nutritional content, 249
processing, 248
quality, 249
raising, 246
Chlorine, 80
Cholesterol, 37
Cis, 36
Clam, 269, 270
Cl. botulinum, 147, 149, 155, 159, 172, 202, 217, 314
Cl. butyricum, 240
Cleaning the plant, 79
Cod, 257, 262
Code of Federal Regulations, 224
Codex Alimentarius Commission, 106
Coffee, 310
decaffeinated coffee, 312
instant coffee, 311
instant tea, 322
Cold point, 147
Color Additive Amendments, 205
Colorants, 204
apocarotenal, 204
beets, 204
beta-carotene, 204
burnt sugar, 204
canthaxanthin, 204
carotene, 204
chlorophyll, 204
cochineal insect, 204
FD&C dyes, 204
FD&C lakes, 204
nature-identical, 204
tomatoes, 204
Colorings, 110
Composition and nutritional value of foods, 11
Consumer sensory testing, 97
hedonic rating, 98
triangle test, 97, 98
Cooling of foods, 72
Corn, 287
milled, 287
modified starch, 288
popcorn, 288
sweeteners, 288
Corn syrup and sugar, 328
high-fructose corn syrup (HFCS), 329
hydrolysis, 328
Cottonseeds, 293
Coxiella burnetti, 228
Crabs, 274
blue crab, 274
dungeness crab, 275
Jonah crab, 276
king crab, 276
red crab, 276
snow or tanner crab, 276
Crackers, 300
Crayfish, 277
cross-contamination, 351
Crustaceans, 272
Cucumbers, 316
pickles, 316
relishes, 316
Cutting boards, 73
Cutting equipment, 349
chefs knife, 349
choppers, 351
commercial food processors, 351
cutting boards, 351
cutting surface, 351
knives, 349
paring knife, 349
serrated slicing knife, 350
slicers, 351
steel, 350
Cyanide, 45
Cysteine, 22

Dairy products, 224, 360
butterfat, 360
foam, 361
half and half, 361
heavy whipping cream, 360
light cream, 360
light whipping cream, 360

Dairy product substitutes, 244
Degree of lethality, 149
Delaney Clause, 184, 205
Denaturation, 20
 enzymes, 20
 proteins, 20
Denature, 20
Desirable changes in foods, 66
Detergents, 79
Detergents in removing soil, 80
Dicarbonyls, 134
Dietary fiber, 31
Dietary Guidelines for Americans, 12
Doughnuts, 299
Dry ice, 175
Drying, 160
 deterioration of dried foods, 167
 enzymatic changes in dried foods, 168
 nonenzymatic browning, 168
 oxidative spoilage, 167
 packaging of dried foods, 166
 pretreatment, 160
 reconstitution of dried foods, 166
 the effect of drying on microorganisms, 167
Ducks, 245
Ducks and geese, 250

Educational Foundation of the National Restaurant Association (EF), 107, 182
Eggs, 245, 250, 361, 362
 chemical composition, 250
 chemistry, 362
 classified, 252
 coagulation, 362
 consumption, 255
 custards, 363
 egg substitutes, 255
 ferrous sulfide, 363
 fried eggs, 363
 nonenzymatic browning, 254
 omelet, 363
 ovalbumen, 364
 pasteurization, 254
 pasteurized, 254
 poached eggs, 363
 processing, 253
 scrambled eggs, 363
 soft boiled, 362
 spray-dried, 254
 whipped, 364
 yolk, 251
Emergency permit control, 159
Environmental Protection Agency (EPA), 105, 184
Enzymes, 20, 28, 123, 194
 alpha-amylase, 28
 amylases, 194
 amyloglucosidase, 28
 applications, 128
 beta-amylase, 28, 29
 catalase, 195
 cellulases, 32, 194
 enzymes that decompose carbohydrates (carbohydrases), 127
 fat-splitting enzymes (lipases), 126
 glucose oxidase, 195
 invertase, 194
 lipases, 195
 maltase, 28
 oxidizing enzymes (oxidases), 126
 pectinase, 194
 proteases, 195
 proteolytic enzymes (proteases), 125
 pullulanase, 28
Equipment, 76, 349
Escherichia coli, 240
Essential amino acids, 284
Ethyl alcohol, 20, 66

Fats, 12, 337, 373
Fats and oils, 33
 chaulmoogric acid, 338
 coconut oil, 339
 deterioration, 373
 diet, 338
 fatty acids, 338
 fire-point, 338
 flash-point, 338
 iodine number, 338
 melting points, 373
 oxidation, 338
 saponification, 338
 smoke-point, 338
 smoke points, 373
 unsaturated fatty acids, 35
Fatty acids, 34
 cis-form, 36
 oleic acid, 36
 saturated fatty acid, 34
 trans-, 36
 trans-form, 36
 unsaturated fatty acid, 34
FDA, 159, 182
FDA Regulations and Safety, 159
Federal Food and Drug Act, 100
Federal Trade Commission (FTC), 105
Fellers, Carl, 4, 5
Fibers, 31
Fish, 256, 364
 baking, 364
 broiling, 364
 frying, 364
 poaching, 364
 steaming, 364
Flatfish, 267
 blackback flounder, 267
 flounder, 268
 fluke, 267
 halibut, 267, 268

lemon sole, 267
plaice, 267
sand dab, 267
turbot, 267, 268
yellowtail, 268
yellowtail flounder, 267
Flavor enhancers, 188
disodium inosinate and disodium GMP, 189
monosodium glutamate (MSG), 188
Flavorings, 188
Flavor profile, 95
score sheet, 94
Flexible pouches, 156
Flow chart, 87, 88
Fluid milk, 224
Food additives, 183
philosophy of food additives, 185
Food and Drug Act of 1906, 183
Food and Drug Administration (FDA), 9, 100, 109, 183
Food and Drug Aministration (Department of Health and Human Services), 100
Food, Drug and Cosmetic Act, 184
Food Guide Pyramid, 12
Food infections, 54, 61
Food intoxications, 62, 65
Food label
code dating, 121
ingredients, 110
legal symbols, 121
open dating, 120
religious symbols, 122
universal product code, 121
Food plant sanitation, 74
Food poisoning, 53
Food preparation, 349
Food safety, 68
Food Safety and Inspection Service (FSIS), 100
Food science as a profession, 9
Foodservice facility, 74
Foodservice sanitation, 72
Food spoilage, 52
Frankfurters, 220, 222
Freeze drying, 6
Freezing, 172
air-blast freezing, 173
blanching, 178
bulk foods, 179
dehydrofreezing, 175
fluidized-bed freezing, 175
freezing methods, 173
individually quick-frozen (IQF), 175, 176
liquid freezers, 175
packaging, 178
plate freezing, 175
preparation of foods for freezing, 177
preservation effect of freezing, 173
quality changes during frozen storage, 179
quick freezing, 173
shelf-life of frozen foods, 181
slow freezing, 176
thawing, 181
Frozen Food Industry Coordinating Committee, 107
Fruits, 302, 369, 371
appearance, 372
discoloration, 372
freezing, 302
heat-processed, 302
texture, 372

Geese, 245
Generally recognized, 184
Georges Bank, 268, 272
Glass containers, 157
Glucose oxidase, 160
Glycogen, 30
GMPs, *see* Good Manufacturing Practices
Goitrogens, 45
Gonyaulax catenella, 60
Good Manufacturing Practices (GMPs), 159, 204
Gossypol, 45
Grade A Pasteurized Milk Ordinance Recommendations, 224
Grades, 111
Grapefruit, 313
Grapes, 303
brandies, 304
concord, 303
drying, 306
grape juice concentrate, 303
jam, 306
jelly, 306
juice, 303
raisins, 305
vinegar, 304
wines, 303
GRAS (generally recognized as safe), 101
Green and wax beans, 315
Green no. 3, 110
G. tamarensis, 60
Guernsey, 224
Gums, 32, 33, 193

Haddock, 257, 264
Hakes, 264
Hart, Edwin, 4
Harvard Medical School, 288
Hazard Analysis and Critical Control Point System (HACCP), 84
Healthy diet, 12
Heating utensils, 352
aluminum, 353
carbon rolled steel, 353
cast iron, 353
conduction, 352
convection, 352
copper, 353

earthware, 353
glass, 353
pans, 352
porcelain enamel, 353
radiation, 352
stainless steel, 353
teflon, 353
tinware, 352
Heat sources, 354
Hemagglutinins, 45, 46
Hemiacetyl bridge, 24
Herring, 257
Herring family, 259
Heteropolysaccharides, 32
Holstein, 224
HTST process, 134
Hydrogenation, 342
catalyst, 342
cis, 342
elaidic acid, 342
nickel, 342
trans, 342
Hydrolysis of polysaccharides, 29
Hydrolytic rancidity, 132
Hydrolyzed animal protein, 110
Hydrolyzed casein, 110
Hydrolyzed protein, 110
Hydrolyzed soy protein, 110
Hydrolyzed vegetable protein, 110

Ice cream, 232, 234
low- and nonfat ice cream-like products, 235
% overrun, 235
procedures, 235
Imitation, 110
Immobilized enzyme, 129
Institute of Food Technologists, (IFT), 9, 106, 154, 289
Internal Revenue Service (IRS), 105
International Association of Milk, Food and Environmental Sanitarians (IAMFES), 84, 107
International Radiation Logo, 199
Invertase, 128
Iodine, 80
Iodine number, 338
Iodophor, 80
Ionization radiation, 184
Ionizing radiation, 198
Iso-ascorbic acid, 217, 221

Jams and preserves, 32
Jersey, 224
Jungle, The, 183

Ketones, 24
Kramer, Amihud, 5

Lactobacillus acidophilus, 236
Lactobacillus bulgaricus, 236
Lactobacillus cucumeris, 321
Lactobacillus pentoaceticus, 321
Lamb/mutton, 215
graded as prime, choice, good, utility, and cull, 216
slaughtered, 216
Lard, 339
interesterification, 340
leavening agents, 197
ammonium bicarbonate, 197
ammonium carbonate, 197
baking powder, 197
double-acting baking powder, 198
single-acting baking powder, 197
sodium bicarbonate, 197
Lemons, 313
Lethality rates, 149
Lettuce, 316
Leuconostoc mesenteroides, 321
Lipid emulsifiers, 343
mono- and diglycerides, 343
Lipids, 11, 33, 337
Liquid nitrogen, 175
Listeria, 149, 159
Listeria monocytogenes, 172, 225
Little, Arthur D. (ADL), 5
Lobsters, 272
Low- and no-fat bakery items, 300
Lysozyme, 252

Mackerel, 257, 265, 266
Maillard reaction, 160
maple syrup sap, 328
Margarine, 343
Marine crayfish, 276
Massecuite, 325
Meat, 209, 358
appearance, 359
bacon, 218
Canadian bacon, 218
capocollo, 218
chipped beef, 219
color, 217
composition, 358
corned beef, 218
cured meat products, 217
curing, 217
flavor, 360
hams, 217
loaf-type, 222
myoglobin, 359
pastrami, 219
products, 222
slaughter, 210, 212
smoking, 217
storage, 211
stored, 211
tenderization, 210

tenderness, 358
tenderness of beef, 210
wholesome, 211
Meat Inspection Act, 211
Meat Inspection Bureau, 103, 104
Meat Inspection Division and Poultry Inspection Service, 100
Mendel, Gregor, 11
Menhaden, 261
Mercury, 46
Methods of drying, 160
drum drying, 163
fluidized-bed drying, 162
freeze-drying, 164
hot-air drying, 161
microwave drying, 166
puff drying, 166
spray drying, 163
sun drying, 160
Metmyoglobin, 217
Microbes, 47
effect of oxidation–reduction, 52
effect of pH on, 50
effect of temperature, 50
nutritional requirements, 50
oxygen requirements of, 51
reproduction in microbes, 49
size of microbes, 48
structure and shape of microbes, 47
water requirements, 51
Microbial activity, 47
Microorganisms, 224
Microwavable containers, 158
Microwave processing, 156
Milk
antibiotics, 227
bacteriological, 232
butyric acid, 240
certified milk, 225
chemical, 232
direct steam heating, 230
dried milk, 232
flavor treatment, 228
handling, 226
indirect heating, 230
on farms, 226
pasteurize, 228
processing of, 228
pulsed electrical fields, 231
quality control, 235
sources of bacteria, 225
temperature standards, 232
transportation, 226
UHT, 230
ultra-high-temperature, 230
vitamin A, 228
vitamin D, 228
Millet, 293
Minerals, 41
aluminum, 45
boron, 45
cadmium, 45
calcium, 42, 43
chlorine, 42
chromium, 45
cobalt, 44
copper, 44
fluorine, 43
iodine, 43, 45
iron, 41
magnesium, 43
manganese, 44
phosphorus, 43
potassium, 42
selenium, 44, 46
silicon, 44
sodium, 42
sulfur, 43
tin, 45
vanadium, 44
zinc, 44, 46
Misbranding, 101
Model Food Code, 102
Modified-Atmosphere and Controlled Atmosphere Packaging (MAP), 155
Modified atmosphere packaged (MAP) foods, 172
Modified atmosphere packaging, 314
Molasses, 325
refining, 326
Molds, 47, 49, 50, 53, 67
Mussels, 271
Mycotoxins, 66
Myoglobin, 217
Myosin, 211

National Aeronautics and Space Administration, 85
National Bureau of Standards (NBS), 105
National Environmental Health Association (NEHA), 107
National Labeling and Education Act (NLEA), 109
National Marine Fisheries Service, National Oceanic and Atmospheric Administration, 105
National Pest Control Association (NPCA), 108
National Restaurant Association (NRA), 107
National Sanitation Foundation International (NSF), 77, 108, 351
"New generation" gourmet type products, 6
New-generation refrigerated foods, 172
Nitrite, 217
Nitrosamines, 203, 220
Nitrosohemochrome, 217
Nonenzymatic browning, 133
Northwestern University, 288
Nutrient additives, 187
iron, 188
mannitol, 196
propylene glycol, 196
protein concentrate, 187

sorbitol, 196
vitamin D, 187
Nutrition, 11
Nutrition labeling, 112
calories, 120
cholesterol, 120
daily values, 115
DRVs, 115
exemptions and exceptions, 114
fat, 120
fiber, 120
health claims, 119
raw foods, 117
RDIs, 115
serving sizes, 114
sodium, 120
sugar, 119
Nutrition Labeling and Education Act (NLEA), 6

Oats, 288
bran, 288
cholesterol, 288
lowered, 288
milling, 288
Office of Technical Services (OTS), 105
Oils, 337, 340, 373
canola, 341
coconut, 340
corn, 341
cottonseed, 341
expressed oils, 341
extracted or expelled oils, 341
hydrogenation, 342
olive, 341
palm kernel, 340
peanut, 341
refined, 341
sesame, 341
soybean, 341
Oleuropein, 317
Oligosaccharides, 27
stachyose, 27
Olives, 66, 317
Oranges, 309
canned, 309
juice, 309
pectin, 309
Oser, Bernard, 5
Oxalic acid, 45
Oxidation, 22, 130, 180
Oxidative-rancidity, 132
Oxidizing agents, 22
Oxymyoglobin, 217
Oysters, 268, 269

Pasteur, Louis, 4
Peaches, 313
Peanuts, 293
aflatoxins, 293
Pears, 313
Pectic enzymes, 128
Pectin, 32
Pectinases, 129
Peptide bond, 17
Perfringens poisoning, 65
Peroxidases, 126
Personal hygiene, 68
Personnel facilities, 78
Phoenix sylvestris, 328
Phospholipases, 127, 136
Phospholipids, 36, 37
Pickles, 66
Pie crusts, 300
Pilchards, 262
Pillsbury Co., 85
Pineapples, 313
Plant construction, 76
Plastic, flexible containers, 158
Plums, 313
Pollock, 264
Polydextrose, 30, 31
Polyhydric alcohols, 196
glycerol, 196
Polyphenolases, 180
Polysaccharides, 28
Pork, 214
graded, 215
slaughter, 215
Potatoes, 318
canned, 318
chips, 318
frozen, 318
sugar content, 318
Potentially hazardous foods, 86
Poultry, 361
Poultry and eggs, 245
Poultry Inspection Service, 104
Prescott, Samuel, 4
Proctor, Bernard, 5
Profile Attribute Analysis (PAA), 96
Protein denaturation, 180
Protein hydrolysates, 110
Proteins, 11, 13, 18
albumins, 18
collagen, 18, 19
conformation, 19
denaturation, 19, 20, 21
gliadin, 19
glutenin, 19
helical structure, 20
histones, 18
hydrogen bond, 19
hydrolysates, 22
hydrolysis, 22
keratin, 18, 19
lactalbumin, 18
nitrogen, 13
oryzenin, 19

ovalbumin, 18
pleated-sheet structure of a protein, 19, 21
primary structure, 19
quaternary structures, 19
reduction, 22
renaturation, 20
secondary, 19
solubility, 18
structure, 19
tertiary structures, 19
water binding, 19
zein, 19
Proteolysis, 125
Protozoa, 61
Prussic acid, 45
Pseudomonas, 252
Putrefactive Anaerobe 3679 (PA 3679), 147

Q fever, 228
Quaternary ammonia, 80

Rancidification, 168
Rancidity, 132, 265
Raspberries, 313
Ready-to-eat cereals, 284
Recommended dietary allowances (RDAs), 13, 14
Red nos. 40 and 3, 110
"Red Tide" shellfish poisoning, 60
Reducing agent, 22
Refrigeration, 169
chilled seawater (CSW), 172
mechanical refrigeration, 170
refrigerated seawater (RSW), 172
refrigeration practices, 171
Rice, 291
bran, 292
flour, 292
puffed rice cereal, 292
quick-cooking rice, 292
RSW, 266
Rye, 291

Saccharomyces cerevisiae, 296
Saccharum, 324
Sacks, 288
Safrole, 45
Salmon, 257, 266
Atlantic salmon, 266
blueback, 266
chinook, 266
chum or dog, 266
king, 266
pink or humpback, 266
red, 266
silversides or coho, 266
sockeye, 266
spring, 266
Salmonella, 149, 159, 254, 255, 359, 361, 362, 363
Salmonellosis, 54
Salt, 11
Sanitation, 68
Sanitation in retail outlets, 81
Sanitation in the home, 69
Sanitizers, 80
Saturated fat, 12
Saturated fatty acids, 35
Sauerkraut, 321
Sausage
bologna, 222
braunschweiger, 222
cooked salami, 222
dried sausages, 223
fermented sausages, 222
frankfurters or hot dogs, 219
fresh pork sausage, 219
kielbasa, 222
knockwurst, 222
pepperoni, 223
salami, 223
Sausage products, 219
Scallops, 269, 271
Sensory evaluation of food, 92
appearance, 92
color, 92, 93
consistency, 93
electric nose, 99
flavor, 93
texture, 93
viscosity, 93
Sequestrants, 196
citric, 196
malic, 196
tartaric, 196
Sewage disposal, 81
Shad, 257, 261
Shapes of various microorganisms, 48
Shellfish, 256, 268, 364, 365
crustacea, 365
molluscs, 365
Shrimp, 273
Sinclair, Upton, 183
Sjöström, Loren, 5, 95
Sodium erythrobate, 217, 220
Sodium nitrite, 220
sodium or calcium propionate, 298
Soft-shell clam, 270
Solanine, 46, 319
Sorghum, 292
Sour cream, 243
Soured cream, 66
Sous-Vide process, 155
Soybeans, 293
imitation bacon bits, 293
textured protein, 293
SO_2, 39
Sphingolipids, 36
sphingomyelin, 37
Stabilizers, 30

Stamler, Jeremiah, 288
Staphylococcal poisoning, 63
Starches, 28, 29, 193, 332
  amylopectin, 28, 30, 193
  amylose, 28, 193
  freeze–thaw stability, 30, 336
  gel strength, 29
  modified starches, 29, 332
  retrogration, 30
  stabilization, 336
  “waxy” maize, 30
Sterilization value, 149
Sterols, 36, 37
Storage, 79
Strawberries, 313
Strecker Degradation, 134
Streptococcus diacetyllactis, 243
Streptococcus lactis, 236
Streptococcus thermophilus, 236
Sublimation, 164
Sugar from cane, 324
Sugars, 11, 23, 27, 324, 372
  beverages, 329
  brown sugar, 373
  candy or confections, 330
  cellobiose, 26
  confectioner’s sugar, 326
  disaccharides, 26
  fructose, 23, 27
  furanoses, 25
  galactose, 23, 27
  glucose, 23, 25, 27
  high-fructose corn syrup (HFCS), 329
  invert sugar, 327
  lactose, 26, 27
  maltose, 26, 27
  mannose, 23
  monosaccharide, 23
  production of raw sugar, 325
  sorbose, 23
  sucrose, 26, 27
Sulfiting agents, 111
Sulfur dioxide, 39, 303
Sulfurous acid, 168
Surface-active agents, 196
  emulsifiers, 197
  lecithin, 197
  mono- and diglycerides, 197
  tweens, 197
Surf clams, 271
Surimi, 281, 282
Sweeteners, 27, 111, 190
  acesulfame K, 190
  aspartame, 190, 193
  fructose, 27, 190
  galactose, 27
  glucose, 27
  glycyrrhizin, 190
  honey, 190, 191
  lactose, 27, 190, 192
  maltose, 27, 190, 192
  mannitol, 190, 192
  maple sugar, 190, 191
  molasses, 190, 191
  neohesperidin dihydrochalcone, 27
  polyhydric alcohols, 190
  saccharin, 27, 190, 193
  sodium cyclamate, 27
  sorbitol, 190, 192
  sucrose, 27
  xylitol, 190, 192

Tapeworm, 61
  diphyllobothrium latum, 61
  taenia saginatta, 61
  taenia solium, 61
Tea, 322
Temperature zones used for processing, holding, and preserving foods, 70
Thermal processing, 139
  aseptic fill method, 154
  canning of acid foods, 158
  continuous agitating retort, 152
  cooking under pressure, 155
  cooling heat-processed foods, 151
  D-value, 147
  estimating processing times, 147
  filling the cans, 144
  heat process, 146
  high-temperature–short-time (HTST) process, 153
  hydrostatic cooker, 153
  liquid in cans, 143
  pretreatment of foods, 140
  retort, 150
  sealing the cans, 145
  vacuum in cans, 141
Thermophilic bacteria, 152
Thiaminase, 45
Tocopherols (vitamin E), 168
Tomatoes, 319
  canned, 319
  catsup, 320
  chili sauce, 320
  genetic engineering, 320
  ketchup, 320
  paste, 320
  puree, 320
  tomato juice, 320
  tomato soup, 320
Torula, 239
Total Quality Control (TQC), 104
Toxicants, 45
Trans configurations, 36
Trichinella spiralis, 218
Trichinosis, 59, 82
Triticale, 294
Tropomyosin, 211

Tuna, 265
  albacore, 265
  bluefin, 265
  skipjack, 265
  yellowtail, 265
Turkey, 245, 249
  processed, 250
Types of heat
  boiling/simmering/poaching, 354
  broiling/toasting, 355
  deep-frying, 355, 356
  electric appliances, 357
  gas, 357
  microwave, 355
  microwave heating, 356
  microwave ovens, 357
  pan-frying, 355
  pan-frying depends, 356
  pressure-cooking, 354
  roasting/baking, 355
  steaming, 354
Tyramine, 45
Tyrosinase, 126

Underwood, William, 4, 139
Underwriters Laboratories, Inc. (UL), 108
U.S. Army Natick Laboratories, 85
U.S. Centers for Disease Control (CDC), 103
U.S. Department of Agriculture (USDA), 9, 12, 100, 111, 369
U.S. Department of Commerce (USDC), 105, 111, 156
U.S. Public Health Service, 102, 232
Unit of lethality, 149

Vegetables, 313, 369
  anthocyanins, 371
  appearance, 370
  asparagus, 323
  beets, 323
  brussel sprouts, 323
  canned, 315
  carrots, 323
  cauliflower, 323
  celery, 323
  chlorophyll, 371
  cooking, 370
  flavor, 371
  freezing, 315
  green peas, 323
  lima beans, 323
  mushrooms, 323
  onions, 323
  other vegetables, 323
  spinach, 323
  squash, 323
  sweet corn, 323
  sweet potatoes, 323
  texture, 370
  turnips, 323
Viruses, 62
Vitamins, 37
  fat-soluble vitamins, 37
  vitamin A, 37
  water-soluble vitamins, 39
Vitamins, fat soluble
  vitamin A, 45
  vitamin D, 38, 45
  vitamin E, 38, 42, 44
  vitamin K, 38
Vitamins, water soluble
  biotin, 40, 45
  choline, 40
  folacin, 40
  niacin (nicotinic acid), 39
  pantothenic acid, 40
  vitamin B_1 (Thiamin), 39, 45
  vitamin B_2 (Riboflavin), 39
  vitamin B_6 (Pyridoxine), 39
  vitamin B_{12} (Cyanocobalamin), 40, 44
  vitamin C (Ascorbic Acid), 41, 42

Warehouse storage of canned foods, 158
Washington State University, 231
Water activity, 51
Water supply, 81
Waxes, 36, 37
Wheat, 284
  enriched, 287
  flour, 286
  milling, 285
  turbomilling, 286
  wheat flour, 287
Whey, 243
Wiley, Dr. Harvey Washington, 100, 183

Yeasts, 48, 49, 50
Yellow nos. 5 and 6, 110
Yogurt, 236, 66
  L. acidophilus, 237
  L. bulgaricus, 236, 237
  quality problems, 238
  S. thermophilus, 236
  sundae style, 237
  "swiss" style, 237
  symbiosis, 236

Z-value, 149